Fourth International Workshop on Complex Structures and Vector Fields

Aspects of

Complex Analysis, Differential Geometry, Mathematical Physics and Applications

St. Konstantin, Bulgaria
September 3 – 11, 1998

Fourth International Workshop on Complex Structures and Vector Fields

Aspects of Complex Analysis, Differential Geometry, Mathematical Physics and Applications

St. Konstantin, Bulgaria
September 3 – 11, 1998

Editors

Stancho Dimiev
Bulgarian Academy of Sciences, Bulgaria

Kouei Sekigawa
Niigata University, Japan

World Scientific
Singapore • New Jersey • London • Hong Kong

Published by

World Scientific Publishing Co. Pte. Ltd.
5 Toh Tuck Link, Singapore 596224
USA office: 27 Warren Street, Suite 401-402, Hackensack, NJ 07601
UK office: 57 Shelton Street, Covent Garden, London WC2H 9HE

British Library Cataloguing-in-Publication Data
A catalogue record for this book is available from the British Library.

ASPECTS OF COMPLEX ANALYSIS, DIFFERENTIAL GEOMETRY, MATHEMATICAL PHYSICS AND APPLICATIONS
Fourth International Workshop on Complex Structures and Vector Fields

ISBN-13 978-981-02-3868-1
ISBN-10 981-02-3868-1

FOURTH INTERNATIONAL WORKSHOP ON COMPLEX STRUCTURES AND VECTOR FIELDS

September 3–11, 1998, St.Konstantin, Bulgaria

CONTRIBUTED COMMUNICATIONS

1. Toshiaki ADACHI
   Kähler magnetic fields and length spectrum on a complex hyperbolic space
2. Vestilav APOSTOLOV
   Bihermitian surfaces of non-Kähler type
3. Johan DAVIDOV
   Geometry of twistor spaces (with G. GRANCHAROV and O. MUSHKAROV)
4. Stancho DIMIEV
   Holomorphic coordinates on real analytic manifolds (with D. FRENKEV and K. SPALLEK)
5. Stoil DONEV
   Photon-like $3+1$ solution solutions in extended electrodynamics
6. Albert FÄSSLER
   Block structured matrices and numerical methods for partial differential equations
7. Antonio FERNANDEZ
   Belinfante-Rosenfeld-Hilbert formulas in generalized gauge theory
8. Hideya HASHIMOTO
   On CR-submanifolds of nearly Kähler S^6 and related topics
9. Mikihiro HAYASHI
   Bounded analytic functions on Riemann surfaces
10. Milen HRISTOV
    On the integrability conditions for almost complex and almost contact structures
11. Azniv KASPARJAN
    Hermitian symmetry of infinitesimally trivial harmonic images of compact Kähler manifolds

12. Takashi KODA
Nearly Kähler weakly symmetric spaces (with L. VANHECKE and Y. WATANABE)

13. Shozo KOSHI
Ordered topological linear spaces

14. Vladimir KOSTOV
On Deligne-Simpson problem

15. Alexander KYTMANOV
Boundary Morera theorem in classical symmetric domains

16. Julian ŁAWRINOWICZ
An approach to 5-,9-and 13-dimensional complex dynamics (with O. SUZUKI)

17. Sadahiro MAEDA
Isoparametric hypersurfaces in a real space form

18. Sawa MANOFF
Invariant projections and covariant divergency of energy momentum tensors

19. Stefano MARCHIAFAVA
Complex and hypercomplex structures on quaternionic manifolds (with D. ALEKSEEVSKI and M. PONTEKORVO)

20. Ivailo MLADENOV
Geometric quantization of the assymetric rotors

21. Oleg MUSHKAROV
Weak Einstein conditions on complex surfaces (with. V. APOSTOLOV)

22. Simona MYSLIVETS
On one-side holomorphic extension of CR functions along complex curves

23. Gregory NABER
An introduction to topology and classical guage theory (five lectures)

24. Galja NAKOVA
4-dimensional submanifolds of 7-dimensional almost contact manifolds with B-metric

25. Nikolai NIKOLOV
Higher order Caratheodory and Kobayashi metrics

26. Kaoru ONO
Obstruction theory for Floer homology for Lagrangian intersections

27. Kiyoshi OHBA
Riemann surfaces with Abelian differentials (with Y. HASHIMOTO)

28. Yoshio OHNUKI
Quantization on closed manifolds

29. Georgi POPOV
Wave-trace formulae

30. Walter SCHEMPP
Unitary symplectic spinors and their applications to quantum teleportation and magnetic resonance tomography

31. Kouei SEKIGAWA
Goldberg conjecture (4-dimensional case)

32. Armen SERGEEV
Vortex equation on a Riemann surface

33. Tatsuru TAKAKURA
Hamiltonian group action, equivariant index and cobordism

34. Todor TODOROV
Boundary surface properties and commutativity of self-adjoint ectensions of certain operators of mathematical physics.

This volume is dedicated to Professor Shozo Koshi on occasion of his 70th birthday and to Professor Sumio Sawaki, who is one of our teachers.

PREFACE

This workshop continues investigations initiated in the previous Workshops held respectively in Sofia(1992), Pravetz(1994), St.Konstantin(1996). Now, the scope of interests is enlarged including Applications. This is naturally motivated by the presence of many new participants. Mickhael Gromov notices that a new breed of Mathematical Professionals is able to meditatebetween Pure Mathematics and Applied Sciences (cf. Notice AMS, 45, 7(1998), pp. 846-84). In fact,one may find that new mathematics appears in various directions, for example,in magnetic resonance imaging,in crystals,in selfsimilarity for fractacles and in many other fields. Of course, the classical trends in Analysis,Geometry and Mathematical Physics are remaining fundamentally in our work.

The editors express their deepest gratitude to Professor T. Oguro for his outstanding co-operation and efforts in the arrangement of this volume, and to Professors S. Manoff and R. Lazov who helped Professor T. Oguro in the preparation of the Fourth Workshop.

Editors

CONTENTS

NORMAL FORMS OF UNFOLDINGS OF NON-FUCHSIAN SYSTEMS

V.P. KOSTOV

Université de Nice - Sophia Antipolis
U.R.A. du C.N.R.S. N⁰ 168 Jean-Alexandre Dieudonné
Parc Valrose, 06108 Nice Cedex 2, France

We find polynomial with respect to the time holomorphic or meromorphic normal forms of unfoldings with bifurcation of the poles of non-resonant non-fuchsian linear systems on $(\mathbf{C}, 0)$.

We consider meromorphic linear systems with complex time. Such a system in the neighbourhood of a pole (which we consider to be at 0) can be written in the form

$$t^r \dot{X} = A(t)X \tag{1}$$

where $t \in (\mathbf{C}, 0)$ denotes the time, "$\cdot$" $= d/dt$, A is a holomorphic matrix-function, $A(0) \neq 0$, A and X are $n \times n$-matrices. If $r = 1$ ($r > 1$) the system is called *fuchsian* (respectively *non-fuchsian*).

It is natural to consider a non-fuchsian system (1) as the limit of a fuchsian family of the kind

$$(t - a_1) \dots (t - a_r)\dot{X} = \tilde{A}(t, a_1, \dots, a_r)X \tag{2}$$

(as the limit for $a_j \to 0$, $j = 1, \dots, r$; $\tilde{A}(t, 0, \dots, 0) \equiv A(t)$). This approach was proposed to the author by V.I. Arnold; it has been used in the works of J.-P. Ramis as well, see [17]. Family (2) is an *unfolding* (or *deformation*) of system (1).

In the present paper we find holomorphic (in the exceptional case meromorphic) normal forms of unfoldings of non-fuchsian systems. In these forms the matrix $\tilde{A}$ is a polynomial of t of degree $r - 1$ and contains the minimal possible number of non-zero coefficients in every matrix entry, see Theorems 5 and 6 and Corollaries 7 and 8. Their number is the same as the number of parameters necessary to represent a generic fuchsian system on $\mathbf{C}P^1$ with r poles up to conjugacy; it is also the same as the number of analytic invariants of a generic non-fuchsian system in $(\mathbf{C}, 0)$. Another advantage of these forms

---

Key words: (unfolding of a) non-fuchsian system, fuchsian system, monodromy operator, monodromy group.

is that they are defined for all $t \in \mathbf{C}$, i.e. they are global with respect to the time. The normalizing transformation is of the kind

$$X \mapsto W(t, a_1, \ldots, a_r)X. \tag{3}$$

The deformed system is assumed to be *non-resonant* (see Section 1). Transformation (3) is holomorphic if the deformed system is either *irreducible*, i.e. it is impossible to block-triangularize it by means of a holomorphic and holomorphically invertible transformation (3), or with a diagonalizable *monodromy operator* around 0. In the first case it is equivalent to say that there exists no proper subspace of $\mathbf{C}^n$ invariant for all *Stokes multipliers*, see [10] (for their definition see [19], [3] or [1]).

In Section 1 we give the definitions and results (see Subsection 1.1) concerning the normal forms of non-fuchsian systems and of their unfoldings and we give some remarks concerning the monodromy groups of fuchsian systems in Subsection 1.2. We prove Theorems 5 and 6 in Section 2.

1 Normal forms of unfoldings of non-fuchsian systems

1.1 Definitions and results

The theory of normal forms of non-fuchsian systems has been developed in two trends. Formal normal forms have been obtained by Turittin [18], Babbitt and Varadarajan [1] and Levelt [15]. Call *resonant* the case when the matrix $A(0)$ has multiple eigenvalues. The basic result is given by

Theorem 1. *Any non-fuchsian system (1) can be transformed to a normal form*

$$\tau^s \mathrm{d}X/\mathrm{d}\tau = (D_0 + \cdots + D_{s-2}\tau^{s-2} + C\tau^{s-1})X \tag{4}$$

by means of a formally holomorphic and holomorphically invertible transformation $X \mapsto H(t)X$ in the non-resonant case (in this case $s = r$ and $\tau = t$) or by means of a formally fractionally-meromorphic transformation $X \mapsto \tilde{H}(\tau)$, $\tau = t^{1/q}$, $q \in \mathbf{N}$ in the resonant one. The matrices D_j are diagonal and the matrix C is in Jordan normal form. The matrices $D_1, \ldots, D_{s-2}, C$ commute pairwise.

"Formally holomorphic (fractionally-meromorphic)" means that the matrix $H(t)$ (the matrix $\tilde{H}(t)$) is a formal Taylor (Laurent) power series of t (of a root of t); these series are, in general, divergent. Levelt (see [15]) has obtained an exact estimation of the number s; in general, $s \geq r$. The formal classification of non-fuchsian systems is the starting point for the analytic and meromorphic ones, see [3] and [1].

The second trend seems to begin with Birkhoff who obtains analytic normal forms for generic non-fuchsian systems, see [5]. Define the *monodromy operator* of system (1) at 0 as follows: fix a value t_0 of t close to 0 and a non-degenerate matrix G. The solutions to system (1) are, in general, ramified at 0. The monodromy operator at 0 is the linear operator mapping G into the value of the analytic continuation of the solution to the system which equals G for $t = t_0$ along the circle centered at 0 and of radius $|t_0|$. The monodromy operator is defined up to conjugacy due to the freedom to choose G and t_0. Birkhoff proves

Theorem 2. *If the monodromy operator at 0 of system (1) is diagonalizable, then there exists a holomorphic and holomorphically invertible at 0 transformation*

$$X \mapsto W(t)X \tag{5}$$

which transforms the system into one of the kind

$$t^r\dot{X} = (B_0 + B_1 t + \cdots + B_{r-1}t^{r-1})X \tag{6}$$

where B_j are constant $n \times n$-matrices.

Transformation (5) is called an *equivalence*. It changes the matrix $A(t)$ according to the rule

$$A(t) \mapsto -W^{-1}\dot{W} + W^{-1}A(t)W. \tag{7}$$

In fact, Birkhoff omits the condition that the monodromy operator must be diagonalizable without which the proof of the theorem is not true, see the counterexample given by Gantmacher in [9].

Recently A.A. Bolibrukh (A.A. Bolibruch) obtained analytic normal forms of irreducible systems, see [6] (for $n = 2$ see also [11], for $n = 3$ see also [2]). His basic result is

Theorem 3. *An irreducible system (1) can be transformed into one of the kind (6) by means of a transformation (5) holomorphic and homomorphically invertible at 0.*

In the same paper he proves (see Theorem 6 there)

Theorem 4. *A non-resonant non-fuchsian system can be transformed by means of a meromorphic at 0 transformation (5) to the normal form (6) without increasing the order of the pole at 0.*

Remark. In transformation (5) one usually assumes that $W(0) = I \equiv \text{diag}(1,\ldots,1)$. Such transformations are called *Birkhoff* ones. Note that the normal forms of Birkhoff and Bolibrukh are defined for all $t \in \mathbf{C}$. The order of the pole of a meromorphic system is often called its *Poincaré rank*.

Consider a *deformation* of system (1) of the kind

$$f(t,\lambda)\dot{X} = \tilde{A}(t,\lambda,u)X \tag{8}$$

where $\lambda \in (\mathbf{C}^{r-1},0)$, $u \in (\mathbf{C}^k,0)$, $f(t,\lambda) := t^r+\lambda_1 t^{r-2}+\ldots+\lambda_{r-1}$, $\tilde{A}$ is a holomorphic in (t,λ,u) $n\times n$-matrix-function, $\tilde{A}(t,0,0) \equiv A(t)$. Each deformation (2) is *induced* from some deformation (8), i.e. there exist functions $\lambda(a)$, $u(a)$ such that deformation (2) coincides with some deformation (8) in which we substitute $\lambda = \lambda(a)$, $u = u(a)$. In this sense deformation (8) is more general than deformation (2). Further we consider (8) to be a deformation of the normal form (6) of Birkhoff-Bolibrukh, i.e. $\tilde{A}(t,0,0) = B_0+B_1t+\cdots+B_{r-1}t^{r-1}$.

Call *exceptional* the case in which there are at least two eigenvalues of the matrix B_{r-1} whose difference is a non-zero integer. If system (6) is fuchsian, i.e. $B_0 = \ldots = B_{r-2} = 0$, then this is the definition of the resonant case; we use the adjective "exceptional" because "resonant" for non-fuchsian systems means that B_0 has (a) multiple eigenvalue(s).

Theorem 5. *If (8) is a deformation of a non-exceptional system (6), then there exists a holomorphic and holomorphically invertible transformation*

$$X \mapsto W(t,\lambda,u)X, \qquad W(0,\lambda,u) \equiv I \tag{9}$$

called equivalence and defined in $(\mathbf{C},0)\times(\mathbf{C}^{r-1},0)\times(\mathbf{C}^k,0)$ *which brings deformation (8) to the form*

$$f(t,\lambda)\dot{X} = (B_0(\lambda,u) + B_1(\lambda,u)t + \ldots + B_{r-1}(\lambda,u)t^{r-1})X \tag{10}$$

with holomorphic at 0 matrices $B_0, \ldots, B_{r-1}$.

Remark. It would be more correct to write about germs at 0 of systems, deformations and transformations; transformation (9) is, in general, defined in a smaller neighbourhood of zero than deformation (8).

Let system (6) be exceptional. Consider its deformation

$$g(t,\varepsilon)\dot{X} = \tilde{A}'(t,\varepsilon,u)X \tag{11}$$

$g(t,\varepsilon) = t^r + \varepsilon_1 t^{r-1} + \cdots + \varepsilon_{r-1} t$. Deformation (11) is more general than deformation (2) (with $a_1 = 0$) but is less general than deformation (8).

Theorem 6. *For every exceptional non-resonant system (6) there exists a transformation*

$$X \mapsto M(t)X \tag{12}$$

holomorphic and holomorphically invertible in a punctured neighbourhood of 0 and meromorphic at 0 which brings system (6) to the form (6) again (without increasing r), the new system being non-exceptional.

Corollary 7. *For any deformation (11) of an exceptional non-resonant system (6) there exists a meromorphic transformation*

$$X \mapsto (W(t,\varepsilon,u)M(t))X \tag{13}$$

which brings the deformation to the form

$$g(t,\varepsilon)\dot{X} = (B_0(\varepsilon,u) + B_1(\varepsilon,u)t + \cdots + B_{r-1}(\varepsilon,u)t^{r-1})X \tag{14}$$

with holomorphic at 0 matrices $B_0, \ldots, B_{r-1}$. The matrices W and M are defined as in (9) and (12).

Indeed, one has to combine Theorems 6 and 5 to obtain the corollary. The less general form of deformation (11) compared to (8) is necessary to allow the meromorphic transformation (12) whose divisor of the poles is $\{t = 0\}$.

Corollary 8. *Any unfolding (2) of a non-resonant non-fuchsian system (1) can be transformed to one of the kind (2) in which $\tilde{A}$ is a matrix of polynomials of t of degree $r-1$. The transformation is of the kind*

$$X \mapsto (W_j(t,a)M_j(t-a_j))X, \qquad 1 \le j \le r$$

where $X \mapsto W_j(t,a)X$ is an equivalence and the matrix $M_j(t-a_j)$ is holomorphic and holomorphically invertible in a punctured neighbourhood of 0 and holomorphic and holomorphically invertible or meromorphic at 0. The corollary is also true if the matrix $\tilde{A}$ in (2) depends holomorphically on additional parameters (in this case W_j depends holomorphically on them and M_j doesn't depend on them).

1.2 Some remarks on fuchsian systems

A fuchsian linear system on $\mathbf{C}P^1$ with $r+1$ poles one of which is at ∞ can be written in the form

$$\dot{X} = \left[\sum_{j=1}^{r} A_j/(t-a_j)\right] X.$$

The only analytic invariant of such a system is its *monodromy group*. It is defined like this: fix a point $a \in \mathbf{C}$ different from the poles of the system. Fix a non-degenerate matrix G. Fix r contours which generate the fundamental group $\pi_0(\mathbf{C}P^1 \backslash \{a_1, \ldots, a_{r+1} = \infty\})$; each of them begins at a, approaches one of the poles $a_1, \ldots, a_r$ (say, a_j), circumvents it anticlockwise and then goes back to a again. If one considers the analytic continuation of the solution to the system which equals G for $t = a$ along such a contour, then one finds that the continuation has a value $G_j = GM_j$ which is in general different from G. The matrix M_j is the matrix of the *monodromy operator* corresponding to the contour. The operators $M_1, \ldots, M_r$ generate the *monodromy group*. One has $M_{r+1} = (M_1 \ldots M_r)^{-1}$ for a suitable ordering of the points $a_1, \ldots, a_{r+1} = \infty$.

The Riemann-Hilbert problem (or Hilbert's XXI problem) is formulated as follows:

Prove that for any set of points $a_1, \ldots, a_{r+1} \in \mathbf{C}P^1$ and for any set of matrices $M_1, \ldots, M_p \in GL(n, \mathbf{C})$ there exists a fuchsian linear system with poles at $a_1, \ldots, a_{r+1}$ and only there for which the corresponding monodromy operators are $M_1, \ldots, M_r$, $M_{r+1} = (M_1 ... M_r)^{-1}$.

After the negative answer to the Riemann-Hilbert problem (A.A.Bolibrukh, see [7]) it is reasonable to state it as follows:

Give necessary and sufficient conditions for the sets $\{M_1, \ldots, M_r\}$ and $\{a_1, \ldots, a_{r+1}\}$ so that there exist a fuchsian system with these and only these poles which has this monodromy group.

The Riemann-Hilbert problem has a positive answer (for any set of poles) if one of the monodromy operators M_j is diagonalizable. This can be deduced from the correct part of Plemelj's wrong proof of the initial version of the problem, see [16]. It is worth pointing out the similarity of this fact and Theorem 2. Another case in which the new version of the problem has a positive answer for any set of poles is when the monodromy group of the system is irreducible (see [8], [13] and [14]). It is interesting to compare this result with Theorem 3.

2 Proof of Theorems 5 and 6

We prove Theorem 5 in Subsections 2.1 and 2.2 and Theorem 9 in Subsection 2.3. In Subsection 2.1 we explain the homotopy method and the plan of the proof and we end the proof of Theorem 5 in Subsection 2.2.

2.1 The homotopy method

Consider for the sake of simplicity the case when there is no dependence on the parameters u; if there is such, then the proof is the same and only the convergence of the series with respect to u has to be proved additionally; its proof is trivial.

Set $\tilde{A}(t,\lambda) = A_0(\lambda) + \cdots + t^{r-1}A_{r-1}(\lambda) + t^r A^*(t,\lambda)$. We construct a one-parameter family of transformations $\mathcal{D}(h) : X \mapsto (I + P(t,\lambda,h))X$, $h \in [0,1]$ such that $\mathcal{D}(h)$ brings family (8) to the form

$$\begin{aligned} f(t,\lambda)\dot{X} =& B(t,\lambda,h)X \equiv B_0(\lambda,h) + B_1(\lambda,h)t + \cdots + B_{r-1}(\lambda,h)t^{r-1} \\ &+ (1-h)t^r A^*(t,\lambda))X. \end{aligned} \tag{15}$$

For $h = 0$ we have $B_j(\lambda, 0) \equiv A_j(\lambda)$, for $h = 1$ we obtain (10).

We don't find the matrix-function P explicitly. We find the matrix-function $Q(t,\lambda,h) := P_h(t,\lambda,h)(I + P(t,\lambda,h))^{-1}$ which for all $h \in [0,1]$ is a convergent matrix-series of (t,λ) (in one and the same polydisc, with uniform in h estimations for the absolute values of the coefficients). Here and further the index h denotes $\partial/\partial h$.

Then we find the matrix-function P from the system of linear differential equations

$$P_h = Q(t,\lambda,h)(I+P), \qquad P|_{h=0} \equiv 0. \tag{16}$$

The matrix Q will satisfy for all $h \in [0,1]$ the equalities $Q|_{t=0} \equiv Q|_{\lambda=0} \equiv 0$. Hence, P will satisfy the conditions $P|_{t=0} \equiv P|_{\lambda=0} \equiv 0$. This implies that for small (t,λ) and for all $h \in [0,1]$ the matrix $I + P$ will be holomorphic and holomorphically invertible.

Proposition 9. *The matrix-function $Q(t,\lambda,h)$ satisfies the equation*

$$-f(t,\lambda)\dot{Q} + [B(t,\lambda,h), Q] = B_h(t,\lambda,h). \tag{17}$$

Proof. Consider two values of h – h^0, h^1 – and set $\delta = h^1 - h^0$. Denote by $(15)|_{h=h^0}$ equation (15) for $h = h^0$. The transformation

$$\Delta(h^0,\delta) \equiv \mathcal{D}(h^1) \circ \mathcal{D}^{-1}(h^0)$$

transforms system $(15)|_{h=h^0}$ into system $(15)|_{h=h^1}$. A direct computation shows that

$$\partial\tilde{P}/\partial\delta|_{\delta=0} = Q(t,\lambda,h^0)$$

where

$$\tilde{P} := ((I+P(t,\lambda,h^1))((I+P(t,\lambda,h^0))^{-1}.$$

Hence, $\tilde{P} = I + \delta Q + o(\delta)$. We have $B = B(t,\lambda,h^0) + \delta B_h(t,\lambda,h^0) + o(\delta)$. Perform transformation Δ in $(15)|_{h=h^0}$ in first approximation with respect to δ:

$$\begin{aligned} f(t,\lambda)\dot{X} &= (B(t,\lambda,h^0) + \delta(-f(t,\lambda)\dot{Q} + [B(t,\lambda,h),Q]) + o(\delta))X \\ &= (B(t,\lambda,h^0) + \delta B_h(t,\lambda,h^0) + o(\delta))X. \end{aligned}$$

Comparing the terms linear in δ we obtain the proof of the proposition. □

Denote by (17;k), $k \in \mathbf{N} \cup 0$ the equation obtained from (17) when one leaves only the coefficients before t^k considered as matrices depending on (λ,h). We first solve system $(\mathcal{A})$ =(17;r)∪ (17;$r+1$)∪ ⋯, setting $Q = tQ^1 + t^2Q^2 + \cdots$. In this system we consider the entries of the matrices B_j as parameters $\tilde{b}$ varying in fixed neighbourhoods $\mathcal{U}_j$ of the matrices $A_j(0)$, $j = 0, \ldots, r-1$.

Thus the matrices Q^s are holomorphic matrix-functions of (λ, b) where b are the differences between the parameters $\tilde{b}$ and the corresponding entries of the matrices $A_j(0)$. Hence, $b \in (\mathbf{C}^{rn^2}, 0)$. For $B_j \in \mathcal{U}_j$, $h \in [0,1]$ we obtain uniform estimations for the coefficients of the series $Q(t,\lambda,b,h)$ which verifies the condition $Q(0,\lambda,b,h) \equiv Q(t,0,b,h) \equiv 0$.

Then we solve system $(\mathcal{B})$ =(17;0)∪ ⋯ ∪(17;$r-1$). This is a system of differential equations of first order (the time is h). We show that if the coefficients of $Q(t,\lambda,b,h)$ satisfy the estimations mentioned above, then for λ sufficiently small the matrices B_j which are solutions to system $(\mathcal{B})$ remain in the neighbourhoods $\mathcal{U}_j$ for all λ sufficiently small and all $h \in [0,1]$. This follows from $Q(t,0,b,h) \equiv 0$. Solving $(\mathcal{B})$ we obtain its solutions $b(t,\lambda)$, then we substitute them in $Q(t,\lambda,b,h)$. This gives the solution Q to (17).

2.2 End of the proof of Theorem 8

1^0. Choose the neighbourhood $\mathcal{U}_{r-1}$ so small that for all $B_{r-1} \in \mathcal{U}_{r-1}$ all the differences between the eigenvalues of B_{r-1} should be outside the d-neighbourhoods of all the non-zero integers for some small but fixed $d > 0$. Such a choice is possible because the deformed system (6) is non-exceptional.

2^0. The changes of the scope of the t-axis $t \mapsto \rho t$ for $\rho > 1$ transform system (6) into the system

$$t^r \dot{X} = (B_0/\rho^{r-2} + (B_1/\rho^{r-3})t + \cdots + B_{r-1}t^{r-1})X.$$

Hence, choosing a greater ρ one can make the matrices $B_0, \ldots, B_{r-2}$ arbitrarily small (in any matrix norm). These changes correspond to shrinking with respect to t of the polydisc in which deformation (8) is defined. The choice of ρ (and of the neighbourhoods $\mathcal{U}_j$, $j = 0, \ldots, r-2$) depends on the choice of d, see below.

3^0. Let $||.||$ be some fixed matrix norm in $gl(n, \mathbf{C})$. Define two matrix norms in the space of convergent matrix power series of t: if $N(t) = N_0 + N_1 t + \cdots$ then set

$$||N||_0 = \sup_{k \in \mathbf{N} \cup 0} (\max_{0 \le j \le k} ||N_j||),$$
$$||N||_1 = \sup_{k \in \mathbf{N} \cup 0} (\max_{0 \le j \le k} ||N_j||/(j+1)).$$

Hence, $||N||_1 \le ||N||_0$. For matrices depending on (λ, b, h) all norm estimations are assumed uniform in all these parameters.

4^0. Set $f(t, \lambda) = t^r + \varphi(t, \lambda)$, $\varphi = \lambda_1 t^{r-2} + \cdots + \lambda_{r-1}$. Hence, $\varphi = O(\lambda)$. Represent equation (17) in the form

$$-t^r \dot{Q} + [B_{r-1}t^{r-1}, Q] = B_h + \varphi \dot{Q} - [B_0 + \cdots + B_{r-2}t^{r-2}, Q].$$

Define the linear operators $\mathcal{P}$, $\mathcal{Q}$, $\mathcal{R}$ as follows:

$$\mathcal{P}[Q] = -t^r \dot{Q} + [B_{r-1}t^{r-1}, Q],$$
$$\mathcal{Q}[Q] = \varphi \dot{Q},$$
$$\mathcal{R}[Q] = -[B_0 + \cdots + B_{r-2}t^{r-2}, Q].$$

Equation (17) can be written in the form

$$\mathcal{P}[Q] = B_h + \mathcal{Q}[Q] + \mathcal{R}[Q].$$

Denote by $\mathcal{S}[.]$ the $(r-1)$-jet with respect to t at 0 of the matrix-function $(.)$ and set $\mathcal{M}[.] = (\mathrm{id} - \mathcal{S})[.]$. The last form of equation (17) allows us to represent the solution to system $(\mathcal{A})$ as

$$Q = \mathcal{P}^{-1}(\mathcal{M}[B_h] + \mathcal{M}(\mathcal{Q} + \mathcal{R})[Q]), \qquad \text{i.e.}$$

$$Q = \mathcal{P}^{-1}(\mathcal{M} + \mathcal{M}(\mathcal{Q}+\mathcal{R})\mathcal{P}^{-1}\mathcal{M} + \mathcal{M}(\mathcal{Q}+\mathcal{R})\mathcal{P}^{-1}\mathcal{M}(\mathcal{Q}+\mathcal{R})\mathcal{P}^{-1}\mathcal{M} + \cdots)[B_h]$$
$$= \mathcal{P}^{-1}\mathcal{M}(\mathrm{id} + \mathcal{V} + \mathcal{V}^2 + \cdots)[B_h], \qquad \mathcal{V} = (\mathcal{Q}+\mathcal{R})\mathcal{P}^{-1}\mathcal{M}.$$

Remark 1. We assume that the solution Q to the equation $\mathcal{P}^{-1}\mathcal{M}[Q] = T$ has no constant term with respect to t.

5^0. Evidently, if $||Q||_j = a$, then $||\mathcal{M}[Q]||_j \leq a$, $j = 0, 1$. Estimate the norms of the operators $\mathcal{P}^{-1}\mathcal{M}$, $\mathcal{Q}$ and $\mathcal{R}$.

Lemma 10. *There exists a constant $c > 0$ depending only on d, see 1^0, such that if $||N||_1 = a$, then $||\mathcal{P}^{-1}\mathcal{M}[N]||_0 \leq ca$.*

Proof. The eigenvalues of the linear operator

$$\mathcal{T}_k : gl(n, \mathbf{C}) \to gl(n, \mathbf{C}), \qquad \mathcal{T}_k : C \mapsto -kC + [B_{r-1}, C]$$

are equal to $\lambda_i - \lambda_j - k$, see [9], where (λ_i, λ_j) are all the pairs of eigenvalues of B_{r-1} (this operator shows the action of $\mathcal{P}^{-1}\mathcal{M}$ upon the matrix-coefficient before t^k). Hence, for $k \in \mathbf{N}^*$ and $B_{r-1} \in \mathcal{U}_{r-1}$ one has $|\lambda_i - \lambda_j - k| \geq d > 0$. For $k \to \infty$ one has $-k\mathcal{T}_k^{-1}(C) \to C$ uniformly for $||C|| = 1$. This implies the existence of the constant $c > 0$. It depends only on d.

The lemma is proved. □

Fix $d > 0$ and choose $\rho > 1$ so great and the neighbourhoods $\mathcal{U}_0, \ldots, \mathcal{U}_{r-2}$ such that if $||N||_j = a$, then $||\mathcal{R}[N]||_j \leq a/4c$, c is from Lemma 10, $j = 0, 1$.

Lemma 11. *There exists a constant $c' > 0$ such that $||\mathcal{Q}[.]||_1 \leq c' \max(|\lambda_1|, \ldots, |\lambda|_{r-1})||.||_0$.*

Indeed, if $Q = Q^0 + tQ^1 + \cdots$, then $\dot{Q} = Q^1 + 2tQ^2 + \cdots$ and

$$\begin{aligned}\lambda_j t^{r-1-j}\dot{Q} &= \lambda_j \sum_{k=1}^{\infty} t^{r-2-j+k} kQ^k \\ &= \lambda_j \sum_{k=1}^{\infty} t^{r-2-j+k} \left(\frac{k}{r-1-j+k}\right)(r-1-j+k)Q^k.\end{aligned}$$

There exists $c'' > 0$ such that $|k/(r-1-j+k)| \leq c''$ for all $j = 1, \ldots, r-1$; $k = 1, 2, \ldots$. Hence, one can set $c' = (r-1)c''$.

6^0. Let $\lambda_1, \ldots, \lambda_{r-1}$ vary in a polydisc of a radius so small that $||\mathcal{Q}[.]||_1 \leq (1/4c)||.||_0$. Then we have

$$\begin{aligned}&||(\mathcal{Q} + \mathcal{R})\mathcal{P}^{-1}\mathcal{M}[.]||_1 \\ \leq &||\mathcal{Q}\mathcal{P}^{-1}\mathcal{M}[.]||_1 + ||\mathcal{R}\mathcal{P}^{-1}\mathcal{M}[.]||_1 \\ \leq &(1/4c)||\mathcal{P}^{-1}\mathcal{M}[.]||_0 + (1/4c)||\mathcal{P}^{-1}\mathcal{M}[.]||_1 \\ \leq &(1/2c)||\mathcal{P}^{-1}\mathcal{M}[.]||_0 \leq (1/2)||.||_1.\end{aligned}$$

The inequalities follow from 5^0. Hence, the operator $\mathcal{V}$ involved in the definition of Q is contracting and the series $Q = \mathcal{P}^{-1}\mathcal{M}(\mathrm{id} + \mathcal{V} + \mathcal{V}^2 + \cdots)[B_h]$ is convergent. This series doesn't depend on all coefficients of B_h as a matrix-series of t but only on B_h^r, B_h^{r+1}, ... because of the presence of $\mathcal{M}$ in the definition of $\mathcal{V}$.

Hence, the estimations for its coefficients are uniform in h as $\mathcal{M}[B_h]$ doesn't depend on h.

The definition of the operator $\mathcal{P}^{-1}\mathcal{M}$ implies that $Q|_{t=0} \equiv 0$. If $\lambda = 0$, then $\mathcal{M}[B_h] \equiv 0$, hence, $Q|_{\lambda=0} \equiv 0$.

The estimations are uniform in b as well because the neighbourhoods $\mathcal{U}_j$ are fixed.

7^0. System $(\mathcal{B})$ can be written in the form

$$\mathcal{S}[B_h] = \mathcal{S}[-f(t,\lambda)\dot{Q} + [B, Q]].$$

Its right hand-side is identically equal to 0 for $\lambda = 0$. Fix $\rho > 1$ as above and the neighbourhoods $\mathcal{U}_j$, i.e. the bounds for the parameters b and shrink with respect to λ the polydisc in which the system is defined. For λ sufficiently small, ρ fixed and b bounded, the right hand-side of the last equation is $O(\max(|\lambda_1|, \ldots, |\lambda_{r-1}|)$. Hence, for such (λ, ρ, b) and for all $h \in [0,1]$ the solution to this equation $b(\lambda, h)$ $(b(\lambda, 0) = 0)$ belongs to the neighbourhood $\mathcal{U}_0 \times \cdots \times \mathcal{U}_{r-1}$ of $0 \in \mathbf{C}^{rn^2}$. Substituting $b = b(\lambda, h)$ in $Q(t, \lambda, b, h)$ found in $4^0 - 6^0$, we obtain $Q(t, \lambda, h)$ – the solution to (6) possessing all the properties claimed in the theorem.

The theorem is proved.

2.3 Proof of Theorem 6

1^0. Let in system (6) the matrix B_{r-1} be in upper-triangular Jordan normal form. Let Σ denote a set of eigenvalues of B_{r-1} differing from one another by integers and let there be at least one non-zero integer among them. Let $\sigma \in \Sigma$ be the eigenvalue with greatest real part. Assume that (one of) its Jordan block(s) is placed in the right lower corner of B_{r-1}. The "shearing" transformation

$$X \mapsto \mathrm{diag}(1, \ldots, 1, t)X \tag{18}$$

brings the matrix B_{r-1} to the form $\begin{pmatrix} K & 0 \\ * & \sigma - 1 \end{pmatrix}$ where the $(n-1) \times (n-1)$-block K hasn't changed and the terms $*$ come into B_{r-1} from B_{r-2}. Using such transformations one can make all integer differences equal to 0. However, the order of the pole at 0 of the system, in general, increases. In order to avoid this we use more complicated transformations but based on the same idea – to decrease by 1 the real part of some eigenvalue from Σ for all such sets Σ.

2^0. Partition the matrices B_{r-1} and B_0 into blocks:

$$B_{r-1} = \begin{pmatrix} K & L \\ 0 & \sigma \end{pmatrix}, \qquad B_0 = \begin{pmatrix} \Lambda & D \\ C & q \end{pmatrix}$$

where K and Λ are $(n-1)\times(n-1)$-blocks. We assume that Λ is in Jordan normal form (this can be achieved by means of a transformation $X \mapsto \begin{pmatrix} R & 0 \\ 0 & 1 \end{pmatrix} X$ with a constant matrix $R \in GL(n-1, \mathbf{C})$). Therefore from now on K and B_{r-1} are not assumed to be in Jordan normal form.

3^0.

Lemma 12. *There exists a transformation*

$$X \mapsto \begin{pmatrix} S_1 & S_2 \\ 0 & 1 \end{pmatrix} X, \qquad S_1 \in GL(n-1, \mathbf{C})$$

after which in the matrix B_0 we have $D = 0$; B_0 is assumed to be non-resonant.

Hence, if one performs transformation (18) after the one described in the lemma, then the order of the pole at 0 of the system doesn't increase and $\mathrm{Re}\sigma$ is decreased by 1.

The theorem is proved.

Proof of Lemma 12. **A)** Call a Jordan block of Λ *covered* if in the block C and the columns of B_0 where this block is situated there is at least one non-zero entry. Assume that all non-covered Jordan blocks of Λ are situated in its left upper corner. Hence, B_0 has a block upper-triangular structure and all eigenvalues of the non-covered blocks are eigenvalues of B_0 as well, which implies that they are distinct. Represent B_0 in the form

$$B_0 = \begin{pmatrix} \Lambda' & 0 & D' \\ 0 & \Lambda'' & D'' \\ 0 & C'' & q \end{pmatrix}, \quad C = (0, C''), \quad \Lambda = \begin{pmatrix} \Lambda' & 0 \\ 0 & \Lambda'' \end{pmatrix}$$

(Λ' contains the non-covered Jordan blocks of Λ). Set $\begin{pmatrix} \Lambda'' & D'' \\ C'' & q \end{pmatrix} = F$, $(0, D') = G$, $q \in \mathbf{C}$. Hence, F and Λ' have no common eigenvalues as B_0 is non-resonant.

B) Perform the transformation $\mathcal{T}_1 : X \mapsto \begin{pmatrix} I & Q \\ 0 & I \end{pmatrix} X$ where the block Q is of the size of G. This transformation changes the matrix B_0 as follows:

$$B \mapsto \begin{pmatrix} \Lambda' & G + \Lambda' Q - QF \\ 0 & F \end{pmatrix}.$$

The eigenvalues of the linear operator $\mathcal{S} : Q \mapsto \Lambda' Q - QF$ are all the differences between an eigenvalue of Λ' and an eigenvalue of F, see [9]. Hence, they are all non-zero.

Choose Q such that $G + \Lambda' Q - QF = 0$. This makes the matrix B_0 block-diagonal. All further transformations will be of the kind $X \mapsto \begin{pmatrix} I & 0 \\ 0 & R \end{pmatrix} X$ where the sizes of I and R are the same as the ones of Λ' and F. Therefore we omit in what follows the block I and assume that the block decomposition of the block R is the same as the one of F, see **A)**.

C) Perform a transformation $\mathcal{T}_2 : X \mapsto RX$, $R = \begin{pmatrix} I & T \\ 0 & 1 \end{pmatrix}$, T being of the size of D''. Hence, $R^{-1} = \begin{pmatrix} I & -T \\ 0 & 1 \end{pmatrix}$ and

$$R^{-1} F R = \begin{pmatrix} \Lambda'' - TC & D'' - Tq + \Lambda'' T - TC'' T \\ C'' & q + C'' T \end{pmatrix}.$$

In the rest of the proof we show that one can choose a matrix T such that

$$D'' - Tq + \Lambda'' T - TC'' T = 0. \tag{19}$$

The transformation claimed by the lemma is $\mathcal{T}_2 \circ \mathcal{T}_1$.

D) Set $T = \begin{pmatrix} t_1 \\ \vdots \\ t_k \end{pmatrix}$, $D'' = \begin{pmatrix} d_1 \\ \vdots \\ d_k \end{pmatrix}$, $C'' = (c_1, \ldots, c_k)$ and $l = C''T \equiv c_1 t_1 + \cdots + c_k t_k$. By the letter J (indexed or not) we denote Jordan blocks with eigenvalue λ_0.

Consider first the private case when $\Lambda'' = J$. Then equation (19) can be written in the form

$$\left.\begin{aligned} (\lambda_0 - q) t_1 + t_2 - l t_1 &= d_1 \\ &\cdots \\ (\lambda_0 - q) t_{k-1} + t_k - l t_{k-1} &= d_{k-1} \\ (\lambda_0 - q) t_k - l t_k &= d_k \end{aligned}\right\} \tag{20}$$

Hence,

$$t_1 = \frac{d_1 - t_2}{(\lambda_0 - q) - l}, \ t_2 = \frac{d_2 - t_3}{(\lambda_0 - q) - l}, \dots, \ t_k = \frac{d_k}{(\lambda_0 - q) - l}$$

so

$$t_j = \sum_{\nu=j}^{k} \frac{(-1)^{\nu-j} d_\nu}{((\lambda_0 - q) - l)^{\nu-j+1}} \tag{21}$$

and

$$l = \sum_{j=1}^{k} \sum_{\nu=j}^{k} \frac{(-1)^{\nu-j} c_j d_\nu}{((\lambda_0 - q) - l)^{\nu-j+1}}. \tag{22}$$

If $d_1 = \dots = d_k = 0$, then we set $T = 0$. If there exists m, $1 \le m \le k$ such that $d_m = \dots = d_k = 0$, then we set $t_m = \dots = t_k = 0$ and we reduce the number of equations of system (20) from k to $m - 1$ without changing its form. Let $d_k \neq 0$. The coefficient before $1/((\lambda_0 - q) - l)^s$ in (22) equals

$$\begin{aligned} a_s = & \sum_{\nu-j+1=s \ , \ j \le \nu \le k \ , \ 1 \le j \le k} (-1)^{\nu-j} c_j d_\nu = (-1)^{s-1} \sum_{\nu=s}^{k} c_{\nu+1-s} d_\nu \\ = & (-1)^{s-1}(c_1 d_s + c_2 d_{s+1} + \dots + c_{k+1-s} d_k). \end{aligned}$$

Hence, the system of linear equations $a_1 = \dots = a_k = 0$ with unknown variables c_j has a unique solution $c_1 = \dots = c_k = 0$ (because its matrix is triangular, with $d_k \neq 0$ on the diagonal). Hence, as $C'' \neq 0$, at least one of the coefficients a_s is not zero.

Consider the function $\varphi = l - \sum_{s=1}^{k} a_s/((\lambda_0 - q) - l)^s$, $l \in \mathbf{C}P^1$. It has poles at ∞ and at $\lambda_0 - q$. Hence, it has at least one zero – l_0 – different from them. Formulas (21) with $l = l_0$ give a solution to system (20).

E) *The matrix* Λ'' *from* **B)** *for its every eigenvalue* λ_0 *contains no more that two Jordan blocks with this eigenvalue.*

Indeed, if it does, then $\mathrm{rk}(F - \lambda_0 I) \le \mathrm{rk}F - 2$ which implies that λ_0 is a multiple eigenvalue of F and, hence, of B_0 as well.

Let

$$F = \begin{pmatrix} H & 0 & 0 & D_1'' \\ 0 & J_1 & 0 & D_2'' \\ 0 & 0 & J_2 & D_3'' \\ C_1'' & C_2'' & C_3'' & q \end{pmatrix}. \tag{23}$$

The block H contains the set of Jordan blocks of F with eigenvalues different from λ_0, the eigenvalue of the Jordan blocks J_1 and J_2 is λ_0 (one of them might not be present). Set $\det(F - \lambda I) = g(\lambda - \lambda_0) + o(\lambda - \lambda_0)$. We find necessary and sufficient conditions for the coefficient g to be non-zero; this is equivalent to the fact that λ_0 is a simple eigenvalue of F. Denote the size of the block H in (23) by p.

Let λ be close to λ_0. Then $\det(H - \lambda I) \neq 0$. Subtract the first p rows of F multiplied by suitably chosen germs of holomorphic functions of $\lambda - \lambda_0$ to achieve the condition $c_1 = \cdots = c_p = 0$. From the other entries of $F - \lambda I$, this subtraction can change only q; it preserves $\det(F - \lambda I)$. After the subtraction one obtains a block upper-triangular matrix with diagonal blocks $(H - \lambda I)$ and $(P - \lambda I)$ where $P = \begin{pmatrix} J_1 & 0 & D_2'' \\ 0 & J_2 & D_3'' \\ C_2'' & C_3'' & \tilde{q} \end{pmatrix}$. Hence, $\det(F - \lambda I)=\det(H - \lambda I)\times\det(P - \lambda I)$ and $g \neq 0$ if and only if $\det(P - \lambda I)/(\lambda - \lambda_0) \neq 0$. Therefore in what follows we consider the case when the blocks H, C' and D' in F are missing, i.e. $F = P$.

If both blocks J_1 and J_2 are present, then one of them is of size 1.

Indeed, let both blocks be of size > 1. Develop $\det(P - \lambda I)$ with respect to the first column:

$$\det(P - \lambda I) = (\lambda - \lambda_0) \det F^* \pm c_1 \det F^{**}$$

for self-evident matrices F^* and F^{**}. F^* looks exactly as $F - \lambda I$ but the sizes of the blocks J_1, C'' and D'' are decreased by 1. Hence, $\lambda - \lambda_0$ is an eigenvalue of F^* and $(\lambda - \lambda_0) \det F^* = o(\lambda - \lambda_0)$. Develop F^{**} with respect to the first row:

$$\det F^{**} = \pm d_1 (\lambda - \lambda_0) \det F^{***} \pm \det F^{****}.$$

$\det F^{***}$ is divisible by $\det(J_2 - \lambda I)$. Hence, $\det F^{***} = o(\lambda - \lambda_0)$. If the size of J_1 is 2, then one finds that $\det F^{****} = \pm d_2 \det(J_2 - \lambda I) = o(\lambda - \lambda_0)$. If it is > 2, then one continues to develop the determinant in the same way and obtains only terms which are $o(\lambda - \lambda_0)$.

Let J_2 be of size 1, i.e. $P = \begin{pmatrix} \lambda_0 & 0 & d_{k-1} \\ 0 & \lambda_0 & d_k \\ c_{k-1} & c_k & \tilde{q} \end{pmatrix}$. One checks directly that $g \neq 0$ if and only if $c_{k-1} d_{k-1} + c_k d_k \neq 0$.

Let now J_2 be of size > 1. Develop $\det(P - \lambda I)$ with respect to the first column:

$$\det(P - \lambda I) = (\lambda_0 - \lambda)\det F^* \pm c_1 \det F^{**}, \ \det F^{**} = \pm d_1 \det(J_2) = o(\lambda - \lambda_0)$$

$$\det F^*|_{\lambda = \lambda_0} = \pm c_2 d_k.$$

Hence, in this case $g \neq 0$ if and only if $c_2 d_k \neq 0$.

F) Let $\Lambda'' = \begin{pmatrix} \lambda_0 & 0 \\ 0 & \lambda_0 \end{pmatrix}$. Then equation (19) can be written in the form

$$\left.\begin{aligned} ((\lambda_0 - q) - l)t_1 = d_1 \\ ((\lambda_0 - q) - l)t_2 = d_2 \end{aligned}\right\} \tag{24}$$

Hence,

$$t_j = d_j/((\lambda_0 - q) - l), \qquad j = 1, 2, \tag{25}$$

$$l = (c_1 d_1 + c_2 d_2)/((\lambda_0 - q) - l). \tag{26}$$

According to **E)** $c_1 d_1 + c_2 d_2 \neq 0$. Hence, equation (26) has at least one solution – l_0 – different from ∞ and $\lambda - q$, see **D)**. Setting $l = l_0$ we find from (24) t_1 and t_2.

G) Let $\Lambda'' = \begin{pmatrix} \lambda_0 & 0 \\ 0 & J^* \end{pmatrix}$ where J^* is of size > 1. System (19) takes the form

$$\left.\begin{aligned} ((\lambda_0 - q) - l)t_1 &= d_1 \\ ((\lambda_0 - q) - l)t_2 + t_3 &= d_2 \\ &\cdots \\ ((\lambda_0 - q) - l)t_{k-1} + t_k &= d_{k-1} \\ ((\lambda_0 - q) - l)t_k &= d_k \end{aligned}\right\} \tag{27}$$

Hence,

$$\begin{aligned} t_1 &= d_1/((\lambda_0 - q) - l), \\ t_j &= \sum_{\nu=j+1}^{k+1} (-1)^{\nu-j-1} d_\nu/((\lambda_0 - q) - l)^{\nu-j}, \quad j = 2, \ldots, k, \end{aligned} \tag{28}$$

$$l = \Phi(l) \equiv \frac{c_1 d_1}{(\lambda_0 - q) - l} + \sum_{j=2}^{k} \sum_{\nu=j+1}^{k+1} \frac{(-1)^{\nu-j-1} c_j d_\nu}{((\lambda_0 - q) - l)^{\nu-j}}.$$

The coefficient before $1/((\lambda_0 - q) - l)^{k-1}$ is equal to $(-1)^{k-2} c_2 d_k$; it is non-zero according to **E)**. Hence, the function $l - \Phi(l)$ has poles at ∞ and $\lambda - q$ and at least one zero – l_0 – different from them. Setting $l = l_0$ in (27), we find t_1, $\ldots$, t_k.

H) Consider the general case, i.e. Λ'' has different eigenvalues which satisfy the conditions obtained in **E)**. Equation (19) becomes a direct sum of systems of the kinds (20), (24) and (25). For everyone of the direct summands express each t_j as in (21), (25) or (28). Using these expressions, represent $l = c_1t_1 + \cdots + c_kt_k$ in the form

$$l = \tilde{\Phi} \equiv \sum_{\mu}\sum_{\nu} a_{\mu\nu}/((\lambda_\mu - q) - l)^\nu, \qquad a_{\mu\nu} \in \mathbf{C}$$

where λ_μ are all the eigenvalues of Λ''. The coefficients before $1/((\lambda_\mu - q) - l)^\nu$ for μ fixed depend only on d_i, c_i in the lines and columns of F in which the one or two Jordan blocks with eigenvalue λ_μ are situated (this can be deduced easily from the form of the right upper block of $R^{-1}FR$, see **C**)).

As in **D**) and **F**) we show that for every λ_μ at least one of the coefficients before $1/((\lambda_\mu - q) - l)^\nu$ is non-zero. Hence, the function $l - \tilde{\Phi}$ has poles at ∞ and at all points $\lambda_\mu - q$, i.e. it has at least one zero – l_0 – different from them. Setting $l = l_0$ in (19) we obtain for each direct summand the values of the unknowns t_j. The lemma is proved. □

Acknowledgement

V.I. Arnold and Yu.S. Il'yashenko introduced me into the analytic theory of the linear systems. Most part of this paper was finished during my post doctoral study in Utrecht where my supervisor was D. Siersma. The final impetus for writing the paper was given by the discussion with A.A. Bolibrukh. I would like to thank them all most sincerely.

References

1. D.G. Babbitt and V.S. Varadarajan, *Local moduli for meromorphic differential equations*, Astérisque (1989), 169–170.
2. W. Balser, *Meromorphic transformations to Birkhoff standard form in dimension three*, J. Fac. Sci. Univ. of Tokyo, Sec. IA **36**(1989), 233–246.
3. W. Balser, W.B. Jurkat and D.A. Lutz, *A general theory of invariants for meromorphic differential equations. Part 1. Formal invariants*, Funkcialaj Ekwacioj **22**(1979), 197–221.
4. ———, W. Balser, W.B. Jurkat and D.A. Lutz, *A general theory of invariants for meromorphic differential equations. Part 2. Proper invariants*, Funkcialaj Ekwacioj **22**(1979), 257–283.
5. G.D. Birkhoff, *A theorem of matrices of analytic functions*, Math. Ann. **74**(1913), 122–133.
6. A.A. Bolibruch, *Fuchsian systems and holomorphic vector bundles on the Riemann sphere*, Université de Nice – Sophia Antipolis, U.R.A. du

C.N.R.S. N^0 168 Jean-Alexandre Dieudonné, Prépublication N^0 347, juin 1993.

7. ______, *The Riemann-Hilbert problem*, Russian Mathematical Surveys **45**(1990), 1–49.
8. ______, *On the sufficient conditions for the positive resolution of the Riemann-Hilbert problem*, Mathematical Notes, Acad. Sci. of the U.S.S.R. (Mathematics of the USSR – Zametki) **51**(1992).
9. F.R. Gantmacher, *Theory of matrices*, Chelsea, New York, 1959.
10. W.B. Jurkat, *Meromorphe Differentialgleichungen*, in Lecture Notes in Mathematics, 637, 1978.
11. W.B. Jurkat, D.A. Lutz and A. Peyerimhoff, *Birkhoff invariants and effective calculations for meromorphic linear differential equations. Part 1*, J. Math. Anal. Appl. **53**(1976), 438–470.
12. ______, W.B. Jurkat, D.A. Lutz and A. Peyerimhoff, *Birkhoff invariants and effective calculations for meromorphic linear differential equations. Part 2*, Houston J. Math. **2**(1976), 207–238.
13. V.P. Kostov, *Fuchsian linear systems on* $\mathbf{C}P^1$ *and the Riemann-Hilbert problem*, C. R. Acad. Sci. Paris **1**(1992), 143–148.
14. ______, *Fuchsian linear systems on* $\mathbf{C}P^1$ *and the Riemann-Hilbert problem*, Prepublication de l'Université de Nice – Sophia Antipolis, P.U.M.A. N^0 303, November 1991.
15. A.H.M. Levelt, *Jordan decomposition for a class of singular differential operators*, Arkiv för Matematik **13**(1975), N^0 1.
16. J. Plemelj, *Problems in the sense of Riemann and Klein*, Inter. Sc. Publ. New York-Sydney, 1964.
17. J.-P. Ramis, Confluence et résurgence, manuscript.
18. H.L. Turrittin, *Convergent solutions of ordinary linear homogeneous differential equations in the neighbourhood of an irregular singular point*, Acta Mathematica **93**(1955), 27–66.
19. W.R. Wasow, *Asymptotic expansions for ordinary differential equations*, Huntington, New York, Krieger, 1976.

BLOCKSTRUCTURE AND EQUIVALENCE OF MATRICES

EUGENE L. ALLGOWER
Colorado State University, Department of Mathematics,
Ft. Collins, Colorado 80523 USA

ALBERT F. FÄSSLER
Berner Fachhochschule, University of Applied Sciences,
Biel School of Engineering and Architecture,
CH-2501 Biel/Bienne Switzerland

1 Introduction

Often, a problem of linear algebra as a final step of the solution of an applied problem manifests itself in a matrix which has a block-structure or equivariance properties, that is to say symmetries. Here we show how to handle these patterns when dealing with eigenvalue problems or systems of linear equations. With the use of the tensor product calculus and the use of group theory the complexity as well as the quality of the results can be essentially improved. These tools are discussed for symbolic computations (parameters) as well as for purely numerical ones. Basically, the above goals are attained by considering a block diagonalization of the linear operator.

2 Block Patterns

We consider block structured matrices which can be expressed as a tensor product or as a sum of them. Let the matrix P be a tensor product of an $m \times m$-matrix A and $n \times n$-matrix B, which is defined by the $mn \times mn$-matrix

$$P = A \otimes B = \begin{pmatrix} a_{11}B & a_{12}B & \cdots & a_{1m}B \\ a_{21}B & a_{22}B & \cdots & a_{2m}B \\ \vdots & \vdots & \vdots & \vdots \\ a_{m1}B & a_{m2}B & \cdots & a_{mm}B \end{pmatrix}. \tag{1}$$

We assume in this Section, that we know a block diagonalization for A as well as for B, i.e. we know a nonsingular $m \times m$-matrix S such that

$$\widetilde{A} = S^{-1}AS = \bigoplus_{j=1}^{J} A^{(j)} = \begin{pmatrix} A^{(1)} & & & \\ & A^{(2)} & \text{O} & \\ & \text{O} & \ddots & \\ & & & A^{(j)} \end{pmatrix},$$

where the $A^{(j)}$ are $m_j \times m_j$-blocks and similarly we know a nonsingular $n \times n$-matrix T such that

$$\widetilde{B} = T^{-1}BT = \bigoplus_{j=1}^{K} B^{(j)} = \begin{pmatrix} B^{(1)} & & & \\ & B^{(2)} & \text{O} & \\ & \text{O} & \ddots & \\ & & & B^{(K)} \end{pmatrix},$$

where the $B^{(j)}$ are $n_j \times n_j$-blocks. Methods for obtaining block diagonalizations are discussed in the following sections.

If we have $m \times m$-matrices M_1, M_2 and $n \times n$-matrices N_1, N_2, then the familiar multiplication rule is

$$(M_1 \otimes N_1)(M_2 \otimes N_2) = (M_1 M_2) \otimes (N_1 N_2).$$

With

$$(S \otimes T)^{-1} = S^{-1} \otimes T^{-1}$$

we get the following block diagonalization for our tensor product P:

$$\widetilde{P} = (S \otimes T)^{-1}(A \otimes B)(S \otimes T) = (S^{-1}AS) \otimes (T^{-1}BT) = \widetilde{A} \otimes \widetilde{B} \quad (2)$$

$$= \bigoplus_{j=1}^{J} (A^{(j)} \otimes \widetilde{B}) = \begin{pmatrix} A^{(1)} \otimes \widetilde{B} & & & \\ & A^{(2)} \otimes \widetilde{B} & & \text{O} \\ & \text{O} & \ddots & \\ & & & A^{(J)} \otimes \widetilde{B} \end{pmatrix}. \quad (3)$$

Hence eigenvalue problems as well as systems of linear equations concerning the tensor product $P = A \otimes B$ can be decomposed into $m_j n$-dimensional subproblems with sparse matrices $A^j \otimes \widetilde{B}$.

In the special case, where both factors A and B can be completely diagonalized by the transformation matrices S and T we get also the large matrix $\widetilde{P}$ diagonal.

If

$$a = (a_1, a_2, \dots, a_m)^T \quad \text{and} \quad b = (b_1, b_2, \dots, b_n)^T$$

are eigenvectors of A and B respectively, then their tensor product

$$a \otimes b = (a_1 b_1, a_1 b_2, \ldots, a_1 b_n, a_2 b_1, a_2 b_2, \ldots, a_2 b_n, a_m b_1 . a_m b_2, \ldots, a_m b_n)^T$$

is an eigenvector of P. Taking all possible tensor products gives us the mn desired eigenvectors of P.

A block diagonalization can also be obtained for a *sum* of tensor products

$$\sum_{i=1}^{N} A_i \otimes B_i$$

if we assume, that the $m \times m$-matrices A_i and the $n \times n$-matrices B_i can be transformed by S and T respectively *simultaneously* into the same-sized block diagonalization, because of

$$(S \otimes T)^{-1} (\sum_{i=1}^{N} A_i \otimes B_i)(S \otimes T) = \sum_{i=1}^{N} (S^{-1} A_i S) \otimes (T^{-1} B_i T). \tag{4}$$

3 Equivariance, symmetry properties

Besides all the classical algorithms for block diagonalizations in linear algebra and numerical analysis we want to emphasize a group theoretically based algorithm if a factor A or B, or even both have symmetry properties, that is so to say equivariance. This Section describes this briefly and gives examples. For a detailed discussion of the theory in Subsection 3.1 see [8].

3.1 Commutator algebra of a representation

A representation θ of a group G is a homomorphism $\theta : s \mapsto D(s)$, $s \in G$, where the $D(s)$ are nonsingular linear transformations of a vector space V, that is to say we have the properties: $D(s)D(t) = D(st)$ for $\forall s, t \in G$. For the identity $e \in G$ it follows that $D(e)$ is the identity transformation.

Let us assume that a given linear operator A commutes with a representation of a finite group G in an d-dimensional vector space V, that is to say $AD(s) = D(s)A$ holds for $\forall s \in G$.

Representation theory gives a proof for the existence of a suitable transformation matrix S, such that

$$\widetilde{D}(s) = S^{-1} D(s) S = \bigoplus_{i=1}^{N} c_i D_i(s), \tag{5}$$

where all the invariant subspaces are irreducible: there are no proper invariant subspaces left under *all* transformations. This is the theorem about the so-called *complete reducibility*. The $c_i \in \{1, 2, 3, \dots\}$ are the multiplicities of the irreducible representations $\theta_i : s \mapsto D_i(s)$, $s \in G$, of dimensions d_i. For a finite group G, there exists only a *finite* number of irreducible and inequivalent representations θ_i. For many groups, these are tabulated or implemented into appropriate software, for example in [10]. We call the set of the columns of the matrix S in (5) a *symmetry adapted basis.*

Finally, by renumbering the symmetry adapted basis vectors suitably, our linear operator described by the matrix A in the original basis has the following block diagonal form

$$\widetilde{A} = d_1 A_1 \otimes d_2 A_2 \otimes \cdots \otimes d_K A_K \tag{6}$$

with $c_i \times c_i$-blocks A_i. Here A_i is repeated d_i times on the diagonal. Therefore, we have an *exchange of the dimensions and multiplicities* between the block diagonalization of the representing matrices $D(s)$ in (5) on one hand and the block diagonalization of our equivariant linear operator described by $\widetilde{A}$ in (6) on the other hand.

We shell demonstrate the technique mentioned here by use of a representative example in the next subsection. The symmetry adapted basis is taken from the tabular description in [8,Chapter 3].

For an analysis of the complexity of the equivariance techniques, see for example, [6] or [2].

The crucial question is of course, how to get a symmetry adapted basis in general. In principle, there are two possibilities:

- With the knowledge of the representation θ and the list of the irreducible representations θ_i of the group G under consideration, there exists an *algorithm based* on *projectors* which *generates a symmetry adapted basis.*
- For any permutation representation on the finite point groups in the plane or in the space, see [9], where there are symmetry adapted bases tabulated.

3.2 Matrix with pentagonal symmetry

Many applied problems exhibit a symmetry structure which manifests itself in a group of transformations. Among these there are engineering structures, brusselators, finite element and boundary element discretizations. Let us consider an example that involves the symmetry of the regular pentagon that represent these.

We consider a 15×15-matrix $A = (a_{jk})$, whose structure is defined by the directed graph of the figure 1.

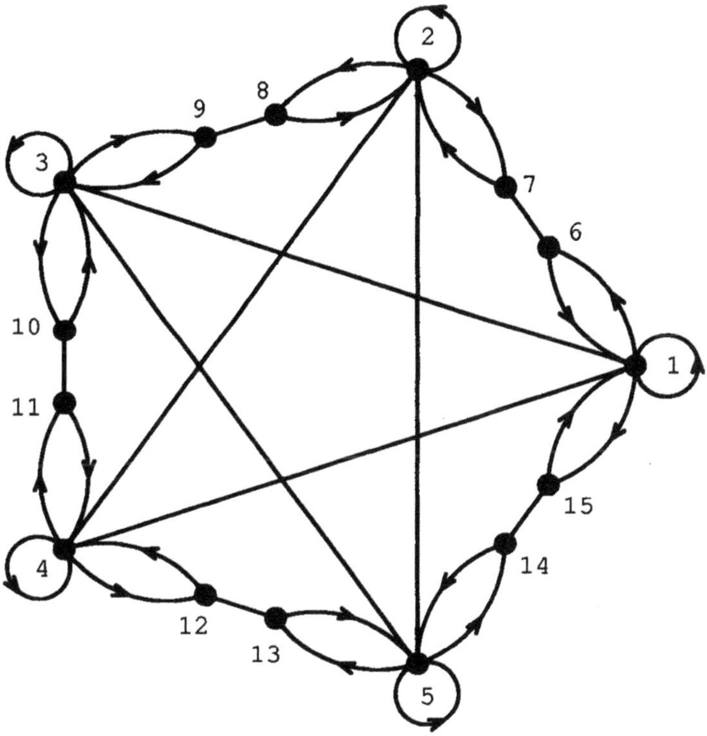

Figure 1.

- An existing edge pointing from nod j to node k means that the matrix element $a_{jk} \neq 0$.
- A strait line between the nodes j and k means $a_{jk} = a_{kj} \neq 0$.
- If there is no edge between the nodes j and k, then $a_{jk} = 0$.

We assume that the graph as well as the values of the matrix elements are invariant under the five rotations around the centre with angles $k.\frac{2\pi}{5}$, $k = 0, \ldots, 4$ and five reflections with respect to the symmetry axes. Those ten elements form the so called *dihedral group* D_5. As a consequence, for the matrix elements of A the following equations hold:

$$\begin{aligned} a_{11} = a_{22} = \cdots = a_{55} &= a \\ a_{16} = a_{27} = a_{28} = \cdots = a_{1\,15} &= b \\ a_{61} = a_{72} = a_{82} = \cdots = a_{15\,1} &= c \\ a_{67} = a_{76} = a_{89} = \cdots = a_{14\ 15} = a_{15\ 14} &= d \\ a_{13} = a_{31} = a_{14} = \cdots = a_{35} = a_{53} &= e. \end{aligned}$$

So, the non-symmetric matrix A contains the parameters a, b, c, d, e. Each element s of the group D_5 corresponds to a permutation $\pi(s)$ of the fifteen nodes. Our linear operator A is acting on vectors, whose coordinates can best be visualized by assigning them to the different nodes in the figure. Let us denote the canonical basis with $e_1, e_2, \ldots, e_{15}$: all coordinates of the vector e_i are 0, except the i the coordinate which is 1.

Now, the permutations of the nodes correspond to a linear transformation of our 15-dimensional complex vector space, which can be described by the following permutation matrices:

$$s \mapsto \Pi(s) = (e_{\pi(s)(1)}, e_{\pi(s)(2)}, \ldots, e_{\pi(s)(15)}), \tag{7}$$

where the different items represents the columns of the matrix $\Pi(s)$ and where $\pi(s)(i)$ denotes the image of element i under the permutation $\pi(s)$.

Equivariance manifests itself in the property

$$A\Pi(s) = \Pi(s)A, \tag{8}$$

that is to say $A = \Pi(s)^{-1}A\Pi(s)$.

$\Pi(s)$ is orthogonal, therefore $\Pi(s)^{-1} = \Pi(s)^T$. The product $A' = A\Pi(s)$ is the matrix whose columns are the permuted columns of A with respect to the permutation $\pi(s)$ and $\Pi(s)^{-1}A'$ is the matrix whose rows are the permuted rows of A and also with respect to π.

Therefore, an equivalent statement of property (8) for the matrix $A = (a_{jk})$ is:

$$a_{jk} = a_{\pi(s)(j),\pi(s)(k)} \quad \text{for} \quad j, k = 1, 2, \ldots, 15. \tag{9}$$

All irreducible representations of the group D_5 have dimension 1 or 2. It turns out that for the complete decomposition of the representation $\theta : s \mapsto \Pi(s)$ we have

$$\theta = 2\theta_1^{(1)} \oplus \theta_2^{(1)} \oplus 3\theta_1^{(2)} \oplus 3\theta_2^{(2)}, \tag{10}$$

where $\theta_1^{(1)} : s \mapsto 1$ for $\forall s$ is the unit representation,

$$\theta_2^{(1)} : s \mapsto \begin{cases} -1 & \text{if } s \text{ is a reflection} \\ 1 & \text{if } s \text{ is a rotation} \end{cases},$$

is the alternating representation and

$$\theta_k^{(2)} : rot \mapsto \begin{pmatrix} \cos k\phi & -\sin k\phi \\ \sin k\phi & \cos k\phi \end{pmatrix}, \quad \mathit{refl} \mapsto \begin{pmatrix} 0 & 1 \\ 1 & 0 \end{pmatrix}, \quad k = 1, 2$$

with *rot* being the rotation with angle $\phi = 2\pi/5$ and *refl* is the reflection with respect to the symmetry axis through point numbered 1 in figure 1.

The upper index of the symbol θ in parentheses indicates the dimension of the representation. Equation (10) will be confirmed with an explicitly given symmetry adapted basis. to make sure that its structure is clear, we have to introduce some notions first:

Let us consider any point P of the 15 points in the plane of the figure. The *orbit of* P under D_5 is the set of all its images $\{s(P) \mid s \in D_5\}$. To simplify the notation, let us just write down the labels of those points belonging to an orbit in a counterclockwise sequence in brackets. In our example there are just two different orbits:

$$orbit_1 = [1, 2, 3, 4, 5] \quad \text{and} \quad orbit_2 = [6, 7, \ldots, 15].$$

We define the following product, where the left factor has to be an orbit and the right factor has to be a sequence of complex numbers (of the same length as the orbit) as a vector $v \in \mathbf{C}^{15}$

$$[i, j, \ldots, l] \odot [\alpha, \beta, \ldots, \rho] = v = [v_1, v_2, \ldots, v_{15}]^T$$

such that $v_i = \alpha$, $v_j = \beta$, ... , $v_l = \rho$, and where all the remaining coordinates of v are 0.

We need the following sequences of complex numbers, where $\omega_k = e^{2\pi ik/5}$ with $k = 1, 2$:

$$c = [1, 1, \ldots, 1] \text{ of length 5 and 10},$$
$$\widetilde{c} = [1, -1, 1, -1, 1, -1, 1, -1, 1, -1],$$
$$f_k = [1, \omega_k, \omega_k^2, \omega_k^3, \omega_k^4],$$
$$g_k = [1, 1, \omega_k, \omega_k, \omega_k^2, \omega_k^2, \omega_k^3, \omega_k^3, \omega_k^4, \omega_k^4],$$
$$\widetilde{g}_k = [-1, 1, -\omega_k, \omega_k, -\omega_k^2, \omega_k^2, -\omega_k^3, \omega_k^3, -\omega_k^4, \omega_k^4].$$

By (10), the representation θ has the following basis of symmetry adapted vectors:

1. Two invariant 1-dimensional subspaces for $\theta_1^{(1)}$. They are spanned by the vectors $x = orbit_1 \odot c$ and $y = orbit_2 \odot c$.

2. One invariant 1-dimensional subspaces for $\theta_2^{(1)}$. It is spanned by the vector $z = orbit_2 \odot \tilde{c}$.

3. The vectors

$$\begin{aligned} p_k &= orbit_1 \odot f_k, & \widehat{p}_k &= \overline{p_k}, \\ q_k &= orbit_2 \odot g_k, & \widehat{q}_k &= \overline{q_k}, \\ r_k &= orbit_2 \odot \widetilde{g}_k, & \widehat{r}_k &= -\overline{r_k}, \end{aligned} \tag{11}$$

span the three irreducible 2-dimensional subspaces invariant under $\theta_k^{(2)}$: $\{p_k, \widehat{p}_k\}$, $\{q_k, \widehat{q}_k\}$, $\{r_k, \widehat{r}_k\}$ for $k = 1, 2$.

Remark. By consequence, there is no proper subspace that is invariant under all transformations of the representation.

Therefore, we have a total of six invariant and irreducible 2-dimensional subspaces.

Now (6) manifests itself for our representation (10) in the fact, that the following sets of vectors span invariant subspaces of the commuting linear operator A:

- One subspace of dimension 2:

$$\{x, y\} \tag{12}$$

- One subspace of dimension 1:

$$\{z\} \tag{13}$$

- Two subspaces of dimension 3:

$$\{p_1, q_1, r_1\} \qquad \{\widehat{p}_1, \widehat{q}_1, \widehat{r}_1\} \tag{14}$$

- Two subspaces of dimension 3:

$$\{p_2, q_2, r_2\} \qquad \{\widehat{p}_2, \widehat{q}_2, \widehat{r}_2\}. \tag{15}$$

So our operator A has the following block diagonal structure

$$\widetilde{A} = S^{-1}AS = \begin{pmatrix} A_2 & & & & & \\ & A_1 & & & \mathrm{O} & \\ & & A_3 & & & \\ & & & A_3 & & \\ & \mathrm{O} & & & B_3 & \\ & & & & & B_3 \end{pmatrix},$$

where the index i is associated with an $i \times i$-block.

3.3 Symmetric Toeplitz matrices

Toeplitz matrices are square matrices characterized by the property that they have constant diagonals from top left to bottom right. They are used in many fields of applications, such as signal processing, coding theory, antenna theory, speech analysis, probability and statistics.

Let us consider the general *symmetric* $n \times n$ Toeplitz matrices denoted by T_n.

So for $n = 5$ we have

$$T_5 = \begin{pmatrix} a & b & c & d & e \\ b & a & b & c & d \\ c & b & a & b & c \\ d & c & b & a & b \\ e & d & c & b & a \end{pmatrix}. \tag{16}$$

For the permutation $\pi = (1 \quad n)(2 \quad n-1)(3 \quad n-3)\cdots$, property (9) obviously holds for T_n. The corresponding permutation matrix for $n = 5$ has the form

$$\begin{pmatrix} & & & & 1 \\ & O & & 1 & \\ & & 1 & & \\ & 1 & & O & \\ 1 & & & & \end{pmatrix}.$$

Because $\pi^2 = e$ (identity), we have to deal with the cyclic group $\{\pi, e\}$.

There are just two different irreducible representations, both of dimension 1:

$$\theta_1^{(1)} : \pi \mapsto 1 \quad \text{and} \quad \theta_2^{(1)} : \pi \mapsto -1.$$

Obviously, the invariant 1-dimensional subspaces under:

1. the unit representation $\theta_1^{(1)}$ are spanned by the vectors

$$\begin{matrix} (1 & 0 & 0 & 0 & \cdots & 0 & 0 & 0 & 1)^T \\ (0 & 1 & 0 & 0 & \cdots & 0 & 0 & 1 & 0)^T \\ (0 & 0 & 1 & 0 & \cdots & 0 & 1 & 0 & 0)^T \\ & & & & \cdots\cdots\cdots & & & & \end{matrix} \tag{17}$$

Therefore, the multiplicity c_1 is

$$c_1 = \begin{cases} n/2 & \text{for } n \text{ even} \\ (n+1)/2 & \text{for } n \text{ oddd} \end{cases}.$$

2. the alternating representation $\theta_2^{(1)}$ are spanned by the vectors

$$\begin{array}{l} (1\ 0\ 0\ 0\ \cdots\ 0\ \ 0\ \ 0\ -1)^T \\ (0\ 1\ 0\ 0\ \cdots\ 0\ \ 0\ -1\ \ 0)^T \\ (0\ 0\ 1\ 0\ \cdots\ 0\ -1\ \ 0\ \ 0)^T \\ \quad\quad\cdots\cdots\cdots \end{array} \tag{18}$$

Therefore, the multiplicity c_2 is:

$$c_2 = \begin{cases} n/2 & \text{for } n \text{ even} \\ (n-1)/2 & \text{for } n \text{ oddd} \end{cases}$$

Because of

$$\theta = c_1\theta_1^{(1)} \oplus c_2\theta_2^{(1)},$$

we know from (6) that the symmetric Toeplitz problem decomposes into two invariant problems of dimensions $n/2$ in the even case and into a $(n+1)/2$-dimensional and a $(n-1)/2$-dimensional subproblem in the odd case.

Consequently, and together with (17) and (18), all eigenvectors are of the form either

$$(\alpha, \beta, \gamma, \ldots, \gamma, \beta, \alpha)^T \quad \text{or} \tag{19}$$

$$(\alpha, \beta, \gamma, \ldots, -\gamma, -\beta, -\alpha)^T. \tag{20}$$

4 Equivariance in tensor products

In a number of applications, for example, difference approximations and fast Fourier transforms, tensor structures have been observed and exploited. This exploitation can be carried out even further under inclusion of equivariance structures. Let us demonstrate this here with several illustrative examples.

To prepare some of the examples, let us just mention the known equivariance of the general circulant matrix C_n of order n:

$$C_n = \begin{pmatrix} c_1 & c_2 & c_3 & \cdots & c_n \\ c_n & c_1 & c_2 & \cdots & c_{n-1} \\ \vdots & \vdots & \vdots & \vdots & \vdots \\ c_2 & c_3 & c_4 & \cdots & c_1 \end{pmatrix},$$

which fulfils (9) for the cyclic permutation $\pi = (1\ 2\ 3 \cdots\ n)$ and its powers π^k. In other words: C_n commutes with the permutation matrix

$$\Pi = \begin{pmatrix} 0\,0 \cdots 0\,1 \\ 1\,0 \cdots 0\,0 \\ 0\,1 \cdots 0\,0 \\ 0\,0 \cdots \vdots\ \vdots \\ \vdots\ \vdots \qquad 0\,0 \\ 0\,0 \cdots 1\,0 \end{pmatrix}$$

and its powers. C_n has the equivariance of the cyclic group $\{e, \pi, \pi^2, \ldots, \pi^{n-1}\}$ of order n. All their irreducible representations θ_k have dimension 1 and in total there are n different ones. Their generating element π is represented by

$$\theta_k : \pi \mapsto \omega^k \quad \omega = e^{2\pi i/n} \quad k = 1, 2, \ldots, n-1.$$

Therefore, with Π also C_n is diagonalized due to the fact, that all multiplicities and dimensions are 1. The Fourier matrix

$$F_n = \frac{1}{\sqrt{n}} \begin{pmatrix} 1 & 1 & 1 & \cdots & 1 \\ 1 & \omega & \omega^2 & \cdots & \omega^{n-1} \\ 1 & \omega^2 & \omega^4 & \cdots & \omega^{2(n-1)} \\ \vdots & \vdots & \vdots & & \vdots \\ 1 & \omega^{n-1} & \omega^{2(n-1)} & \cdots & \omega^{(n-1)^2} \end{pmatrix} \tag{21}$$

which is symmetric and unitary, does the diagonalization

$$diag(\lambda_1, \lambda_2, \ldots, \lambda_n) = \overline{F_n} C_n F_n.$$

An interesting generalization of the circulant matrix is given in [7]: Diaconis introduces *G-circulant* $|G| \times |G|$-matrices of a finite group G. Its i, j entry is $f(s_j s_i^{-1})$, with f a function on the group elements s_i, which is constant on conjugacy classes of G.

Several cases are given with an explicit diagonalization, among them the example for the permutation group $G = S_3$. The corresponding 6×6 matrix has the following pattern with 3 parameters (= $\sharp$ of conjugacy classes):

$$\begin{pmatrix} a & b & b & c & c & b \\ b & a & c & b & b & c \\ b & c & a & b & b & c \\ c & b & b & a & c & b \\ c & b & b & c & a & b \\ b & c & c & b & b & a \end{pmatrix}. \tag{22}$$

Its diagonalization is given by:

Eigenvalue	Eigenvectors
$a+3b+2c$	$\{1,1,1,1,1,1\}$
$a-3b+2c$	$\{1,-1,-1,1,1,-1\}$
$a-c$	$\{0,-1,0,0,0,1\}$
$a-c$	$\{-1,0,0,0,1,0\}$
$a-c$	$\{-1,0,0,1,0,0\}$
$a-c$	$\{0,-1,1,0,0,0\}$

In the examples that follow, let A be any matrix respecting the D_5-symmetry of the figure and $y_1, y_2, \dots, y_n$ the eigenvectors of the circulant matrix C_n (the columns of the Fourier matrix F_n).

4.1 Example $P = B \otimes C_n$

Let

$$B = \begin{pmatrix} a & b & c \\ b & a & c \\ d & d & e \end{pmatrix}.$$

B fulfils (9) for the permutation $(1\ \ 2)$, therefore B is equivariant with respect to the cyclic group $\{e, (1\ 2)\}$. There are three 1-dimensional subspaces invariant under the permutation $(1\ \ 2)$ and spanned by the vectors

$$u = (1,1,0)^T \quad v = (0,0,1)^T \quad z = (1,-1,0)^T.$$

The vectors u, v are fixed, and $z \mapsto -z$ when subjected to the permutation $(1\ \ 2)$. Therefore, we have

$$\theta = 2\theta_1^{(1)} \oplus \theta_2^{(1)}$$

and u, v span an invariant 2-dimensional subspace under B.

Each pair of the n tensor products

$$\{u \otimes y_1, v \otimes y_1\}, \{u \otimes y_2, v \otimes y_2\}, \dots, \{u \otimes y_n, v \otimes y_n\}$$

is invariant under P and $\{z \otimes y_1\}$, $\{z \otimes y_2\}$, $\dots$, $\{z \otimes y_n\}$ are eigenvectors of P.

4.2 Example $P = A \otimes C_n$

Because of (11)÷(15), the $15n$-dimensional matrix P can be split up into

- n problems of dimension 2 with the basis vectors $\{x \otimes y_i, y \otimes y_i\}$,
- n problems of dimension 1 with the basis vectors $\{z \otimes y_i\}$,

- $2n$ problems of dimension 3 with the basis vectors $\{p_1 \otimes y_i, q_1 \otimes y_i, r_1 \otimes y_i\}$, $\{\widehat{p}_1 \otimes y_i, \widehat{q}_1 \otimes y_i, \widehat{r}_1 \otimes y_i\}$,

- $2n$ problems of dimension 3 with the basis vectors $\{p_2 \otimes y_i, q_2 \otimes y_i, r_2 \otimes y_i\}$, $\{\widehat{p}_2 \otimes y_i, \widehat{q}_2 \otimes y_i, \widehat{r}_2 \otimes y_i\}$.

4.3 Example $P = A \otimes A$

With Section 3.2 in mind, the 15^2-dimensional problem decomposes into subproblems of dimension 30, 15, 45, 45, 45, 45. The symmetry adapted basis consists of the tensor products of the vectors described in (11)÷(15).

4.4 Block G-circulants with G-circulant blocks

We consider $mn \times mn$-matrices consisting of n blocks of size $m \times m$ in each row of blocks with the following pattern:

Each block is G-circulant to a specific group G with $m = |G|$ elements. And the global structure of the different blocks is arranged according to a specific group H with $n = |H|$ elements. As an example, we could take $n = 6$ in general different $m \times m$-circulants according to (22). With the knowledge of a diagonalization or a partial diagonalization for the blocks as well as for the global structure, we get a diagonalization or a partial diagonalization of the large $nm \times nm$-matrix.

The special case of block circulants with circulant blocks is treated in the book [5]. It deals with $mn \times mn$-matrices of the form:

$$M = \begin{pmatrix} M_1 & M_2 & M_3 & \cdots & M_n \\ M_n & M_1 & M_2 & \cdots & M_{n-1} \\ \vdots & \vdots & \vdots & & \vdots \\ M_2 & M_3 & M_4 & \cdots & M_1 \end{pmatrix},$$

where the $m \times m$-matrices M_i are circulant.

We have a complete diagonalization of M because with

$$M = \sum_{k=1}^{n} \Pi^{k-1} \otimes M_k$$

each of the tensor products can be diagonalized simultaneously by the Fourier matrices F_n (left factors) and F_m (right factors).

5 Partial differential equations

Let us conclude with two sample examples concerning partial differential equations.

5.1 Poisson equation

We consider the Poisson problem on a square Q with general Dirichlet boundary conditions on ∂Q:

$$\Delta u(x_1, x_2) = f(x_1, x_2) \text{ on } Q \text{ and } \quad u(x_1, x_2) = b(x_1, x_2) \text{ on } \partial Q.$$

No symmetry property is required for the given functions f and b.

Discretizing this problem via central differences of a uniform mesh in Q with n^2 interior points leads to an equation of the form

$$A_{n^2} \cdot \tilde{u} = \tilde{b},$$

where A_{n^2} is a $n^2 \times n^2$-matrix which can be expressed as the Kronecker sum

$$A_{n^2} = A_n \otimes I_n + I_n \otimes A_n.$$

I_n is the $n \times n$-identity matrix and A_n is the tridiagonal matrix

$$A_n = \begin{pmatrix} 2 & -1 & & & O \\ -1 & 2 & -1 & & \\ & \ddots & \ddots & \ddots & \\ & & -1 & 2 & -1 \\ O & & & -1 & 2 \end{pmatrix}. \tag{23}$$

Since the identity matrix is invariant under any coordinate transformation, only a diagonalization of A_n needs to be done to obtain a diagonalization of A_{n^2}. Since A_n itself is a symmetric Toeplitz matrix, it too falls under the reduction results given in Section 2.3.

The analogous Poisson problem on a cube in three dimensions has an even greater reduction. In this case, A_{n^2} is replaced by the matrix A_{n^3}, which is given by the Kronecker sum

$$A_{n^3} = A_n \otimes I_n \otimes I_n + I_n \otimes A_n \otimes I_n + I_n \otimes I_n \otimes A_n.$$

5.2 Biharmonic equation

Consider the biharmonic boundary value problem on a square Q

$$\Delta^2 u(x_1, x_2) = f(x_1, x_2) \quad \text{on} \quad Q$$

with the boundary conditions

$$u(x_1, x_2) = \alpha(x_1, x_2) \quad \text{and} \quad \frac{\partial^2}{\partial n^2} u(x_1, x_2) = \beta(x_1, x_2) \text{ on } \partial Q.$$

No symmetry property is required for the given functions f, α, and β.

Discretizing this problem via central differences on a uniform mesh in Q with n^2 interior points leads to an equation of the form

$$B_{n^2} \cdot \widetilde{u} = \widetilde{b},$$

where B_{n^2} is a $n^2 \times n^2$-matrix which can again be expressed as a Kronecker sum:

$$B_{n^2} = A_n^2 \otimes I_n + 2A_n \otimes A_n + I_n \otimes A_n^2.$$

Here A_n again is the tridiagonal $n \times n$ Toeplitz matrix described in (23). The term $A_n \otimes A_n$ fits into the reduction case discussed in Section 4.3.

Remark. For other problems concerning the Laplacian see also [3,4,11-13] and for equivariant integral equations see [1].

References

1. E.L. Allgower, K. Böhmer, K. Georg and R. Miranda, *Exploiting Symmetry in Boundary Element Methods*, SIAM J. Numer. Anal. **29**(1992), 534–552.
2. E.L. Allgower, K. Georg and R. Miranda, *Exploiting Permutation Symmetries with Fixed Points in Linear Equations*, in Proc. of the AMS-SIAM 1992 Summer Seminar "Exploiting Symmetry in Applied and Numerical Analysis". Lectures in Applied Mathematics (AMS, Providence, 1993).
3. E.L. Allgower, K. Böhmer and M. Zhen, *On a Problem Decomposition for Semilinear Nearly Symmetric Elliptic Problems*, Proc. of the 6th GAMM-Seminar "Parallel Algorithms for Partial Differential Equations", Kiel 1990. Book series: Notes on Numerical Fluid Mechanics, vol. 31, (Vieweg, Braunschweig, 1991).
4. S.-L. Chan, *Application de la théorie des représentations dans certains domaines sans symétrie classique*, C. R. Acad. Sc. Paris **280A**(1975), 389–391.
5. P.J. Davis, *Circulant Matrices*, Wiley, New York, 1979.
6. P. Diaconis and D. Rockmore, *Efficient Computation of the Fourier Transform on finite Groups*, Technical Report No. 292, Dept. of Statistics, Stanford University, California, 1988.

7. P. Diaconis, *Group Representations in Probability and Statistics*, Vol. 11, S. S. Gupta Series Editor, Institute of Mathematical Statistics, Hayward, California, 1988.
8. A. Fässler and E. Stiefel, *Group Theoretical Methods and Their Applications*, Birkhäuser, Boston, 1992.
9. A. Fässler and R. Mäder, *Symmetriegerechte Basisvektoren für Permutationsdarstellungen aller endlicher Punktgruppen*, Zeitschrift für angewandte Mathematik und Physik ZAMP **31**(1980), 277–292.
10. K. Gaterman, *Symmetry*, Ein Paket für lineare Darstellungstheorie endlicher Gruppen. Reduce network library, netlib@rand.org, 1991.
11. J. Hersch, *On Harmonic Measures, Conformal Moduli and some Elementary Symmetry Methods*, Journal d'analyse mathématique **42**(1982–83), 211–228.
12. J. Hersch, *Sur les fonctions propres des membranes vibrantes couvrants un secteur symétrique de polygone régulier ou de domaine périodique*, Comment. Math. Helv **41**(1966–67), 222-236.
13. J.R. Kuttler and V.G. Sigillito, *Eigenvalues of the Laplacian in two dimensions*, SIAM Rev. **26**(1984), 163–193.

PARTIALLY ORDERED TOPOLOGICAL LINEAR SPACES

SHOZO KOSHI

Hokkaido Institute of Technology, Teineku, Maede 7–15, Sapporo, Japan

We introduce a generalized supremum (supset) and generalized infimum (infset) for subsets A of a partially ordered topological linear space E. They generalize the notions of supremum and infimum in Riesz spaces. We shall investigate properties of supsets and infsets in this paper. We also consider distributive law in a partially ordered linear spaces.

1 Partially ordered linear space and the supset of a subset

Let E be a linear space over the real field. Let us consider a convex cone P in E which is generating and proper. Namely, the following two conditions are satisfied:

(a) $E = P - P$.

(b) $P \cap (-P) = \{0\}$.

We say that $x \geq y$ (or equivalently, $y \leq x$) if $x - y \in P$. It is well-known that conditions (a) and (b) are equivalent to the following five conditions for a given subset P of E:

(1) $x \geq y$ and $y \geq x$ implies $x = y$.

(2) $x \geq y$ and $y \geq z$ implies $x \geq z$.

(3) $x \geq y$ implies $x + z \geq y + z$ for all z in E.

(4) $x \geq 0$ implies $\alpha x \geq 0$ for all positive scalars α.

(5) For every x in E, there exist $x_1 \geq 0$ and $x_2 \geq 0$ such that $x = x_1 + x_2$.

E is called a *partially ordered linear space* and P is called an *order* of E provided conditions (a) and (b) (or equivalently, (1), (2), (3), (4) and (5)) are satisfied. Elements of P are said to be *positive* in E.

Let $A = \{ a_\lambda : \lambda \in \Lambda \}$ be a subset of a partially ordered linear space E with order P. We define the *supset* of A (or sup A) to be $\bigvee A = \bigvee_{\lambda \in \Lambda} a_\lambda = \{z \in E : z \geq a_\lambda$ for all λ in Λ and $z = w$ whenever $z \geq w$ and $w \geq a_\lambda$ for all λ in $\Lambda \}$. Sometimes, we use the notation $\sup A$ instead of $\bigvee A$. Hence, $\bigvee A$ is the set of all minimal elements of $U\{A\} = \{ z \in E : z \geq a, \ \forall a \in A \}$.

Elements of $U\{A\}$ are called *upper bounds* of A. Similarly, we define the *infset* $\bigwedge A$ (or $\inf A$) of A to be $\bigwedge A = \{\, z \in E : z$ is a maximal element of $L\{A\}\,\}$, where $L\{A\} = \{z \in E : z \le a,\ \forall a \in A\}$ is the set of *lower bounds* of A. A is said to be *upper bounded* (resp. *lower bounded*) if $U\{A\} \ne \emptyset$ (resp. $L\{A\} \ne \emptyset$). We shall discuss when an upper bounded (resp. lower bounded) set has a nonvoid supset (resp. infset).

A partially ordered linear space E is said to satisfy *Condition* (A) if for every upper bound a of a subset A of E there exists a minimal element x in $U\{A\}$ such that $x \le a$. E satisfies Condition (A) if and only if $\sup A + P = U\{A\}$ for all upper bounded subsets A of E. Later, we shall show that every finite dimensional partially ordered linear space with a closed order always satisfies Condition (A). From here to the end of this section, we state some elementary observations, in which $A = \{\, a_\lambda : \lambda \in \Lambda \,\}$ is a subset of a partially ordered linear space E satisfying Condition (A).

Proposition 1.1. *If* $\sup A$ *is a singleton* $\{u\}$ *then* u *is the least upper bound of* A. *If* $\inf A$ *is a singleton* $\{l\}$ *then* l *is the greatest lower bound of* A.

The least upper bound of A is called the *supremum* of A. The greatest lower bound of A is called the *infimum* of A.

Proposition 1.2. *1.* $-\bigvee_{\lambda\in\Lambda} a_\lambda = \bigwedge_{\lambda\in\Lambda}(-a_\lambda)$; *or equivalently,* $-\sup A = \inf(-A)$.

2. For every positive number α, $\alpha \bigvee_{\lambda\in\Lambda} a_\lambda = \bigvee_{\lambda\in\Lambda} \alpha a_\lambda$; *or equivalently,* $\alpha \sup A = \sup \alpha A$.

3. For every positive number α, $\alpha \bigwedge_{\lambda\in\Lambda} a_\lambda = \bigwedge_{\lambda\in\Lambda} \alpha a_\lambda$; *or equivalently,* $\alpha \inf A = \inf \alpha A$.

Proposition 1.3. *For every* b *in* E, *we have*

1. $\bigvee_{\lambda\in\Lambda} a_\lambda + b = \bigvee_{\lambda\in\Lambda}(a_\lambda + b)$; *or equivalently,* $\sup A + b = \sup\{A + b\}$.

2. $\bigwedge_{\lambda\in\Lambda} a_\lambda + b = \bigwedge_{\lambda\in\Lambda}(a_\lambda + b)$; *or equivalently,* $\inf A + b = \inf\{A + b\}$.

Proposition 1.4. $\sup A = \sup(coA)$, *where* coA *is the convex hull of* A.

Proposition 1.5. *If* x *is a positive element of* E *(i.e.* $x \in P$*) then* $x \vee 0 = \{x\}$ *and* $x \wedge 0 = \{0\}$.

Theorem 1.6. *If* $a \vee b \ne \emptyset$ *for some* a *and* b *in* E *then* $a \wedge b \ne \emptyset$ *and*

$$a + b - (a \wedge b) = a \vee b.$$

Proof. Since $a \wedge b = -\{(-a) \vee (-b)\}$, we have by Propositions 1.2 and 1.3 that

$$a + b - (a \wedge b) = a + b + \{(-a) \vee (-b)\} = b \vee a = a \vee b.$$

The following example shows that $\sup\{\sup A\} \neq \sup A$, in general.

Example 1.7. Let E be a 3-dimensional Euclidean space $\mathbf{R}^3$ with order P generated by four points $(1,0,0)$, $(1,1,0)$, $(1,0,1)$ and $(1,1,1)$. Let $z = (0,0,1)$ and $0 = (0,0,0)$. Let $A = \{z, 0\}$. Then $\sup A = \{a \in \mathbf{R}^3 : a = (1, \alpha, 0), 0 \leq \alpha \leq 1\}$. and $\sup\{\sup A\} = \{a \in \mathbf{R}^3 : a = (2,2,\beta), 1 \leq \beta \leq 2\}$. Hence, $\sup A \neq \sup\{\sup A\}$ in this case.

It may also happen that $\inf\{\inf A\} \neq \inf A$. However, we have the following

Theorem 1.8.

1. $\sup \inf \sup A = \sup A$.
2. $\inf \sup \inf A = \inf A$.

2 Monotone complete order and dual order

Let E be a partially ordered linear space. If E satisfies Condition (A), it is easy to see that the supset $\sup A$ of an upper bounded set A is not empty. Similarly, the infset $\inf A$ of a lower bounded subset A of E is not empty as well. In this section, we shall consider a sufficient condition to ensure that E satisfies Condition (A). To this end, we recall the notion of a monotone complete order.

A subset A of E is a *linear set* if every two elements x and y of A is comparable, *i.e.* $x \leq y$ or $y \leq x$. We say that E is *monotone complete* (in an order P), if every upper bounded linear subset A of E has the least upper bound, *i.e.* $\sup A \neq \emptyset$ consisting of a single element.

Let E be a topological linear space with a linear topology τ. An order of E determined by a convex cone P is called a *topologically continuous order* (or equivalently, τ is called a *order continuous topology*) if every directed linear subset $\{a_\lambda\}$ with $\inf a_\lambda = 0$ converges to 0 in τ.

Theorem 2.1. *Let E be partially ordered linear space with order continuous topology and ordered by a closed convex cone P. If E is monotone complete,* *then the supset* $\sup A$ *of every non-empty upper bounded subset* A *of* E *is not empty. Moreover,* E *has Condition (A). (However,* $\sup A$ *does not necessarily consist of a single element in this case.)*

Proof. Let A be a non-empty upper bounded subset of E. Hence $U\{A\}$ is not empty. Let $a \in U\{A\}$. Then we can find a maximal linear subset B of E which contains a and is contained in $U\{A\}$ by Zorn's maximal theorem. By monotone completeness of the order P, $\inf B = \{b\}$ is a singleton. Since the linear topology τ of E is order continuous and P is closed, $b \in U\{A\}$ and $b \leq a$. Hence b is a minimal element of $U\{A\}$ and thus $\sup A$ is not empty.

In the following, E^* denotes the dual of a partially ordered normed space E. Let P^* be the positive cone dual to P in E^*, *i.e.* $P^* = \{ f \in E^* : f(x) \geq 0, \forall x \in P \}$. By definition, $P^* \cap \{-P^*\} = \{0\}$. But P^* might not necessarily be an order in E^*, in general. In fact, $P^* - P^*$ is not necessarily the whole of E^*. When E is a Banach space with closed order P, T. Ando 2 gave several equivalent conditions to ensure that P^* is an order in E^*, *i.e.* $P^* - P^* = E^*$.

Theorem 2.2. *Let* E *be a Banach space with closed order* P. *If* $P^* - P^* = E^*$ *then* E^* *is monotone complete in the order determined by* P^*. *Moreover, the weak* topology of* E^* *is order continuous with respect to* P^*. *Hence,* $\sup A^*$ *is non-empty for every non-empty upper bounded subset* A^* *of* E^*. *In this case,* E^* *satisfies Condition (A) in the order* P^*.

Proof. By the definition of P^* and the theorem of Banach-Steinhaus, we see that E^* is monotone complete. Since weak* topology of E^* is order continuous, the assertion follows from Theorem 2.1.

It is shown in [2] that for a closed order P in a Banach space E, $E^* = P^* - P^*$ if and only if every order interval $[x, y] = \{ z \in E : x \leq z \leq y \}$ in E is norm bounded.

Corollary 2.3. *Every finite dimensional Euclidean space* E *with a closed order* P *always satisfies Condition (A).*

3 Norm and order

Let E be a partially ordered normed space. A norm $\|\cdot\|$ of E is called an *ordered norm* if $0 \leq x \leq y$ implies $\|x\| \leq \|y\|$. We shall investigate when a norm is equivalent to an order norm.

For a symmetric absorbing convex subset V of E, we shall define the P-*envelop* of V by

$$E_P(V) = (V - P) \cap (V + P).$$

Lemma 3.1.

$$E_P(E_P(V)) = E_P(V).$$

Proof. Let $U = (V+P)\cap(V-P)$. Then, $V \subset U$. Since $U+P \subset V+P+P = V+P$ and $U-P \subset V-P-P = V-P$, $(U+P)\cap(U-P) \subset (V+P)\cap(V-P) = U$. But, we always have $U \subset E_P(U)$. The assertion follows.

Theorem 3.2. *If $U = E_P(V)$ is the unit ball of E in a norm $\|\cdot\|_U$ then this norm is an ordered norm.*

Proof. We shall show that if $x_1 \in U$ and $x_1 \geq x_2 \geq 0$ then $\|x_1\|_U \geq \|x_2\|_U$. In fact, the norm $\|\cdot\|_U$ is the Minkowski functional of E defined by U. It thus suffices to show that $\alpha x_1 \in U$ for some $\alpha \geq 1$ implies $\alpha x_2 \in U$. Since $x_1 = x_2 + p$ for some p in P, $\alpha x_2 = \alpha x_1 - \alpha p \in U - P$. On the other hand, $\alpha x_2 = x_2 + (\alpha - 1)x_2 \in U + P$. This means that $\alpha x_2 \in E_P(U) = U$, by Lemma 3.1. Therefore, $\|x_1\|_U \geq \|x_2\|_U$, as asserted.

Theorem 3.3. *Let E be a partially ordered normed linear space with an order P. The norm of E is equivalent to an order norm if and only if $(V + P) \cap (V - P) \subset \alpha V$ for some $\alpha > 0$, where V is the unit ball of E.*

Proof. Suppose, without loss of generality, that the norm $\|\cdot\|$ of E with unit ball V is an ordered norm. Let $U = (V + P) \cap (V - P)$. Since $U = \bigcup\{ z \in E : x_1 \leq z \leq x_2$ for some x_1, x_2 in V with $x_1 \leq x_2 \}$, by the order interval relation $[x_1, x_2] = x_1 + [0, x_2 - x_1]$, we have

$$\|z\| \leq \|x_1\| + \|x_2 - x_1\| \leq 3, \quad \forall z \in U.$$

Hence $U \subset 3V$.

Conversely, if $U = (V + P) \cap (V - P) \subset \alpha V$ for some $\alpha > 0$ then the norm $\|\cdot\|_U$ of E with unit ball $U = (V + P) \cap (V - P)$ is equivalent to the original norm of E with unit ball V. Moreover, $\|\cdot\|_U$ is an order norm by Theorem 3.2.

4 Riesz spaces and distributive law

In this section, we shall consider the distributive law in a partially ordered linear space E. If E is a Riesz space, it is known that the distributive law holds in E. We shall consider when a partially ordered linear space E becomes a Riesz space. The following provides some criteria.

Proposition 4.1. *Let E be an n-dimensional Euclidean space with a closed order P. If P is generated by a set of n linearly independent elements of E, then E is a Riesz space.*

Corollary 4.2. *Let E be a 2-dimensional Euclidean space with a closed order P. Then E is a Riesz space.*

Theorem 4.3. *Let E be a Hausdorff topological linear space. Let P be an order in E such that $P \setminus \{0\}$ is open in E. If E is has dimensioan greater than 2, then E cannot be a Riesz space in the order P.*

We shall make a remark here that if E is one-dimensional then E is a Riesz space in any order P and $P \setminus \{0\}$ is open in this case.

Proof. Suppose on the contrary that E were a Riesz space in P. At first we shall notice that the topological boundary of P relative to $P \setminus \{0\}$ is equal to $P^- \setminus P$. Since E is has dimension greater than 2 and so $E \setminus \{0\}$ is connected, we can conclude that $P^- \setminus P \neq \emptyset$. So, there exists $0 \neq x \in P^- \setminus P$. Let $y = x \vee 0 \in P$. Then $x < y$ and $0 < y$. Since $P \setminus \{0\}$ is open and $y \in P \setminus \{0\}$, there is a positive number α with $0 < \alpha < 1$ such that

$$z = \alpha x + (1 - \alpha) y \in P.$$

It is easy to see that $0 < z, x < z$ and $z = \alpha x + (1 - \alpha) y < y$. But this is a contradiction to $y =$ least upper bound for x and 0. This establishes our assertion.

We shall present an example of a closed order P in which a 3-dimensional Euclidean space is not a Riesz space.

Example 4.4. Let E be a 3-dimensional Euclidean space $\mathbf{R}^3$. Let P be a generating proper convex cone in E generated by 4 elements $(1,0,0)$, $(1,0,1)$, $(1,1,0)$ and $(1,1,1)$. Then, for two elements $0 = (0,0,0)$ and $z = (0,0,1)$ of E, there is no least upper bound for them. For example, both $a = (1,0,1)$ and $b = (1,1,1)$ are greater than 0 and z in the order P. But, a and b are not comparable. This says that E is not a Riesz space in the order P.

We now consider the distributive law in partially ordered linear spaces.

Proposition 4.5. *Let E be a Riesz space. Then for all x_1, x_2 and y in E, we have*

$$(x_1 \vee x_2) \wedge y = (x_1 \wedge y) \vee (x_2 \wedge y).$$

Similarly, we also have

$$(x_1 \wedge x_2) \vee y = (x_1 \vee y) \wedge (x_2 \vee y).$$

However, the distributive law does not hold, in general, in a partially ordered linear space E. For every pair of subsets A and B of E, we define

$$A \vee B = \sup\{A \cup B\}.$$

There are many possible ways to define $A \vee B$, other than the one stated above. In this paper, though, we consider only the above condition.

Proposition 4.6. *1. $A \cup B = C \cup D$ implies $A \vee B = C \vee D$.*

2. $A \vee B = B \vee A$.

3. $(A \vee B) \vee C = A \vee (B \vee C)$.

We can also define $A \wedge B = \inf(A \cup B)$. With these definitions, we shall provide an example of a partially ordered linear space in which the distributive law holds for some elements x_1, x_2 and y.

Proposition 4.7. *Let E be a partially ordered space and $z \in E$. Then, $y \in z \vee (-z)$ implies $y \geq 0$.*

Proof. Let $y \in z \vee (-z) = (2z \vee 0) - z$. Then, we can find an a from $2z \vee 0$ such that $y = a - z = (1/2)a - z + (1/2)a$. Since $(1/2)a \geq z$ and $a \geq 0$, we conclude that $y \geq 0$.

Example 4.8. Let E be a 3-dimensional Euclidean space with an order as in Examples 1.7 and 4.4. We shall show that the distributive law is true for some elements and false for others in E. Let $z = (0, 0, 1)$. Then

$$(z \wedge 0) \vee (-z \wedge 0) = (z \vee -z) \wedge 0 = 0.$$

But, if we take z, $(1/2)z = (0, 0, 1/2)$ and 0, then the distributive law fails to hold for these three elements.

In the following, we shall verify the converse of Proposition 4.5.

Theorem 4.9. *If the distributive law*

$$(x_1 \vee x_2) \wedge y = (x_1 \wedge y) \vee (x_2 \wedge y)$$

holds for all elements x_1, x_2 and y in a partially ordered linear space E then E is a Riesz space.

Proof. Suppose that $x \wedge 0$ is a subset consisting of more than two elements and $y \leq x$ and $y \neq x$. We shall show that

$$((x \vee y) \wedge 0) \cap ((x \wedge 0) \vee (y \wedge 0)) = \emptyset.$$

In particular,

$$(x \vee y) \wedge 0 \neq (x \wedge 0) \vee (y \wedge 0).$$

Let $z \in ((x \vee y) \wedge 0) \cap ((x \wedge 0) \vee (y \wedge 0))$. Since $y \leq x$, $(x \vee y) \wedge 0 = x \wedge 0$. On the other hand, $z \in (x \wedge 0) \vee (y \wedge 0)$. It follows that $z \geq w$ for all w in $x \wedge 0$. Hence, z is the maximum of the subset $x \wedge 0$. It says that $x \wedge 0 = \{z\}$. This conflicts with the assumption that $x \wedge 0$ contains more than two elements.

5 Supsets for two non-comparable elements

In this section, we shall consider $a \vee b$ for any non-comparable pair a and b of elements of a partially ordered Hausdorff topological linear space E. It is not easy to determine the exact form for $a \vee b$. In some cases, we can present $a \vee b$ in terms of boundary sets. To this end, we need to introduce some definitions.

Throughout this section, E is always an Euclidean space in a closed order P. A subset F of the order P is called a *face* if there exists a supporting hyperplane H of P such that $F = P \cap H$. We shall use the notation $\dim F$ as the dimension of the affine hull of F for a convex subset F of P.

Theorem 5.1. *Let a and b be any non-comparable pair of elements of a partially ordered Hausdorff topological linear space E in the order P. If $\dim F \leq 1$ for all faces F of P then we have*

$$a \vee b = \partial(a + P) \cap \partial(b + P),$$

where ∂ means boundary, i.e. ∂C is the topological boundary of a subset C of E.

We first prepare the following

Lemma 5.2. *If $x \in \partial P$ and $0 \leq y \leq x$ then $y \in \partial P$.*

Proof. If y belongs to the interior P° of P then $x+(x-y) \in P$. It follows that $[y, x+x-y) \subset P^\circ$ since P is convex. Hence $x = (1/2)(y+x+x-y) \in P^\circ$, a contradiction.

Proof of Theorem 5.1. Assume that $x_0 \in a \vee b$ and $x_0 \in (a+P)^\circ$, the interior of $a+P$. Since the affine hull of $a+P = E$ equals E, there exists a number $\lambda > 0$ such that $z = (1-\lambda)b + \lambda x_0 \in a+P$. Hence, $z < x_0$ and $z \in (a+P) \cap (b+P)$. As a result, x_0 is not a minimal element. This contradiction establishes that $x_0 \in \partial(a+P) \cap \partial(b+P)$.

Conversely, let $x_0 \in \partial(a+P) \cap \partial(b+P)$ and suppose $y_0 \leq x_0$ for some $y_0 \in U\{a, b\}$, the set of upper bounds of $\{a, b\}$. By virtue of the fact that $a \leq y_0 \leq x_0$ and Lemma 5.2, we have $y_0 \in [a, x_0] \subset \partial(a+P)$. Similarly, we have $y_0 \in [b, x_0] \subset \partial(b+P)$. Hence, we have $[a, x_0] \cap (a+P)^\circ = \emptyset = [b, x_0] \cap (a+P)^\circ$. By the separation theorem, there exist a closed hyperplane H_1 such that $[y_0, x_0] \subset H_1$ separating $a+P$, and another closed hyperplane H_2 such that $[y_0, x_0] \subset H_2$ separating $b+P$. By assumption, $H_1 \cap (a+P)$ is a half line which contains a and x_0. Also $H_2 \cap (b+P)$ is a half line which contains b and x_0. Since a and b are not comparable in the order P, these two half lines have different directions. This means that $y_0 = x_0$ and so $x_0 \in a \vee b$.

To illustrate Theorem 5.1, we consider the following

Example 5.3. Let E be the set of all Hermitian operators on a 2-dimensional Euclidean space. Let P be the set of all positive semi-definite operators. More precisely, E is considered as a 3-dimensional space in the order $P = \{ (a, b, c) : a \geq 0, b \geq 0, ab \geq c^2 \}$. It is easy to see that the dimension of every face of P is less than or equal to 1 and so the assumption of Theorem 5.1 is satisfied. For any $p = (a, b, c)$ in E with real coordinates a, b and c, we have

$$\begin{aligned} p \vee 0 = \{ (x, y, z) : x \geq 0,\ y \geq 0,\ x-a \geq 0, \\ y-b \geq 0,\ xy = z^2,\ (x-a)(y-b) = (z-c)^2 \}. \end{aligned}$$

References

1. I. Amemiya, *A generalization of Riesz-Fisher's theorem*, J. Math. Soc. Japan **5**(1953), 353–354.
2. T. Ando, *On fundamental properties of a Banach space with cone*, Pacific J. Math. **12**(1962), 1163–1169.
3. S. Koshi, *Lattice structure of partially ordered linear space*, Memoirs of the Hokkaido Institute of Technology **25**(1997), 1–7.

4. ———, *Partially ordered normed linear space with weak Fatou property*, Taiwanese J. Math. **1**(1997), 1–9.
5. W. A. J. Luxemburg and A. C. Zaanen, *Riesz spaces I*, North-Holland Math. Lib. 1, 1971.
6. H. Nakano, *Linear topologies on semi-ordered linear space*, J. Faculty of Sciences, Hokkaido Univ. **12**(1953), 87–104.
7. I. Namioka, *Partially ordered linear topological spaces*, Mem. Amer. Math. Soc. **24**, 1957.
8. H. H. Schaefer, *Banach lattices and positive operators*, Grundlehre der Math. Wiss. 215, Springer-Verlag, Berlin-Heidelberg-New York, 1974.
9. Y. C. Wong, *The topology of uniform convergence on order-bounded sets*, Lecture Notes in Math. 531, Springer-Verlag, Berlin-Heidelberg-New York, 1976.
10. A. C. Zaanen, *Riesz Spaces II*, North-Holland Math. Lib. 30, 1983.
11. ———, *Introduction to operator theory in Riesz spaces*, Springer-Verlag, Berlin-Heidelberg-New York, 1997.

BOUNDED ANALYTIC FUNCTIONS ON RIEMANN SURFACES

MIKIHIRO HAYASHI
Hokkaido University, Sapporo 060-0810, Japan
e-mail: hayashi@math.sci.hokudai.ac.jp

1 The Algebra $H^\infty(R)$

Let R be a Riemann surface (or any domain with analytic structure). We denote by $H^\infty(R)$ the algebra of all bounded analytic functions on R.

With the pointwise operations and the sup-norm

$$\|f\| := \|f\|_\infty := \|f\|_R := \sup_{p\in R} |f(p)|,$$

$H^\infty(R)$ is a commutative Banach algebra over $\mathbf{C}$, where $\mathbf{C}$ is the complex number field. Identifying scalars with constant functions, we assume

$$\mathbf{C} \subseteq H^\infty(R).$$

The algebra $H^\infty(R)$ may have a different character according to the choice of the base Riemann surface R. Therefore, we have to investigate both

$$\begin{cases} \text{a theory (theorems)} \\ \qquad\quad \text{and} \\ \text{examples (\& counter examples)} \end{cases}$$

in order to answer a problem completely. In many cases, unfortunately, we merely have partial answers.

The *maximal ideal space* of the algebra $H^\infty(R)$, which is identified with the set

$$\mathcal{M}(R) = \{\,\phi : H^\infty(R) \to \mathbf{C},\ \text{nonzero algebra homomorphism}\},$$

because

$$\text{every maximal ideal} \quad \overset{1:1}{\longleftrightarrow} \quad \text{kernel } \phi^{-1}(0) : \phi \in \mathcal{M}(R).$$

The *Gelfand transform* $\hat{f}$ of $f \in H^\infty(R)$ is defined by

$$\hat{f}(\phi) := \phi(f) \quad (\phi \in \mathcal{M}(R)).$$

The *Gelfand topology* of $\mathcal{M}(R)$ is the weakest topology among those by which all Gelfand transforms $\hat{f}$ are continuous on $\mathcal{M}(R)$.

Since $H^\infty(R)$ is a *Banach* algebra, each $\phi \in \mathcal{M}(R)$ turns out to be bounded ([Gm69b,Gr81b,St71b] etc.), i.e.,

$$\|\phi\| = \phi(1) = 1.$$

Hence, $\mathcal{M}(R)$ is a weak* compact subset of the closed unit ball $\{ \psi \in H^\infty(R)^* : \|\psi\| \leq 1 \}$, where $H^\infty(R)^*$ denotes the dual of the Banach space $H^\infty(R)$.

Note that

$$\text{the weak* topology of } \mathcal{M}(R) = \text{the Gelfand topology.}$$

For each $p \in R$, define a functional ϕ_p by

$$\phi_p(f) := f(p) \quad (f \in H^\infty(R)).$$

Then, ϕ_p is a nonzero algebra homomorphism of $H^\infty(R)$. Hence, $\phi_p \in \mathcal{M}(R)$.

This yields a natural 'inclusion' mapping

$$\tau : R \ni p \mapsto \phi_p \in \mathcal{M}(R),$$

which is continuous but may not be injective. Identifying $p \in R$ with ϕ_p, we often regard R as a subset of $\mathcal{M}(R)$.

Trivially, a Riemann surface R determines $H^\infty(R)$, and the algebra $H^\infty(R)$ determines $\mathcal{M}(R)$. Thus, if one can distinguish the points of $\tau(R)$ with other points in $\mathcal{M}(R)$, then $H^\infty(R)$ determines R.

Even if it is not the case, the maximal ideal space $\mathcal{M}(R)$ is full of information on $H^\infty(R)$.

Note that $H^\infty(R)$ has the algebraic and analytic characters and $\mathcal{M}(R)$ the geometric characters.

2 On the 'inclusion' $R \subset \mathcal{M}(R)$

If $H^\infty(R) = \mathbf{C}$, we have nothing to do with the algebra $H^\infty(R)$.

A hard problem is to

Problem 1. *Characterize R with $H^\infty(R) \neq \mathbf{C}$.*

If, for any pair of distinct points p, q in R, there is a function $f \in H^\infty(R)$ with $f(p) \neq f(q)$, we say that $H^\infty(R)$ is *(point) separating* ($\Longleftrightarrow$the 'inclusion' mapping τ is injective).

In case $H^\infty(R) \neq \mathbf{C}$, one may ask:

Problem 2 (The point separation problem). *Is the algebra $H^\infty(R)$ separating?*

Problem 3 (The open homeomorphic problem). *Is R embedded in $\mathcal{M}(R)$ as its open subset homeomorphically?*

On Problem 2 there is a general theory (Theorem 2.1).
For Problem 3, we have a satisfactory answer (Theorem 2.2).

Theorem 2.1 (Royden [Ry65]). *If $\mathcal{A}$ is an algebra of analytic functions on R with $\mathcal{A} \supsetneqq \mathbf{C}$, then there exists a triple $(\widetilde{R}, \varphi, \widetilde{\mathcal{A}})$ of*

$$\widetilde{R} : \text{a (connected) Riemann surface}$$
$$\varphi : R \to \widetilde{R}, \text{ an analytic mapping}$$
$$\widetilde{\mathcal{A}} : \text{an algebra of analytic functions on } \widetilde{R}$$

such that

(a) $\mathcal{A} = \widetilde{\mathcal{A}} \circ \varphi$.

(b) $\widetilde{\mathcal{A}}$*: separating except for at most a countable number of points of* $\widetilde{R}$ *(i.e., weakly separating).*

(c) $\widetilde{R}$*:* $\widetilde{\mathcal{A}}$*-maximal, i.e.,* $\widetilde{\mathcal{A}}$ *has no analytic continuation to a Riemann surface containing* $\widetilde{R}$ *properly.*

The surface $\widetilde{R}$ is determined uniquely up to the conformal equivalence, and is called Royden's resolution (w.r.t. the algebra $\mathcal{A}$).

[Construction of $\widetilde{R}$]: Let $f \in \mathcal{A}$ be nonconstant. Choose a point p with $f'(p) \neq 0$ and a small neighborhood U of p such that f is a coordinate for U. Simultaneous analytic continuations of the function elements $\{ g \circ (f|U)^{-1} : g \in \mathcal{A} \}$ gives $\widetilde{R}$.

In case $\mathcal{A} = H^\infty(R)$: It follows that

$$\widetilde{H^\infty(R)} = H^\infty(\widetilde{R}).$$

Therefore, we may restrict our consideration to the case $R = \widetilde{R}$ if we are interested in only the structures of an algebra $H^\infty(R)$.

For a concrete example, however, Theorem 2.1 is little helpful to determine whether it is separating or not. In §5, we consider this problem for a two-sheeted discs.

Set

$M^\infty(R) := \{f$: a meromorphic function bounded off a compact subset of $R\}$

$\mathcal{P}(R) := \{a \in R : \text{some } f \in M^\infty(R) \text{ has a pole at } a\}$

We call $\mathcal{P}(R)$ the *pole set* of R. Trivially,

- The poles of $f \in M^\infty(R)$ are at most finite.
- If $f \in M^\infty(R)$ has no poles, then $f \in H^\infty(R)$.

Theorem 2.2 ([H87]). *Suppose $H^\infty(R) \neq \mathbf{C}$ and $\widetilde{R} = R$. The followings are equivalent:*

(1) the mapping $\tau : R \to \mathcal{M}(R)$ is open and homeomorphic in a neighborhood of a.

(2) $a \in \mathcal{P}(R)$.

(3) $\exists h \in M^\infty(R)$: only one and simple pole at a.

(4) $\exists$ bounded linear operator T on $H^\infty(R)$:

$$T(fg) = Tf \cdot g + f(a)Tg, \quad (Tf_0)(a) \neq 0 \text{ for some } f_0.$$

(5) Let $\Delta : |z| < 1$. $\exists$ homeomorphic mapping $\Phi : \Delta \overset{onto}{\to}$ (an open subset of $\mathcal{M}(R)$) s.t.

$$\Phi(0) = \phi_a, \text{ and } \hat{f} \circ \Phi\text{: analytic } (f \in H^\infty(R)).$$

The set $\Phi(\Delta)$ is called a *weak* open analytic disc* with center ϕ_a.

The pole set $\mathcal{P}(R)$ has nice properties. Many of the results known for plane domains can be generalized to Riemann surfaces as far as they concern the pole sets $\mathcal{P}(R)$ (cf. [H87], [GmH87]).

For examples of Riemann surfaces, we summarize them in §4 together with other examples.

3 Riemann surfaces with $M^\infty(R) \neq \mathbf{C}$

In connection with Theorem 2.2, we are interested in the class P_{M^∞} of Riemann surfaces R satisfying

$$M^\infty(R) \neq \mathbf{C}.$$

Note that class P_{M^∞} includes the class P_{H^∞} of Riemann surfaces R satisfying $H^\infty(R) \neq \mathbf{C}$.

As mentioned earlier, the characterization of the class P_{H^∞} is a hard problem. Nevertheless, we can classify the class P_{M^∞} in terms of P_{H^∞}.

The Royden's resolution of R with respect to the algebra $M^\infty(R)$ can be defined in the same way as the case $H^\infty(R)$, and it is denoted also by $\widetilde{R}$ in this section.

Theorem 3.1 ([H97]). *Let R be a Riemann such that $M^\infty(R) \neq \mathbf{C}$. Then, either one of the following two occurs:*

(1) $H^\infty(R) \neq \mathbf{C}$.

(2) $\widetilde{R}$ *is a closed Riemann surface.*

The following lemma is crutial.

Lemma 3.2. *Let f, $g \in M^\infty(R)$. Then, there is a nonconstant polynomial P of two variables such that*

$$P(f, g) = 0.$$

Proof. We may assume that both f, g are nonconstants. Let L be the number of poles $f^{-1}(\infty) \cup g^{-1}(\infty)$, M the maximum order of these poles, and $\mathcal{L}_N$ the linear span of $f^j g^k$ $(1 \leq j, k \leq N)$.

For each $F \in \mathcal{L}_N$, we consider all coefficients of the principal terms of the Laurent expansion of F around all its poles. Then, we have a linear mapping $\mathcal{L}_N \to \mathbf{C}^{2LMN}$. Since $f^j g^k$ $(1 \leq j, k \leq N)$ are N^2 elements, we can find a nontrivial linear combinaion $F = \sum c_{jk} f^j g^k \in \mathcal{L}_N$ in the kernel of this linear mapping. Since F has no poles, $F \in H^\infty(R)$, and hence, $F = c$ (constant). Therefore, the disired polynomial is given by

$$P(z, w) = c - \sum c_{jk} z^j w^k.$$

4 Examples of Riemann surfaces

Let us begin with the famous examples due to P. Myrberg: Let

$$\Omega = \mathbf{C} \setminus \{0\}, \text{ a sequence: } 0 < a_{n+1} < a_n < 1,\ a_n \searrow 0.$$

1. Ω_+, Ω_-: two copies of Ω, and let $I_n = [a_{2n}, a_{2n-1}]$.

2. Cutting both sheets Ω_+ and Ω_- along all I_n, we past the two sides of I_n in Ω_+ with those in Ω_- crosswise.

Let $\widetilde{\Omega}$ be the resulting surface and $\pi : \widetilde{\Omega} \to \Omega$ the natural projection. This is an unlimited covering map with the ramification points $\pi^{-1}(a_n)$ $(n = 1, 2, \ldots)$ and the conformal structure of $\widetilde{\Omega}$ is induced from Ω by π. (see Figure 1; the left).

Example 1. There exists a Riemann surface R with $H^\infty(R) \neq \mathbf{C}$, but not separating.

Set $\Delta = \{z : |z| < 1\}$, $\Delta_0 = \Delta \setminus \{0\}$, $\widetilde{\Delta_0} = \pi^{-1}(\Delta_0)$. We see that

$$H^\infty(\widetilde{\Delta_0}) = H^\infty(\Delta_0) \circ \pi = H^\infty(\Delta) \circ \pi. \tag{4.1}$$

The proof is simple. For $f \in H^\infty(\widetilde{\Delta_0})$, set

$$g(z) = (f(z_+) - f(z_-))^2 \quad (\pi^{-1}(z) = \{z_+, z_-\}).$$

Then, $g \in H^\infty(\Delta_0)$. By Riemann's removable singularity theorem, $g(z)$ extends analytically to $z = 0$. Thus, $g \in H^\infty(\Delta)$. Since $(a_n)_+ = (a_n)_-$ at a ramification point, $g(a_n) = 0 \quad (n = 1, 2, \ldots)$. The uniqueness theorem implies $g(z) \equiv 0$, and

$$f(z_+) = f(z_-) \ (z \in \Delta_0).$$

This shows that $F(z) = f(z_+) = f(z_-)$ defines a (bounded analytic function on Δ, and $f = F \circ \pi$. This proves (4.1).

Note. Deleting a closed disc from one of two sheets of $\widetilde{\Omega}$, Myrberg gave a Riemann surface with a large boundary but $H^\infty(R) = \mathbf{C}$. ([My49]; see also Ahlfors [Ah51]). These examples show that pathological phenomena may occur for bounded analytic functions of Riemann surfaces. Here one might aware that the algebras $H^\infty(R)$ on these surfaces are isomorphic to ones on plane domains, and hence, these algebras themselves are not pathological. This suggested me to study Riemann surfaces from the point of view of the algebra $H^\infty(R)$ and led me to finding Theorem 2.2 and the next example.

Example 2 **([H87]).** There exists a Riemann surface R with $\mathcal{P}(R) \subsetneq R$. Set

$$R = \bigcup_{k=0}^{\infty} D_k, \quad D_k = \begin{cases} \Delta & (k = 0) \\ \Delta \setminus \bigcup_{n \geq 1, n \neq k} J_n & (k \neq 0). \end{cases}$$

Here, J_n is any sequence of mutually disjoints closed interval, which converges to a boundary point in Δ. For each k, we choose I_k be a finite union of closed subintervals of J_k, and past the sheets D_0 and D_k along the subintervals I_k crosswise. The resulting surface is R. (Figure 2).

Let R_n be the surface obtained by pasting only finite sheets $D_1, \ldots,$ D_n with D_0 in the same way as R. Since we removed intervals J_n, there is an analytic mapping $\pi_n : R \to R_n$ with $\pi_n(D_k) = D_k \ (1 \leq k \leq n), = D_0$ (otherwise). Since $H^\infty(R_n) \circ \pi_n \subset H^\infty(R)$ and since R_n is a finite bordered

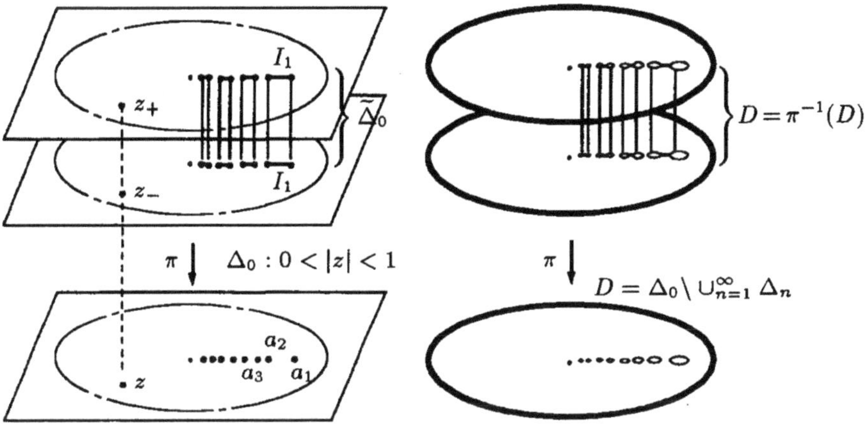

Figure 1. Myrberg's Example & the Myrberg phenomenon

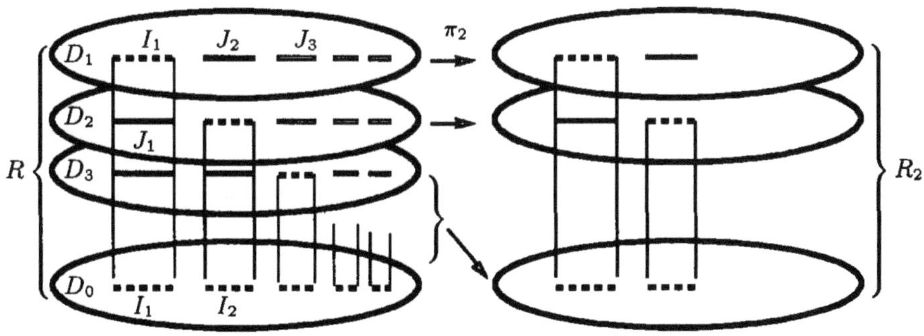

Figure 2. Example 2

surface, $H^\infty(R_n)$ is separating, $H^\infty(R)$ turns out to be separating. If we choose the numbers of subintervals I_k in J_k sufficiently large, the sheets D_n converges to the sheet D_0 as $n \to \infty$ (in the maximal ideal space $\mathcal{M}(R)$). As to its pole set we have

$$\mathcal{P}(R) = \bigcup_{n=1}^{\infty} D_n = R \setminus D_0.$$

Observing this example, I have posed the following conjecture ([H97]).

Conjecture. *Let R be a Riemann surface. Suppose that there are an open subset U and a compact subset K of R such that $H^\infty(U)$ is weakly separating and such that $R = K \cup U$ and $\partial K \subset \mathcal{P}(U)$. Then, $M^\infty(R)$ is weakly separating.*

The examples in Figure 3 supports this conjecture, where we use the complex plane in place of the unit disc to define the sheet D_n and choose intervals J_n so as to converges to the infinity of the complex plane.

Note. As a modification of this example, I constructed also in the same paper ([H87]) a Riemann surface such that $H^\infty(R)$ is separating and such that $\mathcal{P}(R) = \emptyset$.

On the other hand, there are Riemann surfaces with $\mathcal{P}(R) = R$ (e.g., Riemann surfaces of Parreau-Widom type [St75]). Therefore, Theorem 2.2 gives a complete answer to the open homeomorphic problem.

5 The Myrberg phenomenon

Recall Example 1: $\pi : \widetilde{\Delta_0} \to \Delta_0$, $\Delta_0 : 0 < |z| < 1$, which has the ramification points $\{\pi^{-1}(a_n)\}$.

Let $\Delta_n = \{z : |z - a_n| \leq r_n\}$ be a small closed disc with center a_n, and

$$\widetilde{D} = \pi^{-1}(D), \quad D = \Delta_0 \setminus \bigcup_{n=1}^{\infty} \Delta_n \qquad \text{(see Figure 1; the right).}$$

If we have

$$H^\infty(\widetilde{D}) = H^\infty(D) \circ \pi,$$

then we say that a *Myrberg phenomenon* occurs. It was a bit surprising when we found that the Myrberg phenomena can occur for $(\widetilde{D}, D, \pi)$ in spite of the complete lack of branch points.

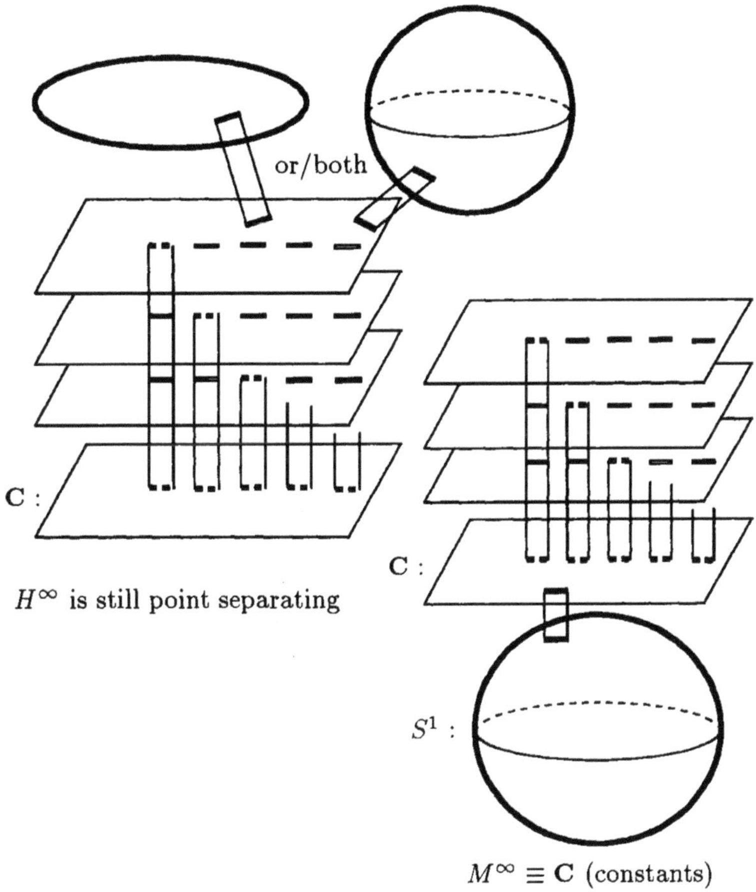
or/both
C :
H^∞ is still point separating
C :
S^1 :
$M^\infty \equiv \mathbf{C}$ (constants)

Figure 3.

Our next purpose is to clarify why such a phenomenon can occur. For instance, what kind of conditions are related to the Myrberg phenomena? Can we write such a condition in terms of the radii r_n? Since all ramification points are extracted with the discs $\pi^{-1}(\Delta_n)$ from $\widetilde{D}$, we can no longer make use of the simple argument used in Example 1.

In place of Riemann's removable singularity theorem, we find that a generalization of the uniqueness theorem can be a sufficient condition.

We say that the *uniqueness theorem* holds if for $f \in H^\infty(D)$

$$f^{(n)}(0) := \lim_{z<0, z\to 0} f^{(n)}(z) = 0 \ (n = 0, 1, 2, \ \ldots) \text{ implies } f(z) \equiv 0.$$

Theorem 5.1 ([HN98]). *In Example 1, let* $a_n = 2^{-n}$ *and* $\Delta_n = \{ z : |z - a_n| \le 2^{-nN(n)} \}$. *Set* $\widetilde{D} = \pi^{-1}(D)$, *where* $D = \Delta_0 \setminus \bigcup_{n=1}^{\infty} \Delta_n$.

(1) The uniqueness theorem implies the Myrberg phenomenon, i.e., $H^\infty(\widetilde{D}) = H^\infty(D) \circ \pi$.

(2) $\sum_{n=1}^{\infty} 1/N(n) < \infty$ *(Wiener criterion) implies the uniqueness theorem.*

(3) Set $M(l) := \{ m \in \mathbf{N} : N(\mu) \ge l + 2 \ (\mu \ge m) \}$. *Then,* $\sum_{l=1}^{\infty} 2^{-M(l)} = \infty$ *(Carleman condition) implies the uniqueness theorem.*

(4) If $\sup_{n\in\mathbf{N}} n \left(N(n) - \frac{n+1}{2}\right) < \infty$, *then Myrberg phenomenon does not occur, i.e.,* $H^\infty(\widetilde{D})$ *is separating.*

No necessary and sufficient conditions are not known.

For the Myrberg phenomena/the point separation problem of 2-sheeted disc, several other investigations have been made in different situations ([HK98,HN88,HNS92]).

6 Corona problem and isomorphic problem

The corona problem asks

Problem 4. *Is* R *dense in* $\mathcal{M}(R)$*?*

The Royden's resolution $\widetilde{R}$ looks a 'natural' domain for $H^\infty(R)$. Thus, one might expect that the algebra $H^\infty(R)$ determine $\widetilde{R}$.

More precisely,

Problem 5 (The isomorphic problem). *If $H^\infty(R_1) \cong H^\infty(R_2)$ as Banach algebras, then $\widetilde{R}_1 \cong \widetilde{R}_2$ as Riemann surfaces?*

Surprisingly (for me), this was false.

A counter example to the corona problem constructed by E. Bishop (cf. [Gm70]) in the case of R one-dimensional analytic space. B. Cole (cf. [Gm78b]) constructed a counter example for a Riemann surface and Nakai ([Nk82]) constructed one which is of Parreau-Widom type. A counter example to the isomorphic problem constructed by me ([H91]) and this at the same time is a counter example to the corona problem.

The construction of this example starts with the following modification of the Bishop's counter example. Let $\{r_n\}_{n=1}^\infty$ be any increasing positive numbers with $r_n \nearrow 1$ and $\{N_n\}_{n=1}^\infty$ a sequence of positive integers (fixed later). Set

$$V = \{\,(z,w) \in \Delta^2 \;:\; z \in E \text{ or } w \in E\,\} \quad \text{(see Figure 4)},$$

where

$$E = \bigcup_{n=1}^\infty E_n, \; E_n = \{a_{nk} : k = 1, \ldots, N_n\}, \; a_{nk} = r_n e^{2\pi i k/N_n}.$$

If N_n are chosen sufficiently large, then every $f \in H^\infty(V)$ extends uniquely to a bounded analytic function on the bidisc Δ^2.

$$\text{i.e.} \quad H^\infty(V) \cong H^\infty(\Delta^2).$$

By an arbitrariness in the above construction of V, there are many non isomorphic V.

If we make smoothing deformation to V (see Figure 4), we have a Riemann surface R. A geometric deformation is easy, but we have to encounter a new difficulty; namely, it is difficult to make bounded analytic function on V. Thus, we invent an analytic deformation Δ^2.

Set

$$h_n(z) = \prod_{a \in E_n} (z - a), \qquad H_n(z,w) = h_n(z)h_n(w).$$

If we choose positive numbers $\{\delta_n\}$ sufficiently small,

$$H(z,w) = \frac{\delta_1}{H_1} + \frac{\delta_2}{H_1 H_2} + \cdots + \frac{\delta_n}{H_1 \cdots H_n} + \cdots$$

converges and defines a meromorphic function on Δ^2.

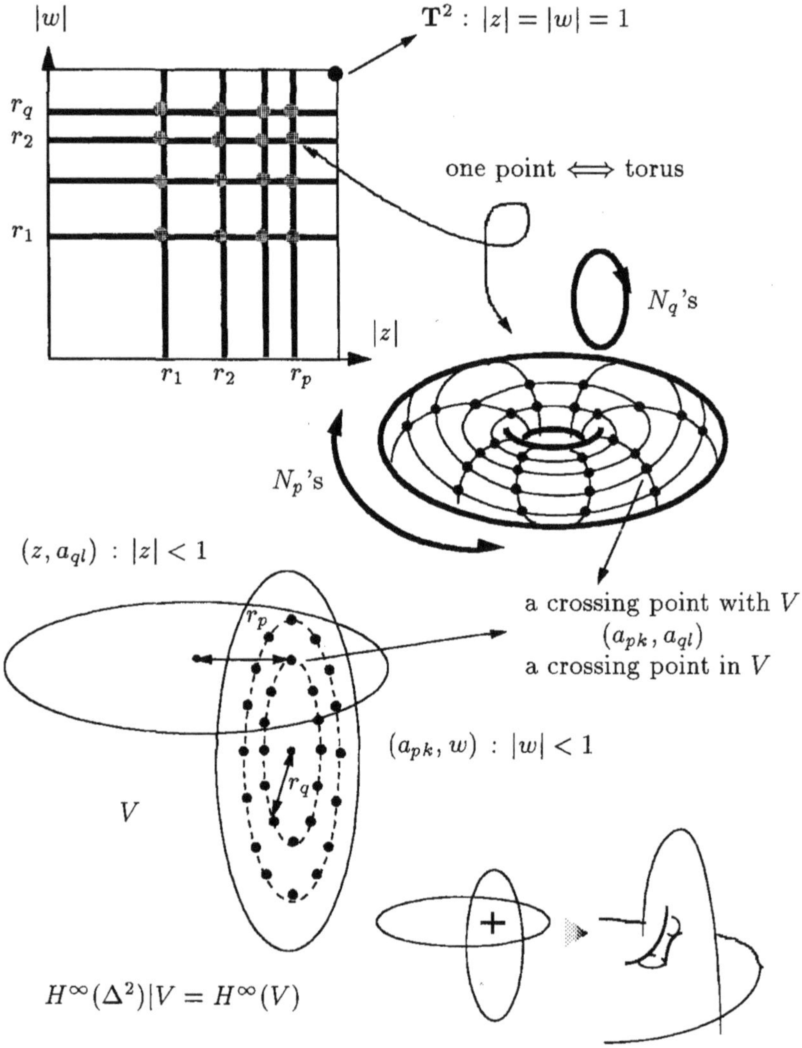

Figure 4. V & its deformation

Example 3. There exits a Riemann suface R such that $H^\infty(R) \cong H^\infty(\Delta^2)$. Set $R = \{ (z,w) \in \Delta^2 : 1 = H(z,w) \}$. If we choose the positive integers sufficiently large and positive numbers δ_n sufficiently small, then every $f \in H^\infty(R)$ extends uniquely to a bounded analytic function on Δ^2.

As a variation of this example, we can also construct a example of a Riemann surface R such that R is weakly separating but not separating, or such that R is not dense in $\widetilde{R}$ in the maximal ideal space $\mathcal{M}(R)$.

There is a sufficient condition in order that the isomorphic problem is solved affirmatively:

Theorem 6.1. *If the pole set $\mathcal{P}(R)$ is not empty, then the isomorphic theorem holds, that is, if $H^\infty(R_1) \cong H^\infty(R_2)$ (as algebras), and if either of the sets $\mathcal{P}(R_j)$ $(j = 1, 2)$ is not empty, then $\widetilde{R}_1 \cong \widetilde{R}_2$ (conformally).*

References

[AS75] P.R. Ahern and R. Schneider, *Isometries of H^∞*, Duke Math. J. **42**(1975), 321–326.

[Ah51] L.V. Ahlfors, *Remarks on the classification of open Riemann surfaces*, Ann. Acad. Sci. Fenn. Ser. A.I. **87**(1951), 8pp.

[AW98b] H. Alexander, J. Wermer, *Several Complex Variables and Banach Algebras* (Third edition), Springer-Verlag, 1998.

[BS49] H. Behnke and K. Stein, *Entwicklung analytischer Funktionen auf Riemannschen Flächen*, Math. Ann. **120**(1949), 430–461.

[Bs62] E. Bishop, *Analyticity in certain Banach algebras*, Trans. Amer. Math. Soc. **102**(1962), 507–544.

[Fs83b] S. D. Fisher, *Function Theory of Planar domains*, John Wiley & Sons, 1983.

[Gm69b] T.W. Gamelin, *Uniform Algebras* (Second edition), Chelsea Press, 1984. (First edition was published from Prentice-Hall, Englewood Cliffs, N. J., 1969).

[Gm70] ———, *Localzation of the corona problem*, Pacific J. Math. **34**(1970), 73–81.

[Gm78b] ———, *Uniform Algebras and Jensen Measures*, London Math. Soc. Lecture Note Series. No.32, 1978.

[GmH87] T. W. Gamelin and M. Hayashi, *The algebra of bounded analytic functions on a Riemann surface*, J. Reine Angew Math. **382**(1987), 49–73.

[Gr81b] J.B. Garnett, *Bounded Analytic Functions*, Academic Press, 1981.

[GJ85] J.B. Garnett and P.W. Jones, *The corona theorem for Denjoy domains*, Acta Math. **155**(1985), 27–40.

[Hr82] M. Hara, *The corona problem on 2-sheeted disks*, Proc. Japan Acad. **58**, Ser. A (1982), 256–257.

[Hs83b] M. Hasumi, *Hardy Classes on Infinitely Connected Riemann Surfaces*, Lecture Notes in Math. **1027**, Springer, 1983.

[H79] M. Hayashi, *Hardy Classes on Riemann Surfaces*, Univ. of California, Los Angeles, thesis. (1979).

[H83] ———, *Invariant subspaces on Riemann surfaces of Parreau-Widom type*, Trans. Amer. Math. Soc. **279**(1983), 737–757.

[H86] ———, *An example of a domain of Parreau-Widom type*, Complex Variables **6**(1986), 73–80.

[H87] ———, *The maximal ideal space of the bounded analytic functions on a Riemann surface*, J. Math. Soc. Japan **39**(1987), 337–344.

[H91] ———, *An example of a Riemann surface such that the algebra of all bounded analytic functions on it isomorphic to that on the bidisc* (Japanese), at the Fall Conference of Math. Soc. Japan, 1991.

[H97] ———, *Classification of Riemann surfaces with a nonconstant meromorphic function bounded at infinity*, Izvestiya Nat. Akad. Nauk Armenii. Mat. 32, No. 4 (1997), 37–44.

[HK98] M. Hayashi and T. Kato, *Point separation of a two-sheeted disc by bounded analytic functions*, Hokkaido MathJ. (1998) (to appear).

[HN88] M. Hayashi and M. Nakai, *Point separation by bounded analytic functions of a covering Riemann surface*, Pacific J. Math. **134**(1988), 261–273.

[HN98] ———, *A unicity theorem and the Myrberg phenomenon*, J. d'Analyse Math. (1998) (to appear).

[HNS92] M. Hayashi, M. Nakai and S. Segawa, *Bounded analytic functions on two sheeted discs*, Trans. Amer. Math. Soc. **333**(1992), 799–819.

[My49] P.J. Myrberg, *Über die analytische Fortsetzung von beschränkten Funktionen*, Ann. Acad. Sci. Fenn. Ser. A.I. **58**(1949), 7pp.

[Nk82] M. Nakai, *Corona problem for Riemann surfaces of Parreau-Widom type*, Pacific J. Math. **103**(1982), 103–109.

[Nk83] ———, *The corona problem on finitely sheeted surfaces*, Nagoya Math. J. **92**(1983), 163–173.

[Nr98] J. Narita, *Separation and weak separation on Riemann surfaces*, Hokkaido Math. J. (1998) (to appear).

[Ry65] H.L. Royden, *Algebras of bounded analytic functions on Riemann surfaces*, Acta Math. **114**(1965), 113–142.

[Rd78] W. Rudin, *Some theorems on bounded analytic functions*, Trans.

Amer. Math. Soc. **78**(1955), 333–342.

[Rd69b] W. Rudin, *Function Theory in Polydiscs*, W. A. Benjamin, Inc., 1969.

[St75] C.M. Stanton, *Bounded analytic functions on a class of open Riemann surfaces*, Pacific J. Math. **59**(1975), 557–565.

[St71b] E.L. Stout, *The Theory of Uniform Algebras*, Bogden & Quigley, Inc., 1971.

ON THE ANTI-PARALLEL IGETA CONSTRUCTION OF RIEMANN SURFACES

YOSHITAKE HASHIMOTO

Department of Mathematics and Physics, Graduate School of Science, Osaka City University, Sugimoto, Sumiyoshi-ku, Osaka 558-8585, Japan
e-mail: hashimot@sci.osaka-cu.ac.jp

KIYOSHI OHBA

Department of Mathematics, Faculty of Science, Ochanomizu University, Otsuka 2-1-1, Bunkyo-ku, Tokyo 112-8601, Japan
e-mail: ohba@math.ocha.ac.jp

In this paper we introduce a method of constructing a holomorphic family of once punctured Riemann surfaces of any genus. The parameter space has the same dimension as the moduli space of such objects. We show the fact that the classifying map of this family is open, which implies that our construction gives a good visualization of the deformation of complex structures of them. All the hyperelliptic curves can be obtained by our construction.

1 Introduction

In this paper we introduce a method of constructing once punctured Riemann surfaces using "Igeta", which we introduced in our previous paper [HO]. By an Igeta we mean a pair of disjoint parallel line segments in the Gauss plane $\mathbb{C}$ with the same length. Given a set of g Igeta, we first cut the plane $\mathbb{C}$ along the segments of each Igeta, and then we paste the edges appropriately to obtain a once punctured Riemann surface of genus g. For each Igeta we have two natural ways of pasting the edges: pasting each side of one segment and the opposite side of the other segment by a parallel translation, and pasting each side of one segment and the same side of the other segment by a rotation of angle π. (See § 2, Figure 1.) We shall call an Igeta "a parallel Igeta" when its edges are pasted with a parallel translation, "an anti-parallel Igeta" when its edges are pasted with a rotation of angle π. We considered in [HO] the case when the set of Igeta consists of only parallel Igeta, and we consider in this paper the case when the set consists of not only parallel but also anti-parallel Igeta.

The set of configurations of g disjoint Igeta yields a family of once punctured Riemann surfaces of genus g. Let η be the set of parallel and anti-parallel Igeta in $\mathbb{C}$, and let $I_g\eta$ be the collection of sets Γ of g disjoint elements of η. (The notations η and $I_g\eta$ are different from the ones in [HO], where η was the set of only parallel Igeta in $\mathbb{C}$, and $I_g\eta$ was the collection of sets of g disjoint parallel Igeta.) $I_g\eta$ turns out to be a $3g$-dimensional complex manifold. For $\Gamma \in I_g\eta$ we denote by $(R(\Gamma), p_\infty)$ the resulting once punctured Riemann surface of genus g. We then obtain a holomorphic family of once punctured Riemann surfaces of genus g over $I_g\eta$ whose fiber on $\Gamma \in I_g\eta$ is $(R(\Gamma), p_\infty)$.

It is clear that two once punctured Riemann surfaces $(R(\Gamma), p_\infty)$ and $(R(\Gamma'), p_\infty)$ are biholomorphic if there exists an affine automorphism of $\mathbb{C}$ that transform Γ to Γ'. The quotient space $I_g\eta/\mathrm{Aut}(\mathbb{C})$ turns out to be a $3g-2$-dimensional complex V-manifold, whose dimension is the same as the dimension of the moduli space $\mathcal{M}_{g,1}$ of once punctured Riemann surfaces of genus g. We thus expect to have a visual image of the moduli space by using this construction.

We considered in [HO] the Kodaira-Spencer map of this family, and proved that the classifying map from $I_g\eta/\mathrm{Aut}(\mathbb{C})$ to $\mathcal{M}_{g,1}$ is locally surjective at any $[\Gamma]$ if Γ consists of only parallel Igeta. In this paper we obtain the following result:

Theorem 1. *The classifying map*

$$F : I_g\eta/\mathrm{Aut}(\mathbb{C}) \to \mathcal{M}_{g,1}$$

is an open map.

This result implies that our construction gives a good visualization of the deformation of complex structures of once punctured Riemann surfaces. However, we do not know whether any once punctured Riemann surface can be obtained by the Igeta-construction.

On the other hand we obtain the following result:

Theorem 2. *Any hyperelliptic curve of genus g is obtained from a set of g disjoint anti-parallel Igeta each of which is symmetric with respect to the origin.*

Given such a symmetric configuration of anti-parallel Igeta, we easily obtain a defining equation of the resulting hyperelliptic curve. We can then tell when two such symmetric configurations represent the same hyperelliptic curve. We have some other examples of configurations of Igeta which represent the same once punctured Riemann surface.

The authors wishes to express their gratitude to Professor Akio Hattori and Professor Yukio Matsumoto for continuous advice and encouragement. They are also grateful to Professor Koei Sekigawa and Professor Hideya Hashimoto who gave the second author an opportunity to speak at the workshop held at St. Constantine in Bulgaria 1998.

2 Parallel and anti-parallel Igeta

In this section we give definitions of "parallel Igeta" and "anti-parallel Igeta", and then construct a holomorphic family on the configuration space of Igeta. (See also [HO, § 2].)

The Gauss plane is the complex affine line $\mathbb{A}^1$ with a fixed global coordinate $\zeta : \mathbb{A}^1 \to \mathbb{C}$. Consider the set η consisting of pairs $((\sigma^+, \sigma^-), \psi)$ where (σ^+, σ^-) is an unordered pair of disjoint parallel line segments in the Gauss plane $(\mathbb{A}^1, \zeta)$ with the same length, and ψ is either the identity map *id* of the Gauss plane $(\mathbb{A}^1, \xi)$ or the affine automorphism τ of $(\mathbb{A}^1, \xi)$ defined by $\tau(\xi) = -\xi$. We call such a pair $((\sigma^+, \sigma^-), id)$ a *parallel Igeta*, and a pair $((\sigma^+, \sigma^-), \tau)$ an *anti-parallel Igeta*. We denote by $I_g\eta$ the collection of unordered sets Γ of g elements of η where

$$\Gamma = (((\sigma_j^+, \sigma_j^-), \psi_j) \in \eta \; ; \; j = 1, \ldots, g)$$

such that $\sigma_j^\pm$ are pairwise disjoint. Let $\phi_j^\pm = \phi_j^\pm[\Gamma]$ be the affine map from the Gauss plane $(\mathbb{A}^1, \xi)$ to $(\mathbb{A}^1, \zeta)$ given by

$$\zeta = a_j \xi + b_j^\pm, \quad a_j \in \mathbb{C}^\times, \; b_j^\pm \in \mathbb{C}$$

such that $\sigma_j^\pm = \phi_j^\pm([-1, 1])$. The space $I_g\eta$ is a $3g$-dimensional open complex manifold with local coordinates $(a_j, b_j^\pm \; ; \; j = 1, \ldots, g)$ for a fixed order of line segments.

We construct a holomorphic family of once punctured Riemann surfaces of genus g over $I_g\eta$ as follows. Let B be an open and relatively compact subset of $I_g\eta$. Set

$$E_\Gamma^0 = \mathbb{A}^1 - \bigcup_{j=1}^{g} (\sigma_j^+ \cup \sigma_j^-)$$

for $\Gamma = ((\sigma_j^\pm), \psi_j); \; j = 1, \ldots, g) \in I_g\eta$ and set

$$E^0 = \bigsqcup_{\Gamma \in I_g\eta} E_\Gamma^0 \subset I_g\eta \times \mathbb{A}^1,$$

$$E_B^0 = E^0 \cap (B \times \mathbb{A}^1).$$

Let U_∞ be the disk $\{w \in \mathbb{C} \; ; \; |w| < \epsilon\}$ and V_j $(j = 1, \ldots, g)$ copies of the annulus

$$\{z \in \mathbb{C} \; ; \; (1 + \epsilon)^{-1} < |z| < 1 + \epsilon\}$$

for a sufficiently small $\epsilon > 0$, and let

$$V_j^+ = \{z \in V_j\ ;\ |z| > 1\},\quad V_j^- = \{z \in V_j\ ;\ |z| < 1\}.$$

We paste the patches

$$E_B^0,\quad B \times U_\infty,\quad B \times V_j\ (j = 1, \ldots, g)$$

by the attaching maps

$$\begin{aligned} B \times (U_\infty - \{0\}) \ni (\Gamma, w) &\longmapsto (\Gamma, w^{-1}) \in E_B^0, \\ B \times V_j^+ \ni (\Gamma, z) &\longmapsto (\Gamma, \phi_j^+[\Gamma] \circ J(z)) \in E_B^0, \\ B \times V_j^- \ni (\Gamma, z) &\longmapsto (\Gamma, \phi_j^-[\Gamma] \circ \psi_j \circ J(z)) \in E_B^0 \end{aligned}$$

where J is the Joukowski transform

$$J(z) = \frac{1}{2}(z + z^{-1}).$$

We then obtain a complex manifold E_B, which is the total space of a holomorphic family of once punctured Riemann surfaces of genus g over B. As $I_g\eta$ is locally compact, we can construct the holomorphic family $\pi : E \to I_g\eta$ such that $\pi^{-1}(B) = E_B$ for any open and relatively compact subset B of $I_g\eta$. For a point Γ of $I_g\eta$ the Riemann surface $R(\Gamma) = \pi^{-1}(\Gamma)$ is constructed by pasting the patches

$$E_\Gamma^0,\quad U_\infty,\quad V_j\ (j = 1, \ldots, g)$$

through the attaching map

$$\begin{aligned} U_\infty \ni w &\longmapsto \zeta = w^{-1} \in E_\Gamma^0, \\ V_j^+ \ni z &\longmapsto \zeta = \phi^+ \circ J(z) \in E_\Gamma^0, \\ V_j^- \ni z &\longmapsto \zeta = \phi^- \circ \psi_j \circ J(z) \in E_\Gamma^0. \end{aligned}$$

We denote by p_∞ the puncture on $R(\Gamma)$ corresponding to $0 \in U_\infty$.

We call the construction mentioned above the *Igeta-construction*. Roughly speaking, the Igeta-construction is as follows: Given g disjoint Igeta $((\sigma_j^\pm, \psi_j);\ j = 1, \ldots, g)$, cut the Gauss plane along the $2g$ line segments $(\sigma_j^\pm;\ j = 1, \ldots, g)$, paste each side of σ_j^+ and the opposite side of σ_j^- by a parallel translation if $((\sigma_j^+, \sigma_j^-), \psi_j)$ is a parallel Igeta, and then paste each side of σ_j^+ and the same side of σ_j^- by a rotation of angle π if $((\sigma_j^+, \sigma_j^-), \psi_j)$ is an anti-parallel Igeta. (Figure 1. The numbers (1), ... , (6) in Figure 1 indicate where to paste.)

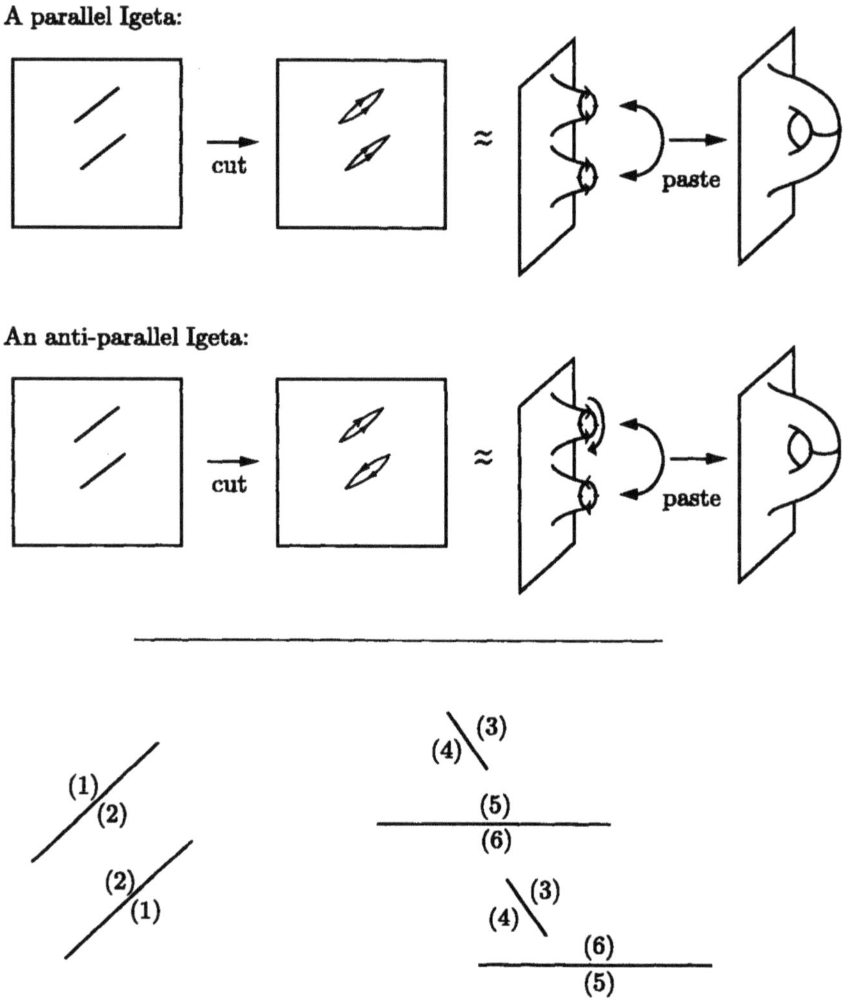

Figure 1. The Igeta-construction

3 Abelian differentials of the second kind and the classifying map from $I_g\eta$.

When we construct a once punctured Riemann surface $(R(\Gamma), p_\infty)$ from $\Gamma \in I_g\eta$ as in the previous section, we obtain g disjoint circles C_j $(j = 1,$

$\ldots, g)$ on $R(\Gamma)$ each of which comes from the unit circle of V_j. The circles C_j's generate a Lagrangian lattice Λ_Γ — a subgroup Λ of $H_1(R;\mathbb{Z})$ is called a Lagrangian lattice of a closed Riemann surface R if Λ coincides with its orthogonal complement with respect to the intersection form. If $\Gamma \in I_g\eta$ consists of only parallel Igeta in $(\mathbb{A}^1, \zeta)$, the 1-form $d\zeta$ on E^0_Γ extends to a meromorphic 1-form ω_Γ on $R(\Gamma)$, which played an important role in our previous paper [HO]. The Abelian differential ω_Γ of the second kind is characterized uniquely up to multiple scalars by the following properties: (1) ω_Γ has a pole of order 2 at p_∞ and has no other poles, (2) the integral of ω_Γ along any element of Λ_Γ equals zero. (See [HO].)

When Γ contains anti-parallel Igeta, the Lagrangian lattice Λ_Γ and the point p_∞ also determine an Abelian differential $\omega_{\Lambda_\Gamma, p_\infty}$ uniquely up to scalars satisfying the condition (1) and (2) above. However $\omega_{\Lambda_\Gamma, p_\infty}$ does not come from the differential $d\zeta$ on E^0_Γ because that rotations of angle π does not preserve $d\zeta$. Instead, we hence construct a natural Abelian differential $\widetilde{\omega_\Gamma}$ on a certain double covering space of $R(\Gamma)$, which plays an important role in showing that the classifying map from $I_g\eta/\mathrm{Aut}(\mathbb{A}^1)$ to the moduli space $\mathcal{M}_{g,1}$ of Riemann surfaces of genus g is an open map.

Let Γ contain at least one anti-parallel Igeta. Set that $(\sigma_1^\pm, \tau), \ldots, (\sigma_s^\pm, \tau)$ are the anti-parallel Igeta of Γ and that $(\sigma_{s+1}^\pm, id), \ldots, (\sigma_g^\pm, id)$ are the parallel Igeta of Γ. We then obtain from Γ a once punctured Riemann surface $(R(\Gamma), p_\infty)$, a Lagrangian lattice Λ_Γ, and a non-trivial element $c_\Gamma = [C_1 + \cdots + C_s]$ in $\Lambda_\Gamma \otimes \mathbb{Z}_2$. We construct $\widetilde{R(\Gamma)}$ and $\widetilde{\omega_\Gamma}$ from these ingredients.

The element c_Γ determines a double covering space $p : \widetilde{R(\Gamma)} \to R(\Gamma)$ via the following homomorphism:

$$\pi_1(R(\Gamma)) \ni \gamma \longmapsto \gamma \cdot c_\Gamma \in \mathbb{Z}_2$$

where $\gamma \cdot c_\Gamma$ is the $\mathbb{Z}_2$-intersection number of γ and c_Γ. The double covering space $\widetilde{R(\Gamma)}$ is obtained explicitly from two Gauss planes $\mathbb{C}^+ = (\mathbb{A}^1, \zeta^+)$ and $\mathbb{C}^- = (\mathbb{A}^1, \zeta^-)$ by the following steps: (1) cut each of the two Gauss planes $\mathbb{C}^+$ and $\mathbb{C}^-$ along Γ, (2) paste each side of $\sigma_j^\pm$ $(j = 1, \ldots, s)$ in $\mathbb{C}^+$ and the same side of $\sigma_j^\mp$ in $\mathbb{C}^-$, (3) paste each side of σ_j^+ $(j = s+1, \ldots, g)$ in $\mathbb{C}^+$ and the opposite side of σ_j^- in $\mathbb{C}^+$, (4) paste each side of σ_j^+ $(j = s+1, \ldots, g)$ in $\mathbb{C}^-$ and the opposite side of σ_j^- in $\mathbb{C}^-$, (5) compactify by adding two points p_∞^+ and p_∞^-. (See Figure 2. Γ in Figure 2 has one parallel Igeta and one anti-parallel Igeta. The numbers on the figure indicate where to paste.) The involution ι of $\widetilde{R(\Gamma)}$ is induced by the map

$$\mathbb{C}^+ \longrightarrow \mathbb{C}^-$$
$$\zeta^+ \longmapsto \zeta^- = \zeta^+$$

Note that $\widetilde{R(\Gamma)}$ is of genus $2g-1$.

Let $\widetilde{\Lambda_\Gamma}$ be the inverse image of Λ_Γ by $p_* : H_1(\widetilde{R(\Gamma)} - \{p_\infty^+, p_\infty^-\}; \mathbb{Z}) \to H_1(R(\Gamma); \mathbb{Z})$. Note that $\widetilde{\Lambda_\Gamma}$ is isomorphic to $\mathbb{Z}^{2g}$. It follows from the Riemann-Roch formula and the bilinear relations of Riemann that

$$\dim H^0(\widetilde{R(\Gamma)};\ \Omega^1(2p_\infty^+)) = 2g,$$
$$\dim H^0(\widetilde{R(\Gamma)};\ \Omega^1(2p_\infty^+ + p_\infty^-)) = 2g+1,$$
$$\dim H^0(\widetilde{R(\Gamma)};\ \Omega^1(2p_\infty^+ + 2p_\infty^-)) = 2g+2,$$

and that the homomorphism by Abelian integrals

$$f:\ H^0(\widetilde{R(\Gamma)};\ \Omega^1(2p_\infty^+ + 2p_\infty^-)) \longrightarrow \operatorname{Hom}(\widetilde{\Lambda_\Gamma}, \mathbb{C})\ (\cong \mathbb{C}^{2g})$$

is surjective. Hence the kernel $\widetilde{Z_\Gamma}$ of the homomorphism f is two-dimensional.

The Abelian differentials $d\zeta^+$ on $\mathbb{C}^+$ and $-d\zeta^-$ on $\mathbb{C}^-$ induce an Abelian differential $\widetilde{\omega_\Gamma}$ of the second kind. $\widetilde{\omega_\Gamma}$ has poles of order 2 at p_∞^+ and p_∞^- and has no other poles. In addition, the integral of $\widetilde{\omega_\Gamma}$ along any element of $\widetilde{\Lambda_\Gamma}$ vanishes. Namely $\widetilde{\omega_\Gamma}$ is in $\widetilde{Z_\Gamma}$. In fact the Abelian differential $\widetilde{\omega_\Gamma}$ and the pullback $p^*(\omega_{\Lambda_\Gamma, p_\infty})$ generate $\widetilde{Z_\Gamma}$: $\widetilde{\omega_\Gamma}$ is an eigenvector corresponding to the eigenvalue -1 of the involution ι^* of $\widetilde{Z_\Gamma}$, and $p^*(\omega_{\Lambda_\Gamma, p_\infty})$ is an eigenvector corresponding to the other eigenvalue 1 of the involution ι^* of $\widetilde{Z_\Gamma}$. Therefore $\widetilde{\omega_\Gamma}$ is characterized uniquely up to scalars by the following three properties: (1) the Abelian differential $\widetilde{\omega_\Gamma}$ is of the second kind, has poles of order 2 at p_∞^+ and p_∞^-, and has no other poles, (2) the integral of $\widetilde{\omega_\Gamma}$ along any element of $\widetilde{\Lambda_\Gamma}$ vanishes, (3) $\iota^*(\widetilde{\omega_\Gamma}) = -\widetilde{\omega_\Gamma}$. We remark here that we later use the following property: the set of zeros of $\widetilde{\omega_\Gamma}$ coincides with the set of points corresponding to the endpoints of Igeta in Γ on $\mathbb{C}^\pm$.

We next construct a holomorphic invariant $I(R,p)$ of once punctured Riemann surfaces (R,p).

When a once punctured Riemann surface (R,p) is given a Lagrangian lattice Λ and a non-trivial element c of $\Lambda \otimes \mathbb{Z}_2$, we obtain a double covering space $\widetilde{R}$ and an Abelian differential $\widetilde{\omega}_{R,p,\Lambda,c}$ uniquely up to scalars on $\widetilde{R}$ as in the case of $(R(\Gamma), p_\infty, \Lambda_\Gamma, c_\Gamma)$. Let $I(\widetilde{\omega}_{R,p,\Lambda,c})$ be the image of the following period map:

$$\int \widetilde{\omega}_{R,p,\Lambda,c} :\ H_1(\widetilde{R}, O(\widetilde{\omega}_{R,p,\Lambda,c}); \mathbb{Z}) \longrightarrow \mathbb{C},$$

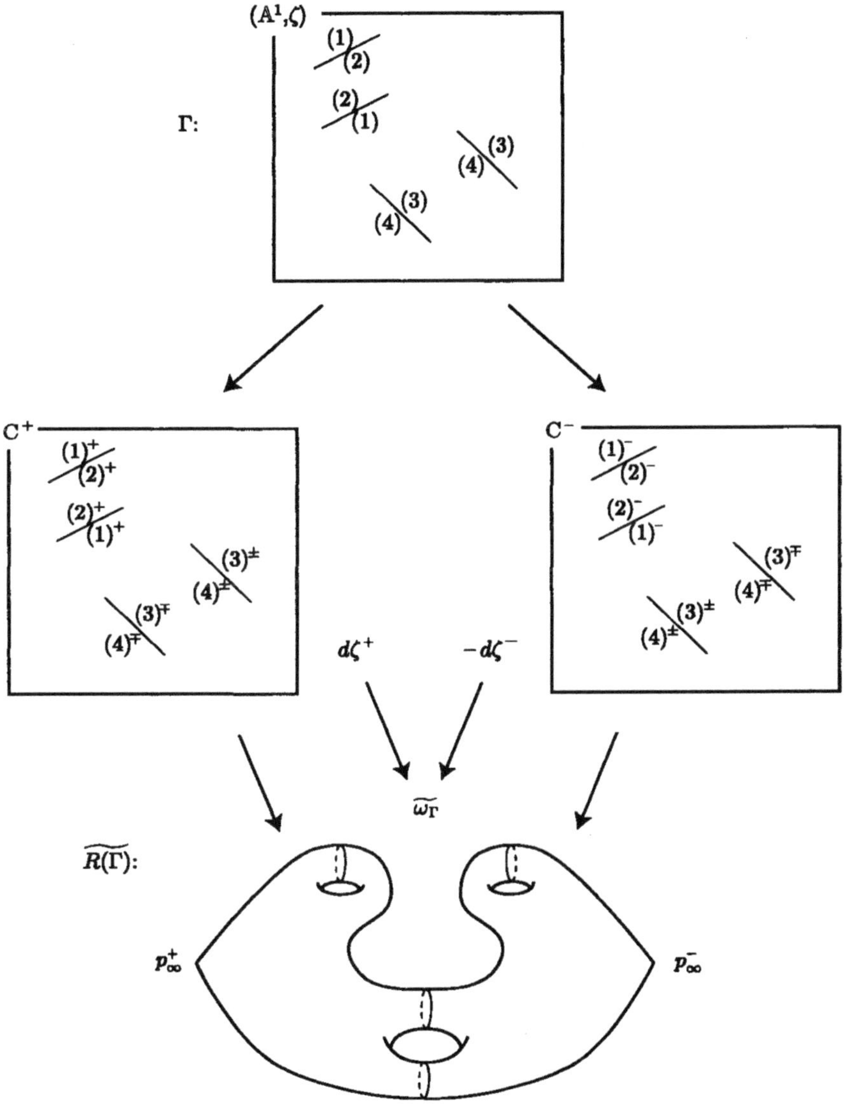

Figure 2. The double covering space $\widetilde{R(\Gamma)}$.

where $O(\widetilde{\omega}_{R,p,\Lambda,c})$ is the set of zeros of $\widetilde{\omega}_{R,p,\Lambda,c}$. Obviously the set $I(\widetilde{\omega}_{R,p,\Lambda,c})$ is a countable subgroup of $\mathbb{C}$.

We denote by $\Omega_{R,p,\Lambda,c}$ the set of those $\widetilde{\omega}_{R,p,\Lambda,c}$ such that $I(\widetilde{\omega}_{R,P,\Lambda,c}) \ni 1$. The set $\Omega_{R,p,\Lambda,c}$ is countable. Set

$$I(R,p,\Lambda,c) := \bigcup_{\widetilde{\omega}_{R,p,\Lambda} \in \Omega_{R,p,\Lambda,c}} I(\widetilde{\omega}_{R,p,\Lambda,c}).$$

The set $I(R,p,\Lambda,c)$ is also a countable subgroup of $\mathbb{C}$, and is an invariant of (R,p,Λ,c).

Furthermore let C_Λ be the set of non-trivial elements of $\Lambda \otimes \mathbb{Z}_2$, $L(R)$ the set of Lagrangian lattices of R, and set

$$I(R,p) := \bigcup_{\Lambda \in L(R)} \bigcup_{c \in C_\Lambda} I(R,p,\Lambda,c).$$

Then $I(R,p)$ is a countable subgroup of $\mathbb{C}$, and is a holomorphic invariant of once punctured Riemann surfaces, which we use to prove the following theorem:

Theorem 1. *The classifying map*

$$F : I_g\eta/\mathrm{Aut}(\mathbb{A}^1) \to \mathcal{M}_{g,1}$$

is an open map.

Proof. Let $I_g\eta_0$ be the subset of $I_g\eta$ consisting of those Γ having the interval $\zeta^{-1}([0,1])$ as one of its $2g$ line segments. Note that the $3g-2$ dimensional submanifold $I_g\eta_0$ of $I_g\eta$ gives a local manifold cover of the V-manifold $I_g\eta/\mathrm{Aut}(\mathbb{A}^1)$ at any point. We consider the classifying map

$$F' : \ I_g\eta_0 \to \mathcal{M}_{g,1}.$$

Assume that F' is not open map. Since F' is complex analytic there exists a simple path

$$\begin{aligned} \gamma : \ [0,1] &\longrightarrow I_g\eta_o \\ t &\longmapsto \Gamma_t \end{aligned}$$

such that $F' \circ \gamma : \ [0,1] \to \mathcal{M}_{g,1}$ is constant. We set

$$(R_0, p_0) = (R(\Gamma_t), p_\infty) \quad (\forall t \in [0,1]).$$

We consider the invariant $I(R_0,p_0)$. Let $V(\Gamma_t) \subset \mathbb{C}$ denote the image of the endpoints of Igeta in Γ_t by ζ. Then $V(\Gamma_t)$ is included in $I(R(\Gamma_t), p_\infty) =$

$I(R_0, p_0)$ because $\widetilde{\omega_{\Gamma_t}}$ belongs to $\Omega_{R(\Gamma_t), p_\infty, c_{\Gamma_t}}$. Hence

$$\bigcup_{t \in [0,1]} V(\Gamma_t) \subset I(R_0, p_0)$$

The set $\bigcup_{t\in[0,1]} V(\Gamma_t)$ is not countable because γ is injective, however the set $I(R_0, p_0)$ is countable. This is a contradiction. □

This result means that the complex structures vary necessarily as we move the position of Igeta. The Kodaira-Spencer map of the family on $I_g\eta_0$ is not known yet to be isomorphic at Γ when Γ has at least one anti-parallel Igeta.

Remark 1. When we construct a once punctured Riemann surface $(R(\Gamma), p_\infty)$ from a collection Γ of g disjoint Igeta on $(\mathbb{A}^1, \zeta)$, the quadratic differential $(d\zeta)^2$ on $\mathbb{A}^1$ induces a quadratic differential ω'_Γ on $R(\Gamma)$, which has a pole at p_∞ of order 4. We obtain the double covering space $\widetilde{R(\Gamma)}$ of $R(\Gamma)$ by resolving the singularities of the image of double valued section

$$\begin{array}{rccc} \sqrt{\omega'_\Gamma}: & R(\Gamma) - \{p_\infty\} & \longrightarrow & H^0(R(\Gamma) - \{p_\infty\}; \Omega^1) \\ & [\zeta] & \longmapsto & \pm\sqrt{\omega'_\Gamma([\zeta])} \end{array}$$

4 Hyperelliptic curves and anti-parallel Igeta.

We call a Riemann surface X a hyperelliptic curve if it can be realized as a ramified double covering space of the projective line $\mathbb{C}P_1$, i.e., if X has a meromorphic function with either two simple poles or a single pole of order 2 — we shall call such a meromorphic function ρ a hyperelliptic quotient map:

$$\rho: X \longrightarrow \mathbb{C}P_1.$$

When X has genus $g \geq 1$, there are $2g+2$ branch points $a_1, \ldots, a_{2g+2}$ of ρ. X is determined by the configuration of its branch point set $\mathcal{B} = \{a_1, \ldots, a_{2g+2}\} \subset \mathbb{C}P_1$: X is biholomorphic to a hyperelliptic curve X' if and only if X' has a hyperelliptic quotient map $\rho': X' \to \mathbb{C}P_1$ with branch point set $\mathcal{B}'$ which is transformed to $\mathcal{B}$ by some automorphism of $\mathbb{C}P_1$. We then obtain an arbitrary hyperelliptic curve R of genus g from a set of $2g-1$ distinct points $a_1, \ldots, a_{2g-1}$ on $\mathbb{C}$ each of which is neither 0 nor 1. R is obtained as the resolution of the curve defined by the following equation:

$$y^2 = x(x-1)(x-a_1)(x-a_2)\ldots(x-a_{2g-1}).$$

Note that R has a hyperelliptic quotient map whose branch point set is $\{0, 1, \infty, a_1, \ldots, a_{2g-1}\}$.

In this section we construct an arbitrary hyperelliptic curve R using anti-parallel Igeta. Namely we show the following:

Theorem 2. *Any hyperelliptic curve of genus g is obtained from a set of g disjoint anti-parallel Igeta each of which is symmetric with respect to the origin.*

Proof. Let Γ be a set of g disjoint anti-parallel Igeta in the Gauss plane $(\mathbb{A}^1, \zeta)$ each of which is symmetric with respect to the origin:

$$\Gamma = ((([\alpha_j, \beta_j], [-\alpha_j, -\beta_j]), \tau);\ j = 1, \ldots, g),$$

where $[\alpha, \beta]$ is the line segment on $(\mathbb{A}^1, \zeta)$ connecting the points $\zeta = \alpha$ and $\zeta = \beta$ for complex numbers α, β. Then the involution

$$\begin{aligned} i_\Gamma :\ \mathbb{A}^1 - \bigcup_{j=1}^{g}([\alpha_j, \beta_j] \cup [-\alpha_j, -\beta_j]) &\longrightarrow \mathbb{A}^1 - \bigcup_{j=1}^{g}([\alpha_j, \beta_j] \cup [-\alpha_j, -\beta_j]) \\ \zeta &\longmapsto -\zeta \end{aligned}$$

extends to a holomorphic involution of the resulting Riemann surface $R(\Gamma)$:

$$\hat{i}_\Gamma :\ R(\Gamma) \longrightarrow R(\Gamma).$$

The fixed point set of $\hat{i}_\Gamma$ is $\{p_\infty, [0], [\alpha_1], \ldots, [\alpha_g], [\beta_1], \ldots, [\beta_g]\}$, where $[\alpha]$ is the point on $R(\Gamma)$ corresponding to the point $\zeta = \alpha$ on $\mathbb{A}^1$. (See Figure 3.)

It is easy to see that the map

$$\begin{aligned} \rho_\Gamma :\ (\mathbb{A}^1 - \bigcup_{j=1}^{g}([\alpha_j, \beta_j] \cup [-\alpha_j, -\beta_j])) &\longrightarrow \mathbb{C}P_1 \\ \zeta &\longmapsto \zeta^2. \end{aligned}$$

extends to a hyperelliptic quotient map of $R(\Gamma)$:

$$\hat{\rho}_\Gamma :\ R(\Gamma) \longrightarrow \mathbb{C}P_1.$$

The branch point set of $\hat{\rho}_\Gamma$ is $\{0, \infty, \alpha_1^2, \ldots, \alpha_g^2, \beta_1^2, \ldots, \beta_g^2\}$. Hence the following equation is a defining equation of $R(\Gamma)$:

$$y^2 = x(x - \alpha_1^2) \cdots (x - \alpha_g^2)(x - \beta_1^2)(x - \beta_g^2).$$

Therefore from any set of g disjoint anti-parallel Igeta each of which is symmetric with respect to the origin we obtain a hyperelliptic curve of genus g.

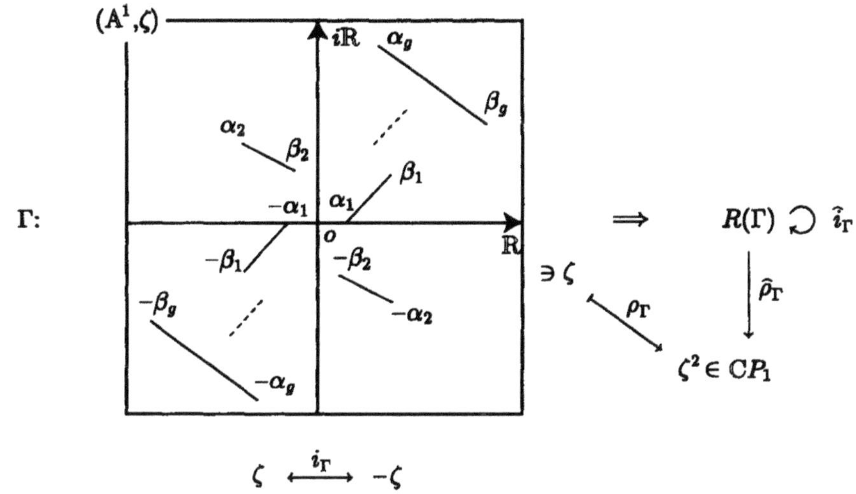

Figure 3. A symmetric configuration of g disjoint anti-parallel Igeta

Let R be an arbitrary hyperelliptic curve of genus g, and let the following equation be a defining equation of R:

$$y^2 = x(x-1)(x-a_1)\cdots(x-a_{2g-1}),$$

where $0, 1, a_1, \ldots, a_{2g-1}$ are pairwise distinct. For R we put the $4g$ points $\pm 1, \pm\sqrt{a_1}, \ldots, \pm\sqrt{a_{2g-1}}$ on the Gauss plane $(\mathbb{A}^1, \zeta)$. We find a set Γ of g disjoint anti-parallel Igeta satisfying the following conditions:

(1) each element of Γ is symmetric with respect to the origin,

(2) the endpoint set of Γ is equal to $\{\pm 1, \pm\sqrt{a_1}, \ldots, \pm\sqrt{a_{2g-1}}\}$.

A number of such configurations of anti-parallel Igeta exist, and we construct one of them as follows.

Let $\sqrt{a_1}, \ldots, \sqrt{a_{2g-1}}$ be in $\mathbb{H} \sqcup \mathbb{R}_+$, where $\mathbb{H}$ is the upper half plane and $\mathbb{R}_+$ is the set of positive real numbers. Let $\mathcal{P}$ be the collection of sets P of g disjoint unordered pairs

$$P = \{(b_1, b_1'), \ldots, (b_g, b_g')\}$$

where each pair (b_i, b_i') of P consists of two distinct elements of

$\{1, \sqrt{a_1}, \ldots, \sqrt{a_{2g-1}}\}$. We then consider the following function:

$$\Psi : \mathcal{P} \longrightarrow \mathbb{R}_+$$
$$P \longmapsto ||b_1 - b'_1|| + \cdots + ||b_g - b'_g||.$$

The function Ψ has the minimal value. We assume that Ψ takes the minimal value at

$$P_0 = \{(c_1, c'_1), \ldots, (c_g, c'_g)\} \in \mathcal{P}.$$

Then the g line segments $[c_1, c'_1], \ldots, [c_g, c'_g]$ on $\mathbb{A}^1$ do not intersect each other because $||c_i - c'_i|| + ||c_j - c'_j||$ is larger than $||c_i - c_j|| + ||c'_i - c'_j||$ if $[c_i, c'_i]$ intersects $[c_j, c'_j]$. Therefore we obtain a desired set of g disjoint anti-parallel Igeta

$$\Gamma = ((([c_j, c'_j], [-c_j, -c'_j]), \tau);\ j = 1, \ldots, g). \qquad \square$$

Remark 2. Given a set Γ of disjoint anti-parallel Igeta each of which is symmetric with respect to the origin, we easily obtain a defining equation of $R(\Gamma)$ as above. We can then tell when two such sets represent the same hyperelliptic curve.

Remark 3. Let Γ be a set of g disjoint Igeta each of which is symmetric with respect to the origin. Even if some of the Igeta in Γ are parallel, we also obtain a hyperelliptic curve of genus g because we obtain a hyperelliptic involution of $R(\Gamma)$ by extending the involution i_Γ. However the map ρ_Γ does not extend to a hyperelliptic quotient map of $R(\Gamma)$. So we do not know a defining equation of $R(\Gamma)$ yet.

5 Examples of configurations which represent the same Riemann surface.

From distinct elements of $I_g\eta/\mathrm{Aut}\mathbb{A}^1$ we occasionally obtain the same Riemann surface. In the previous section we obtained any hyperelliptic curve from a set of g disjoint anti-parallel Igeta each of which is symmetric with respect to the origin. Let Γ and Γ' be such elements of $I_g\eta$. If the set of endpoints of Igeta in Γ is equal to the one of Igeta in Γ', we obtain from Γ and Γ' the same hyperelliptic curve. Note that Γ and Γ' do not represent the same class in $I_g\eta/\mathrm{Aut}\mathbb{A}^1$ in general. In this section we give three other examples of elements of $I_g\eta$ which represent the same Riemann surface. (We

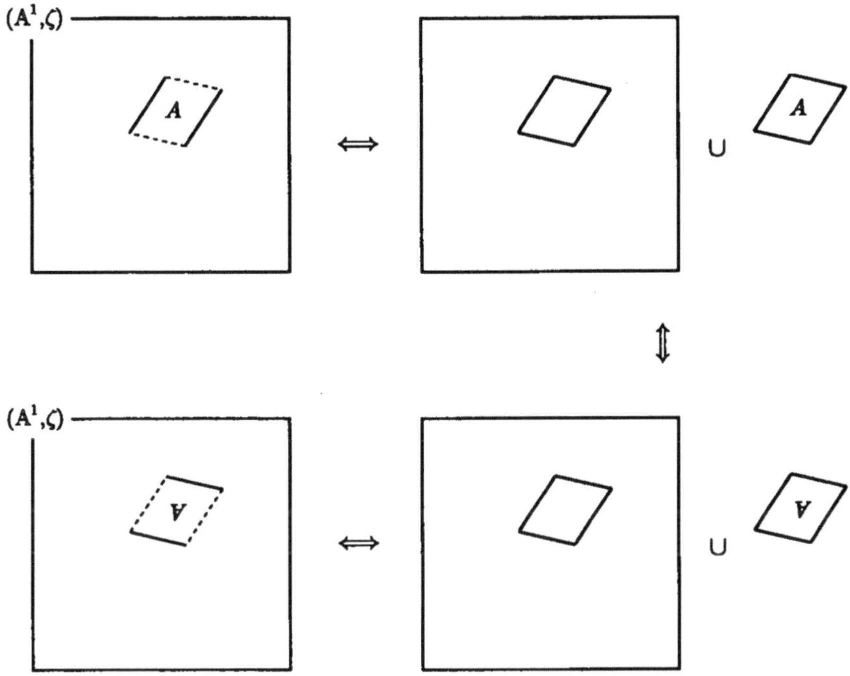

Figure 4. A half-rotation of a parallelogram

do not know yet any necessary and sufficient condition to judge when two distinct elements of $I_g\eta$ represent the same Riemann surface.)

We shall denote by $[a, b]$ the line segment on $(\mathbb{A}^1, \zeta)$ connecting the points $\zeta = a$ and $\zeta = b$ for complex numbers a, b as before, and by $[a, b, c, d]$ the quadrilateral bounded by the line segments $[a, b]$, $[b, c]$, $[c, d]$, and $[d, a]$. We shall also denote by $|\Gamma|$ the set of $2g$ line segments included in $\Gamma \in I_g\eta$.

Example 1. A half-rotation of a parallelogram. (Figure 4.)

Assume that an anti-parallel Igeta $(([a^+, a^+ + b], [a^-, a^- + b]), \tau)$ in $\Gamma \in I_g\eta$ satisfies the condition that the line segments $[a^+, a^-]$ and $[a^+ + b, a^- + b]$ do not intersect any line segment in $|\Gamma| - \{[a^+, a^+ + b], [a^-, a^- + b]\}$. Let us denote by A the parallelogram $[a^+, a^+ + b, a^- + b, a^-]$.

We cut off the parallelogram A from the Gauss plane $(\mathbb{A}^1, \zeta)$, rotate A with angle π, paste the edge $[a^+, a^+ + b]$ on A and the edge $[a^- + b, a^-]$ on

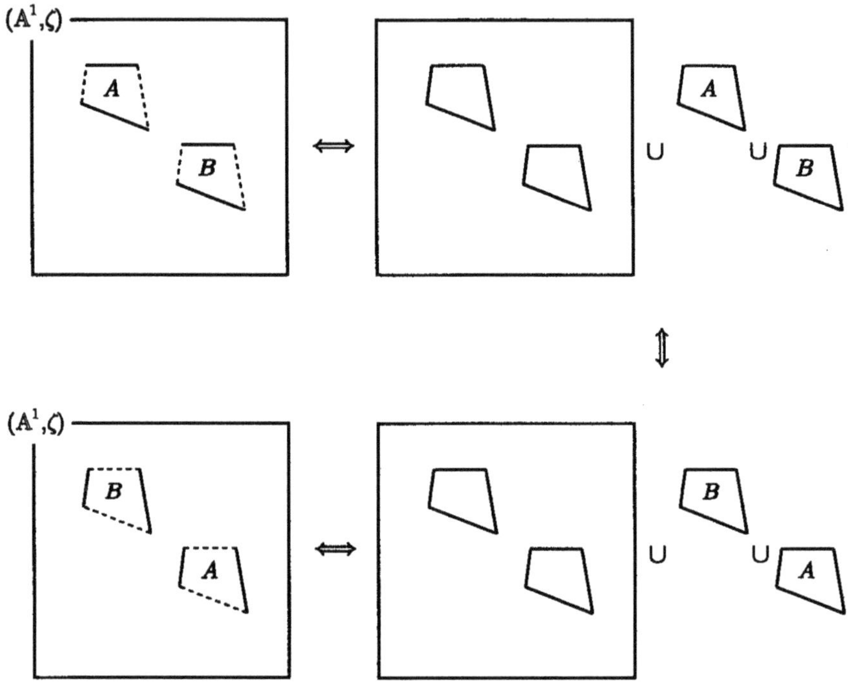

Figure 5. Exchange of two congruent quadrilaterals, I

$\mathbb{A}^1 - \mathrm{Int}A$, and paste the edge $[a^-, a^- + b]$ on A and the edge $[a^+ + b, a^+]$ on $\mathbb{A}^1 - \mathrm{Int}A$. We then obtain another element Γ' of $I_g\eta$, which obviously represent the once punctured Riemann surface $(R(\Gamma), p_\infty)$. (Note that the anti-parallel Igeta $(([a^+, a^-], [a^+ + b, a^- + b]), \tau)$ is in Γ', and that the anti-parallel Igeta $(([a^+, a^+ + b], [a^-, a^- + b]), \tau)$ is not in Γ'.)

Example 2. Exchange of two congruent quadrilaterals, I. (Figure 5.)

Assume that two parallel Igeta $(([a_1^+, a_1^+ + b_1], [a_1^-, a_1^- + b_1]), id)$ and $(([a_2^+, a_2^+ + b_2], [a_2^-, a_2^- + b_2]), id)$ in $\Gamma \in I_g\eta$ satisfy the following three conditions:

(1) The line segments $[a_1^+, a_2^+]$ and $[a_1^+ + b_1, a_2^+ + b_2]$ do not intersect any line segment in $|\Gamma| - \{[a_1^+, a_1^+ + b_1], [a_2^+, a_2^+ + b_2]\}$.

(2) The line segments $[a_1^-, a_2^-]$ and $[a_1^- + b_1, a_2^- + b_2]$ do not intersect any line

segment in $|\Gamma| - \{[a_1^-, a_1^- + b_1], [a_2^-, a_2^- + b_2]\}$.

(3) Two quadrilaterals $[a_1^+, a_1^+ + b_1, a_2^+ + b_2, a_2^+]$ and $[a_1^-, a_1^- + b_1, a_2^- + b_2, a_2^-]$ are congruent.

Let us denote by A the quadrilateral $[a_1^+, a_1^+ + b_1, a_2^+ + b_2, a_2^+]$, and by B the quadrilateral $[a_1^-, a_1^- + b_1, a_2^- + b_2, a_2^-]$.

We cut off the quadrilaterals A and B from the Gauss plane $(\mathbb{A}^1, \zeta)$, paste the edge $[a_1^+, a_1^+ + b_1]$ on A and the edge $[a_1^-, a_1^- + b_1]$ on $\mathbb{A}^1 - (\mathrm{Int}A \cup \mathrm{Int}B)$, paste the edge $[a_2^+, a_2^+ + b_2]$ on A and the edge $[a_2^-, a_2^- + b_2]$ on $\mathbb{A}^1 - (\mathrm{Int}A \cup \mathrm{Int}B)$, paste the edge $[a_1^-, a_1^- + b_1]$ on B and the edge $[a_1^+, a_1^+ + b_1]$ on $\mathbb{A}^1 - (\mathrm{Int}A \cup \mathrm{Int}B)$, and paste the edge $[a_2^-, a_2^- + b_2]$ on B and the edge $[a_2^+, a_2^+ + b_2]$ on $\mathbb{A}^1 - (\mathrm{Int}A \cup \mathrm{Int}B)$. We then obtain another element Γ' of $I_g\eta$, which obviously represent the once punctured Riemann surface $(R(\Gamma), p_\infty)$. (Note that both the parallel Igeta $(([a_1^+, a_2^+], [a_1^-, a_2^-]), id)$ and $(([a_1^+ + b_1, a_2^+ + b_2], [a_1^- + b_1, a_2^- + b_2]), id)$ are in Γ', and that neither the parallel Igeta $(([a_1^+, a_1^+ + b_1], [a_1^-, a_1^- + b_1]), id)$ nor $(([a_2^+, a_2^+ + b_2], [a_2^-, a_2^- + b_2]), id)$ is in Γ'.)

Example 3. Exchange of two congruent quadrilaterals, II. (Figure 6.)

Assume that two anti-parallel Igeta $(([a_1^+, a_1^+ + b_1], [a_1^-, a_1^- + b_1]), \tau)$ and $(([a_2^+, a_2^+ + b_2], [a_2^-, a_2^- + b_2]), \tau)$ in $\Gamma \in I_g\eta$ satisfy the following three conditions:

(1) The line segments $[a_1^+, a_2^+]$ and $[a_1^+ + b_1, a_2^+ + b_2]$ do not intersect any line segment in $|\Gamma| - \{[a_1^+, a_1^+ + b_1], [a_2^+, a_2^+ + b_2]\}$.

(2) The line segments $[a_1^-, a_2^-]$ and $[a_1^- + b_1, a_2^- + b_2]$ do not intersect any line segment in $|\Gamma| - \{[a_1^-, a_1^- + b_1], [a_2^-, a_2^- + b_2]\}$.

(3) Two quadrilaterals $[a_1^+, a_1^+ + b_1, a_2^+ + b_2, a_2^+]$ and $[a_1^- + b_1, a_1^-, a_2^-, a_2^- + b_2]$ are congruent.

Let us denote by A the quadrilateral $[a_1^+, a_1^+ + b_1, a_2^+ + b_2, a_2^+]$, and by B the quadrilateral $[a_1^- + b_1, a_1^-, a_2^-, a_2^- + b_2]$.

We cut off the quadrilaterals A and B from the Gauss plane $(\mathbb{A}^1, \zeta)$, rotate A and B with angle π, paste the edge $[a_1^+, a_1^+ + b_1]$ on A and the edge $[a_1^- + b_1, a_1^-]$ on $\mathbb{A}^1 - (\mathrm{Int}A \cup \mathrm{Int}B)$, paste the edge $[a_2^+, a_2^+ + b_2]$ on A and the edge $[a_2^- + b_2, a_2^-]$ on $\mathbb{A}^1 - (\mathrm{Int}A \cup \mathrm{Int}B)$, paste the edge $[a_1^-, a_1^- + b_1]$ on B and the edge $[a_1^+ + b_1, a_1^+]$ on $\mathbb{A}^1 - (\mathrm{Int}A \cup \mathrm{Int}B)$, and paste the edge $[a_2^-, a_2^- + b_2]$ on B and the edge $[a_2^+ + b_2, a_2^+]$ on $\mathbb{A}^1 - (\mathrm{Int}A \cup \mathrm{Int}B)$. We then obtain another element Γ' of $I_g\eta$, which obviously represent the once

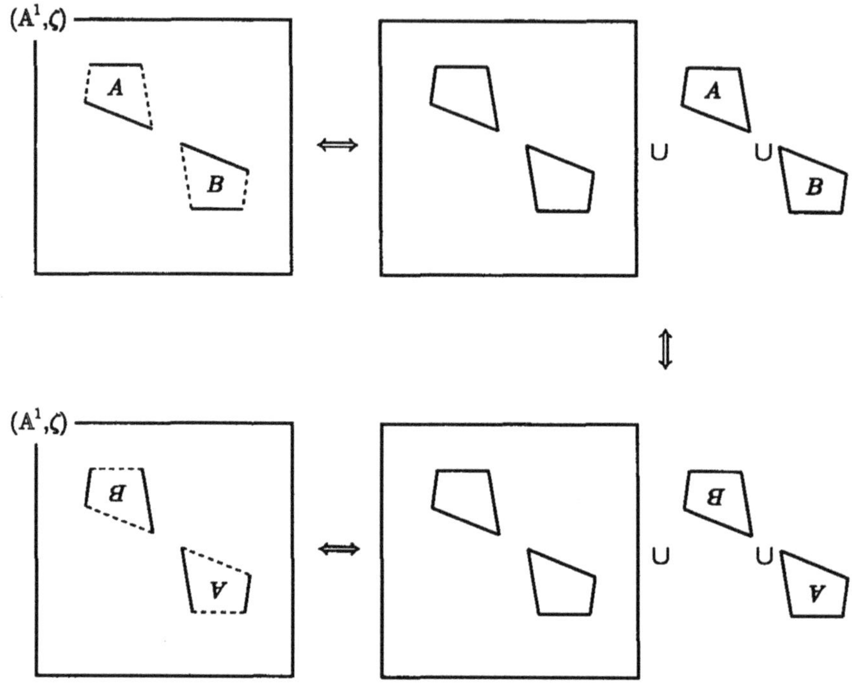

Figure 6. Exchange of two congruent quadrilaterals, II

punctured Riemann surface $(R(\Gamma), p_\infty)$. (Note that both the anti-parallel Igeta $(([a_1^+, a_2^+], [a_2^- + b_2, a_1^- + b_1]), \tau)$ and $(([a_1^+ + b_1, a_2^+ + b_2], [a_2^-, a_1^-]), \tau)$ are in Γ', and that neither the anti-parallel Igeta $(([a_1^+, a_1^+ + b_1], [a_1^-, a_1^- + b_1]), \tau)$ nor $(([a_2^+, a_2^+ + b_2], [a_2^-, a_2^- + b_2]), \tau)$ is in Γ'.)

References

[HO] Y. Hashimoto and K. Ohba, *Cutting and Pasting of Riemann Surfaces with Abelian Differentials, I*, Int. J. Math., to appear.

[M] D. Mumford, *Tata Lectures on Theta II*, Progress in Math. vol. 43, Birkhäuser, 1984.

[N] R. Narasimhan, *Compact Riemann Surfaces*, Lectures in Math. ETH Zürich, Birkhäuser, 1992.

ON THE BOUNDARY MORERA THEOREM FOR CLASSICAL SYMMETRIC DOMAINS

S. KOSBERGENOV AND A. KYTMANOV
Nukus, Krasnoyarsk

In last years the interest of specialists in complex analysis is attracted by multidimensional boundary analogues of the Morera theorem (see [1]-[7]). All of them deal with whole boundary of a domain and the opportunity of holomorphic continuation of function f from the boundary ∂D of a domain D in $\mathbf{C}^N$ in this domain consists in vanishing of integrals from f on the boundaries of analytical disks lying on ∂D. So the work of E.Grinberg [2] considers the case of a ball, and the disks are the intersections of a ball with complex lines. In article of J. Globevnik and E.L. Stout [3] the Grinberg result is transferred on arbitrary bounded domains with a smooth boundary. The local variant of given theorem is obtained in [4]. The other proof and some generalization of Globevnik and Stout theorem is obtained in [5]. Complex lines are replaced by some classes complex curves in [6,7].

In given paper we consider the boundary variant of the Morera theorem for classical domains. The starting point of it is the Nagel and Rudin result [1] (see also [8]) which says that if the function f is continuous on the boundary of a ball in $\mathbf{C}^N$ and integral

$$\int_0^{2\pi} f(\psi(e^{i\varphi}, 0 \dots, 0))\, e^{i\varphi}\, d\varphi = 0$$

for all (holomorphic) automorphisms ψ of the ball then function f holomorphically extends in the ball.

We generalize the given statement on the case of classical domains, replacing the boundary of a domain by its Shilov boundary (distinguished boundary). And our proof, based on other ideas, than in [1], permits in case of the ball to strengthen the Nagel and Rudin theorem.

We remind some definitions and we enter necessary to us notations. Under classical domain $D \subset \mathbf{C}^N$ (see [9]) we shall understand the irreducible bounded symmetric domain in many complex variables one of the following four types:

1. A domain D_I is formed by matrices Z from m lines and n columns (elements of matrixes are complex numbers) satisfying the condition

$$I^{(m)} - ZZ^* > 0.$$

Here $I^{(m)}$ is unit matrix of order m, $Z^* = \overline{Z'}$ is a complex conjugate matrix to transposed matrix Z', notation $H > 0$ means, as usually, for Hermitian matrix H that the given matrix is positive definite.

2. A domain D_{II} is formed by symmetric (square) n-matrices Z satisfying the condition

$$I^{(n)} - Z\overline{Z} > 0.$$

3. A domain D_{III} is formed by antisymmetric matrices Z of order n satisfying the condition

$$I^{(n)} + Z\overline{Z} > 0.$$

4. A domain D_{IV} is formed by n-vectors $z = (z_1, \dots, z_n)$ satisfying the conditions

$$|zz'|^2 + 1 - 2\overline{z}z' > 0, \qquad |zz'| < 1.$$

The complex dimension of these four types of domains is equal, accordingly, mn, $n(n+1)/2$, $n(n-1)/2$, n. The given domains are complete circular convex domains. In our article the domain D means the domain of one of above-stated types.

Let S be the Shilov boundary for a domain D (see [9]).

1. S_I is formed by matrices U from m lines and n columns with condition, that

$$UU^* = I^{(m)}.$$

In particular, at $m = n$ the manifold S_I coincides with set of all unitary matrices $U(n)$.

2. S_{II} consists of all symmetric unitary matrices of order n.

3. S_{III} is defined by various manner in dependence on evenness or oddness n. If n is even, S_{III} consists of all antisymmetric unitary matrices of order n. If n is odd, S_{III} consists of matrices of a kind UFU', where U is arbitrary unitary matrix and

$$F = \begin{pmatrix} 0 & 1 \\ -1 & 0 \end{pmatrix} \dotplus \begin{pmatrix} 0 & 1 \\ -1 & 0 \end{pmatrix} \dotplus \dots \dotplus 0.$$

4. S_{IV} consists of the vectors of a kind $e^{i\varphi}x$, where x is real n-vector, satisfying to condition $xx' = 1$, $0 \leqslant \varphi \leqslant 2\pi$.

The manifolds S_I, S_{II}, S_{III}, S_{IV} have the real dimension $m(2n - m)$, $n(n+1)/2$, $n(n-1)/2 + (1 + (-1)^n)(n-1)/2$ and n, accordingly. All these manifolds are CR manifolds, and in case, when their dimension is equal to (complex) dimension of space $\mathbf{C}^N$, given manifolds are totally real (i.e. not possessing of complex tangent vectors).

We define the class $H^1(D)$ as the class of all holomorphic functions F in D, for which

$$\sup_{0<r<1} \int_S |F(r\zeta)| d\mu < +\infty.$$

Here $r\zeta = (r\zeta_1, \dots, r\zeta_N)$, and $d\mu$ is normalized Lebesgue measure on the manifold S, which is Haar measure.

For any function F in D and for any $\zeta \in S$ we consider slice-function F_ζ in $\Delta = \{ t \in \mathbf{C} : |t| < 1 \}$ of the following form: $F_\zeta(t) = F(t\zeta)$. This slice-function enables to connect some N-dimensional properties of function F with one-dimensional properties of F_ζ.

We need the following properties of functions from $H^1(D)$ (see [10,11], as well as review [12]).

1. In order that $F \in H^1(D)$ it is necessary and sufficient, that the function $F_\zeta \in H^1(\Delta)$ for almost all $\zeta \in S$ and radial boundary values of function F_ζ (at $r \to 1-0$) were function of class $\mathcal{L}^1(S)$ ([11]).

2. Each function $F \in H^1(D)$ has the radial boundary values on S of class $\mathcal{L}^1$. We shall designate it through f. The function $F \in H^1(D)$ is restored in D by its radial boundary values f with the help of Cauchy-Szegö integral (or invariant Poisson integral) ([10]).

We fix the point $\lambda_0 \in S$ $(\lambda_0 = (\lambda_1^0, \dots, \lambda_N^0))$. And we consider the following embedding of a disk Δ in domain D

$$\{ \zeta \in \mathbf{C}^N : \zeta_j = t\lambda_j^0,\ j = 1, \dots, N,\ |t| < 1 \}. \tag{1}$$

The boundary T of a disk Δ thus turn into a circle lying on S. If ψ is arbitrary (holomorphic) automorphism of a domain D (i.e. biholomorphic mapping of a domain D on itself) then the set of the form (1) under action of its automorphism turns into some analytic disk with the boundary on S.

Theorem 1. *If the function $f \in C(S)$ satisfies the condition*

$$\int_T f(\psi(t\lambda_0))\, dt = 0 \tag{2}$$

for all automorphisms ψ of a domain D then function f holomorphically extends in D up to function F of class $C(\overline{D})$.

Proof. The subgroup of automorphisms leaving fixed 0 acts transitively on S. We name there unitary transformations, as far as they are linear and for the case of domains, consisting from square matrixes, are set unitary matrixes. As S is invariant under unitary transformations (also, as well as a domain D), condition (2) will be replaced by arbitrary points $\lambda \in S$.

We enter parametrization of the manifold S by the following manner

$$\zeta = t\lambda, \quad t = e^{i\varphi}, \quad 0 \leqslant \varphi \leqslant 2\pi, \quad \lambda \in S',$$

if $\zeta \in S$. The manifold S' is defined by different for different domains D (see [11]). For domains of the first type and $m = n$ it is presented by the group $SU(n)$ of special unitary matrices, for domains of fourth type it is the sphere and etc. As is shown in [11] the measure $d\mu$ can be written in the form

$$d\mu = \frac{d\varphi}{2\pi} \wedge d\mu_0(\lambda) = \frac{1}{2\pi i}\frac{dt}{t} \wedge d\mu_0(\lambda),$$

where $d\mu_0(\lambda)$ is the differential form determining the positive measure on S'.

Multiplying the equality (2) on $d\mu_0$ and integrating on S', we from (2) obtain

$$\int_S f(\psi(\zeta))\,\zeta_k\,d\mu(\zeta) = 0, \tag{3}$$

where ζ_k are the components of vector ζ, $k = 1, \ldots, N$.

We consider automorphism ψ_A transforming the point A from D in 0. It is defined to within unitary transformation.

Substituting then in condition (3) instead of ψ the automorphism ψ_A^{-1} and doing the replacement of variable $W = \psi_A^{-1}(\zeta)$, we receive

$$\int_S f(W)\,\psi_k^A(W)\,d\mu(\psi_A(W)) = 0, \tag{4}$$

where ψ_k^A are the components of automorphism ψ_A.

As is shown in [10] (Lemma 3.4) (for the case of square matrices see the proof of Theorem 4.6.3 from [9])

$$d\mu(\psi_A(W)) = P(W, A)\,d\mu(W),$$

where $P(W, A)$ is the invariant Poisson kernel of a domain D.

Thus, from condition (4) we get that

$$\int_S f(W)\psi_k^A(W)\,P(W, A)\,d\mu(W) = 0 \tag{5}$$

for all points A from D and all $k = 1, \ldots, N$.

Thus, taking into account the Theorem 5.7.1 from [9] about properties of the Poisson integral for continuous functions, Theorem 1 will succeed from the following statement

Theorem 2. *If the function $f \in \mathcal{L}^1(S)$ and for it the equality (5) holds for all automorphisms ψ_A of a domain D transforming the point A from D in 0 and all $k = 1, \ldots, N$ then function f is radial boundary value of some function $F \in H^1(D)$.*

Let the classical domain D is such, that the dimension S is strictly more than N. We remind function $f \in \mathcal{L}^1(S)$ is CR function if

$$\int_S f(\zeta)\overline{\partial}(\omega \wedge d\zeta) = 0$$

for all exterior differential forms ω with coefficients from class C^∞ on S of appropriate dimension, $d\zeta$ is the exterior product of all differentials $d\zeta_k$.

The Rossi and Vergne theorem [13] says that any CR function $f \in \mathcal{L}^1(S)$ is radial boundary value on S of function $F \in H^1(D)$. The Poisson kernel in classical domain D is expressed through Cauchy-Szegö kernel $C(W, A)$

$$P(W, A) = \frac{C(W, A)C(A, W)}{C(A, A)}.$$

As expression $C(W, A)d\mu(W)$ (as is shown in [14]) is the restriction on S of some Cauchy-Fantappie kernel $\Omega(W, A)$ being $\overline{\partial}$-closed differential form. Therefore in condition (5) of Theorem 2 we get the product of the Cauchy-Fantappie kernel Ω and the holomorphic function vanishing at the point A. Then (see [15]) given product is some $\overline{\partial}$-exact form with coefficients of class $^\infty$ on S. Thus, Theorem 2 is the generalization of Rossi and Vergne theorem from [13].

The proof of Theorem 2 shows that it remains true if the condition (5) holds only for that automorphisms ψ_A, for which the point A lies in some open set $V \subset D$. Therefore it is true the following generalization of Theorem 2.

Theorem 3. *If the function $f \in \mathcal{L}^1(S)$ satisfies to condition (5) for all points A lying in some open set $V \subset D$ and all components of automorphism ψ_A then f is radial boundary value on S of some function $F \in H^1(D)$.*

Hence, Theorem 1 also admits generalization.

Theorem 4. *Let the function $f \in C(S)$ and condition (2) holds for all automorphisms ψ transforming the point 0 in point from some open set $V \subset D$ then f holomorphically extends in D up to some function $F \in C(\overline{D})$.*

This theorem in the case of a ball D generalizes the Nagel and Rudin theorem from [1].

We note through Δ_ψ the analytical disk of a kind

$$\Delta_\psi = \{\zeta : \zeta = \psi(t\lambda_0),\ |t| < 1\},$$

where λ_0 is fixed point from the Shilov boundary S and ψ is the automorphism of a domain D. Then the boundary T_ψ of this analytical disk lies on S.

Corollary 1. *If the function $f \in C(S)$ holomorphically extends (by t) in analytical disks Δ_ψ for all automorphisms ψ (or for all automorphisms ψ transforming the point 0 in point of some fixed open set $V \subset D$) then function f holomorphically extends in D.*

Corollary 1 is analogue for classical domains of the Stout theorem [16] on functions with one-dimensional property of holomorphic continuation (see also [3,17,18]). Besides it generalizes for given class of manifolds S the Tumanov theorem by which if the smooth function f holomorphically extends in all analytical disks with boundary on S then f is CR function on S ([19]).

Acknowledgment

This work is supported by RFFI, grant 96-01-00080.

References

1. A. Nagel and W. Rudin, *Moebius-invariant function spaces on balls and spheres*, Duke Math. J. **43**(1976), 841–865.
2. E. Grinberg, *A boundary analogue of Morera's theorem on the unit ball of* $\mathbf{C}^n$, Proc. Amer. Math. Soc. **102**(1988), 114–116.
3. J. Globevnik and E.L. Stout, *Boundary Morera theorems for holomorphic functions of several complex variables*, Duke Math. J. **64**(1991), 571–615.
4. J. Globevnik, *A boundary Morera theorem*, J. Geometric Anal. **3**(1993), 269–277.
5. A.M. Kytmanov and S.G. Myslivets, *On an application of the Bochner–Martinelli operator*, Contemporary Math. **212**(1998), 133–136.
6. ———, *On functions with one-dimensional property of holomorphic continuations and boundary analogues of the Morera theorem*, in Complex

Analysis and Mathematical Physics, Krasnoyarsk State Univ., Krasnoyarsk, 1998, pp. 88–96 (in Russian).

7. S.G. Myslivets, *On holomorphic continuation of functions along complex curves*, in Topics in Complex Analysis, Differential Geometry and Mathematical Physics, Third Int. Workshop on Complex Structures and Vector Fields. Varna, 1996, World Scientific, Singapore, 1997, pp. 39–45.
8. W. Rudin, *Function theory in the unit ball of* $\mathbf{C}^n$, Springer-Verlag, Berlin, Heidelberg, New York, 1980.
9. Hua Loken, *Harmonic analysis of functions of many complex variable in classical domains*, IL, Moscow, 1959.
10. A. Koranyi, *The Poisson integral for generalized half-planes and bounded symmetric domains*, Ann. Math. **82**(1965), 332–350.
11. A.M. Kytmanov and T.N. Nikitina, *Analogues of Carleman formula for classical domains*, Math. Notes **45**(1989), 243–248.
12. V.S. Vladimirov and A.G. Sergeev, *Complex analysis in future tube*, Sovremennye problemy matematiki. Fundamentalnie napravlenija. VINITI **8**(1985), 191–266 (in Russian).
13. H. Rossi and M. Vergne, *Equations de Cauchy-Riemann tangentiells, assosies a un domaine de Siegel*, Ann. Ecole Norm. Super. Pisa. **9**(1976), 31–80.
14. A. Airapetyan and G.M. Khenkin, *Integral representations of differential forms on Cauchy-Riemann manifolds and the theory of CR functions*, Russian Math. Surveys **39**(1984), 39–106.
15. A.M. Kytmanov and T.N. Nikitina, *Holomorphic continuation of CR functions with singularities on generic manifold*, Ross. Acad. Sci. Izv. Math. **40**(1993), 623–635.
16. E.L. Stout, *The boundary values of holomorphic functions of several complex variables*, Duke Math. J. **44**(1977), 105–108.
17. M.L. Agranovskii and A.M. Semenov, *Boundary analog of Hartogs theorem*, Siberian Math. J. **32**(1991), 137–139.
18. A.M. Kytmanov and S.G. Myslivets, *On holomorphness of functions represented by the formula of logarithmic residue*, Sib. Mat. Zh. **38**(1997), 351–361 (in Russian).
19. A.E. Tumanov, *Extension of CR functions into a wedge from a manifold of finite type*, Math. USSR Sb. **64**(1989), 129–140.

ON ONE-SIDE HOLOMORPHIC CONTINUATION OF CR FUNCTIONS ALONG COMPLEX CURVES

S. MYSLIVETS

Krasnoyarsk State University, Krasnoyarsk, Russia

This is a short report on joint results with A. Kytmanov

The well-known Hartogs theorem says that if the function f which is holomorphic in some domain of space $\mathbf{C}^n$ holomorphically extends (as the function of one complex variable) along parallel complex lines in a domain D then it holomorphically extends in D as the function of many complex variables.

The limit case of this Theorem when the function f is CR function on some hypersurface was proven by L.A. Aĭzenberg and C. Rea in [4].

In given work we generalize the result from [4] replacing the class of complex lines some class of complex curves. For it we get the set of assertions about one-side holomorphic continuation of CR functions in fixed domain (in spirit of work [2], [3]) using the formula of multidimensional logarithmic residue.

Let Ω be a domain in $\mathbf{C}^n$ and let Γ be a smooth (class $\mathcal{C}^\infty$) hypersurface in Ω of the form

$$\Gamma = \{ z \in \Omega : \rho(z) = 0 \},$$

where $\rho \in \mathcal{C}^\infty(\Omega)$ and $d\rho \neq 0$ on Γ.

We denote through $\Omega^+ = \{ z \in \Omega : \rho(z) > 0 \}$ and through $\Omega^- = \{ z \in \Omega : \rho(z) < 0 \}$. The orientation Γ is agreed with Ω^+.

Let the function $f \in \mathcal{C}(\Omega^\pm)$. If for any function $\varphi \in \mathcal{D}(\Gamma)$ exists limit

$$\lim_{\varepsilon \to +0} \int_\Gamma f(z \pm \varepsilon\nu(z))\varphi(z)\, d\sigma,$$

where $\nu(z)$ is the field of unit normals to Γ at the points $z \in \Gamma$ directed in the side of the growth of function ρ ($d\sigma$ is Lebesgue measure on Γ), this limit defines the distribution on Γ. We denote it through $[f]_0$ and we name it *generalized boundary value of function f*. If the function f is continuous up to Γ the generalized boundary value $[f]_0$ coincides with usual restriction of f on Γ.

Theorem 1. *Let the functions $f^{\pm}$ be continuous on sets $\Omega^{\pm}$ accordingly. We assume that they have the generalized boundary values $[f^{\pm}]_0$ and the equality $[f^{+}]_0 = [f^{-}]_0$ holds on Γ. If there exists the differential form ω of type $(0,1)$ with coefficients from class $C^m(\Omega)$, $0 \leqslant m \leqslant \infty$ or real analytic in Ω for which*

$$\overline{\partial} f^{+} = \omega \quad on \quad \Omega^{+} \quad and \quad \overline{\partial} f^{-} = \omega \quad on \quad \Omega^{-}, \tag{1}$$

then the functions $f^{\pm}$ are restrictions on sets $\Omega^{\pm}$ of one function F, given on Ω and it is the same class of smoothness that ω.

If $m = 0$ (i.e. function $f^{\pm}$ and form ω are continuous) the equality (1) is fulfilled in generalized sense, i.e. for any differential form $\gamma \in \mathcal{D}_{n,n-1}(\Omega^{\pm})$

$$\int \omega \wedge \gamma = -\int f^{\pm} \overline{\partial} \gamma. \tag{2}$$

We consider the problem of existence of holomorphic continuation of CR function f on Γ in the domain Ω^{+}. We get the result in the spirit of works [2,3] only instead of the Bochner–Martinelli and Cauchy–Fantappie kernels (as in [2,3]) we use the kernels being in the formula of multidimensional logarithmic residue.

Let $\Psi = (\psi_1, \dots, \psi_n)$ be a vector-function in $\mathbf{C}^n$ consisting from functions holomorphic on set

$$E_{\Omega} = \{ w \in \mathbf{C}^n \ : \ w = \zeta - z, \ z, \zeta \in \Omega \},$$

and having unique zero at the point $w = 0$ of multiplicity μ.

We consider the differential form $U(\Psi(\zeta - z))$ where

$$U(w) = \frac{(n-1)!}{(2\pi i)^n} \sum_{k=1}^{n} (-1)^{k-1} \frac{\overline{w}_k}{|w|^{2n}} d\overline{w}[k] \wedge dw$$

is the Bochner–Martinelli kernel, $dw = dw_1 \wedge \ldots \wedge dw_n$, form $d\overline{w}[k]$ is obtained from the form $d\overline{w}$ by deletion of differential $d\overline{w}_k$, the point z is fixed. The form $U(\Psi(\zeta - z))$ is $\overline{\partial}$-closed form of type $(n, n-1)$ with real analytic coefficients at $\zeta \neq z$. It is the kernel in formula of multidimensional logarithmic residue (see, for example, [5]).

We consider the sequence of bounded domains Ω_s with smooth boundary such, that $\overline{\Omega}_s \subset \Omega_{s+1}$, $\bigcup_{s=1}^{\infty} \Omega_s = \Omega$ and $\partial\Omega_s$ intersect Γ transversally, $s = 1, 2 \ldots$

Let $f \in C(\Gamma)$, we denote

$$M_s(f)(z) = \int_{\Omega_s \cap \Gamma} f(\zeta) U(\Psi(\zeta - z)), \quad z \notin \Gamma.$$

If $z \in \Omega_s^+$ we shall write $M_s^+(f)$ and if $z \in \Omega_s^-$ we shall write $M_s^-(f)$.

If $f \in \mathcal{L}^1(\Gamma)$ then

$$M(f)(z) = \int_\Gamma f(\zeta) U(\Psi(\zeta - z)), \quad z \notin \Gamma.$$

Theorem 2. *Let $f \in \mathcal{C}^k(\Gamma)$ $(0 \leqslant k \leqslant \infty)$ be CR function on Γ. In order that the function f holomorphically extends in Ω^+ up to function F of class $\mathcal{C}^k$ up to Γ it is necessary and sufficient that the integrals $M_s^-(f)$ real analytically extend from sets Ω_s^- on sets Ω_s for all $s = 1, 2, \ldots$*

If besides the function $f \in \mathcal{L}^1(\Gamma)$ for holomorphic continuation of it in a domain Ω^+ it is necessary and sufficient that the integral $M^-(f)$ has real analytic continuation in Ω.

The given statement for Bochner-Martinelli kernel $U(\zeta - z)$ and the special choosing of Cauchy–Fantappie kernels was proven in [2,3]. We here consider the case when $f \in \mathcal{C}^k$ though this theorem can be proven for the other classes of functions $\mathcal{L}^p$, $\mathcal{D}'$, as well as in [2,3].

The *Proof* of the theorem uses the following lemma:

Lemma 3. *If the function $f \in \mathcal{C}^k(\Gamma)$ $(k \geqslant 0)$ and f is integrable on Γ then the integrals $M^\pm(f)$ have the generalized boundary values on Γ and the equality*

$$[M^+(f)]_0 - [M^-(f)]_0 = \mu f \tag{3}$$

holds on Γ in the sense of the theory of generalized functions.

Let the point 0 belong Ω^-. We consider the decomposition of functions $M_s^-(f)$ in Teylor series at the point 0. We select from it the holomorphic terms and denote their sums through $M_s^h(f)$. The sum of rest terms we denote through $M_s^g(f)$. Then

$$M_s^-(f) = M_s^h(f) + M_s^g(f)$$

in a neigbourhood of the point 0.

Assume the domain Ω is such, that all its sections of complex lines passing through 0 are connected and simply connected (for example, the domain Ω is star-like with respect to zero) and hypersurface Γ is connected also. The domains Ω_s choose so that they have the same property.

Corollary 4. *In conditions of theorem 2 in order that CR function f holomorphically extends in Ω^+ it is necessary and sufficient that for all $s =$*

1, 2, ... *the functions* $M_s^h(f)$ *have the holomorphic continuation from neighbourhood of zero on sets* Ω_s.

We pass to continuation of functions along complex curves.

We consider the mapping $\Phi(z) = (\varphi_1(z), \dots, \varphi_n(z))$ and $\Phi(0) = 0$ consisting from functions

$$w_j = \varphi_j(z), \qquad j = 1, \dots, n$$

holomorphic in a domain D for which $\Phi^{-1}(D) = \Omega$ and such, that there exists $\Psi(w) = (\psi_1(w), \dots, \psi_n(w))$ consisting from holomorphic on E_Ω functions with property

$$(\Psi \circ \Phi) = (z_1^{p_1}, \dots, z_n^{p_n}),$$

where p_j $(j = 1, \dots, n)$ are some natural numbers. Thus, $\Psi(0) = \Phi(0) = 0$ and $\Psi(z) \neq 0$, $\Phi(z) \neq 0$ at $z \neq 0$.

The class of such mappings is enough wide. For example, in the capacity of Φ it can be taken a biholomorphic mapping, then Ψ will be equal to 1. In more general case the mapping Φ can be the composition of mapping $(z_1^{p_1}, \dots, z_n^{p_n})$ and biholomorphic.

With the help of the mapping Φ we define the next class of complex curves L_b

$$L_b = \{\zeta \in \mathbf{C}^n : \zeta_j = \varphi_j(b_1 t^{k_1}, \dots, b_n t^{k_n}),\ j = 1, \dots, n\},$$

where $t \in \mathbf{C}$ is complex parameter, natural numbers k_j are chosen so that

$$p_j k_j = p, \qquad j = 1, \dots, n,$$

and vector $b = (b_1, \dots, b_n)$ is element of weighted projective space $\mathbf{CP}^{n-1}$ $(k_1, \dots, k_n)$. We remind that the weighted projective space $\mathbf{CP}^{n-1}$ $(k_1, \dots, k_n)$ is obtained from $\mathbf{C}^n \setminus \{0\}$ by identifying of points, lying on the curve of kind

$$\{\zeta \in \mathbf{C}^n : \zeta_j = b_j t^{k_j},\ j = 1, \dots, n,\ t \in \mathbf{C}\}.$$

For any point ζ is found the curve L_b (at suitable choosing of vector b) passing through ζ. All curves L_b are intersected at the point 0. If they are intersected in other point, it is easy to show that j-coordinates of vectors b are obtained from each other by turn on the angle multiplied by $\dfrac{2\pi}{p_j}$. Therefore for unequivocal definition the vector b we can assume that the argument b_j satisfies to condition:

$$0 \leqslant \arg b_j < \frac{2\pi}{p_j}, \qquad j = 1, \dots, n.$$

Then the curves form the fibering of the sets $\mathbf{C}^n \setminus \{0\}$.

The Sard theorem shows that for almost all $b \in \mathbf{CP}^{n-1}(k_1, \ldots, k_n)$ the intersection $\Gamma \cap L_b$ is the union of finite number of piece-wise smooth curves (if it is not empty).

Hereinafter assume the domain Ω is such that the intersections $\Omega \cap L_b$ are connected and simply connected (domain Ω_s are chosen with the same condition) and hypersurface Γ is connected also and *regular* , i.e. intersections $\Omega^+ \cap L_b$ are connected (if they are not empty) and complex curves L_b intersect the surface Γ transversally for all b.

Remark 1. Corollary 4 is true for given class of domains Ω.

Theorem 5. *Let CR function f be continuous on Γ and holomorphically extends (by t) on intersections $\Omega^+ \cap L_b$ for all $b \in \mathbf{CP}^{n-1}(k_1, \ \ldots, k_n)$ (for which the intersection $L_b \cap \Gamma$ is not empty). Then the function f holomorphically extends on Ω^+ up to function F, continious up to Γ.*

For the case of complex lines the given assertion is obtained in [4].

For proof of this theorem we need the following auxiliary statement.

Lemma 6. *In conditions of theorem 5 we consider the function F being continuation of f along curves L_b. Then the function F is continuous in Ω^+ and continuously extends on the set $\Omega^+ \cup \Gamma$.*

Proof. see in [4]. □

Proof of the Theorem 5. On the strength of Remark 1 it is necessary to show that the holomorpic part $M_s^h(f)$ holomorphically extends in Ω_s^+ for each s. It is enough to prove that the holomorphic part of integrals $M_s^-(f)$ and

$$B_s(F) = \int\limits_{\partial\Omega_s^+ \setminus \Gamma} F(\zeta) U(\Psi(\zeta - z))$$

coincide. Here F is a function defined by Lemma 6. For this it is enough to show that derivatives by z at the point 0 from functions $M_s^-(f)$ and $B_s(f)$ coincide.

We notice that holomorphic and non-holomorphic parts of integral $B_s(f)$ is determined in all domains Ω_s. To prove this it is possible the same way as in Corollary 4 restoring the non-holomorphic part $B_s(f)$ by values of derivative from $B_s(f)$ by $\overline{z}_j$.

We denote

$$D^\alpha = \frac{\partial^{\alpha_1+\ldots+\alpha_n}}{\partial z_1^{\alpha_1}\ldots\partial z_n^{\alpha_n}},$$

where $\alpha = (\alpha_1, \ldots, \alpha_n)$ is multi-index.

We consider the kernel $U(\Psi(\zeta))$ in new coordinates $\zeta_j = \varphi_j(b_1 t^{k_1}, \ldots, b_n t^{k_n})$.

We denote

$$\Psi^*(t,b) = (\psi_1^*, \ldots, \psi_n^*) = (\Psi\circ\Phi)(b_1 t^{k_1}, \ldots, b_n t^{k_n}) = (b_1^{p_1} t^p, \ldots, b_n^{p_n} t^p).$$

Then the form $U(\Psi(\zeta - z))$ takes the kind

$$U^*(t,b) = U(\Psi^*(t,b)) = \frac{(n-1)!}{(2\pi i)^n}\cdot\frac{\sum\limits_{k=1}^{n}(-1)^{k-1}\overline{\psi_k^*}\, d\overline{\psi^*}[k]}{|\Psi^*|^{2n}}\wedge d\psi^*.$$

We write this formula in variable t and b. We have

$$|\Psi^*|^2 = |t|^{2p}\sum_{j=1}^{n}|b_j|^{2p_j},$$

and

$$\begin{aligned} d\psi^* &= d\psi_1^*\wedge\ldots\wedge d\psi_n^* \\ &= d(b_1^{p_1}t^p)\wedge\ldots\wedge d(b_n^{p_n}t^p) = t^{p(n-1)}p\, dt\wedge\sum_{k=1}^{n}(-1)^{k-1}b_k^{p_k}\, db^p[k], \end{aligned}$$

where

$$db^p[k] = db_1^{p_1}\wedge\ldots\wedge db_{k-1}^{p_{k-1}}\wedge db_{k+1}^{p_{k+1}}\wedge\ldots\wedge db_n^{p_n}.$$

Similarly we get

$$\begin{aligned} d\psi^*[k] &= d(b_1^{p_1}t^p)\wedge\ldots[k]\ldots\wedge d(b_n^{p_n}t^p) \\ &= (t^p\, db_1^{p_1} + b_1^{p_1}pt^{p-1}\, dt)\wedge\ldots[k]\ldots\wedge(t^p\, db_n^{p_n} + b_n^{p_n}pt^{p-1}\, dt) \\ &= t^{p(n-1)}db^p[k] \\ &\quad + \frac{pt^{p(n-1)}}{t}dt\wedge\left(\sum_{j=1}^{k-1}(-1)^{j-1}b_j^{p_j}\, db^p[j,k] + \sum_{j=k+1}^{n}(-1)^j b_j^{p_j}\, db^p[k,j]\right). \end{aligned}$$

Then

$$\sum_{k=1}^{n}(-1)^{k-1}\psi_k^* \, d\psi^*[k] = t^{pn}\sum_{k=1}^{n}(-1)^{k-1}b_k^{p_k}\, db^p[k] + pt^{pn-1}dt \wedge$$
$$\sum_{k=1}^{n}(-1)^{k-1}b_k^{p_k}\left(\sum_{j=1}^{k-1}(-1)^{j-1}b_j^{p_j}\, db^p[j,k] + \sum_{j=k+1}^{n}(-1)^{j}b_j^{p_j}\, db^p[k,j]\right). \tag{4}$$

As far as

$$\sum_{j>k}(-1)^{j+k-1}b_k^{p_k}b_j^{p_j}\, db^p[k,j] = \sum_{j<k}(-1)^{j+k-1}b_j^{p_j}b_k^{p_k}\, db^p[j,k],$$

therefore second term in (4) vanishs.

From here the differential form

$$\sum_{k=1}^{n}(-1)^{k-1}\overline{\psi}_k d\overline{\psi}[k]$$

in coordinates (t, b) takes the form

$$\overline{t^{pn}\sum_{k=1}^{n}(-1)^{k-1}b_k^{p_k}\, db^p[k]}.$$

Hence

$$U^*(t,b) = \frac{(n-1)!}{(2\pi i)^n}\times$$
$$\frac{\bar{t}^{pn}\sum_{k=1}^{n}(-1)^{k-1}\bar{b}_k^{p_k}\, d\bar{b}^p[k]\wedge t^{pn-1}p\, dt\wedge \sum_{k=1}^{n}(-1)^{k-1}b_k^{p_k}\, db^p[k]}{|t|^{2pn}\left(\sum_{k=1}^{n}|b_k|^{2p_k}\right)^n}$$
$$= \frac{|t|^{2pn}\, dt\wedge\lambda(b)}{t|t|^{2pn}} = \frac{dt}{t}\wedge\lambda(b).$$

Where

$$\lambda(b) = \frac{(n-1)!}{(2\pi i)^n}\frac{p\overline{\sum_{k=1}^{n}(-1)^{k-1}b_k^{p_k}\, db^p[k]}\wedge\sum_{k=1}^{n}(-1)^{k-1}b_k^{p_k}\, db^p[k]}{\left(\sum_{k=1}^{n}|b_k|^{2p_k}\right)^n}.$$

Finding $D^\alpha U(\Psi(\zeta - z))$ at $z = 0$ we obtain that they will consist of terms of the form

$$\frac{\overline{\psi_1^{\beta_1} \cdots \psi_n^{\beta_n} \sum_{k=1}^{n} (-1)^{k-1} \psi_k \, d\psi[k]}}{|\Psi|^{2n+2\beta_1+\ldots+2\beta_n}} \wedge \gamma_\beta,$$

where the differential form γ_β is the form of type $(n, 0)$ with holomorphic coefficients.

Considering $D^\alpha U$ in variable (t, b) we get that they will consist of terms of the kind

$$\frac{dt}{t^{pn+\beta_1+\ldots+\beta_n}} \wedge \gamma_\beta^*(t, b),$$

in which the form $\gamma_\beta^*(t, b)$ is $(n-1, n-1)$-form and has the holomorphic coefficients by t.

Then on the strength of the condition of theorem

$$\int\limits_{(\partial\Omega_s^+ \cap L_b)\setminus\partial\Omega_s} f^*(t, b)\frac{dt}{t^{pn+m}} = \int\limits_{(\partial\Omega_s^+ \cap L_b)\setminus\Gamma} F^*(t, b)\frac{dt}{t^{pn+m}}$$

for all b and all multi-indeces β.

Therefore the values of derivatives D^α at the point $z = 0$ from integrals $M_s^-(f)$ and $B_s(f)$ coincide by Fubini theorem. □

Acknowledgment

This work is supported by RFFI, grant 96-01-00080.

References

1. L.A. Aĭzenberg and Sh.A. Dautov, *Differential forms orthogonal to holomorphic functions or forms, and their properties*, AMS, Providence, RI, 1983.
2. L.A. Aĭzenberg and A.M. Kytmanov, *On the possibility of holomorphic extension into a domain of functions defined on a connected piece of its boundary*, Math. USSR Sb. **72**(1992), 467–483.
3. ———, *On the possibility of holomorphic extension into a domain of functions defined on a connected piece of its boundary, II*, Russian Acad. Sci. Sb. Math. **78**(1994), 1–10.

4. L.A. Aĭzenberg and C. Rea, *The moment-conditions for the free boundary problem for CR functions*, Annali Scuola Norm. Super. Pisa. Ser. IV **20**(1993), 313–322.
5. L.A. Aĭzenberg and A.P. Yuzhakov, *Integral representations and residues in multidimensional complex analysis*, AMS, Providence, RI, 1983.
6. A.M. Kytmanov, *The Bochner–Martinelli integral and its applications*, Birkhäuser Verlag, Basel, Boston, Berlin, 1995.
7. A.M. Kytmanov and S.G. Myslivets, *On holomorphness of functions represented by the formula of logarithmic residue*, Sib. Mat. Zh. **38**(1997), 351–361 (in Russian).
8. E.J. Straube, *Harmonic and analytic functions admitting a distribution boundary value*, Annali Scuola Norm. Super. Pisa. Ser. IV **11**(1984), 559–591.

SYMPLECTIC AND ALMOST COMPLEX STRUCTURES ON DIFFERENTIABLE SPACES

URSULA OESING
Hattinger Straße 421, 44795 Bochum

KARLHEINZ SPALLEK
Ruhr-Universität Bochum, Fakultät für Mathematik,
Universitätsstraße 150, 44780 Bochum

Symplectic and almost complex structures on vectorspacebundles and their properties of compatibility are well known. But in most cases the tangent spaces of (differentiable) spaces (for example quotient spaces) are not vectorspacebundles. So if one wants to define those structures on spaces, one first has to introduce vectorfiberbundles, then one can define symplectic and almost complex structures on these bundles. In the following article the properties of these structures on vectorfiberbundles (especially on two kinds of tangent spaces) are discussed. The following is part of the thesis of U. Oesing, which was written under the guide of K. Spallek.

1 Some introducing facts

In the first section we introduce the most important definitions which we need for the following examples and theorems and present first results. All (differentiable) spaces are proposed to be paracompact and locally compact.

1.1 Definition

α) A *vectorfiberbundle* (FX, π, X), short FX, over a space X is a pair of two spaces FX, X together with a morphism $\pi : FX \to X$, such that the following conditions hold:

i) Each fiber $F_xX := \pi^{-1}(x)$, $x \in X$ is a $\mathbb{R}$-vectorspace.

ii) To each $x \in X$ exist $x \in U \subset_{open} X$ and diffeomorphisms $\psi : U \to \psi(U) =: V \subset \mathbb{R}^{s_u}$ and $\varphi : \pi^{-1}(U) \to \varphi(\pi^{-1}(U)) \subset V \times \mathbb{R}^{n_u}$, such that for each $y \in U$ the mapping $\varphi_{|F_yX} : F_yX \to \{\psi(y) \times \mathbb{R}^{n_u}\}$ is a vectorspace homomorphism. (Especially $\varphi_{|F_yX}(F_yX)$ must be a vectorspace.) Write $FX_U := \pi^{-1}(U)$. $(FX_U, \varphi, \varphi(\pi^{-1}(U)))$ is called a *bundlemap*, $\{(FX_{U_i}, \varphi_i, \varphi_i(FX_{U_i})) \mid i \in I\}$ with $X = \cup_{i \in I} U_i$ is called a *bundleatlas*.

β) A *vectorspacebundle* (VX, π, V), short VX, over X is a vectorfiberbundle with the property: $\exists n \in \mathbb{N}$ and a bundleatlas $\{ (VX_{U_i}, \varphi_i, V_i \times \mathbb{R}^n) \mid i \in I \}$.

γ) $(FX, \pi_{FX}, X) = FX$ is called a *subbundle* of $\tilde{F}X := (\tilde{F}X, \pi, X)$, short $FX \subset_\pi \tilde{F}X$, if $FX \subset \tilde{F}X$ and $\pi_{FX} = \pi_{|FX}$.

δ) Define a *riemannian metric* d_r, a *symplectic structure* d_s, a *bundlehomomorphism* L, exspecially an *almost complex structure* J, and an *hermitean metric* d_h analogous to those well known structures on vectorspacebundles on vectorfiberbundles FX respectively $F^2X := FX \bigoplus FX$. (FX, d_r) {respectively (FX, d_s), (FX, d_h, J), (FX, J)} is called a *riemannian* {*symplectic, hermitean, almost complex*} *space*.

ε) J *hermitesates* (is *compatible* with) (FX, d_s) and d_s *hermitesates* (is *compatible* with) (FX, J), if $d_s(J, \sim)$ is a riemannian metric.

Analogous to vectorspacebundles on manifolds one can prove:

1.2 Theorem

Each symplectic vectorspacebundle (VX, d_s) ist hermitesable.

The examples at the end of this section show, that this is wrong for vectorfiberbundles. Together with those examples and the following note one concludes, that on vectorfiberbundles there exist in general "more" symplectic structures than almost complex structures:

1.3 Note

Each almost complex space (FX, J) is hermitesable.

Proof. Take an arbitrary riemannian metric d_r' on F^2X, then $d_r := d_r'(J-, J\sim) + d_r'(-, \sim)$ is riemannian, too, and J-invariant. By simple calculation one can show that $d_s := -d_r(J-, \sim)$ is a symplectic structure, which is compatible with J.

One can draw some conclusions about the existence of extensions onto VX of the structures above, given on FX, where $FX \subset_\pi VX$:

1.4 Note

Let $FX \subset_\pi VX$.

α) A bundlehomomorphism $L : FX \to FX$ is extendable to a bundlehomomorphism $\tilde{L} : VX \to VX$.

β) A morphism $d : F^2X \to \mathbb{R}$, which is $\mathbb{R}$-bilinear on the fibers, is extendable to a morphism $\tilde{d} : V^2X \to \mathbb{R}$, which is $\mathbb{R}$-bilinear on the fibers.

γ) Not all d_s and d_r on FX are extendable to symplectic structures $\tilde{d}_s$ respectively riemannian metrics $\tilde{d}_r$ on VX.

Proof.

to α) *First step*:

To $x \in X$ $\exists x \in U \subset_{open} X$ and a bundlemap of $VX = (VX_U, \varphi, V \times \mathbb{R}^n)$, $V \times \mathbb{R}^n \subset \mathbb{R}^{s+n}$. Let $FW := \varphi(FX_U)$, then $\hat{L} := \varphi \circ L \circ \varphi^{-1} : FW \to FW$ is a bundlehomomorphism. For U chosen small enough there exists $\tilde{V} \subset \mathbb{R}^s$ with $V \subset \tilde{V}$ and a (differentiable) extension $f : \tilde{V} \times \tilde{U}(0) \subset_{open} \mathbb{R}^{s+n}$ of $\hat{L}_{|(\tilde{V} \times \tilde{U}(0)) \cap FW}$ with $f(y, w) = (y, f_2(y, w))$ $\forall (y, w) \in \tilde{V} \times \tilde{U}(0)$. One gets: $F : \tilde{V} \times \mathbb{R}^n \to \tilde{V} \times \mathbb{R}^n$, $F(y, w) := (y, \sum_{i=1}^n \frac{\partial f_2(y,0)}{\partial v_i} w_i)$ is a bundlehomomorphism and an extension of $\hat{L}$. Hence $\tilde{L}_U : VX_U \to VX_U$, $\tilde{L}_U := \varphi^{-1} \circ F \circ \varphi$ is a bundlehomomorphism and an extension of $L_{|FX_U}$.

Second step:

Because of the paracompactness of VX there exists a locally finite open covering $\{ VX_{U_i} \mid i \in I \}$ of VX with extensions $\tilde{L}_i$ of $L_{|FX_{U_i}}$. It follows the existence of a locally finite open covering $\{ VX_{U_i'} \mid i \in I \}$ of VX with $\bar{U_i} \subset U_i$ with partition of 1: $\{ \lambda_i \mid i \in I \}$ respecting $\cup_{i \in I} U_i'$. Hence $\tilde{L} : VX \to VX$, $\tilde{L} := \sum_{i \in I} \lambda_i \circ \pi \tilde{L}_i$ is an extension of L like above.

to β) Analogous to α) for d with $F(y, v, w) := (y, \sum_{i,j=1}^n \frac{\partial^2 f_2(y,0,0)}{\partial v_i \partial w_j} v_i w_j)$ in the first step. (F and f_2 are analogously defined for d instead of L.)

to γ) *Conterexamples*:

$X := (\mathbb{R}^2 \times \{0\}) \cup (\{0\} \times \mathbb{R}^2) \subset \mathbb{R}^4$, $VX := X \times \mathbb{R}^4$, $FX := T^i X := \{ (p, v) \mid p \in X, \ \exists V : X \to TX$ locally integrable with $V(p) = v \}$. d_r and d_s are given by the bilinear extensions $\tilde{d}_r(v, w) := w^t D_r v$ respectively $\tilde{d}_s(v, w) := w^t D_s v$ with $D_r := \begin{pmatrix} k & 0 & 0 & 0 \\ 0 & k & 0 & 0 \\ 0 & 0 & 1 & 0 \\ 0 & 0 & 0 & 1 \end{pmatrix}$ and $D_s :=$

$\begin{pmatrix} 0 & k & 0 & 0 \\ -k & 0 & 0 & 0 \\ 0 & 0 & 0 & 1 \\ 0 & 0 & -1 & 0 \end{pmatrix}$, $k := {x_1}^2 + {x_2}^2$. Then extensions of d_r and d_s cannot be positive definite respectively non degenerated.

1.5 Theorem

Let $FX \subset_\pi VX$ and d_s be symplectic on FX with a symplectic extension $\tilde{d}_s$ on VX. Then α) and β) are equivalent:

α) There exists $\tilde{d}_r$ on VX, which is compatible with $\tilde{d}_s$, i.e.: $w \perp_{\tilde{d}_r} FX \Leftrightarrow w \perp_{\tilde{d}_s} FX \ \forall w \in VX$.

β) There exists $J : FX \to FX$ with an almost complex extension $I : VX \to VX$, such that J hermitesates (FX, d_s), and I hermitesates $(VX, \tilde{d}_s)$.

Proof.

"$\Leftarrow$" Define $\tilde{d}_r : V^2X \to \mathbb{R}$; $\tilde{d}_r := \tilde{d}_s(I-, \sim)$. For $w \in VX$, $w \perp_{\tilde{d}_r} FX$ one gets: $w \perp_{\tilde{d}_r} FX \Leftrightarrow \tilde{d}_r(v, w) = 0 \ \forall\ v \in FX \Leftrightarrow \tilde{d}_s(Jv, w) = 0 \ \forall\ v \in FX \Leftrightarrow \tilde{d}_s(v, w) = 0 \ \forall\ v \in FX \Leftrightarrow w \perp_{\tilde{d}_s} FX$.

"$\Rightarrow$" $\exists!\ L : VX \to VX$ with $\tilde{d}_s(-, \sim) = \tilde{d}_r(L-, \sim)$. For $v \in F_qX$ and $V_qX \ni L\ v =: w_1 + w_2$, $w_1 \in F_qX$ and $w_2 \perp_{\tilde{d}_r} F_qX$ one gets $0 = \tilde{d}_s(v, w_2) = \tilde{d}_r(Lv, w_2) = \tilde{d}_r(w_1 + w_2, w_2) = \tilde{d}_r(w_2, w_2) \Rightarrow w_2 = 0 \Rightarrow L\ v \in F_qX \Rightarrow -L \circ L_{|F_qX} : F_qX \to F_qX$ and $-L \circ L$ is selfadjoined and positive. $\Rightarrow \exists!\ P : VX \to VX$ bundlehomomorphism, positive, with $P \circ P = -L \circ L$ and $P \circ L = L \circ P$ and $P_{|F_qX} : F_qX \to F_qX$. Hence $I : VX \to VX$, $I := -L \circ P^{-1}$ and $J : FX \to FX$, $J := I_{|FX}$ have the properties of β).

The following examples give a further impression about the theory:

1.6 Example

$X := (\mathbb{R}^2 \times \{0\}) \cup (\{0\} \times \mathbb{R}^2)$, $FX := TX$, d_s with $D_s := \begin{pmatrix} 0 & k & 1 & 0 \\ -k & 0 & 0 & -1 \\ -1 & 0 & 0 & 1 \\ 0 & 1 & -1 & 0 \end{pmatrix}$,

$k := {x_1}^2 + {x_2}^2$. (TX, d_s) is a symplectic space, which is not hermitesable, because there does not exist $J : TX \to TX$, such that $D_s(0)J(0)$ is positive definite.

1.7 Examples

$X := (\mathbb{R}^{m_1}\times\mathbb{R}^s\times\{0\})\cup(\{0\}\times\mathbb{R}^s\times\mathbb{R}^{m_2}) =: X_1\cup X_2 \subset \mathbb{R}^m$; $m := m_1+m_2+s$; m, m_1, $m_2 > 0$ even.

α) Let (T^iX, d_s) be a symplectic space with an extension $\tilde{d}_s : X\times\mathbb{R}^m\times\mathbb{R}^m \to \mathbb{R}$ (1.4)β)). (One can always choose $\tilde{d}_s$ screw symmetric.) If $\tilde{d}_{s|T^{i2}X_1}$ and $\tilde{d}_{s|T^{i2}X_2}$ are not degenerated, then (T^iX, d_s) is hermitesable.

β) In the case $m_1 = m_2 = 2$, (T^iX, d_s) is hermitesable for all d_s.

γ) β) is wrong for $m_1 > 2$.

Proof.

to α) One has to show that there exists $U \subset_{open} X$ with $X_1 \cap X_2 \subset U$ and $(T^iX_U, d_{s|T^{i2}X_U})$ hermitesable. Then α) follows with 2.1 and 1.2. Let $\tilde{d}_s$ be an extension of d_s with $\tilde{D}_s := \begin{pmatrix} D_{s_1} & D_{s_2} & 0 \\ -D_{s_2}^t & D_{s_4} & D_{s_5} \\ 0 & -D_{s_5}^t & D_{s_6} \end{pmatrix}$. $D_{s_4}(x)$ is invertible $\forall x \in U$ for an U like above, which is small enough. Hence d_{s_1}, d_{s_2} and d_{s_3} with matrices $D_{s_1} + D_{s_2}D_{s_4}^{-1}D_{s_2}^t$, D_{s_4} respectively $D_{s_6} + D_{s_5}^tD_{s_4}^{-1}D_{s_5}$ are symplectic on $U\times(\mathbb{R}^{m_1}\times\{0\}\times\{0\})$, $U\times(\{0\}\times\mathbb{R}^{m_2}\times\{0\})$ respectively $U\times(\{0\}\times\{0\}\times\mathbb{R}^{m_3})$. With 1.2 one has compatible almost complex structures J_1, J_2 and J_3 with matrices I_{11}, I_{22} and I_{33}. Let $I_{21} := D_4^{-1}(I_{22}^tD_{s_2}^t + D_{s_2}^tI_{11})$ and $I_{23} := -D_4^{-1}(D_{s_5}I_{33} + I_{22}^tD_{s_5})$, then $I := \begin{pmatrix} I_{11} & 0 & 0 \\ I_{21} & I_{22} & I_{23} \\ 0 & 0 & I_{33} \end{pmatrix}$ is a matrix of an almost complex structure on (U, T^iU), which is compatible with $d_{s|\,T^{i\,2}U}$.

to β) Analogous to α), in the case $m_1 = m_2 = 2$ one always finds I_{11} and I_{33}, take $\begin{pmatrix} 0 & 1 \\ -1 & 0 \end{pmatrix}$ or $\begin{pmatrix} 0 & -1 \\ 1 & 0 \end{pmatrix}$.

to γ) $X := (\mathbb{R}^4\times\{0\})\cup(\{0\}\times\mathbb{R}^2)$, d_s with $D_s := \begin{pmatrix} 0 & k & 0 & l & 0 & 0 \\ -k & 0 & l & 0 & 0 & 0 \\ 0 & -l & 0 & k & 0 & 0 \\ -l & 0 & -k & 0 & 0 & 0 \\ 0 & 0 & 0 & 0 & 0 & 1 \\ 0 & 0 & 0 & 0 & -1 & 0 \end{pmatrix}$, $k := x_1^2 + x_2^2$, $l := x_3^2 + x_4^2$. Suppose that the almost complex structure J hermitesates (T^iX, d_s). For the matrix I of an arbitrary extension (1.4)α)) one has $I = \begin{pmatrix} I_{11} & I_{12} \\ I_{21} & I_{22} \end{pmatrix}$ with $I_{11}(0)^2 = -E$, and

$D_1(x)I_{11}(x)$ positive definite, symmetric $\forall x \in (\mathbb{R}^4 \times \{0\})\backslash\{0\}$, with

$D_1 := \begin{pmatrix} 0 & k & 0 & l \\ -k & 0 & l & 0 \\ 0 & -l & 0 & k \\ -l & 0 & -k & 0 \end{pmatrix}$. But such I_{11} does not exist (direct calculation).

1.8 Examples

α) Let $X := \{ (x_1, x_2, x_3) \mid x_3 = \sqrt{x_1^{\ 2} + x_2^{\ 2}} \} \subset \mathbb{R}^3$ be the upper half cone, $FX := T^i X$, then (FX, d_s) with $D_s := \begin{pmatrix} 0 & 1 & 0 \\ -1 & 0 & 0 \\ 0 & 0 & 0 \end{pmatrix}$ is a symplectic space, which is not hermitesable, because there does not exist an almost complex structure J on FX.

β) The same is true for $X := X_1 \cup X_2 := \{ (x_1, x_2, x_3) \mid x_3 = 0 \} \cup \{ (x_1, x_2, x_3) \mid x_3 = x_1^{\ 2} + x_2^{\ 2} \} \subset \mathbb{R}^3$, $FX := T^i X$ or $FX := T^g X$, d_s given like in α).

Proof. By calculation: Suppose there exists J on FX then one can extend J to $L : X \times \mathbb{R}^3 \to X \times \mathbb{R}^3$ (1.4)α)). But because of $J: \ FX \to FX$ and $J \circ J(v) = -v \ \forall v \in FX$, L cannot be differentiable.

2 Properties of symplectic and almost complex structures

In the following second section some further properties of symplectic and almost complex structures are presented. The first note gives some information about the connection of local and global hermitesation.

2.1 Note

Let (FX, d_s) be a symplectic space, $X := \cup_{i \in I} U_i$, and J_i hermitesate $(FX_{U_i}, d_{s|\ F^2X_{U_i}})$ $\forall i \in I$. Let $\tilde{X} := \{ x \in X \mid \exists x \in U_x \subset_{open} X,\ FX_{U_x}$ vectorspacebundle$\}$, and $\forall q \in X\backslash\tilde{X}$ there exists $q \in \tilde{U_{iq}} \subset U_i \backslash \cup_{\binom{j \in I}{j \neq i}} U_j$. Then (FX, d_s) is hermitesable.

Proof. Let $X := \cup_{i \in I} U_i$ be a locally finite open covering with $X := \cup_{i \in I} U_i{}'$; $\bar{U_i}{}' \subset U_i$ and $\{\lambda_i | \ i \in I\}$ a partition of 1 respecting $\cup_{i \in I} U_i{}'$. Then $\tilde{L}: \ FX_{\tilde{X}} \to FX_{\tilde{X}}$, $\tilde{L}(v) := \sum_{i \in I} \lambda_i \circ \pi(v) J_i(v)\ \forall v \in FX_{\tilde{X}}$ is a bundlehomomorphism with the property $d_s(\tilde{L}-, \sim) = \sum_{i \in I} \lambda_i \circ \pi.\ d_{r_i}(-, \sim) =: d_r(-, \sim)$.

Hence $d_s(-,\sim) = d_r(L-,\sim)$ for $L := \tilde{L}^{-1}$. So $\tilde{J} : FX_{\tilde{X}} \to FX_{\tilde{X}}$ (like in the proof of 1.5) "$\Rightarrow$") exists, compatible with $d_{s|F^2X_{\tilde{X}}}$, and $\tilde{J}_{|F_{\tilde{q}}X} = J_{i|F_{\tilde{q}}X}$ $\forall \tilde{q} \in \tilde{X} \cap \tilde{U_{iq}}$, $q \in X\backslash\tilde{X}$. Hence $J := \left\{ \begin{array}{l} \tilde{J} \text{ on } FX_{\tilde{X}} \\ J_i \text{ on } \tilde{U_{iq}},\ q \in X\backslash\tilde{X} \end{array} \right\}$ is an almost complex structure on FX, compatible with d_s.

The next note and theorem give some information about the existence of almost complex structures on vectorfiberbundles and their product bundles.

2.2 Note

α) Let (FX, J_1) and $(\tilde{F}X, J_2)$ be almost complex spaces. Then $(FX \bigoplus \tilde{F}X, J)$ is an almost complex space, where $J(v) := (J_1 \circ p_{r_1}(v), J_2 \circ p_{r_2}(v))$ $\forall v \in FX \bigoplus \tilde{F}X$, and p_{r_1} and p_{r_2} are the projections onto FX and $\tilde{F}X$.

β) The same is true for (FX, d_{s_1}), $(\tilde{F}X, d_{s_2})$ and $(FX \bigoplus \tilde{F}X, d_s)$, where $d_s(v,w) := d_{s_1}(p_{r_1}(v), p_{r_1}(w)) + d_{s_2}(p_{r_2}(v), p_{r_2}(w))$ $\forall v, w \in FX \bigoplus \tilde{F}X$.

γ) J_1 is compatible with (FX, d_{s_1}) and J_2 with $(\tilde{F}X, d_{s_2})$ iff J is compatible with $(FX \bigoplus \tilde{F}X, d_s)$.

Proof. Direct calculation.

2.3 Theorem

Let (FX, J) be an almost complex space, $FX := \tilde{F}X \bigoplus VX$, let the dimension of the fibers of VX be even. Then almost complex structures exist locally on $\tilde{F}X$.

Proof. Let $p \in X$, $p \subset U \subset_{open} X$ sufficiently small.
First step:
One can suppose $U \subset \mathbb{R}^m$, $FX_U = F\tilde{X} \bigoplus VX_U \subset \mathbb{R}^m \times \mathbb{R}^{2s}$, $V_qX = \mathbb{R}^{2s}$ $\forall q \in U$. Choose vectorfields $(V_1, JV_1, \ldots, V_k, JV_k) =: \begin{pmatrix} W_{11} \\ W_{12} \end{pmatrix}$ linear independent $\forall x \in U$, k maximal. One can suppose: $\text{span}\left\{ \begin{pmatrix} W_{11}(p) \\ W_{12}(p) \end{pmatrix} \right\} = \{0\} \times \mathbb{R}^{2k}$, $k \supseteq s$ and $\forall q \in U$ there exist vectors $V^q_{2k+1}, \ldots, V^q_{2\,m_q}$ with $F_qX_U := \text{span}\{V_1(q), \ldots, JV_k(q), V^q_{2k+1}, \ldots, V^q_{2\,m_q}\} =: \text{span}\left\{ \begin{pmatrix} W_{11} \\ W_{12} \end{pmatrix}, \begin{pmatrix} W_{21} \\ 0 \end{pmatrix} \right\} \subset \mathbb{R}^n$. (U must be chosen small enough such that W_{12} is an invertible $(2k \times 2k)$-matrix $\forall x \in U$. $2m_q$ is the fiber dimension of F_qX_U.)

Second step:

For each extension I of J, I with matrix $\begin{pmatrix} I_{11} & I_{12} \\ I_{21} & I_{22} \end{pmatrix}$ and J with matrix $\begin{pmatrix} J_{11} & J_{12} \\ J_{21} & J_{22} \end{pmatrix}$ respecting the basis $\begin{pmatrix} W_{21} & W_{11} \\ 0 & W_{12} \end{pmatrix}$, one has $WJ :=$ $\begin{pmatrix} W_{21} & W_{11} \\ 0 & W_{12} \end{pmatrix} \begin{pmatrix} J_{11} & J_{12} \\ J_{21} & J_{22} \end{pmatrix} = \begin{pmatrix} I_{11} & I_{12} \\ I_{21} & I_{22} \end{pmatrix} \begin{pmatrix} W_{21} & W_{11} \\ 0 & W_{12} \end{pmatrix}$, where $J^2 = -E$ and $I \in C^\infty$, and there exists an invertible $(2k \times 2k)$-matrix X with $\begin{pmatrix} W_{21}J_{12} + W_{11}J_{22} \\ W_{12}\, J_{22} \end{pmatrix} = \begin{pmatrix} W_{11} \\ W_{12} \end{pmatrix} X$. Hence $X = J_{22}$, $J_{12} = 0$ and $J_{11}^2 = -E$. With $B := \begin{pmatrix} E & -W_{11}\, W_{12}^{-1} \\ 0 & W_{12}^{-1} \end{pmatrix}$ one gets: $BWJ = BIB^{-1}BW =: \tilde{I}BW$. $\Rightarrow$ $W_{21}J_{11} = \tilde{I_{11}}W_{21} \Rightarrow$ There exists $J_1{}'$ on $U \times \operatorname{span}\left\{\begin{pmatrix} W_{21} \\ 0 \end{pmatrix}\right\}$.

Third step:

Let $\begin{pmatrix} A_1 \\ A_2 \\ 0 \end{pmatrix}$ be a matrix, such that the columns built a basis of $\tilde{F}_qX$ and $\begin{pmatrix} W_{11} \\ W_{12} \end{pmatrix} =: \begin{pmatrix} B_{11} & B_{12} \\ B_{21} & B_{22} \\ B_{31} & B_{32} \end{pmatrix} \subset C^\infty$. One can suppose that B_{31} is an invertible $(2s \times 2s)$-matrix. One gets $F_qX = \operatorname{span}\left\{\begin{pmatrix} A_1 & 0 \\ A_2 & 0 \\ 0 & E \end{pmatrix}\right\} =$ $\operatorname{span}\left\{\begin{pmatrix} W_{21} & B_{11} & B_{12} \\ 0 & B_{21} & B_{22} \\ 0 & B_{31} & B_{32} \end{pmatrix}\right\}$. Hence there exist matrices X_1, X_3 with $\begin{pmatrix} A_1 \\ A_2 \end{pmatrix} =$ $\begin{pmatrix} W_{21} & -B_{11}B_{31}^{-1}B_{32} + B_{12} \\ 0 & -B_{21}B_{31}^{-1}B_{32} + B_{22} \end{pmatrix} \begin{pmatrix} X_1 \\ X_2 \end{pmatrix} =: \begin{pmatrix} W_{21} & \tilde{W_{11}} \\ 0 & \tilde{W_{12}} \end{pmatrix} \begin{pmatrix} X_1 \\ X_2 \end{pmatrix}$ with $\tilde{W_{11}}, \tilde{W_{12}} \in$ C^∞. It follows that the columns of $\begin{pmatrix} W_{21} & \tilde{W_{11}} \\ 0 & \tilde{W_{12}} \\ 0 & 0 \end{pmatrix}$ built a basis of $\tilde{F}_qX$ $\forall q \in U$, and with 1.2 there exists an almost complex structure on $U \times \operatorname{span}\left\{\begin{pmatrix} \tilde{W_{11}} \\ \tilde{W_{12}} \end{pmatrix}\right\}$. Now the theorem follows with 2.2.

3 Weak differentiable structures

The examples in 1.8 show, that there are many symplectic spaces without an existing almost complex structure. Because of this I explore the problem with a weaker kind of differentiability.

3.1 Definition

Let $\vartheta(X)$ be a sheaf of germs of vectorfields $V : X \to FX$ on X; $FX := \{ (x, v) \mid x \in X,\ v \in \vartheta(X)(x) \}$.

α) A map (bundlehomomorphism) $L : FX \to FX$, which is linear on the fibers is called *weakly differentiable*, if L induces a map $\vartheta(X)_x \to \vartheta(X)_x$.

β) A map $d : F^2X \to \mathbb{R}$ respectively $\mathbb{C}$, which is bilinear on the fibers, is called *weakly differentiable*, if d induces a map $\vartheta(X)_x \times \vartheta(X)_x \to D^\infty(X)_x$ respectively $D^{\mathbb{C}\infty}(X)_x$.

   It is obvious now, how to define *weakly differentiable* symplectic and almost complex structures and *weakly hermitesable.*

γ) If $\vartheta(X) := \vartheta^i(X) :=$ sheaf of germs of locally integrable fields $V : X \to TX$, then $FX = T^iX$, and d respectively J are called *i-differentiable* and weakly differentiable becomes *i-differentiable*.

It is obvious, that a differentiable structure is weakly differentiable and that in the case of manifolds i-differentiable and differentiable are equivalent. Further it is obvious (or analogously to be shown) that 1.2, 1.3, 2.1, 2.2 and 2.3 are true in the weakly differentable case, if $\vartheta(X)$ is the sheaf of germs of all vectorfields $V : X \to FX$ respectively $V : X \to VX$.

3.2 Example

Let X be as in 1.7 and $FX := T^iX$. Then one obtains:

α) All i-differentiable almost complex structures on X are differentiable.

β) All i-differentiable symplectic structures on X are differentiable.

Proof.
to α) For the matrix I of J one has

$$I(x) := (i_{ql}(x)) := \left\{ \begin{array}{ll} \begin{pmatrix} I_{11} & I_{12} & * \\ I_{21} & I_{22} & * \\ 0 & 0 & * \end{pmatrix} & x \in X_1 \backslash Sg(X) \\ \begin{pmatrix} * & 0 & * \\ * & I_{22} & * \\ * & 0 & * \end{pmatrix} & x \in Sg(X) \\ \begin{pmatrix} * & 0 & 0 \\ * & I_{22} & I_{23} \\ * & I_{32} & I_{33} \end{pmatrix} & x \in X_2 \backslash Sg(X) \end{array} \right\}.$$

For all $j,\ v,\ w \in \{1, \ldots, m_1\}$, $k \in \{m_1+1, \ldots, m_1+s\}$, $l,\ r \in \{m_1+1+s, \ldots, m\}$ and $t \in \{1, \ldots, m\}$ one gets:

(1) $e_k \in \Gamma^\infty(X, T^iX) \Rightarrow J \circ e_k \in \Gamma^\infty(X, T^iX) \Rightarrow i_{tk} \in C^\infty(\mathbb{R}^m)$.

(2) $x_v\ e_j \in \Gamma^\infty(X, T^iX) \Rightarrow x_v J \circ e_j \in \Gamma^\infty(X, T^iX)$
$\Rightarrow$ There exist extensions $g_{vtj} \in C^\infty(\mathbb{R}^{m_1+s})$ of $x_v i_{tj}$ on $X_1 \backslash Sg(X)$
$\Rightarrow x_v g_{wtj|X_1\backslash Sg(X)} = x_v x_w i_{tj|X_1\backslash Sg(X)} = x_w g_{vtj|X_1\backslash Sg(X)}$
$\Rightarrow \exists \tilde{g_{vtj}} \in C^\infty(\mathbb{R}^{m_1+s})$ with $g_{vtj} = \tilde{g_{vtj}} x_v$ and $x_v x_w \tilde{g_{vth}} = x_v x_w \tilde{g_{wtj}}$
$\Rightarrow \tilde{g_{tj}} := \tilde{g_{vtj}} = \tilde{g_{wtj}}$ on $X_1 \backslash \{x \in X_1 \mid x_v x_w = 0\}$, hence on X_1.
Hence $i_{tj} = \tilde{g_{vtj}} = \tilde{g_{tj}}$ on $X_1 \backslash \{x \in X_1 \mid x_v = 0\}$
$\Rightarrow i_{tj} = \tilde{g_{tj}}$ on $X_1 \backslash Sg(X) \Rightarrow \tilde{g_{tj}}$ is a C^∞-extension of i_{tj} on X_1.
Now extend i_{tj} to $i_{tj} \in C^\infty(\mathbb{R}^m)$.

(3) $x_r e_l \in \Gamma^\infty(X, T^iX) \Rightarrow x_r\ J \circ e_l \in \Gamma^\infty(X, T^iX)$. Continue analogously to (2).
Hence there exists a C^∞-extension of J.

to β) analogous to α).

The examples 1.7)γ) and 3.2) show, that there exist symplectic spaces, that are not weakly hermitesable. The following example shows, that "i-differentiable" and "differentiable" are not equivalent. There exist more i-differentiable almost complex structures than differentiable ones (compare 1.8 with 3.3).

3.3 Example

α) $X := \{(x_1, x_2, x_3) \mid x_3 = 0\} \cup \{(x_1, x_2, x_3) \mid x_3 = {x_1}^2 + {x_2}^2\} =: X_1 \cup X_2 \subset \mathbb{R}^3$. J, given by $\begin{pmatrix} 0 & 1 \\ -1 & 0 \end{pmatrix}$ respecting $\left\{ \begin{pmatrix} 1 \\ 0 \\ 0 \end{pmatrix}, \begin{pmatrix} 0 \\ 1 \\ 0 \end{pmatrix} \right\}$ on $X_1 \backslash \{0\}$ and $\left\{ \begin{pmatrix} 1 \\ 0 \\ 2x_1 \end{pmatrix}, \begin{pmatrix} 0 \\ 1 \\ 2x_2 \end{pmatrix} \right\}$ on $X_2 \backslash \{0\}$ is an i-differentiable almost complex structure.

β) Let d_s be given by $d \cdot \begin{pmatrix} 0 & 1 \\ -1 & 0 \end{pmatrix}$, $d(x) \neq 0$ on $Mf(X)$ respecting the basis of α), then $(T^i X, d_s)$ is hermitesable iff sign $d_{|Mf(X)}$ is constant.

Proof.

to α) For all $\tilde{V} \in \vartheta(X)_o$ exist $v_1, v_2 : U(0) \to \mathbb{R}$, such that $\tilde{V}$ is represented by $V \in \Gamma^\infty(U(0), T^i X_{U(0)})$, where

$$V(x) := \left\{ \begin{array}{cc} \begin{pmatrix} v_1 \\ v_2 \\ 0 \end{pmatrix} & x \in (X_1 \backslash \{0\}) \cap U(0) \\ \begin{pmatrix} 0 \\ 0 \\ 0 \end{pmatrix} & x = 0 \\ \begin{pmatrix} v_1 \\ v_2 \\ 2x_1 v_1 + 2x_2 v_2 \end{pmatrix} & x \in (X_2 \backslash \{0\}) \cap U(0) \end{array} \right\}.$$

One has to show: $JV \in \Gamma^\infty(U(0), T^i X_{U(0)})$. This is true iff $f(x) := \left\{ \begin{array}{cc} 0 & x \in X_1 \\ -2x_2 v_1 + 2x_1 v_2 & x \in X_2 \backslash \{0\} \end{array} \right\} \in D^\infty(X)_o$. Because $V \in \Gamma^\infty(U(0), T^i X_{U(0)})$ there exist extensions $\tilde{v_j}$ of v_j, $j \in \{1, 2\}$ and a_j, b_j, c_j with $\tilde{v_j} = a_j x_1 + b_j x_2 + c_j x_3 \in C^\infty(U(0))$. One can choose a_j, b_j, c_j independent from x_3. Therefore $2x_1 \tilde{v_1} + 2x_2 \tilde{v_2} = 2{x_1}^2 a_1 + 2x_1 x_2 (b_1 + a_2) + x_3(2x_1 c_1 + 2x_2 c_2) + 2{x_2}^2 b_2 \in C^\infty(U(0))$ is an extension of $2x_1 v_1 + 2x_2 v_{2|X_2 \backslash \{0\}}$, so $-2{x_2}^2 a_1 + 2{x_2}^2 b_2 + 2x_1 x_2 (b_1 + a_2) + x_3(2x_1 c_1 + 2x_2 c_2 + 2a_1)$, too. It follows that there exist g_1, $g_2 \in D^\infty(\mathbb{R}^3)_o$ with $x_2(-x_2 a_1 + x_2 b_2 + x_1 b_1 + x_1 a_2) + x_3(x_1 c_1 + x_2 c_2 + a_1) = g_1 x_3 + g_2({x_1}^2 + {x_2}^2 - x_3)$. Let $x_3 = 0$, then $x_2 b_2 = x_2 a_1 - x_1 b_1 - x_1 a_2 + \hat{g_2}({x_1}^2 + {x_2}^2)$ with $x_2 \hat{g_2}(x_1, x_2) := g_2(x_1, x_2, 0)$,

$\hat{g_2} \in D^\infty(\mathbb{R}^2)_0$. Hence $-2x_2\tilde{v_1} + 2x_1\tilde{v_2} \in C^\infty(U(0))$ is an extension of $-2x_2v_1+2x_1v_{2|X_2\setminus\{0\}}$ and $-2x_2\tilde{v_1}+2x_1\tilde{v_2}-0 = ({x_1}^2+{x_2}^2-x_3)(2x_1\hat{g_2}-2b_1) + x_3(-2\,b_1 + 2x_1\hat{g_2} - 2x_2c_1 + 2x_1c_3) =: ({x_1}^2 + {x_2}^2 - x_3)\tilde{g_1} + x_3\tilde{g_2}$. Now one can see that $F := x_3\tilde{g_2}$ is an extension of f, so $f \in D^\infty(X)_o$.

to β) With the help of the vectorfields $V := \left\{\begin{matrix} \begin{pmatrix}0\\0\\0\end{pmatrix} & x \in X_1 \\ \begin{pmatrix}x_3\\0\\2x_1x_3\end{pmatrix} & x \in X_2 \end{matrix}\right\}$,

$W := \left\{\begin{matrix} \begin{pmatrix}0\\0\\0\end{pmatrix} & x \in X_1 \\ \begin{pmatrix}0\\x_3\\2x_2x_3\end{pmatrix} & x \in X_2 \end{matrix}\right\}$, $\tilde{V} := \left\{\begin{matrix} \begin{pmatrix}{x_1}^2+{x_2}^2\\0\\0\end{pmatrix} & x \in X_1 \\ \begin{pmatrix}0\\0\\0\end{pmatrix} & x \in X_2 \end{matrix}\right\}$, $\tilde{W} :=$ $\left\{\begin{matrix} \begin{pmatrix}0\\{x_1}^2+{x_2}^2\\0\end{pmatrix} & x \in X_1 \\ \begin{pmatrix}0\\0\\0\end{pmatrix} & x \in X_2 \end{matrix}\right\}$, $Z := \left\{\begin{matrix} \begin{pmatrix}x_1\\x_2\\0\end{pmatrix} & x \in X_1 \\ \begin{pmatrix}x_1\\x_2\\2({x_1}^2+{x_2}^2)\end{pmatrix} & x \in X_2 \end{matrix}\right\}$ and the method of α) one gets $JV, JW, J\tilde{V}, J\tilde{W}, JZ \in \Gamma^\infty(X, TX)$ and therefore J with matrix $\begin{pmatrix} -j_{22} & j_{12} \\ \frac{1+j_{22}^2}{-j_{12}} & j_{22} \end{pmatrix}$ respecting the basis in α) can only be i-differentiable, if sign $j_{12|Mf(X)}$ is constant. This and α) prove β).

4 Structures on quotient spaces

In the last section I explore this theory on special quotient spaces. Let (for example) a finite group G operate on X. One knows (see [Hi]), that locally integrable fields V^* on the quotient X/G are exactly those fields which can be lifted to G-invariant fields V on X. One can also lift riemannian metrices (see [Hi]).

I take $X = \mathbb{R}^{2n}$ and a finite group $G \subset O(2n)$, which operates on X. Hence X/G exists, and there exist a polynomial quotient map $Q : \mathbb{R}^{2n} \to \mathbb{R}^l$ with $Q(0) = 0$, which determines X/G (see [Re]). So one has $V^* \circ Q = TQ \circ V$, where V^* is locally integrable and $V \circ g = Tg \circ V$, V is determined by V^* and the other way round; and $TQ(x,-) : T_x{}^G X \to T^i_{Q(x)} Q(X)$ is bijective, $T_x{}^G X$ like in 4.1) α).

4.1 Definition

L and d of 3.1)α)β) are called *G-differentiable*, if $FX := T^G X := \{(x,v) \mid \exists V : X \to TX,$ G-invariant with $V(x) = v\}$, and $\vartheta(X) := \vartheta^G(X) :=$ sheaf of germs of G-invariant fields $V : X \to TX$.

(Note, that the representants $\tilde{\vartheta}$ of a germ of G-invariant fields are exactly those that can be extended to G-invariant fields $V : X \to TX$).

4.2 Theorem

α) How to push down J:
Let $J : T^G X \to T^G X$ be a G-invariant, G-differentiable almost complex structure, then there exists exactly one i-differentiable almost complex structure $J^* : T^i(X/G) \to T^i(X/G)$ with $TQ \circ J = J^* \circ TQ$.

β) How to lift J^*:
Let $J^* : T^i(X/G) \to T^i(X/G)$ be an i-differentiable almost complex structure, then there exists exactly one G-differentiable almost complex structure $J : T^G X \to T^G X$ with $TQ \circ J = J^* \circ TQ$, and J is G-invariant.

γ) How to push down d_s (and analogously d_r):
Let $d_s : T^{G\,2} X \to \mathbb{R}$ be a G-invariant, G-differentiable symplectic structure, then there exists exactly one i-differentiable symplectic structure $d_s{}^* : T^{i\,2}(X/G) \to \mathbb{R}$ with $d_s = Q^* \circ d_s{}^*$.

δ) How to lift $d_s{}^*$ (and analogously $d_r{}^*$):
Let $d_s{}^* : T^{i\,2}(X/G) \to \mathbb{R}$ be an i-differentiable symplectic structure, then there exists exactly one G-differentiable symplectic structure $d_s : T^{G\,2} X \to \mathbb{R}$ with $d_s = Q^* \circ d_s^*$, and d_s is G-invariant.

ε) *"lift $\circ$ push = id" and "push $\circ$ lift = id".*

ζ) *$d_s(J, \sim)$ is riemannian $\Leftrightarrow d_s{}^*(J^* -, \sim)$ is riemannian.*

Proof.
to α) Define pointwise: $J^*(y, v^*) := TQ \circ J \circ TQ^{-1}(y, v^*)$, where $TQ^x(x,-) := TQ(x,-) : T_x{}^G X \to T_y{}^i Q(X)$ is given for an arbitrary chosen $x \in Q^{-1}(y)$.

(1) J^* is *well defined*:
Let $\tilde{x} \in Q^{-1}(y) \Rightarrow \exists g \in G$ with $g(x) = \tilde{x}$ and $\exists! v$ and $\tilde{v}$ with $TQ^x(x,v) = (y,v^*) = TQ^{\tilde{x}}(\tilde{x},\tilde{v})$ and $Tg(x,v) = (\tilde{x},\tilde{v}) \Rightarrow TQ^{\tilde{x}} \circ J \circ TQ^{\tilde{x}\ -1}(y,v^*) = TQ^{\tilde{x}} \circ J(\tilde{x},\tilde{v}) = TQ^{\tilde{x}} \circ J \circ Tg(x,v) = TQ^{\tilde{x}} \circ Tg \circ J(x,v) = TQ^x \circ J(x,v) = TQ^x \circ J \circ TQ^{x\ -1}(y,v^*)$

(2) J^* is *i-differentiable*:
$TQ \circ J \circ V : X \to T^iQ(X)$ is G-invariant for each $V \in \vartheta^G(X)_x : TQ \circ J \circ V(g(x)) = TQ \circ J \circ Tg \circ V(x) = TQ \circ Tg \circ J \circ V(x) = TQ \circ J \circ V(x)$.
It follows with [Re] for $V^* \in \vartheta^i(X/G)_y$: $\exists! \tilde{h} \in \Gamma^\infty(Q(x), T^iQ(x))$ with $\tilde{h} \circ Q(x) = TQ \circ J \circ V(x) = J^* \circ V^* \circ Q(x) \Rightarrow J^* \circ V^* = \tilde{h} \in \vartheta^i(Q(x))_y$. The rest is obvious.

to β) Define pointwise: $J(x,v) = TQ^{-1} \circ J^* \circ TQ(x,v)$.

(1) J is *G-invariant*:
$J \circ Tg(x,v) = TQ^{\tilde{x}\ -1} \circ J^* \circ TQ^{\tilde{x}} \circ Tg(x,v)$
$= Tg \circ TQ^{x\ -1} \circ J^* \circ TQ^x(x,v) = Tg \circ J(x,v)$.

(2) J is *G-differentiable*:
Let $V \in \vartheta^G(X)_x$. For $J^* \circ V^* \in \Gamma^\infty(Q(X), T^iQ(X))$ exists $W \in \Gamma^\infty(X, T^GX)$ with $TQ \circ W(x) = J^* \circ V^*(Q(x)) = TQ \circ J \circ TQ^{-1} \circ TQ \circ V(x) = TQ \circ J \circ V(x) \Rightarrow J \circ V = W \in \vartheta^G(X)_x$. ($J \circ V$ is obviously G-invariant). The rest is obvious.

to γ) analogous to α) with $d_s{}^* := Q^{-1\ *} \circ d_s$ (TQ^{-1} pointwise).
to δ) analogous to β) with $d_s := Q^* \circ d_s{}^*$.
to ε) obvious
to ζ) $d_s(J, \sim)$ is riemannian $\Leftrightarrow d_s(J \circ TQ^{-1}-, TQ^{-1} \sim)$ is riemannian $\Leftrightarrow$ $d_s(TQ^{-1} \circ J^*-, TQ^{-1} \sim)$ is riemannian $\Leftrightarrow d_s^*(J^*-, \sim)$ is riemannian.

Now we have a simple and powerful tool to explore the problem of i-hermitesation on quotient spaces. We also see, that these definitions of differentiability and "tangentspaces" fit together very well.

The next question that arises is that about the relationship of differentiability and G-differentiability of J and d_s on T^GX. It is solved in the following special cases:

4.3 Theorem

α) *Let $G := \{id, x \to -x\}$ operate on $X := \mathbb{R}^{2n}$ Then all G-differentiable d and L like in 4.1)α) are differentiable.*

β) *Let an arbitrary G operate on $X := \mathbb{R}^2$, J be almost complex and d_s be symplectic, both G-differentiable. Then J and d_s are differentiable.*

Proof.

to α) Analogous to the proof of 3.2 with the G-invariant fields $x_i e_j$ for $i, j \in \{1, \dots, 2n\}$.

to β) $Q(x_1, x_2) = (x_1^2 + x_2^2, \ \mathrm{Re}((x_1 + i\, x_2)^m), \ \mathrm{Im}((x_1 + i\, x_2)^m))$ [St], where $\begin{pmatrix} \cos\frac{2\pi}{m} & \sin\frac{2\pi}{m} \\ -\sin\frac{2\pi}{m} & \cos\frac{2\pi}{m} \end{pmatrix}$, $m \in \mathbb{N}$, generates G. (Note, that dim $T_x^G X$ has to be even $\forall x \in X$, otherwise there cannot exist J or d_s.) $V_1 := \begin{pmatrix} x_1 \\ x_2 \end{pmatrix}$, $V_2 := \begin{pmatrix} -x_2 \\ x_1 \end{pmatrix}$ and $V_3 := \begin{pmatrix} m\,\mathrm{Re}((x_1 + i\, x_2)^{m-1}) \\ -m\,\mathrm{Im}((x_1 + i\, x_2)^{m-1}) \end{pmatrix} =: \begin{pmatrix} V_{31} \\ V_{32} \end{pmatrix}$ are G-invariant fields. Let $\begin{pmatrix} -j_{22} & j_{12} \\ \frac{1+j_{22}^2}{-j_{12}} & j_{22} \end{pmatrix}$ be the matrix of J on $\mathbb{R}^2 \backslash \{0\}$. $JV_1, JV_2, JV_3 \in \Gamma^\infty(X, T^i X) \Rightarrow$ The following functions g_s, $s \in \{1, \dots, 5\}$ are extendable to $\tilde{g_s} \in D^\infty(\mathbb{R}^2)$, where $g_1 := -j_{22}x_1 + j_{12}x_2$, $g_2 := \frac{1+j_{22}^2}{-j_{12}} x_1 + j_{22}x_2$, $g_3 := j_{22}x_2 + j_{12}x_1$, $g_4 := \frac{1+j_{22}^2}{j_{12}} x_2 + j_{22}x_1$ and $g_5 := -j_{22}V_{31} + j_{12}V_{32}$ on $\mathbb{R}^2 \backslash \{0\}$. On $\mathbb{R}^2 \backslash \{0\}$ one get $g_3 V_{31} + x_2 g_5 = j_{12}(V_{31}x_1 + V_{32}x_2)$, $g_1 V_{31} - x_1 g_5 = j_{12}(V_{31}x_2 - V_{32}x_1)$ and $x_2 g_1 + x_1 g_3 = j_{12}(x_1^2 + x_2^2)$. One has to show: There exists an extension $g \in D^\infty(X)_o$ of j_{12}. Then this is also true for j_{22} and $\frac{1+j_{22}^2}{j_{12}}$ because of g_1 and g_3 respectively g_2 and g_4. One has: $V_{31}x_1 + V_{32}x_2 = m\,\mathrm{Re}((x_1 + i\, x_2)^{m-1})x_1 - m\,\mathrm{Im}((x_1 + i\, x_2)^{m-1})x_2 = m\,\mathrm{Re}[(Re((x_1 + i\, x_2)^{m-1}) + i\,\mathrm{Im}((x_1 + i\, x_2)^{m-1}))(x_1 + i\, x_2)] = m\,\mathrm{Re}[(x_1 + i\, x_2)^{m-1}(x_1 + i\, x_2)] = m\,\mathrm{Re}(x_1 + i\, x_2)^m$ and $V_{31}x_2 - V_{32}x_1 = m\,\mathrm{Re}((x_1 + i\, x_2)^{m-1})x_2 + m\,\mathrm{Im}((x_1 + i\, x_2)^{m-1})x_1 = m\,\mathrm{Re}[(x_1 + i\, x_2)^{m-1}(x_2 - i\, x_1)] = m\,\mathrm{Re}[-i(x_1 + i\, x_2)^{m-1}(x_1 + i\, x_2)] = m\,\mathrm{Re}(-i(x_1 + i\, x_2)^m) = m\,\mathrm{Im}(x_1 + i\, x_2)^m$. Hence there exist extensions h_1 and $h_{2m} \in D^\infty(X)_o$ of $j_{12}(x_1^2 + x_2^2)$ and $j_{12}(x_1 + i\, x_2)^m$. Now

$$
\begin{aligned}
(x_1 + i\, x_2)^m &= \sum_{k=0}^{m} \binom{m}{k} x_1^{m-k} (-i)^k x_2^k \\
&= \sum_{k=0,k\ even} \binom{m}{k} x_1^{m-k} (-1)^{\frac{k}{2}} x_2^k \\
&\quad + i \sum_{k=1,k\ uneven} \binom{m}{k} x_1^{m-k} (-1)^{\frac{k-1}{2}} x_2^k
\end{aligned}
$$

and $j_{12} x_1^k x_2^l$ is extendable to $h_{kl} \in D^\infty(\mathbb{R}^2)_o \Leftrightarrow j_{12} x_1^{k+2} x_2^{l-2}$ is extendable to $h_1 x_1^k x_2^{l-2} - h_{kl} \in D^\infty(\mathbb{R}^2)_o \Leftrightarrow j_{12}\, x_1^{k-2} x_2^{l+2}$ is extendable to $h_1 x_1^{k-2} x_2^l - h_{kl} \in D^\infty(\mathbb{R}^2)_o$. Now one has to show: $h_1 \in D^\infty(\mathbb{R}^2)_o$,

$h_{2m} \in D^{\mathbb{C}\infty}(\mathbb{R}^2)_o \Rightarrow h_1 \in D^\infty(\mathbb{R}^2)_o$, $h_{2\,m-1} \in D^{\mathbb{C}\infty}(\mathbb{R}^2)_o$. (Then after finitely many steps one has $h_{20} = j_{12} \in D^\infty(\mathbb{R}^2)_o$.)

proof.
First case: m is even.

$$\begin{aligned} j_{12}(x_1 + i\,x_2)^m &= j_{12}[\sum_{k=0,k\ even}^{m} \binom{m}{k} x_1^{m-k}(-1)^{\frac{k}{2}} x_2^{\,k}] \\ &\quad + i\,j_{12}[\sum_{k=0,k\ even}^{m-2} \binom{m}{k+1} x_1^{\,m-(k+1)}(-1)^{\frac{k}{2}} x_2^{\,k+1}] \\ &= \tilde{g_1} + [\sum_{k=0,k\ even}^{m} \binom{m}{k}] x_1^{\,m} j_{12} \\ &\quad + i\,\tilde{g_2} + i[\sum_{k=0,k\ even}^{m-2} \binom{m}{k+1}] x_1^{\,m-1} x_2\, j_{12} \end{aligned}$$

for proper $\tilde{g_1}$ and $\tilde{g_2} \in D^\infty(\mathbb{R}^2)_o$. $\Rightarrow x_1^{\,m} j_{12}$ and $x_1^{\,m-1} x_2 j_{12}$ are extendable $\Rightarrow x_1^{\,m-1} j_{12}$ is extendable. Further

$$\begin{aligned} j_{12}(x_1 + i\,x_2)^m &= j_{12}[\sum_{k=0,k\ even}^{m} \binom{m}{k} x_1^{\,k} x_2^{\,m-k}(-1)^{\frac{m-k}{2}}] \\ &\quad + i\,j_{12}[\sum_{k=0,k\ even}^{m-2} \binom{m}{k+1} x_1^{\,k+1}(-1)^{\frac{m-(k+1)}{2}} x_2^{m-(k+1)}] \\ &= g_1{}' + [\sum_{k=0,k\ even}^{m} \binom{m}{k}](-1)^{\frac{m}{2}} x_2^{\,m} j_{12} + i\,g_2{}' \\ &\quad + i[\sum_{k=0,k\ even}^{m-2} \binom{m}{k+1}](-1)^{\frac{m-1}{2}} x_2^{\,m-1} x_1 j_{12} \end{aligned}$$

$\Rightarrow x_2^{\,m} j_{12}$ and $x_2^{\,m-1} x_1\, j_{12}$ are extendable $\Rightarrow x_2^{\,m-1} j_{12}$ is extendable. It follows that $x_1^{\,m-1-k} x_2^{\,k} j_{12} \in D^\infty(\mathbb{R}^2)_o\ \forall k \in \{0, \ldots, m-1\} \Rightarrow h_{2m-1} \in D^\infty(\mathbb{R}^2)_o$.
Second case: m is odd. Analogous proof.
For d_s with matrix $\begin{pmatrix} 0 & d \\ -d & 0 \end{pmatrix}$ one gets $d(x_1{}^2 + x_2{}^2)$, $d(x_1 V_{31} + x_2 V_{32})$ and $d(x_2\, V_{31} - x_1 V_{32}) \in D^\infty(\mathbb{R}^2)_o$. Rest analogous.

The theorems above can be used to calculate the following examples and note:

4.4 Example

$X = \mathbb{R}^2$, $G = \{id, x \to -x\}$, $Q(x_1, x_2) = ({x_1}^2, {x_2}^2, x_1 x_2)$

α) X/G is diffeomorphic to the half cone.

β) There exists an i-differentiable J^* on X/G.

γ) All i-differentiable d_s^* on X/G are i-hermitesable.

δ) There exists an i-differentiable d_s^* on X/G, which is not differentiable.

Proof.

to β) J with matrix $\pm \begin{pmatrix} 0 & 1 \\ -1 & 0 \end{pmatrix}$ is a G-invariant, differentiable almost complex structure on $T^G\mathbb{R}^2$ and 4.2)α).

to γ) With 4.2)δ) one gets d_s G-invariant and G-differentiable on $T^G\mathbb{R}^2$ with matrix $\begin{pmatrix} 0 & d \\ -d & 0 \end{pmatrix}$, sign $d_{|\mathbb{R}^2\setminus\{0\}}$ constant. J in β) is compatible and 4.2).

to δ) For the matrices $D_s := \begin{pmatrix} 0 & d \\ -d & 0 \end{pmatrix}$ of d_s and $D_s^* := \begin{pmatrix} 0 & d_{12}^* & d_{13}^* \\ -d_{12}^* & 0 & d_{23}^* \\ -d_{13}^* & -d_{23}^* & 0 \end{pmatrix}$ of d_s^* one gets: $dQ^t D_s^*(Q(x))\ dQ = D_s \Leftrightarrow d(x) = 4x_1x_2d_{12}^*(Q(x)) - 2{x_2}^2 d_{23}^*(Q(x)) + 2{x_1}^2 d_{13}^*(Q(x))$. For d_s with $d(x) \neq 0\ \forall x \in \mathbb{R}^2$ d_s^* is i-differentiable, but not differentiable.

Compare these results with 1.8)α)!

4.5 Note

All (Y, d_s^*) with $Y := X/G$; $X = \mathbb{R}^2$ and d_s^* i-differentiable are i-hermitesable.

Proof. Let $d_s : T\ {}^2\mathbb{R}^2 \to \mathbb{R}$ be the lift of d_s^*. Then $J:\ T^G\mathbb{R}^2 \to T^G\mathbb{R}^2$ with matrix $\begin{pmatrix} 0 & 1 \\ -1 & 0 \end{pmatrix}$ or $\begin{pmatrix} 0 & -1 \\ 1 & 0 \end{pmatrix}$ is compatible with d_s and G-invariant. The note follows with 4.2)ζ).

4.6 Example

There exists a differentiable, G-invariant $d_s : T\,^2X \to \mathbb{R}$, such that (X, d_s) is not hermitesable.

Proof. $X := \mathbb{R}^4$, $G := \{id, x \to -x\}$, d_s with $D_s := D_1$ like in 1.7)γ). 1.7)γ) and 4.3)α) together prove the example.

References

1. J. Himmelmann, *Zur Geometrie differenzierbarer Quotientenräume*, Dissertation, Bochum (1995).
2. K. Reichhard, *Quotienten analytischer und differenzierbarer Räume nach Transformationsgruppen,*, Habilitationsschrift, Bochum (1978).
3. K. Spallek, *Almost Complex Structures on Spaces, Associated Singularities and Foliations*, in: Almost Complex Structures, Proceedings of the International Workshop, World Scientific, London (1992).
4. R. Strub, *Vollständige Klassifikation der Singularitäten von Quotienten von unendlich oft reell differenzierbaren Mannigfaltigkeiten nach eigentlich diskontinuerlichen Automorphismengruppen*, Dissertation, Mainz (1980).

REMARKS ON SOME FUNCTION THEORIES ON A CLASS OF ALMOST COMPLEX MANIFOLDS

STANCHO DIMIEV
Bulgarian Academy of Sciences,
Institute for Mathematics and Informatics
Acad. G. Bonchev Str., Bl. 8, 1113 Sofia, Bulgaria
e-mail: sdimiev@math.bas.bg,

DIMITAR FRENKEV
Plovdiv University – Branch in Smolian,
ul. Dimo Dichev 32, 4700, Smolian, Bulgaria

KARLHEINZ SPALLEK
Ruhr Universität - Bochum, Fakultät für Mathematik
D-44780 Bochum, Deutschland
e-mail: Karlheinz.Spallek@ruba.rz.ruhr-uni-bohum.dbp.de

The purpose of this note is to describe some facts of elementary geometric character valid on almost complex manifolds admitting non-trivial almost holomorphic functions under the special condition on the constancy of the corresponding type. In the general case of non-constant type, a notion of holomorphic relation is proposed and respectively a notion of a generalized foliation. Some results of more sophisticated character will be exposed somewhere else.

1 Complex self-conjugate co-ordinates

Let (M, J) be an almost complex manifold of real dimension $2n$, and p be a point of M. The set of all bases in the tangent space T_pM is a fibre in the principal fibre bundle $P = P(M, GL(n, \mathbf{C}))$ of all linear frames on M. The restriction of J on T_pM is denoted by J_p. It acts on the fibre of P over the point p.

We can use the local co-ordinates of the underlying smooth manifold M near p, namely $(x^1, \ldots, x^{2n})$, denoting by $J(x)$ the local representation of J in these co-ordinates. The transformation of J under the co-ordinate change $(x^1, \ldots, x^{2n}) \to (y^1, \ldots, y^{2n})$ is given by $J(y) = A^{-1}J(x)A$, $\det A \neq 0$, where $A \in GL(2n, \mathbf{R})$. In any point $p \in M$ we can find local co-ordinates such that

J to be of the following form

$$\begin{pmatrix} 0 & 1 & & & \\ -1 & 0 & & & \\ & & \ddots & & \\ & & & 0 & 1 \\ & & & -1 & 0 \end{pmatrix}.$$

This representation is valid only at the point p, and does not holds in any neighbourhood of p.

It is natural to replace the real co-ordinates $(x^1, \ldots, x^{2n})$ by the self-conjugate complex one $(z^1, \ldots, z^n, \overline{z}^1, \ldots, \overline{z}^n)$, where $z^k := x^{2k-1} - ix^{2k}$, $\overline{z}^k = x^{2k-1} + ix^{2k}$, $k = 1, \ldots, n$. Then each moving base (= local section of P over some open $U \subset M$) $(\partial/\partial x^1, \ldots, \partial/\partial x^{2n})$, can be replaced canonically by a self-conjugated basis $(\partial/\partial z^1, \ldots, \partial/\partial z^n, \partial/\partial\overline{z}^1, \ldots, \partial/\partial\overline{z}^n)$, where $\partial/\partial z^k = (1/2)(\partial/\partial x^{2k-1} - i\partial/\partial x^{2k})$, $\partial/\overline{z}^k = (1/2)(\partial/\partial x^{2k-1} + i\partial/\partial x^{2k})$.

The matrix representation of J_p in the self-conjugate co-ordinates is the following

$$\begin{pmatrix} iE_n & 0 \\ 0 & -iE_n \end{pmatrix},$$

where E_n is the unit $n \times n$ matrix.

Considering the natural splitting $\mathbf{C}T_pM = T_pM^{1,0} \oplus T_pM^{0,1}$, we take $(\partial/\partial z^1, \ldots, \partial/\partial z^n)$ to be a basis of $T_pM^{1,0}$ and $(\partial/\partial\overline{z}^1, \ldots, \partial/\partial\overline{z}^n)$ to be a basis of $T_pM^{0,1}$. In this case, we say that the self-conjugate coordinates $(z^1, \ldots, z^n, \overline{z}^1, \ldots, \overline{z}^n)$ are adapted to J at the point $p \in M$.

2 Almost holomorphic functions

By definition a function $f : U \to \mathbf{C}$, where U is an open subset of M, is called almost holomorphic or almost complex if

$$\overline{\partial} f = 0.$$

The above definition can be reformulated in the following equivalent form:

f is almost holomorphic iff $J^* df = idf$.

By J^* the action of J on differential forms is denoted, i.e. $(J^*\omega)X := \omega(JX)$, where X is a vector field and ω is a form.

For the proof of the equivalence it is enough to take in view that the exterior derivative d is decomposed as $d = \partial + \overline{\partial}$ over the space of smooth functions on M.

Remark. For an almost complex manifold (M, J) with non-integrable J, the above decomposition is not valid over differential (p, q)-forms on (M, J).

The following proposition is well-known.

Proposition 1. *The almost complex structure J of the almost complex manifold (M, J), $\dim_{\mathbf{R}} M = 2n$, is an integrable almost complex structure if and only if for every point $p \in M$, there is a neighbourhood U of p and almost holomorphic functions*

$$f_j : U \to \mathbf{C}, \qquad j = 1, \ldots, n,$$

which differentials at p, i.e. $d_p f_j$, $j = 1, \ldots, n$, are $\mathbf{C}$-linear independent.

Remark. Taking $(U; f_1, \ldots, f_n)$ as local co-ordinate system (as f_j are functionally independent on a neighbourhood of p), we obtain a local complex co-ordinate system $(U; z^1, \ldots, z^n)$, where $z^k = f_k$.

3 Spencer's co-ordinates

3.1. We shall use some mixed vector spaces as $\mathbf{C}^m \times \mathbf{R}^{2n-2m}$, where $0 < m < n$. In complex self-conjugate co-ordinates $\mathbf{C}^m \times \mathbf{R}^{2n-2m}$ is represented as $\mathbf{C}^m \times (\mathbf{C}^{n-m} \times \overline{\mathbf{C}}^{n-m})$. It can be considered as obtained from $\mathbf{C}^n \times \overline{\mathbf{C}}^n$ by complexification of $2m$ real co-ordinates. When we write $\mathbf{C}^m$ without $\overline{\mathbf{C}}^m$ we mean that $\mathbf{C}^m$ is a complex vector space, but on the other hand $\mathbf{C}^m \times \overline{\mathbf{C}}^m$ is a real vector space, isomorphic to $\mathbf{R}^{2m}$.

The affine varieties in $\mathbf{C}^m \times \mathbf{R}^{2n-2m}$ which don't intersect $\mathbf{C}^m$ (called parallel to $\mathbf{C}^m$) are denoted by $A\{x^{2m+1}, \ldots, x^{2n}\}$, i.e. $A\{x^{2m+1}, \ldots, x^{2n}\} = \mathbf{C}^m \times \{x^{2m+1}, \ldots, x^{2n}\}$, where x^j, $j = 2m+1, \ldots, 2n$, are fixed real numbers not all zero. We set $\mathbf{C}^m = \mathbf{C}^m \times \{0, \ldots, 0\} = A\{0, \ldots, 0\}$. Each $A\{x^{2m+1}, \ldots, x^{2n}\}$ is a complex affine variety isomorphic to $\mathbf{C}^m$. So the ambient space is exhausted by all $A\{x^{2m+1}, \ldots, x^{2n}\}$ and can be considered as a smooth family of complex affine spaces.

3.2. In the case $m = n$, if U is an open co-ordinate neighbourhood of M as differentiable manifold, and $f_1, \ldots, f_n$ are functionally independent almost holomorphic functions defined on U then we obtain local coordinate system $(U; z^1, \ldots, z^n)$ with $z^k = f_k$, $k = 1, \ldots, n$. In the next we adopt this remark for m, where $0 < m < n$.

3.3. We say that the almost complex manifold (M, J) is of (constant) type m, if for every point p of M there exists an open self-conjugate neighbourhood U on which there exists m, different functionally independent almost holomorphic functions.

It is to remark that the notion of the constant type is well defined in the category of almost complex manifolds. This follows from the fact that the composition of almost holomorphic mappings and the inverse of almost holomorphic diffeomorphism are almost holomorphic too.

Having in mind that $\mathbf{R}^{2n}$ is canonically isomorphic to $\mathbf{C}^n \times \overline{\mathbf{C}}^n$, we shall consider local co-ordinate systems (local charts) on M with values in $\mathbf{C}^n \times \overline{\mathbf{C}}^n$, i.e. : $\Phi_U : U \to \mathbf{C}^n \times \overline{\mathbf{C}}^n$, where U is defined above, $p \in U$, and $\Phi_U(p) = 0$. Let χ be the canonical isomorphism between $\mathbf{C}^n \times \overline{\mathbf{C}}^n$ and $\mathbf{C}^m \times (\mathbf{C}^{n-m} \times \overline{\mathbf{C}}^{n-m})$ determined by the partial complexification of $2m$ co-ordinates. The composition $\chi \circ \Phi_U$ defines a local chart with values in $\mathbf{C}^m \times (\mathbf{C}^{n-m} \times \overline{\mathbf{C}}^{n-m})$. Especially, we have this one defined by the sequence $(f_1, \dots, f_m)$ of m almost holomorphic functions, namely the chart ψ_U determined by

$$(f_1, \dots, f_m, z^{m+1}, \dots, z^n; \overline{z}^{m+1}, \dots, \overline{z}^n)$$

or

$$\Psi_U(z^j, \overline{z}^j) := (f_1, \dots, f_m) \times (\text{identity of } \mathbf{C}^{n-m} \times \overline{\mathbf{C}}^{n-m}), \qquad j = 1, \dots, n.$$

Setting $w^l = f_l(z^j, \overline{z}^j)$, $l = 1, \dots, m$, $w^k = z^{m+k}$, $k = 1, \dots, n-m$, we obtain new coordinates $(w^1, \dots, w^m, w^{m+1}, \dots, w^n, \overline{w}^{m+1}, \dots, \overline{w}^n)$ on U.

Lemma 1. *The matrix representation of J^* in each local self-conjugate co-ordinate system obtained by the almost holomorphic co-ordinate change*

$$\Psi_U(z^1, \dots, z^n) := (f_1, \dots, f_m, z^{m+1}, \dots, z^m),$$

where

$$f_j = f_j(z^1, \dots, z^n, \overline{z}^1, \dots, \overline{z}^n),$$

are functionally independent almost holomorphic functions, seems as follows

$$\begin{pmatrix} iE_m & * & 0 & * \\ 0 & * & 0 & * \\ 0 & * & -iE_m & * \\ 0 & * & 0 & * \end{pmatrix},$$

E_m being the unit $m \times m$ matrix.

Proof. The action of J^* on the basis $(dz^1, \dots, dz^m, dz^{m+1}, \dots, dz^n, d\overline{z}^1, \dots, d\overline{z}^m, \dots)$ of the cotangent bundle over the self-conjugate co-ordinate neighbourhood $U \subset M$ in which the first m co-ordinates are almost holomorphic functions gives directly the mentioned representation. Indeed,

$$J^* dz^k = i dz^k, \qquad k = 1, \dots, m. \qquad \square$$

Remark. The proof exposed above improves the proof of the same lemma in [2].

Consequences. The first m equations of the system

$$J^* df = i\, df$$

are just the conditions

$$\frac{\partial f}{\partial \overline{z}_j} = 0, \qquad j = 1, \dots, m.$$

3.4. The diffeomorphic image of $A(0, \dots, 0) \cap \Phi_U(U)$ by $\Phi_{U^{-1}}$ is a smooth submanifold of M as a differentiable manifold. We denote it by M_p, $p \in M$, as $\Phi(p) = \mathcal{O} \in A(0, \dots, 0) \cap \Phi_U(U)$.

The restrictions of $f_1, \dots, f_m$ on M_p defines a holomorphic co-ordinates on M_p as the following two lemmas (2 and 3) hold.

Lemma 2. *The restriction on M_p of each almost holomorphic on U function g is a holomorphic function of m complex variables.*

Proof. As $w^j = f_j$, $j = 1, \dots, m$, is a system of smooth functionally independent on U functions, we have $g = G(w^1, \dots, w^m)$ with $G \in C^\infty(U)$. But

$$\overline{\partial} G = \frac{\partial G}{\partial \overline{w}^1} d\overline{w}^1 + \cdots + \frac{\partial G}{\partial \overline{w}^m} d\overline{w}^m,$$

and in view of $\overline{\partial} G = \overline{\partial} g = 0$, we get $\partial G / \partial \overline{w}^j = 0$, $j = 1, \dots, m$. $\square$

Remark. In some more abstract form this lemma had been remarked by L. Apostolova [1] and K. Spallek (private communication).

Lemma 3. *Let $(w^1, \dots, w^m)$ and $(v^1, \dots, w^m)$ be two systems of holomorphic co-ordinates on M_p defined by two different systems of almost holomorphic on*

U systems $(f_1, \ldots, f_m)$ *and* $(g_1, \ldots, g_m)$. *Then there exists a bijective holomorphic transition mapping between the mentioned two co-ordinate systems.*

Proof. According to lemma 2 we have $v_j = G_j(w^1, \ldots, w^m)$, $j = 1, \ldots, n$, where G_j are holomorphic functions of $w^1, \ldots, w^m$. the system $G = (G_1, ..., G_m)$ defines the mentioned transition mapping as the differentials dG_j which are $\mathbf{C}$-linear independent. □

Recapitulating we obtain the following

Proposition 2. *On each paracompact almost complex manifold* (M, J) *of constant Spencer type* m *there exists a locally finite cover* $\{U_j\}$ *by self-conjugated Spencer's co-ordinate system* $(U_j, z_j^1, \ldots, z_j^m, \ldots)$ *such that in every intersection* $U_j \cap U_k$ *the holomorphic co-ordinates* $(z_j^1, \ldots, z_j^m)$ *change holomorphically in the other holomorphic co-ordinates* $(z_k^1, \ldots, z_k^m)$.

4 Local submersions and foliations

Having in mind the same notations as in the Section 3 above, we shall also consider the projection

$$\pi_m : \mathbf{C}^m \times (\mathbf{C}^{n-m} \times \overline{\mathbf{C}}^{n-m}) \to \mathbf{C}^m.$$

The composition $\pi_m \circ \Phi_U$ is a smooth submersion of U in $\mathbf{C}^m$, which is defined by the system of functionally independent functions $(f_1, \ldots, f_m)$. This submersion defines a smooth foliation of U such that each of its leaves can be equipped with holomorphic co-ordinates like for M_p. For this purpose an extension of lemma 2 and lemma 3 is necessary on the foliating leaves different from M_p. So we have smooth foliations which are disjoint families of smooth submanifolds (leaves), all of the same dimension m, which depends smoothly on a multidimensional $(2n - m)$ parameter. On each foliation leaf we have the supplementary structure of complex-analytic manifold defined by m complex ("holomorphic") co-ordinates.

5 Holomorphic functions on the leaves

Here we shall emphasize on the special leaf defined by the set of all submanifolds M_p, $p \in M$. We denote by N the submanifold of M defined by different M_p for all $p \in M$. This is the union of all M_p equipped with hole in which each two composant with non-empty intersection are identified by transition mappings described in lemma 3.

Denoting by f_U the map defined by the almost holomorphic functions $(f_1, \dots, f_m)$ we consider the glued map $f : M \to N$ defined by different f_U. This map is correctly defined according to lemma 3. It is almost holomorphic map as locally we have $f = (f_1, \dots, f_m)$ and $J^* df = (J^* df_1, \dots, J^* df_m) = i(df_1, \dots, df_m) = idf$. Supposing that the complex manifold admits non-trivial holomorphic function we denote by $\mathcal{O}_M$ the sheaf of them. Then we can define the inverse image of $\mathcal{O}_M$ by f, denoted by $\widehat{\mathcal{O}}_M$.

The inverse image of $\mathcal{O}_N$ with respect to the glued mapping $f : M \to N$ is defined as the sheaf determined by the following presheaf on M $(U, \mathcal{O}_M(U))$, where U is an open subset of M and

$$\mathcal{O}_M(U) = \{\, g \circ f_U \mid g \in \mathcal{O}_N(f_U(U)) \,\}.$$

According to general analytic sheaf theory the inverse image of a coherent analytic sheaf is coherent too. In the case considered above we have made more precise an analogous statement in [3]. For instance, on U we have a local foliation each leaf of which is a complex manifold of complex dimension m. So the restriction of $\mathcal{O}_M(U)$ on each of the mentioned leaves must be coherent. We received a smooth family of coherent analytic sheaves. The same is valid when we have a global smooth foliation of M by complex analytic leaves: $\mathcal{O}_M$ coincides with a smooth family of coherent complex analytic sheaves defined on the leaves of the global foliaion.

6 Holomorphic relations

Here we drop the condition about the existence of functionally independent systems of m almost holomorphic functions. A smooth $2n$ dimensional manifold is always denoted by M . Now we assume that for every point $p \in M$ there exists an open co-ordinate neighbourhood U, $p \in U$, with a system $(f_1, \dots, f_m)$ of different almost holomorphic functions.

Lemma 4. *Under the above formulated conditions, for every point $p \in M$ there exists a co-ordinate neighbourhood U, $p \in U$, and a holomorphic function $F_U(w^1, ..., w^m)$ of a number of complex variables, which defines a holomorphic relation for the corresponding system of almost holomorphic functions $f_j(z^k, \overline{z}^k)$, i.e. we have*

$$F_U(f_1, \dots, f_m) \equiv 0 \qquad \text{on } U.$$

Proof. We consider the system $w^j = f_j(z^k, \overline{z}^k)$ of m equations, $j = 1, \dots, m$. Now, as the functions f_j are functionally dependent on U, there exists a function $F_U \in \mathfrak{C}^\infty(U)$ such that $F_U(f_1, \dots, f_m) \equiv 0$ on U.

Taking the inverse functions

$$z^j = \varphi_j(w^1, \dots, w^m, \dots), \qquad j = 1, \dots, m,$$

with respect to the considered system of m equations, we can consider the function F_U as a function of the variables z^j, $j = 1, \dots, m$. But the inverse functions z^j are almost holomorphic with respect to w^j, $j = 1, \dots, m$, and so $\partial z^k / \partial \overline{w}^j = 0$ for $k = 1, \dots, m$. This implies that

$$\frac{\partial F_U}{\partial \overline{w}^j} = \frac{\partial F_U}{\partial z^1} \cdot \frac{\partial z^1}{\partial \overline{w}^j} + \cdots + \frac{\partial F_U}{\partial z^m} \cdot \frac{\partial z^m}{\partial \overline{w}^j} = 0,$$

i.e. F_U is holomorphic with respect to every w_j. □

Recapitulating we say that each almost complex manifold (M, J) equipped with Spencer's co-ordinates (not necessarily of constant type) is covered by a system of couples $(U_i,\ F_i)$, where U_i is a self-co-ordinate neighbourhood on M and F_i is a holomorphic relation on U_i as it is described in Lemma 4.

7 Foliations with singularities

Let p be a point of M and (U_i, F_i) be a couple of a self-conjugate neighbourhood U_i and a holomorphic relation F_i such that $p \in U_i$. We shall consider the corresponding analytic set in $\mathbf{C}^m$

$$\{ (w^1, \dots, w^m) \in \mathbf{C}^m \mid F_i(w^1, \dots, w^m) = 0 \}.$$

We denote this analytic W_i, $W_i \subset \mathbf{C}^m$. Now we suppose that the even natural number m is the minimal one such that the complex vector space $\mathbf{C}^m$, $m < n$, contains all analytic sets of the above described type independently of the index i.

The image of the point $p \in U_i$ by the self-co-ordinate Spencer mapping Φ_{U_j} lies in $W_i \times (\mathbf{C}^{n-m} \times \overline{\mathbf{C}}^{n-m})$, i.e. $\Phi_{U_j}(p) \in W_i$. We set $\Phi_{U_j} = \Phi_j$. Then, by $\widetilde{W_i} \mid \Phi_i(p)$ we denote the germ of W_i at the point $\Phi_i(p) \in W_i$.Denoting further by $\mathcal{F}(\mathbf{C}^m)$ the set of germs of all biholomorphisms of $\mathbf{C}^m$, we shall consider mappings of the kind

$$\Psi_{ij} : U_i \cap U_j \to \mathcal{F}(\mathbf{C}^m),$$

i.e. for each point $p \in U_i \cap U_j$, $\Psi_{ij}(p)$ is a germ of a biholomorphism in $\mathbf{C}^m$.

The following definition is suggested by [9].

Let (M, J) be an almost complex manifold covered by local Spencer self-co-ordinate systems (U_i, Φ_i). We say that an atlas of a generalized m-dimensional foliation is defined on (M, J) if there exists a system of quadruples

$$(U_i, \Phi_i, W_i, \Psi_{ij})$$

such that for each couple (i, j) and for each point $p \in U_i \cap U_j$ we have

$$\Psi_{ij}(p)(\widetilde{W}_i \mid \Phi(p)) = \widetilde{W}_j \mid \Phi_j(p).$$

Two atlases are called equivalent if they are subatlases of a third atlas. Each class of equivalent atlases of the kind mentioned defines a generalized m-dimensional foliation (or foliation with singularities) on (M, J).

8 Problems

The problem of the existence of self-conjugate Spencer co-ordinates on an almost complex manifold is largely open. Negative results are given by Ehresmann, by Herman [5], and others. Some partial results of positive character are given in [6] and [7].

The local co-ordinate expression of the fundamental system of almost holomorphic function $(J^* df = idf)$ contains a number of classical differential equations $(\frac{\partial f}{\partial \overline{z}^j} = 0)$ and some other linear equations in (z^j) and $(\overline{z}^j)$ with non-constant coefficients. We can consider this as a generalization of the classical $\bar{\partial}$-problem. A possible approach is the application of Spencer's theory of overdetermined systems of PDE.

The geometric description in terms of foliations described above leads us to a kind of complex analytic spaces (see [8]) depending on smooth parameters. There are some interconnections with Spallek's theory of vector fields with singularities.

Special attention merits the 4-dimensional case for smooth (M, J) with Spencer's co-ordinates $(z, w, \overline{w})$, where $z \in \mathbf{C}$, and $(w, \overline{w}) \in \mathbf{C} \times \overline{\mathbf{C}} \simeq \mathbf{R}^2$. In the case of constant type we have a smooth foliation of M by complex analytic 1-dimensional surfaces. Here a useful role can play the well known theorem of Thurston [4] according to which if TM admits a 2-plane subbundle F, then F is homotopic to a subbundle FM which is tangent to a smooth foliation of M.

Acknowledgments

One of the authors (S.D.) is too much obliged to his colleagues Lilia Apostolova and Sawa Manoff for their help in preparation of the manuscript. This work is partially supported by the National Science Foundation of Bulgaria under Grant No. $MM - 525/95$.

References

1. L.N. Apostolova, *Nearly Kähler manifolds are holomorphic foliations*, C. R. Acad. Bulg. Sci. **38**(1985), 977–979.
2. S. Dimiev and F. Succi, *Analytic almost complex structures*, in Almost Complex Structures, Proc. of the Intern. Workshop, eds. St. Dimiev, K. Sekigawa, World Scientific, Singapore, 1994, pp. 20–62.
3. S. Dimiev, *Problems in almost analytic geometry*, in Topics in Complex Analysis, Differential Geometry, and Mathematical Physics, eds. St. Dimiev, K. Sekigawa, World Scientific, Singapore, 1997, pp. 46–52.
4. W. Thurston, *Existence of co-dimension-one foliations*, Ann. of Math. **104**(1976), 249–268.
5. R. Hermann, *Complex homogeneous almost complex spaces of positive characteristic*, Trans. Amer. Math. Soc. **83**(1956).
6. S. Dimiev, *Proprietés locales des fonctions presque-holomorphes*, in Analytic Functions (Blazejewsko). Lecture Notes in Mathematics No. 1039, Springer Verlag, Berlin, 1983.
7. O. Mushkarov, *Existence of Holomorphic functions on almost complex manifolds*, Math. Z. **192**(1986), 283–295.
8. H. Grauert and R. Remmert, *Coherent analytic spheres*, Springer Verlag, Berlin, 1984.
9. D. Fuks, *Charakteristic classes of foliations*, Uspehi math. nauk (UMN) **28**(1973), 3–17.

HERMITIAN SYMMETRY OF INFINITESIMALLY TRIVIAL HARMONIC IMAGES OF COMPACT KÄHLER MANIFOLDS

AZNIV KASPARIAN
Kliment Ohridski University of Sofia,
Department of Mathematics and Informatics,
Section of Algebra,
Sofia - Bulgaria
E-mail: kasparia@fmi.uni-sofia.bg

Let $f : M \to \Gamma\backslash G/K$ be a harmonic map of a compact Kähler manifold M in a local Riemannian symmetric space $\Gamma\backslash G/K$ of noncompact type, such that the universal covering of $f(M)$ has trivial tangent bundle, i.e. $f(M)$ is infinitesimally trivial. The work provides sufficient conditions for Hermitian symmetry of $f(M)$. This allows to specify the harmonic images of infinitesimally maximum rank, when $G = SL(2n+1, \mathbf{R})$ or $G = SO(2p+1, n)$.

A real analytic map $f : M \to N$ of compact Riemannian manifolds is harmonic if it attains an extremum for the energy functional $E(f) = \frac{1}{2}\int_M \|df\|^2$, where the norm of $df \in (T^{\mathbf{R}}M)^* \otimes (T^{\mathbf{R}}N)$ is induced by the Riemannian metrics of M and N. Let $N = \Gamma \setminus G/K$ be a local Riemannian symmetric space of noncompact type and $\mathfrak{g} = \mathfrak{k} + \mathfrak{p}$ be the Cartan decomposition of $\mathfrak{g} = \mathrm{Lie}\, G$ with respect to $\mathfrak{k} = \mathrm{Lie}\, K$. One can view df as a section ω of the locally trivial bundle $\mathcal{H} := \mathrm{Hom}(T^{\mathbf{R}}M, f^*T^{\mathbf{R}}(\Gamma \setminus G/K))$ with a flat connection $D : \mathcal{H} \to \mathcal{A}^1_M(\mathcal{H})$, induced by the torsion-free metric connections on $T^{\mathbf{R}}M$ and $T^{\mathbf{R}}(\Gamma \setminus G/K)$. There is a decomposition $D = D' + D''$ into a sum of differential operators of types $(1,0)$ and $(0,1)$, as well as $\omega = \omega' + \omega''$ for $\omega \in \mathcal{A}^1_M(f^*T^{\mathbf{R}}N)$, into a sum of $\mathfrak{p}$-valued $(1,0)$- and $(0,1)$-forms.

Theorem 1 (Sampson [8]). *Let $f : M \to N$ be a harmonic map of a compact Kähler manifold M in a local Riemannian symmetric space $N = \Gamma\backslash G/K$ of noncompact type. Then:*

(i) $d_x f^{\mathbf{C}}(T^{1,0}_x M) \subset \mathfrak{p}^{\mathbf{C}} := \mathfrak{p} \otimes_{\mathbf{R}} \mathbf{C}$ are abelian subspaces for all $x \in M$ and

(ii) there holds the pluriharmonic equation $D''\omega' = 0$.

In general, the restrictions imposed by Theorem 1 allow infinite dimensional families of local pluriharmonic maps $f : U \to \Gamma \setminus G/K$, $U \subset M$.

1991 Mathematics Subject Classification: 58E20, 53C35, 22E46.

However, for f of maximum rank ${}_{\mathbf{C}}d_x f^{\mathbf{C}}$, Theorem 1 *(i)* implies infinitesimal rigidity, i.e., $d_x f^{\mathbf{C}}(T_x^{1,0}U) = k_x d_0 f^{\mathbf{C}}(T_0^{1,0}U)k_x^{-1}$ where U is a neighborhood of $0 \in M$, $f(0) = \breve{o}$, $x \in U$, $k_x \in K$ (cf.[1] and Lemma 4). Further, the pluriharmonic equation *(ii)* from Theorem 1 and the flatness $\nabla^2 = 0$ of the connection $\nabla = D + \omega$ yield local rigidity $k_x = 1, \forall x \in U$ (see [1] and Lemma 5). Due to the nonpositive sectional curvatures of G/K, the harmonic maps $f : M \to \Gamma \setminus G/K$ admit global harmonic liftings $\widetilde{f} : \widetilde{M} \to G/K$ to the universal covering $\widetilde{M}$ of M. Thus, the aforementioned rigidity results imply that $d\widetilde{f}^{\mathbf{C}}(T^{1,0}\widetilde{M}) \simeq \widetilde{f}(\widetilde{M}) \times d_{x_0}\widetilde{f}^{\mathbf{C}}(T_{x_0}^{1,0}\widetilde{M})$ $(x_0\pi_1(M) = 0)$ is a trivial complex vector bundle.

A harmonic image $f(M) \subset \Gamma \setminus G/K$ with a trivial $d\widetilde{f}^{\mathbf{C}}(T^{1,0}\widetilde{M})$, $\breve{o} \in \widetilde{f}(\widetilde{M})$ is called infinitesimally trivial. All $f(M)$ with contractible universal coverings $\widetilde{f}(\widetilde{M})$ are infinitesimally trivial. The present note (Theorem 2) provides sufficient conditions for Hermitian symmetry of infinitesimally trivial $f(M)$.

Let $G_i/K_i \subset G/K$, $i = 1, 2$, be transversally intersecting Hermitian symmetric subspaces with common origins $\breve{o}$, such that

$$d_{x_0}\widetilde{f}^{\mathbf{C}}(T_{x_0}^{1,0}\widetilde{M}) = \sum_{i=1}^{2} T_{\breve{o}}^{1,0}G_i/K_i$$

as a direct sum of $\mathbf{C}$-vector spaces. Recall from [1] that the adjoint action of a Cartan subalgebra $\mathfrak{h} \subset \mathfrak{t}$ splits $\mathfrak{t}^{\mathbf{C}} = \mathfrak{h}^{\mathbf{C}} + \sum_{\alpha \in \Delta_{\mathfrak{t}}} \mathfrak{t}^{\alpha}$ and $\mathfrak{p}^{\mathbf{C}} = \mathfrak{p}^{o} + \sum_{\alpha \in \Delta_{\mathfrak{p}}} \mathfrak{p}^{\alpha}$ into direct sums of 1-dimensional root spaces $\mathfrak{t}^{\alpha}$, $\mathfrak{p}^{\alpha}$ and the centralizer $\mathfrak{p}^{o}$ of $\mathfrak{h}$ in $\mathfrak{p}^{\mathbf{C}}$. If $T \subset K$ is the maximal torus with $\operatorname{Lie} T = \mathfrak{h}$ then $\mathfrak{h}^{\mathbf{C}}$, $\mathfrak{t}^{\alpha}$, $\mathfrak{p}^{\alpha}$, $\mathfrak{p}^{o}$ are $\operatorname{Ad} T$-submodules of $\mathfrak{g}^{\mathbf{C}}$. An equivariant Hermitian symmetric subspace $G_i/K_i \subset G/K$ is said to be $\operatorname{Ad} T$-invariant if the subalgebra $\mathfrak{g}_i^{\mathbf{C}} \subset \mathfrak{g}^{\mathbf{C}}$ is an $\operatorname{Ad} T$-submodule. More precisely, $T_i = T \cap G_i = T \cap K_i$ is a maximal torus of K_i and $\mathfrak{t}_i^{\mathbf{C}} = \operatorname{Lie} T_i \otimes_{\mathbf{R}} \mathbf{C} + \sum_{\alpha \in \Delta_{\mathfrak{t}_i}} \mathfrak{t}^{\alpha}$, $\mathfrak{p}_i^{\mathbf{C}} = \sum_{\alpha \in \Delta_{\mathfrak{p}_i}} \mathfrak{p}^{\alpha}$ are $\operatorname{ad}(\operatorname{Lie} T_i)$-decompositions for appropriate subsets $\Delta_{\mathfrak{t}_i} \subset \Delta_{\mathfrak{t}}$, $\Delta_{\mathfrak{p}_i} \subset \Delta_{\mathfrak{p}}$.

Theorem 2. *Let G/K be a Riemannian symmetric space of noncompact type, T be a maximal torus of K, $G_i/K_i \subset G/K$, $i = 1, 2$, be equivariant $\operatorname{Ad} T$-invariant Hermitian symmetric subspaces and $G_1 \cap G_2 = 1$. If $f : M \to \Gamma \backslash G/K$ is a harmonic map of a compact Kähler manifold M, whose image is infinitesimally trivial with $d\widetilde{f}^{\mathbf{C}}(T^{1,0}\widetilde{M}) \simeq \widetilde{f}(\widetilde{M}) \times (\sum_{i=1}^{2} T_{\breve{o}}^{1,0}G_i/K_i)$, then there is a Hermitian symmetric subspace $\hat{G}/\hat{K} \subset G/K$ and a complex structure on $\hat{G}/\hat{K}$ such that f is a holomorphic map onto a compact discrete quotient $f(M) = \hat{\Gamma}\backslash\hat{G}/\hat{K}$.*

Turning back to the harmonic maps, inducing abelian subspaces $d_x f^{\mathbf{C}}(T_x^{1,0}M) \subset \mathfrak{p}^{\mathbf{C}}$ of maximum dimension, Theorem 1 and Kostant's description [4] of the Lie algebra cohomologies of the holomorphic tangent spaces to the Hermitian symmetric spaces, provide the following:

Theorem 3 (Carlson and Toledo [1]). *Let G/K be an irreducible classical Riemannian symmetric space of noncompact type, $G \neq SL(2n+1, \mathbf{R})$, $G \neq SO(2p+1, n)$, $\operatorname{rank} G/K \geq 3$ and $\beta(G)$ be the maximum complex dimension of an equivariant $\operatorname{Ad} T$–invariant Hermitian symmetric subspace $G^h/K^h \subset G/K$ for some maximal torus $T \subset K$. If a harmonic map $f : M \to \Gamma \setminus G/K$ of a compact Kähler manifold M has constant $\operatorname{rank}_{\mathbf{C}} d_x f^{\mathbf{C}} := \dim_{\mathbf{C}} d_x f^{\mathbf{C}}(T_x^{1,0}M) = \beta(G)$, $\operatorname{rank}_{\mathbf{R}} d_x f := \dim_{\mathbf{R}} d_x f(T_x^{\mathbf{R}}M)$ for all $x \in M$, then f is a holomorphic map onto a compact discrete quotient $f(M) = \Gamma^h \setminus G^h/K^h$ of an equivariant $\operatorname{Ad} T$-invariant Hermitian symmetric subspace of $\dim_{\mathbf{C}} G^h/K^h = \beta(G)$.*

For $G = SL(2n+1, \mathbf{R})$ and $G = SO(2p+1, n)$ Kostant's result does not apply, since the abelian subspaces $d_x f^{\mathbf{C}}(T_x^{1,0}M) \subset \mathfrak{p}^{\mathbf{C}}$ do not produce 1-cocycles. The specific considerations from Sections 2 and 3 derive from Theorem 1 that the harmonic maps of maximum $\operatorname{rank}_{\mathbf{C}} d_x f^{\mathbf{C}}$ are associated with $d\widetilde{f}^{\mathbf{C}}(T^{1,0}\widetilde{M}) \simeq \widetilde{f}(\widetilde{M}) \times W$ for some $\operatorname{ad}\mathfrak{h}$-invariant abelian subspace $W \subset \mathfrak{p}^{\mathbf{C}}$. Combining with Theorem 2, which is elaborated in Section 1, one obtains:

Theorem 4. *Suppose that a harmonic map $f : M \to \Gamma\backslash G/K$ of a compact Kähler manifold M has constant $\operatorname{rank}_{\mathbf{C}} d_x f^{\mathbf{C}} := \dim_{\mathbf{C}} d_x f^{\mathbf{C}}(T_x^{1,0}M) = \alpha(G)$, $\operatorname{rank}_{\mathbf{R}} d_x f := \dim_{\mathbf{R}} d_x f(T_x^{\mathbf{R}}M)$, $\forall x \in M$, where $\alpha(G)$ is the maximum dimension of an abelian subspace of $\mathfrak{p}^{\mathbf{C}}$. If $G = SL(2n+1, \mathbf{R})$, $n \geq 4$, then either f is holomorphic and $f(M) = \hat{\Gamma}\backslash\hat{G}/\hat{K}$ is a compact discrete quotient of a Hermitian symmetric subspace $\hat{G}/\hat{K} \subset G/K$, or f is not holomorphic and $f(M) = \widetilde{\Gamma}\backslash(Sp(n, \mathbf{R})/U(n) \times T_1)$ for an equivariant Siegel upper half-space $Sp(n, \mathbf{R})/U(n) \subset SL(2n+1, \mathbf{R})/SO(2p+1, n)$, invariant under the adjoint action of a maximal torus $T \subset SO(2n+1)$ and a noncompact 1-dimensional torus $T_1 \subset SL(2n+1, \mathbf{R})$, centralizing $Sp(n, \mathbf{R})$. If $G = SO(2p+1, n)$, $\min(2p+1, n) \geq 10$, then f is holomorphic and $f(M) = \hat{\Gamma}\backslash\hat{G}/\hat{K}$ for a Hermitian symmetric $\hat{\Gamma}/\hat{K} \subset SO(2p+1, n)/SO(2p+1) \times SO(n)$.*

Let us note that the presence of equivariant non-holomorphic harmonic images $f(M) = \widetilde{\Gamma} \setminus (Sp(n, \mathbf{R})/U(n) \times T_1) \subset \Gamma \setminus SL(2n+1, \mathbf{R})/SO(2n+1)$ disproves Carlosn and Toledo's conjecture (cf.[1]) on the holomorphy of the harmonic maps $f : M \to \Gamma \setminus G/K$ of maximum complex rank. Theorem 4 extends also to the cases, treated in Theorem 3, as far as $\alpha(G) = \beta(G)$ for the classical simple noncompact groups $G \neq SL(2n+1, \mathbf{R})$, $G \neq SO(2p+1, n)$.

1 Infinitesimal and global Hermitian symmetry

The present section is devoted to the proof of Theorem 2. The crucial steps are the existence of compatible Iwasawa decompositions of G_i, G (Lemma 1) and the integrability of the distribution $G/K \times (\sum_{i=1}^{2} T_{\bar{o}}^{1,0} G_i/K_i)$ by a bounded circular domain (Lemma 2).

Lemma 1. *Under the assumptions of Theorem 2, there exist Iwasawa decompositions $G_i = K_i A_i N_i$ $(i = 1, 2)$, $G = KAN$ with $A_i = A \cap G_i$, $N_i = N \cap G_i$.*

Proof. Let $\mathfrak{g}_i = \mathfrak{t}_i + \mathfrak{p}_i$ be the Cartan decompositions of G_i/K_i, $\mathfrak{a}_i$ be maximal abelian subspaces of $\mathfrak{p}_i$ and $\mathfrak{h}_i$ be Cartan subalgebras of $\mathfrak{g}_i$, containing $\mathfrak{a}_i$. The $\mathrm{Ad}\, T$-invariance of G_i/K_i implies that $\mathfrak{p}_i \subset \mathfrak{p}$, $\mathfrak{t}_i \subset \mathfrak{t}$. Due to $G_1 \cap G_2 = 1$, the spaces $\mathfrak{a}_i$ are disjoint and $\mathfrak{a}_1 + \mathfrak{a}_2 \subset \mathfrak{p}$ extends to a maximal abelian subspace $\mathfrak{a} \subset \mathfrak{p}$. Similarly, $\mathfrak{h}_1 + \mathfrak{h}_2 \subset \mathfrak{g}$ extends to a Cartan subalgebra $\mathfrak{h}_o$ of $\mathfrak{g}$.. With respect to an appropriate ordering, the positive $\mathrm{ad}\, \mathfrak{h}_o$-root vectors of $\mathfrak{g}^{\mathbf{C}}$, which are not Killing orthogonal to $\mathfrak{a}$, determine nilpotent subalgebras $\mathfrak{n}^{\mathbf{C}} \subset \mathfrak{g}^{\mathbf{C}}$ and $\mathfrak{n} := \mathfrak{n}^{\mathbf{C}} \cap \mathfrak{g} \subset \mathfrak{g}$. That provides an Iwasawa decomposition $\mathfrak{g} = \mathfrak{t} + \mathfrak{a} + \mathfrak{n}$ into a direct sum of vector spaces (cf.[3]). Clearly, $\mathfrak{n}_i := \mathfrak{n} \cap \mathfrak{g}_i$ are $\mathrm{ad}\, \mathfrak{h}_i$-invariant nilpotent subalgebras of $\mathfrak{g}_i$. The direct sums $\mathfrak{t}_i + \mathfrak{a}_i + \mathfrak{n}_i$ are claimed to be Iwasawa decompositions of $\mathfrak{g}_i$. To this end, let $Z_{\rho_i} = \sum_{\sigma(\mathfrak{a})=0} z^\sigma Z_\sigma + A + \sum_{\rho(\mathfrak{a}) \neq 0, \rho > 0} z^\rho Z_\rho$ be an $\mathrm{ad}\, \mathfrak{h}_i$-root vector with $\rho_i(\mathfrak{a}_i) \neq 0$, $\rho_i > 0$, expressed by $\mathrm{ad}\, \mathfrak{h}$-root vectors Z_σ, Z_ρ and $A \in \mathfrak{a}$, $z^* \in \mathbf{C}$. If $\rho_i^* = \alpha_i + \beta_i$ with $\alpha_i \in \mathfrak{a}_i$, $\beta_i \in \mathfrak{t}_i \cap \mathfrak{h}_i$, then $\mathrm{ad}\, \alpha_i(Z_{\rho_i}) = \sum_\rho \rho(\alpha_i) z^\rho Z_\rho = \rho_i(\alpha_i) Z_{\rho_i}$ implies the vanishing of z^σ and A. Thus, the Iwasawa complement of $\mathfrak{t}_i + \mathfrak{a}_i$ in $\mathfrak{g}_i$ is contained in $\mathfrak{n}$ and, therefore, coincides with $\mathfrak{n}_i$. □

The global liftings $\widetilde{f}(\widetilde{M}) \subset G/K$ of the infinitesimally trivial harmonic images are studied by the means of the following

Lemma 2. *Under the assumptions of Theorem 2, let $\mathcal{D}_i \subset \mathbf{C}^{\beta_i}$, $\beta_i := \dim_{\mathbf{C}} G_i/K_i$ be the Harish-Chandra realizations of G_i/K_i as bounded symmetric domains. Then there is a bounded circular domain $\mathcal{D}_o \subset \mathcal{D}_1 \times \mathcal{D}_2 \subset \mathbf{C}^{\beta_1+\beta_2}$ with $T^{1,0}\mathcal{D}_o \simeq \mathcal{D}_o \times (\sum_{i=1}^{2} T_{\bar{o}}^{1,0} G_i/K_i)$, diffeomorphic to a subspace $\mathcal{D} \subset G/K$.*

Proof. The Iwasawa decompositions, constructed in Lemma 1, determine connected simply connected solvable subgroups $S_i := A_i N_i \subset G_i$, $S := AN \subset G$.

As far as the sectional curvatures of G/K are nonpositive, Cartan-Hadamard Theorem implies that $\exp_{\check{o}} : \mathfrak{p} \to G/K$ is a covering map. Then the simply connectedness of G/K requires $\exp_{\check{o}}$ to be bijective. The exponential map $\mathrm{Exp} : \mathfrak{p} \to G$ at $1 \in G$ is a global lifting of $\exp_{\check{o}}$, i.e., $\mathrm{Exp}\,(X)K = \exp_{\check{o}}(X)$ for all $X \in \mathfrak{p}$. Bearing in mind that $\mathrm{Exp}\,(\mathfrak{p}) \cap K = 1$, one identifies $\mathrm{Exp}\,(\mathfrak{p})$ with $\mathrm{Exp}\,(\mathfrak{p})K/K = G/K$.

Let us recall form [3] that there is a real analytic diffeomorphism $\Psi : S \to \mathrm{Exp}\,(\mathfrak{p})$. More precisely, if $\theta : G \to G$ is the group homomorphism, induced by the product of the complex conjugations of $G^{\mathbf{C}}$ with respect to the compact and noncompact real forms of $G^{\mathbf{C}}$ with Lie algebras $\mathfrak{t} + \sqrt{-1}\mathfrak{p}$, and, resp., $\mathfrak{t} + \mathfrak{p}$, then $\Psi(s) = \theta(s)s^{-1}$ for $s \in S$. Let $\Pi_{\mathfrak{p}_1+\mathfrak{p}_2} : \mathfrak{p} \to \mathfrak{p}_1 + \mathfrak{p}_2$ be the Killing orthogonal projection and $\mathcal{D} := \mathrm{Exp}\,\Pi_{\mathfrak{p}_1+\mathfrak{p}_2}\mathrm{Exp}^{-1}\Psi(S_1S_2) \subset \mathrm{Exp}\,(\mathfrak{p})$. If P_i are the parabolic subgroups of $G_i^{\mathbf{C}}$ with semisimple parts $K_i^{\mathbf{C}}$ then $\mathfrak{p}_i^+ \ni X \mapsto \mathrm{Exp}\,(X)P_i \in G_i^{\mathbf{C}}/P_i$ are biholomorphisms of $\mathfrak{p}_i^+$ onto dense open subsets of $G_i^{\mathbf{C}}/P_i$, containing $G_iP_i/P_i \simeq G_i/K_i$ (cf.[7]). Their inverses $\chi_i : G_i/K_i \to \mathfrak{p}_i^+$ are the Harish-Chandra realizations as bounded symmetric domains $\mathcal{D}_i = \chi_i(G_i/K_i)$ in $\mathfrak{p}_i^+ \simeq \mathbf{C}^{\beta_i}$. The diffeomorphism $\chi = (\chi_1\mathrm{Exp}\,, \chi_2\mathrm{Exp}\,)\mathrm{Exp}^{-1} : \mathcal{D} \to \mathfrak{p}_1^+ \times \mathfrak{p}_2^+$ is claimed to map onto a bounded circular domain $\mathcal{D}_o = \chi(\mathcal{D}) \subset \mathcal{D}_1 \times \mathcal{D}_2 \subset \mathbf{C}^{\beta_1+\beta_2}$.

Let us observe that $\chi_i\mathrm{Exp}$ are diffeomorphisms, so that $\mathcal{D}_o$ is a domain in $\mathbf{C}^{\beta_1+\beta_2}$ exactly when $\mathrm{Exp}^{-1}(\mathcal{D})$ is open in $\mathfrak{p}_1 + \mathfrak{p}_2$. To this end, it suffices to establish the injectiveness of $d\Pi_{\mathfrak{p}_1+\mathfrak{p}_2} : T^{\mathbf{R}}_{\Phi(s_1s_2)}\Phi(S_1S_2) \to \mathfrak{p}_1 + \mathfrak{p}_2$, where $\Phi := \mathrm{Exp}^{-1}\Psi$, $\forall s_i \in S_i$. For sufficiently small $r \in \mathbf{R}$ the vector fields $rZ_i + O(r^2) \in T^{\mathbf{R}}_{s_i}S_i \simeq \mathfrak{a}_i + \mathfrak{n}_i$ induce $r(\sum_{i=1}^2 Z_i) + O(r^2) \in T^{\mathbf{R}}_{s_1s_2}S_1S_2$. Decomposing $Z_i = Y_i + X_i$ for $Y_i \in \mathfrak{t}_i$, $X_i \in \mathfrak{p}_i$, one has $r(\sum_{i=1}^2 Y_i - X_i) + O(r^2) \in T^{\mathbf{R}}_{\theta(s_1s_2)}\theta(S_1S_2)$, $r(\sum_{i=1}^2 -Y_i - X_i) + O(r^2) \in T^{\mathbf{R}}_{(s_1s_2)^{-1}}(S_1S_2)^{-1}$, whereas $-2r(\sum_{i=1}^2 X_i) + O(r^2) \in T^{\mathbf{R}}_{\Psi(s_1s_2)}\Psi(S_1S_2)$ and $v(r) = -2r(\sum_{i=1}^2 X_i) + O(r^2) \in T^{\mathbf{R}}_{\Phi(s_1s_2)}\Phi(S_1S_2)$. Thus, $d\Pi_{\mathfrak{p}_1+\mathfrak{p}_2}v(r) = -2r(\sum_{i=1}^2 X_i) + O(r^2) = 0$ for all sufficiently small $r \in \mathbf{R}$ imples $X_i = 0$, $i = 1, 2$.

Towards the invariance of $\mathcal{D}_o$ under the circular action on $\mathcal{D}_1 + \mathcal{D}_2$, let us recall that the circular action on a bounded symmetric domain $\mathcal{D}_i$ coincides with the adjoint action of the center $\mathcal{C}_i$ of K_i. According to the root pattern, $\mathrm{Lie}\,\mathcal{C}_i$ is contained in any Cartan subalgebra of K_i. In particular, $\mathrm{Lie}\,\mathcal{C}_i \subset \mathfrak{h}_i$ and $\mathcal{C}_i \subset T$. The $\mathrm{Ad}\,T$-invariance of the Killing form of $\mathfrak{g}$ and the $\mathrm{Ad}\,(T \cap K_i)$-invariance of $\mathfrak{p}_i$ imply the commuting of $\Pi_{\mathfrak{p}_i}$ with the adjoint action of $\mathcal{C}_i$. Further, the group homomorphisms $\mathrm{Ad}\,c_i$, $c_i \in \mathcal{C}_i$ commute with the exponential map and the Harish-Chandra realizations, as well as with the group homomorphism θ. Consequently,

$$\begin{aligned}&\operatorname{Ad} c_i\{\chi_i \operatorname{Exp} \Pi_{\mathfrak{p}_i} \operatorname{Exp}^{-1}[\theta(s_1 s_2)(s_1 s_2)^{-1}]\}\\&= \chi_i \operatorname{Exp} \Pi_{\mathfrak{p}_i} \operatorname{Exp}^{-1}\{\theta[\operatorname{Ad} c_i(s_1)\operatorname{Ad} c_i(s_2)][\operatorname{Ad} c_i(s_1)\operatorname{Ad} c_i(s_2)]^{-1}\}.\end{aligned}$$

Finally, the $\operatorname{Ad}(\operatorname{Exp}\mathfrak{h}_i)$-invariance of S_1 and S_2 reveals that $\mathcal{D}_o$ is closed under the circular action on $\mathcal{D}_1 \times \mathcal{D}_2$. □

Here is the conclusion of the

Proof of Theorem 2. As far as the harmonic lifting $\widetilde{f} : \widetilde{M} \to G/K$ is real, the infinitesimal triviality $d\widetilde{f}^{\mathbf{C}}(T^{1,0}\widetilde{M}) \simeq \widetilde{f}(\widetilde{M}) \times \left(\sum_{i=1}^2 \mathfrak{p}_i^+\right)$ implies that $d\widetilde{f}^{\mathbf{C}}(T^{0,1}\widetilde{M}) \simeq \widetilde{f}(\widetilde{M}) \times \left(\sum_{i=1}^2 \mathfrak{p}_i^-\right)$. Consequently, $d\widetilde{f}^{\mathbf{C}}(T^{\mathbf{C}}\widetilde{M}) \simeq \widetilde{f}(\widetilde{M}) \times \left(\sum_{i=1}^2 \mathfrak{p}_i^{\mathbf{C}}\right)$. On the other hand, complexifying $T^{\mathbf{R}}\widetilde{f}(\widetilde{M}) = d\widetilde{f}(T^{\mathbf{R}}\widetilde{M})$ one gets $T^{\mathbf{R}}\widetilde{f}(\widetilde{M}) \otimes_{\mathbf{R}} \mathbf{C} = d\widetilde{f}^{\mathbf{C}}(T^{\mathbf{C}}\widetilde{M})$. Bearing in mind that the real structures of $\widetilde{f}(\widetilde{M}) \subset G/K$ and $G_i/K_i \subset G/K$ with $T^{\mathbf{R}}_{\check{o}}G_i/K_i = \mathfrak{p}_i$ are determined by the real form $G \subset G^{\mathbf{C}}$, one concludes that $T^{\mathbf{R}}\widetilde{f}(\widetilde{M}) \simeq \widetilde{f}(\widetilde{M}) \times \left(\sum_{i=1}^2 \mathfrak{p}_i\right)$. The invariant complex structures on G_i/K_i give rise to a complex structure J on $\widetilde{f}(\widetilde{M}) \times \left(\sum_{i=1}^2 \mathfrak{p}_i\right)$. That turns $\widetilde{f}(\widetilde{M})$ into a complex manifold. Viewing $T^{1,0}\widetilde{f}(\widetilde{M})$ as the $\sqrt{-1}$-eigen-subbundle of $T^{\mathbf{C}}\widetilde{f}(\widetilde{M})$ with respect to the action of J, one observes that $T^{1,0}\widetilde{f}(\widetilde{M}) \simeq \widetilde{f}(\widetilde{M}) \times \left(\sum_{i=1}^2 \mathfrak{p}_i^+\right)$ coincides with $d\widetilde{f}^{\mathbf{C}}(T^{1,0}\widetilde{M})$ and $\widetilde{f}$ is holomorphic.

In the notations from Lemma 2, the real analytic diffeomorphism $\chi : \mathcal{D} \to \mathcal{D}_o \subset \mathbf{C}^{\beta_1+\beta_2}$ induces a complex structure on $\mathcal{D}$, with respect to which $T^{1,0}\mathcal{D} \simeq \mathcal{D} \times (\sum_{i=1}^2 \mathfrak{p}_i^+)$. As maximal integral manifolds through the origin $\check{o} \in G/K$ for the integrable distribution $G/K \times (\sum_{i=1}^2 \mathfrak{p}_i^+)$, with Lie bracket induced from $\mathfrak{g}^{\mathbf{C}}$, the complex manifolds $\widetilde{f}(\widetilde{M})$ and $\mathcal{D}$ coincide. In particular, $\widetilde{f(M)} = \widetilde{f}(\widetilde{M})$ turns to be a bounded circular domain with a compact discrete quotient $f(M) = \pi_1 f(M) \setminus \widetilde{f(M)}$. The results of Vey from [10] imply that all such $\widetilde{f(M)} = \hat{G}/\hat{K}$ are bounded symmetric domains. □

In the situation from Theorem 2, if the Ricci curvature of M is nonnegative then $\hat{G}/\hat{K} \subset G/K$ is totally geodesic, according to a Bochner-type formula (cf. [2]). Then by the properness of f, this subspace is equivariant, so that G_i are subgroups of $\hat{G}$, $\hat{G} = G_1 \times G_2$ and $\hat{G}/\hat{K}$ is also $\operatorname{Ad} T$-invariant.

2 Abelian subspaces of maximum dimension

The characterization of the harmonic maps $f : M \to \Gamma \setminus G/K$ of maximum rank demands the study of the corresponding abelian subspaces $W = d_x f^{\mathbf{C}}(T_x^{1,0}M) \subset \mathfrak{p}^{\mathbf{C}}$.

For an arbitrary $X \in \mathfrak{g}^{\mathbf{C}}$ the decomposition of the endomorphism $\operatorname{ad} X$ of $\mathfrak{g}^{\mathbf{C}}$ into semisimple and nilpotent parts induces $X = X_\sigma + X_\nu$. According to Kostant and Rallis [5], if $X \in \mathfrak{p}^{\mathbf{C}}$ then $X_\sigma, X_\nu \in \mathfrak{p}^{\mathbf{C}}$. Bearing in mind that X_σ, X_ν are polynomials of X without constant terms, one observes that the abelian subspaces $W \subset \mathfrak{p}^{\mathbf{C}}$ extend to abelian subspaces $W_\sigma + W_\nu$ where $W_\sigma := \{X_\sigma | X \in W\}$, $W_\nu := \{X_\nu | X \in W\}$. Carlson and Toledo [1] establish that for any abelian subspace $W_\nu \subset \mathfrak{p}^{\mathbf{C}}$ of nilpotents, there is a Cartan subalgebra $\mathfrak{h}$ of $\mathfrak{t}$ and root decompositions $\mathfrak{t}^{\mathbf{C}} = \mathfrak{h}^{\mathbf{C}} + \sum_{\alpha \in \Delta_t} \mathfrak{t}^\alpha$, $\mathfrak{p}^{\mathbf{C}} = \mathfrak{p}^o + \sum_{\alpha \in \Delta_p} \mathfrak{p}^\alpha$ such that $W_\nu \subset \mathfrak{p}^+ := \sum_{\alpha \in \Delta_p^+} \mathfrak{p}^\alpha$.

Below are provided explicit matrix realizations of bases of $\mathfrak{h}$ and of $\operatorname{ad} \mathfrak{h}$-root vectors. Let us denote by $E_{k,l}$ the matrix units, i.e., the matrices whose single nonzero entry 1 is in the intersection of the k-th row and the l-th column. For $G = SL(m, \mathbf{R})$, $n := \left[\frac{m}{2}\right]$, $\varepsilon(m) := m - 2\left[\frac{m}{2}\right]$, the semisimple elements $H_i = E_{i,i} - E_{i+n,i+n}$, $1 \leq i \leq n$ form a basis of a Cartan subalgebra $\mathfrak{h}$ of $so(m)$. One choses compact root vectors $Y_{e_i+e_j} = sign(j-i)(E_{i,j+n} - E_{j,i+n})$, $Y_{-e_i-e_j} = sign(j-i)(E_{i+n,j} - E_{j+n,i})$ for $i < j$; $Y_{e_i-e_j} = E_{i,j} - E_{j+n,i+n}$ for $i \neq j$; $Y_{e_i} = \varepsilon(m)(E_{i,2n+1} - E_{2n+1,i+n})$, $Y_{-e_i} = \varepsilon(m)(E_{i+n,2n+1} - E_{2n+1,i})$ for $1 \leq i \leq n$. The centralizer $\mathfrak{p}^o$ of $\mathfrak{h}$ in $\mathfrak{p}^{\mathbf{C}}$ has a basis $X_i^o = \varepsilon(m)(E_{i,i} + E_{i+n,i+n} - 2E_{2n+1,2n+1})$, $1 \leq i \leq n$, of semisimple elements of $\mathfrak{p}$. The noncompact root vectors $X_{e_i+e_j} = E_{i,j+n} + E_{j,i+n}$, $X_{-e_i-e_j} = E_{i+n,j} + E_{j+n,i}$ for $i < j$; $X_{e_i-e_j} = E_{i,j} + E_{j+n,i+n}$ for $i \neq j$; $X_{2e_i} = E_{i,i+n}$, $X_{-2e_i} = E_{i+n,i}$, $X_{e_i} = \varepsilon(m)(E_{i,2n+1} + E_{2n+1,i+n})$, $X_{-e_i} = \varepsilon(m)(E_{i+n,2n+1} + E_{2n+1,i})$, for $1 \leq i \leq n$.

In the case of $G = SO(m,n)$ put $p := \left[\frac{m}{2}\right]$, $q := \left[\frac{n}{2}\right]$, $s := p + q$ and arrange the weights, pertaining to $so(m)$ and $so(n)$ in sets S_0, S_1 of $\operatorname{card} S_0 = p$, $\operatorname{card} S_1 = q$. The matrices $H_i = E_{i,i} - E_{i+s,i+s}, i \in S_0 \cup S_1$ constitute a basis of a Cartan subalgebra $\mathfrak{h}$ of $so(m) + so(n)$. As compact root vectors one choses $Y_{e_i-e_j} = E_{i,j} - E_{j+s,i+s}$ for $i \neq j$; $Y_{e_i+e_j} = sign(j-i)(E_{i,j+s} - E_{j,i+s})$, $Y_{-e_i-e_j} = sign(j-i)(E_{j+s,i} - E_{i+s,j})$ for $i < j$; $Y_{e_i} = \varepsilon_i(E_{i,\delta_i(2s+1)} - E_{\delta_i(2s+1),i+s})$, $Y_{-e_i} = \varepsilon_i(E_{\delta_i(2s+1),i} - E_{i+s,\delta_i(2s+1)})$, where $i,j \in S_0$ or $i,j \in S_1$; $\delta_i := k$ for $i \in S_k$; $\varepsilon_i := \varepsilon(m)$ for $i \in S_0$, $\varepsilon_i := \varepsilon(n)$ for $i \in S_1$. The centralizer $\mathfrak{p}^o$ of $\mathfrak{h}$ in $\mathfrak{p}^{\mathbf{C}}$ is spanned by the semisimple $X^o = \varepsilon(m)\varepsilon(n)(E_{0,2s+1} + E_{2s+1,0}) \in \mathfrak{p}$. As noncompact root vectors serve $X_{e_i-e_\xi} = E_{i,\xi} + E_{\xi+s,i+s}$, $X_{e_i+e_\xi} = E_{i,\xi+s} + E_{\xi,i+s}$, $X_{-e_i-e_\xi} = E_{\xi+s,i} + E_{i+s,\xi}$, $X_{e_i} = \hat{\varepsilon}_i(E_{i,(1-\delta_i)(2s+1)} + E_{(1-\delta_i)(2s+1),i+s})$,

$X_{-e_i} = \hat{\varepsilon_i}(E_{(1-\delta_i)(2s+1),i} + E_{i+s,(1-\delta_i)(2s+1)})$, where $i \in S_0$, $\xi \in S_1$ or $i \in S_1$, $\xi \in S_0$, $\hat{\varepsilon_i} := \varepsilon(n)$ for $i \in S_0$, $\hat{\varepsilon_i} := \varepsilon(m)$ for $i \in S_1$.

According to Malčev's method of the leading root vectors (cf.[6]), any abelian subspace $W_\nu \subset \mathfrak{p}^+$ admits a basis $Z_i = X_{\alpha_i} + T_i$ with $\alpha_1 < \alpha_2 < \cdots < \alpha_d$ and "tails" $T_i \in \mathrm{Span}_{\mathbf{C}}\{X_\alpha | \alpha > \alpha_i, \alpha \neq \alpha_j \text{ for all } j\}$. The lexicographic ordering of the roots is compatible with the addition, so that $[Z_i, Z_j] = 0$ force $[X_{\alpha_i}, X_{\alpha_j}] = 0$. Thus, any abelian subspace $W_\nu \subset \mathfrak{p}^+$ of nilpotents is associated with a root system $C = C(W_\nu) = \{\alpha_1, \ldots, \alpha_d\}$ of $\mathrm{card}\, C = \dim_{\mathbf{C}} W_\nu$ with the property $\alpha, \beta \in C \Longrightarrow \alpha + \beta \notin \Delta_{\mathfrak{t}}$. Such $C \subset \Delta_{\mathfrak{p}}^+$ are called commutative root systems after the corresponding $\mathrm{ad}\,\mathfrak{h}$-invariant abelian subspaces $W(C) = \mathrm{Span}_{\mathbf{C}}\{X_\alpha | \alpha \in C\} \subset \mathfrak{p}^+$.

Lemma 3. *With respect to some maximal torus $T \subset K$ with $\mathfrak{h} = \mathrm{Lie}\, T$, let G^h/K^h be an equivariant $\mathrm{Ad}\, T$-invariant Hermitian symmetric subspace of maximum dimension $\beta(G)$ and $\nu(G)$ be the maximum dimension of an abelian subspace $W_\nu \subset \mathfrak{p}^+$ of nilpotents. Then*

(i) Any commutative root system $\Sigma \subset \Delta_{\mathfrak{p}}^+(SL(m, \mathbf{R}))$ of $\mathrm{card}\, \Sigma \geq \frac{n(n+1)}{2}$, $n := \left[\frac{m}{2}\right]$, is $Aut(\mathfrak{h})$-equivalent to a subset of the commutative root system

$$C_{SL(m,\mathbf{R})} = \{\, e_i + e_j, \varepsilon(m) e_n \mid 1 \leq i \leq j \leq n \,\}$$

of $\mathrm{card}\, C_{SL(m,\mathbf{R})} = \nu(SL(m, \mathbf{R})) = \frac{n(n+1)}{2} + \varepsilon(m)$. All Hermitian symmetric G^h/K^h of $\dim_{\mathbf{C}} G^h/K^h = \beta(SL(m, \mathbf{R})) = \frac{n(n+1)}{2}$ are isomorphic to the Siegel upper half-space $\mathcal{S}_n = Sp(n, \mathbf{R})/U(n)$ with $T_{\check{o}}^{1,0}\mathcal{S}_n = \mathrm{Span}_{\mathbf{C}}\{X_{e_i+e_j} | 1 \leq i \leq j \leq n\} =: W^h_{SL(m,\mathbf{R})}$. The centralizer $\mathcal{C}_{\mathfrak{p}^{\mathbf{C}}}(sp(n, \mathbf{C})) = \mathcal{C}_{\mathfrak{p}^o}(W^h_{SL(m,\mathbf{R})})$ is of dimension $\varepsilon(m)$.

(ii) For $G = SO(m, n)$ with $\min(m, n) \geq 10$, $p := \left[\frac{m}{2}\right]$, $q := \left[\frac{n}{2}\right]$, any commutative root system $\Sigma \subset \Delta_{\mathfrak{p}}^+$ of $\mathrm{card}\, \Sigma \geq pq$ is $Aut(\mathfrak{h})$-equivalent to a subset of the commutative root system

$$C_{SO(m,n)} = \{\, e_i + e_\xi, \varepsilon(n) e_t, \varepsilon(m) e_\tau \mid i \in S_0, \xi \in S_1 \,\},$$

where $t := \max\{i \in S_0\}$, $\tau := \max\{\xi \in S_1\}$ and $\mathrm{card}\, C_{SO(m,n)} = \nu(SO(m, n)) = pq + \varepsilon(m) + \varepsilon(n)$. All Hermitian symmetric G^h/K^h of $\dim_{\mathbf{C}} G^h/K^h = \beta(SO(m, n)) = pq$ are isomorphic to the generalized ball $\mathcal{B}(p, q) = SU(p, q)/S(U(p) \times U(q))$ with $T_{\check{o}}^{1,0}\mathcal{B}(p, q) = \mathrm{Span}_{\mathbf{C}}\{X_{e_i+e_\xi} \mid i \in S_0, \xi \in S_1\} =: W^h_{SO(m,n)}$. The centralizer $\mathcal{C}_{\mathfrak{p}^{\mathbf{C}}}(sl(p+q, \mathbf{C})) = \mathcal{C}_{\mathfrak{p}^o}(W^h_{SO(m,n)})$ is of dimension $\varepsilon(m)\varepsilon(n)$.

Proof. An induction on n reveals that up to a permutation of e_i, one has $\Sigma \subset C_{SL(m,\mathbf{R})}$ if $\mathrm{card}\, \Sigma > \frac{(n-1)(n+2)}{2}$. An induction on $p + q$ implies that up to permutations of S_0, S_1 and some sign changes of e_i, $\Sigma \subset C_{SO(m,n)}$ whenever card $\Sigma \geq pq$.

Viewing G^h/K^h as a universal covering of a compact Kähler manifold $\Gamma^h \setminus G^h/K^h$, immersed in some $\Gamma \setminus G/K$, one applies Theorem 1 *(i)* to conclude that $T^{1,0}_{\bar{o}} G^h/K^h$ is an ad $\mathfrak{h}$-invariant abelian subspace of $\mathfrak{p}^+$. The only such $T^{1,0}_{\bar{o}} G^h/K^h \subset W(C_G)$ with Lie triple systems $T^{\mathbf{R}}_{\bar{o}} G^h/K^h \subset \mathfrak{p}$ are the announced W^h_G.

After observing that the centralizer of the Cartan subalgebra $\mathfrak{h} \cap \mathrm{Lie}\, G^h \subset \mathrm{Lie}\, K^h$ in $\mathfrak{p}^{\mathbf{C}}$ is $\mathfrak{p}^o$, one notes that $\mathcal{C}_{\mathfrak{p}^o}(W^h_G)$ is spanned by $\sum_{i=1}^n X^o_i$ for $G = SL(m,\mathbf{R})$ or, resp., by X^o for $G = SO(m,n)$. □

Now we are ready to deal with the main result of the present section.

Lemma 4. *In the notations from Lemma 3, let $W(C_G) = \{X_\alpha | \alpha \in C_G\} \subset \mathfrak{p}^+$ and $W(G) \subset \mathfrak{p}^{\mathbf{C}}$ be an arbitrary abelian subspace of maximum $\dim_{\mathbf{C}} W(G) = \alpha(G)$, defined over $\mathbf{R}$. Then any $W(SL(m,\mathbf{R}))$ with $m \geq 8$ is $SO(m)$-conjugate to $W(C_{SL(m,\mathbf{R})})$ or $W^h_{SL(m,\mathbf{R})} + \mathcal{C}_{\mathfrak{p}^o}(W^h_{SL(m,\mathbf{R})})$, and any $W(SO(m,n))$ with $\min(m,n) \geq 10$ is $SO(m) \times SO(n)$-conjugate to $W(C_{SO(m,n)})$. In both cases, $\alpha(G) = \nu(G)$.*

Proof. Decomposing $W(G) = W_\nu + W_\sigma$, one computes the reductive centralizer $\mathcal{C}_{\mathfrak{g}}(W^{\mathbf{R}}_\sigma)$ of $W^{\mathbf{R}}_\sigma$, $W_\sigma = W^{\mathbf{R}}_\sigma \otimes_{\mathbf{R}} \mathbf{C}$, in $\mathfrak{g}$ and views $W(G)$ as a subspace of $\mathcal{C}_{\mathfrak{g}}(W^{\mathbf{R}}_\sigma) \otimes_{\mathbf{R}} \mathbf{C}$. There follows $W(G) = W(G)_\nu$ of $\dim_{\mathbf{C}} W(G)_\nu = \nu(G)$ with the only exception of $W(SL(2n+1,\mathbf{R})) = W(SL(2n,\mathbf{R}))_\nu + W(SL(2n+1,\mathbf{R}))^1_\sigma$ for a complex line $W(SL(2n+1,\mathbf{R}))^1_\sigma \subset \mathfrak{p}^{\mathbf{C}}$, generated by a semisimple element.

First, we establish the $K^{\mathbf{C}}$-conjugacy of $W(G)_\nu$ to $W(C_G)$. According to Lemma 3 and Malčev's method of the leading root vectors, $W(G)_\nu$ can be endowed with a basis

$$Z_{e_i+e_j} = X_{e_i+e_j} + \sum_{k<l} z^{k,-l}_{i,j} X_{e_k-e_l} + \sum_{k<n} z^k_{i,j} X_{e_k},$$

$$Z_{e_n} = X_{e_n} + \sum_{k<l} z^{k,-l}_n X_{e_k-e_l} + \sum_{k<n} z^k_n X_{e_k},$$

where $z^*_* \in \mathbf{C}$. Let $Supp Z_\alpha = \{X_\alpha\} \cup \{X_\beta \in \mathfrak{p}^+ | z^\beta_\alpha \neq 0\}$ be the support of Z_α and $\Pi_+ : \mathfrak{t}^{\mathbf{C}} \to \sum_{i<j} \mathfrak{t}^{e_i+e_j}$, $\Pi_- : \mathfrak{t}^{\mathbf{C}} \to \sum_{i\neq j} \mathfrak{t}^{e_i-e_j}$, $\Pi_0 : \mathfrak{t}^{\mathbf{C}} \to \sum_i \mathfrak{t}^{e_i}$, $\Pi_\alpha : \mathfrak{t}^{\mathbf{C}} \to \mathfrak{t}^\alpha$ ($\alpha \in \Delta_{\mathfrak{t}}$) be Killing orthogonal projections.

The adjoint action of $\mathrm{Exp}\,(-\varepsilon(m) \sum_{k<n} z^k_n Y_{e_k-e_n})$ on $W(SL(m,\mathbf{R}))_\nu$ removes $\varepsilon(m) X_{e_i}$, $i < n$ from $Supp(\varepsilon(m) Z_{e_n})$. Gaussian eliminations of $X_\beta \in Supp Z_\alpha$, $\beta \in C_{SL(m,\mathbf{R})} - \{\alpha\}$ preserve the leading root system $C_{SL(m,\mathbf{R})}$.

The vanishing of the projections $\Pi_+[Z_{e_n}, Z_{e_i+e_j}] = 0$, $\Pi_+[Z_{e_i+e_j}, Z_{e_k+e_n}] = 0$ for $i \leq j < n$ and then for $i = j = n$ implies that $Z_\alpha = X_\alpha$, $\forall \alpha \in C_{SL(m,\mathbf{R})}$. If $W(SL(2n+1,\mathbf{R})) = W(SL(2n,\mathbf{R}))_\nu + W(SL(2n+1,\mathbf{R}))^1_\sigma$, one applies the previous considerations to $W(SL(2n,\mathbf{R}))_\nu$ and puts together with Lemma 3 *(i)*, in order to obtain the announced result.

For $G = SO(m,n)$, the conjugation by $\mathrm{Exp}\,(-\sum_{\theta\in\{t,\tau\}}\sum_{i\in S_\delta-\{\theta\}} z^i_\theta Y_{e_i-e_\theta})$ eliminates X_{e_i}, $i \in S_0$ from $SuppZ_{e_t}$ and X_{e_ξ}, $\xi \in S_1$ from $SuppZ_{e_\tau}$. If $\max(t,\tau) \in S_\delta$, for some $\delta \in \{0,1\}$, one lets $\lambda := \min(t,\tau)$ and acts by $\mathrm{Ad\,Exp}\,(\sum_{\lambda<i\in S_\delta} z_\lambda^{\lambda,-i} Y_{-e_i})$. That provides $X_{e_t-e_\zeta} \notin SuppZ_{e_t}$, $\zeta \in S_1$ and $X_{e_\tau-e_k} \notin SuppZ_{e_\tau}$, $k \in S_0$ for the sake of $Z_\alpha \in \mathrm{Span}_{\mathbf{C}}\{X_\beta, X_{-e_u}, X^o | \beta \in \Delta^+_{\mathfrak{p}}, u \in S_{1-\delta}, X^o \in \mathfrak{p}^o\}$. Preserving $C_{SO(m,n)}$, one exploits $\Pi_+[Z_{e_t}, Z_{e_\tau}] = 0$ and $\Pi_+[Z_{e_\theta}, Z_{e_i+e_\xi}] = 0$ with $\theta \in \{t,\tau\}$. Further, $\Pi_+[Z_{e_i+e_\xi}, Z_{j+e_\eta}] = 0$ for $i \neq j$, $\xi \neq \eta$ and then for $i = j$ or $\xi = \eta$ reveals that $Z_{e_i+e_\xi} = X_{e_i+e_\xi} + \sum_{i<\xi} \mathfrak{z}^\zeta_\xi X_{e_i-e_\zeta} + \sum_{\xi<k} \mathfrak{z}^k_i X_{e_\xi-e_k}$ with skew-symmetric $\mathfrak{z}^\zeta_\xi$, $\mathfrak{z}^k_i$. The action of $\mathrm{Ad\,Exp}\left(\sum_{\zeta>\xi} \mathfrak{z}^\zeta_\xi Y_{-e_\xi-e_\zeta} + \sum_{k>i} \mathfrak{z}^k_i Y_{-e_i-e_k}\right)$ produces

$$Z_{e_t} = X_{e_t} + \delta \sum_k z_t^{-k} X_{-e_k}, Z_{e_\tau} = X_{e_\tau} + (1-\delta) \sum_\zeta z_\tau^{-\zeta} X_{-e_\zeta},$$

$$Z_{e_i+e_\xi} = X_{e_i+e_\xi} + \sum_{k,\zeta} z_{i,\xi}^{-k,-\zeta} X_{-e_k-e_\zeta}.$$

Then $\Pi_{e_j-e_k}[Z_{e_i+e_\xi}, Z_{e_j+e_\eta}] = 0$ with $i \neq j$ specify that $Z_{e_i+e_\xi} = X_{e_i+e_\xi}$ and $[Z_{e_\theta}, Z_{e_i+e_\xi}] = 0$ with $\theta \in \{t,\tau\}$ imply that $Z_{e_\theta} = X_{e_\theta}$.

The $K^{\mathbf{C}}$-conjugacy of $W(G)_\nu$ with $W(C_G)$, suffices for their K-conjugacy, according to the coincidence of the $\mathrm{Ad}\,K^{\mathbf{C}}$- and $\mathrm{Ad}\,K$-orbits of $W(C_G)$. That, in turn, is derived from the parabolicity of the $\mathrm{Ad}\,K^{\mathbf{C}}$-stabilizer of $W(C_G)$. □

3 Harmonic images of maximum dimension

Having at hand the infinitesimal rigidity result of Lemma 4, the present section derives the following local rigidity property:

Lemma 5. *In the notations from Lemma 3, let M be a compact Kähler manifold, $\widetilde{f} : \widetilde{M} \to G/K$ be the global lifting of a harmonic map $f : M \to \Gamma\backslash G/K$ of constant* $\mathrm{rank}_{\mathbf{C}} d_x f^{\mathbf{C}} = \alpha(G)$, $\mathrm{rank}_{\mathbf{R}} d_x f$ *for all $x \in M$. If $G = SL(2n+1,\mathbf{R})$,*

$n \geq 4$, then

$$d\widetilde{f}^{\mathbf{C}}(T^{1,0}\widetilde{M}) \simeq \widetilde{f}(\widetilde{M}) \times T^{1,0}_{\breve{o}}\left[\mathcal{S}_n \times SL(2,\mathbf{R})/SO(2)\right] \text{ or}$$

$$d\widetilde{f}^{\mathbf{C}}(T^{1,0}\widetilde{M}) \simeq \widetilde{f}(\widetilde{M}) \times \left[T^{1,0}_{\breve{o}}\mathcal{S}_n \times (\operatorname{Lie} T_1 \otimes_{\mathbf{R}} \mathbf{C})\right]$$

where the torus $T_1 \subset SL(2n+1,\mathbf{R})$ is the 1-dimensional centralizer of $Sp(n,\mathbf{R})$ in the Killing orthogonal complement of $SO(2n+1)$. If $G = SO(m,n)$, $\min(m,n) \geq 10$, then

$$d\widetilde{f}^{\mathbf{C}}(T^{1,0}\widetilde{M}) \simeq \widetilde{f}(\widetilde{M}) \times T^{1,0}_{\breve{o}}\left[\mathcal{B}\left(\left[\frac{m}{2}\right],\left[\frac{n}{2}\right]\right) \times (SL(2,\mathbf{R})/SO(2))^{\varepsilon(m)+\varepsilon(n)}\right].$$

Proof. For a sufficiently small neighborhood $U \subset M$ of $0 \in M$, $f(0) = \Gamma\breve{o}$, one can assume that $f(U) \subset G/K$. Then the real tangent bundle $T^{\mathbf{R}}f(U) \simeq f(U) \times \mathfrak{p}$ and $\mathcal{H}_U := \operatorname{Hom}(T^{\mathbf{R}}U, U \times \mathfrak{p}) \simeq \mathcal{A}^1_U(\mathfrak{p})$. On the bundle $\mathcal{H}_o := \operatorname{Hom}(T^{\mathbf{R}}U, U \times \mathfrak{g}) \simeq \mathcal{A}^1_U(\mathfrak{g})$, containing $\mathcal{H}_U$, there is an action of $\omega \in \mathcal{A}^1_U(\mathfrak{p})$ by exterior multiplication of the forms and Lie bracket of the vector fields. The torsion-free metric connections on $T^{\mathbf{R}}M$ and $T^{\mathbf{R}}G$ give rise to a flat connection $D : \mathcal{H}_o \to \mathcal{A}^1_U(\mathcal{H}_o)$, extending the restriction of $D : \mathcal{H} \to \mathcal{A}^1_M(\mathcal{H})$ from Theorem 1 to U. Their sum $\nabla := D + \omega : \mathcal{H}_o \to \mathcal{A}^1_U(\mathcal{H}_o)$ is a flat G-invariant metric connection. Let $\mathcal{W}(x) := d_x f^{\mathbf{C}}(T^{1,0}_x M) \subset \mathfrak{p}^{\mathbf{C}}$ and $\mathcal{W}^{\mathbf{R}}(x) = d_x f(T^{\mathbf{R}}_x M) \subset \mathfrak{p}$ for $x \in U$. According to Lemma 4, the abelian spaces $\mathcal{W}(x) = k_x W(C_G)k_x^{-1}$ or $\mathcal{W}(x) = k_x(W^h_G + \mathcal{C}_{\mathfrak{p}^o}(W^h_G))k_x^{-1}$ where $k_x \in K$. Towards the local rigidity of $\mathcal{W}$, i.e., $k_x = 1$ for all $x \in U$, sufficiently close to $0 \in U$, it suffices to establish that $\Pi_{\mathfrak{p}} \nabla (\mathcal{W}^{\mathbf{R}}(x))|_{x=0} \subseteq \mathcal{W}^{\mathbf{R}}(0)$.

Let K_1 be the $\operatorname{Ad} K$-stabilizer of $\mathcal{W}(0)$, $\mathfrak{t}_1 = \operatorname{Lie} K_1$ and $\mathfrak{t} = \mathfrak{t}_1 + \mathfrak{m}$, $\mathfrak{p} = \mathcal{W}^{\mathbf{R}}(0) + \mathfrak{q}$ be Killing orthogonal decompositions. The $\mathfrak{t}$-valued connection D splits into a sum $D = D_1 + \mu$ of a $\mathfrak{t}_1$-valued connection D_1 and a real 1-form $\mu = \mu' + \overline{\mu'}$, $\mu' \in \mathcal{A}^{1,0}_U(\mathfrak{m}^{\mathbf{C}})$. The local rigidity follows from the vanishing of μ'. To this end are exploited the pluriharmonic equation $D''\omega' = 0$ and the $(2,0)$-part of the $\mathfrak{p}$ -valued component of the flatness equation $\nabla^2 = 0$, stating $D'\omega' = 0$.

In the case of $G = SL(2n+1,\mathbf{R})$ and $\mathcal{W}(0)_\sigma \neq 0$, one introduces local holomorphic coordinates $x^{ij} \equiv x^{ji}$ $(1 \leq i \leq j \leq n)$, x^o on U. Then $\omega' = \sum_{i \leq j} dx^{ij} \otimes X_{e_i+e_j} + dx^o \otimes (\sum_i X^o_i)$, $K_1 = U(n) \times U(1)$, so that $\mu' = \sum_{i<j} \mu^{i,j} Y_{e_i+e_j} + \sum_{i<j} \mu^{-i,-j} Y_{-e_i-e_j} + \sum_i \mu^i Y_{e_i} + \sum_i \mu^{-i} Y_{-e_i}$ with $\mu^* \in \mathcal{A}^{1,0}_U = \sum_{i \leq j} \mathcal{O}_U dx^{ij} + \mathcal{O}_U dx^o$, where $\mathcal{A}^1_U$ is regarded as a free module over the structure sheaf $\mathcal{O}_U$. The decomposition $\mathfrak{p}^{\mathbf{C}} = \mathcal{W}(0) + \overline{\mathcal{W}(0)} + \mathfrak{q}^{\mathbf{C}}$ induces Killing orthogonal projections $\Pi^*_{\mathfrak{q}} : \mathcal{A}^2_U(\mathfrak{p}^{\mathbf{C}}) \to dx^* \otimes \mathcal{A}^1_U(\mathfrak{q}^{\mathbf{C}})$. There hold $\Pi^*_{\mathfrak{q}} D''\omega' = \Pi^*_{\mathfrak{q}}[\mu'', \omega'] = 0$ and $\Pi^*_{\mathfrak{q}} D'\omega' = \Pi^*_{\mathfrak{q}}[\mu', \omega'] = 0$. In the case under consideration, $\Pi^{jk}_{\mathfrak{q}}[\mu', \omega'] = 0$ implies that $\mu^{-i,-j} \in \mathcal{O}_U dx^{jk}$ for at least

two different $k \notin \{i,j\}$, so that $\mu^{-i,-j} = 0$. Combining with $\Pi^o_{\mathfrak{q}}[\mu'', \omega'] = 0$ one concludes that $\mu' = 0$.

For $G = SL(2n+1, \mathbf{R})$ and $\mathcal{W}(0) = \mathcal{W}(0)_\nu$, one has $\omega' = \sum_{i\leq j} dx^{ij} \otimes X_{e_i+e_j} + dx^n \otimes X_{e_n}$, $K_1 \simeq U(n-1) \times U(1)^2$, $\mu' = \sum_{\sigma\in\{\pm\}} (\sum_{i<j} \mu^{\sigma i, \sigma j} Y_{\sigma e_i + \sigma e_j} + \sum_i \mu^{\sigma i} Y_{\sigma e_i} + \sum_{i<n} \mu^{\sigma i, -\sigma n} Y_{\sigma e_i - \sigma e_n})$. Then $\Pi^n_{\mathfrak{q}}[\mu'', \omega'] = 0$ implies the vanishing of μ^i and $\mu^{-i,n}$, $\Pi^{ij}_{\mathfrak{q}}[\mu'', \omega'] = 0$ forces $\mu^{i,j} = 0$ and $\Pi^{ij}_{\mathfrak{q}}[\mu', \omega'] = 0$ requires μ^{-i}, $\mu^{-i,-k} \in \mathcal{O}_U dx^{ij}$ for at least two different $j \neq i$, whereas $\mu^{-i} = 0$, $\mu^{-i,-k} = 0$. Note that for $\mathfrak{p}^\alpha \subset \mathcal{W}(0)$ with $\alpha \notin \Delta_{\mathfrak{t}_1} + (e_i + e_j)$ the Killing orthogonal projection $\Pi^{ij}_\alpha D''\omega' = \Pi^{ij}_\alpha[\mu'', \omega'] = 0$ does not depend on D_1. Thus, $\Pi^{ij}_{e_j+e_n}[\mu'', \omega'] = 0$ supplies $\mu^{i,-n} = 0$, proving $\mu' = 0$.

If $G = SO(m,n)$ then $\omega' = \sum_{i\in S_\delta} \sum_{\xi \in S_{1-\delta}} dx^{i\xi} \otimes X_{e_i+e_\xi} + dx^t \otimes X_{e_t} + dx^\tau \otimes X_{e_\tau}$, the stabilizer $K_1 = U\left(\left[\frac{m}{2}\right] - \varepsilon(n)\right) \times U\left(\left[\frac{n}{2}\right] - \varepsilon(m)\right) \times U(1)^{2+\varepsilon(n)+\varepsilon(m)}$, and $\mu' = \sum_{\delta\in\{0,1\}} \sum_{\sigma\in\{\pm\}} \{\sum_{i<j\in S_\delta} \mu^{\sigma i, \sigma j} Y_{\sigma e_i + \sigma e_j} + \varepsilon[\delta n + (1-\delta) m] \sum_{i \in S_\delta} \mu^{\sigma i} Y_{\sigma e_i} + \varepsilon[\delta m + (1-\delta) n] \sum_i \mu^{\sigma i, -\sigma\theta} Y_{\sigma e_i - \sigma e_\theta}\}$, where the last summation is over $i \in S_\delta - \{\theta\} = S_\delta - \{t, \tau\}$. The annihilation of $\mu^{i,j}$, μ^i follows from $\Pi^{i\xi}_{\mathfrak{q}}[\mu'', \omega'] = 0$, while $\mu^{-i,\theta} = 0$ is a consequence of $\Pi^\theta_{\mathfrak{q}}[\mu'', \omega'] = 0$. One infers from $\Pi^{i\xi}_{\mathfrak{q}}[\mu', \omega'] = 0$ that $\mu^{-i,-j}, \mu^{-i} \in \mathcal{O}_U dx^{i\xi}$ for at least two different $\xi \neq \theta$, and concludes that $\mu^{-i,-j} = 0$, $\mu^{-i} = 0$. Bearing in mind that $e_\theta + e_\xi \notin \Delta_{\mathfrak{t}_1} + (e_i + e_\xi)$ for $\theta \in S_\delta$, $\xi \in S_{1-\delta}$, one exploits $\Pi^{i\xi}_{e_\theta + e_\xi}[\mu'', \omega'] = 0$ towards $\mu^{i,-\theta} = 0$ and obtains $\mu' = 0$. □

In the case of $d\widetilde{f}^{\mathbf{C}}(T^{1,0}\widetilde{M}) \simeq \widetilde{f}(\widetilde{M}) \times T^{1,0}_{\tilde{o}}[G^h/K^h \times (SL(2,\mathbf{R})/SO(2))^s]$, the Theorem 4 follows immediately from Lemma 5 and Theorem 2. For $G = SL(2n+1, \mathbf{R})$ and $d\widetilde{f}^{\mathbf{C}}(T^{1,0}\widetilde{M}) \simeq \widetilde{f}(\widetilde{M}) \times \left[T^{1,0}_{\tilde{o}} \mathcal{S}_n \times (\mathrm{Lie}\, T_1 \otimes_{\mathbf{R}} \mathbf{C})\right]$, one observes that $\mathcal{S}_n \times T_1$ is a maximal integral manifold of $T^{\mathbf{R}} \widetilde{f}(\widetilde{M})$ and concludes the proof of Theorem 4.

Theorem 3 and Theorem 4 generalize to the following

Corollary 1. *Let G_ρ/K_ρ be an irreducible classical Riemannian symmetric space of noncompact type and* $\mathrm{rank}\, G_\rho/K_\rho = \rho \geq 10$, *$\alpha(G_{\rho-\sigma})$ be the maximum dimension of an abelian subspace of $\mathfrak{p}(G_{\rho-\sigma})^{\mathbf{C}}$ for some $\sigma \in \mathbf{Z}$, $0 \leq \sigma \leq \rho$. If a harmonic map $f : M \to \Gamma \setminus G/K$ of a compact Kähler manifold M has constant* $\mathrm{rank}_{\mathbf{C}} d_x f^{\mathbf{C}} = \alpha(G_{\rho-\sigma}) + \sigma$, $\mathrm{rank}_{\mathbf{R}} d_x f = 2\alpha(G_{\rho-\sigma}) + \sigma$, *then $f(M) \simeq \hat{\Gamma} \setminus \left(\hat{G}/\hat{K} \times \hat{T}\right)$ for a Hermitian symmetric $\hat{G}/\hat{K} \subset G/K$ of* $\dim_{\mathbf{C}} \hat{G}/\hat{K} = \alpha(G_{\rho-\sigma})$ *and a noncompact torus $\hat{T} \subset G$ of* $\dim_{\mathbf{R}} \hat{T} = \sigma$.

Acknowledgements

The author is extremely grateful to Prof. Paul Gauduchon for the discussion of related topics, to her colleaques from the Section of Algebra, Institute of Mathematics and Informatics, Bulgarian Academy of Sciences, and from the Section of Mathematical Analysis, Department of Mathematics and Informatics, Kliment Ohridski University of Sofia, for the encouragement and the computer equipment for the preparation of the present article.

This work is partially supported by Section of Algebra, Institute of Mathematics and Informatics, Bulgarian Academy of Sciences.

References

1. J. Carlson, D. Toledo, *Rigidity of harmonic maps of maximum rank*, Journ. Geom. Anal. **3**(1993), 99–140.
2. J. Eells, J. Sampson, *Harmonic mappings of Riemannian manifolds*, Amer. Journ. Math. **86**(1964), 109–160.
3. S. Helgason, *Differential Geometry, Lie Groups, and Symmetric Spaces*, Academic Press, London, 1978.
4. B. Kostant, *Lie algebra cohomology and the generalized Borel-Weil Theorem*, Ann. Math. **74**(1961), 329–387.
5. B. Kostant, S. Rallis, *Orbits and representations associated with symmetric spaces*, Amer. Jorn. Math. **93**(1971), 753–809.
6. A. Malčev, *Commutative subalgebras of semisimple Lie algebras*, Izv. Akad. Nauk USSR **9**(1945), 125–133 (in Russian).
7. N. Mok, *Metric Rigidity Theorems on Hermitian Locally Symmetric Manifolds*, in World Scientific Series: Pure Math., vol. 6, World Scientific, Singapore, 1989.
8. J. Sampson, *Applications of harmonic maps to Kähler geometry*, Contemp. Math. **49**(1986), 125–133.
9. Y. Siu, *Complex analyticity of harmonic maps and strong rigidity of compact Kähler manifolds*, Ann. Math. **112**(1980), 73–111.
10. J. Vey, *Sur la division des domaines de Siegel*, Ann. Scient. Ec. Norm. Sup. **3**(1970), 479–506.

ON THE INTEGRABILITY CONDITIONS FOR ALMOST CONTACT MANIFOLDS

MILEN J. HRISTOV
Department of Mathematics, "Vasil Levski" Land Forces Academy,
5000 Veliko Tarnovo - Bulgaria
E-mail: milenjh@yahoo.com

A parallel between the conditions of Newlender-Nierenberg for integrability for almost complex structures and the integrability conditions for almost contact structures is given. A class of partially integrable and a class of integrable almost contact metric manifolds in the sence of the classification scheme of Ganchev-Alexiev are given.

1 Almost contact metric manifolds

Almost contact structurte on smooth $2n+1$-dimensional manifold M^{2n+1} is a triple (φ, ξ, η), where φ is a smooth tensor field of type $(1,1)$, ξ a smooth vector field, η a differential form, such that

$$\varphi^2 = -id + \eta \otimes \xi, \quad \eta(\xi) = 1$$

and as consequences $\eta \circ \varphi = 0$, $\varphi(\xi) = 0$, rank $\varphi = 2n$.

$M^{2n+1}(\varphi, \xi, \eta)$ is said to be *almost contact manifold*. If on $M^{2n+1}(\varphi, \xi, \eta)$ there exists Riemmannian definit metric g, compatible with the structure [1]:

$$g \circ \varphi = g - \eta \otimes \eta,$$

then the quadruple (φ, ξ, η, g) is call *almost contact metric structure*, and the manifold $M^{2n+1}(\varphi, \xi, \eta, g)$, *almost contact metric manifold*. On $M^{2n+1}(\varphi, \xi, \eta, g)$ arise the following objects:

— distributions:

Every one fiber T_pM of TM is $2n+1$-dimensional euclidian space with structure $(\varphi_p, \xi_p, \eta_p, g_p)$. The decomposition $T_pM = \{D_p = \mathrm{Ker}\,\eta_p\} \oplus \{\xi_p\}$, orthogonal and $U(n) \times 1$-invariant, implies the existence of the following distributions:

$D = \{\, D_p, p \in M \,\}$, *contact distribution*, and (D_p, φ_p) are Hermitian vector spaces.

$\{\xi\} = \{\, \mathrm{Im}\,(\eta_p).\xi_p, p \in M \,\}$, *vertical distribution*.

— operators:

$$h = -\varphi^2 : hTM = D, \quad v = \eta \otimes \xi : vTM = \{\xi\},$$
$$A : TM \to D : AX = \nabla_X \xi.$$

— tensor fields

$\Phi : \Phi(x, y) = g(x, \varphi y)$, x, $y \in D$, fundamental 2-form for the structure; $\nabla\varphi$; $\nabla\eta$; $\nabla\Phi = -F$ (∇ is the Levi-Civita connection) [2,3]; $d\eta$; $d\Phi$; $\mathcal{L}_\xi g$; $\mathcal{L}\varphi$; $\mathcal{L}\eta$ ($\mathcal{L}$.is the Lie-differentiation);

$N = [\varphi, \varphi] + 2d\eta \otimes \xi$ is the generalized Nijenhuis tensor field of type (1,2), $[\varphi, \varphi]$ is the usual Nijenhuis tensor field of type (1,2), formed with φ :

$$[\varphi, \varphi](X, Y) = [\varphi X, \varphi Y] - \varphi[\varphi X, Y] - \varphi[X, \varphi Y] + \varphi^2[X, Y],$$

and the corresponding to N tensor field of type (0,3): $N(X, Y, Z) = g(N(X, Y), Z)$, X, Y, $Z \in T_pM$.

— differential forms:

The 1-forms f_p, f_p^* and ω_p, associated to F_p, defined by

$$f_p(X) = g^{ij} F_p(e_i, e_j, X), \quad f_p^*(X) = g^{ij} F_p(e_i, \varphi e_j, X), \quad \omega_p(x) = F_p(\xi, \xi, x),$$

where $\{e_1, \dots, e_{2n+1}\}$ is an orthonormal bases of T_pM, define the differential forms f, f^* and ω respectively and

$$\delta\eta : \delta\eta_p = -f_p^*(\xi_p), \quad \delta\Phi : \delta\Phi_p = f_p + \omega_p.$$

The differential 1-form $\theta = -\dfrac{\delta\eta}{2n}.\eta + \dfrac{1}{2(n-1)}.\delta\Phi \circ \varphi$ is said to be *Lee-form* for the structure.

The properties, formulated in the following lemma are well known[3].

Lemma 1.1. *Let $M^{2n+1}(\varphi, \xi, \eta, g)$ be almost almost contact metric manifold. Then:*

(a) $(\nabla\varphi) \circ \varphi = -\varphi \circ (\nabla\varphi) + (\nabla\eta) \otimes \xi + A \otimes \eta$;

(b) $(\nabla_X\eta)Y = g(AX, Y) = F(X, \varphi Y, \xi)$:

(c) $F(X, Y, Z) = g[(\nabla_X\varphi)Y] = -F(X, Z, Y)$
$= -F(X, \varphi Y, \varphi Z) + [\eta \wedge (\nabla_X\eta) \circ \varphi](Y, Z)$;

(d) $F(X, \varphi Y, Z) = F(X, Y, \varphi Z) + \eta(Y)(\nabla_X\eta)Z + \eta(Z)(\nabla_X\eta)Y$;

(e) $(\mathcal{L}_\xi g)(X, Y) = (\nabla_X\eta)Y + (\nabla_Y\eta)X$;

(f) $2d\eta(X, Y) = (\nabla_X\eta)Y - (\nabla_Y\eta)X$;

(g) $2d\eta(\xi, X) = (\mathcal{L}_\xi g)(\xi, X) = -\omega(\varphi X) = g(H, X)$;

(h) $3d\Phi(X, Y, Z) = -\underset{(X,Y,Z)}{\sigma} F(X, Y, Z)$;

(i) $3d\Phi(\xi, Y, Z) = -F(\xi, Y, Z) + (\nabla_Y\eta)\varphi Z - (\nabla_Z\eta)\varphi Y$;

(j) $N(X,Y,Z) = -3d\Phi(\varphi X,Y,Z) + 3d\Phi(\varphi Y,X,Z) + 2F(Z,Y,\varphi X)$
$+ 2[\eta\otimes(\nabla\eta - 2d\eta)](X,Y,Z);$

(k) $f^*_{\bar{d}} = g^{\alpha\beta}(F_{\alpha\bar{\beta}d} - F_{\bar{\alpha}\beta d}) = f_{\bar{d}} - \omega_{\bar{d}} = -3g^{\alpha\beta}\Phi_{\alpha\bar{\beta}d} = 2(n-1)\theta_d + 2\eta_{0d},$

$f^*_{\bar{0}} = g^{\alpha\beta}(F_{\alpha\bar{\beta}0} - F_{\bar{\alpha}\beta 0}) = -3g^{\alpha\beta}\Phi_{\alpha\bar{\beta}0} = 2n\theta_0, \quad f_0 = g^{\alpha\beta}(F_{\alpha\beta 0} + F_{\bar{\alpha}\bar{\beta}0}),$

where $\{e_\alpha,\ e_{\bar{\alpha}} = \varphi e_\alpha,\ e_0 = \xi,\ \alpha \in \{1,\dots,n\}\}$ *is local orthonormal* φ*-bases of* $T_pM,\ p \in M$.

In [2] Ganchev-Alexiev have introduced the 12 basic classes of almost contact metric manifolds, by using orthogonal, $U(n)\times 1$-invariant and irreducible decomposition of the space of tensors of type (0,3) having the symmetries of $F = -\nabla\Phi$, Lemma 1.1. c). These 12 classes play the same rule as the four basic classes of almost Hermitian manifolds, introduced by Gray and Hervella[4].

2 Complexification of the tangential distribution. Characterisation of the twelve basic classes of almost contact metric manifolds

The natural complexification of the tangential distribution TM of almost contact metric manifold $M^{2n+1}(\varphi,\xi,\eta,g)$ is the direct summ:

$$T^CM = D^C \oplus \{\xi\},$$

where D^C is the complexification of the almost Hermitian distribution ($D = \operatorname{Ker}\eta;\varphi$) and $\{\xi\} = (\operatorname{Im}\eta)$. ξ is the vertical distribution. We have $D^C = D^{10}\oplus D^{01}$, where $D^{10}(D^{01})$ is $+i(-i)$-eigenspace of φ, $i=\sqrt{-1}$ and

$$D^{10} = \operatorname{span}\{Z_\alpha = e_\alpha - i\varphi e_\alpha\}_{\alpha\in I}, \quad D^{01} = \operatorname{span}\{Z_{\bar{\alpha}} = e_\alpha + i\varphi e_\alpha\}_{\bar{\alpha}\in\bar{I}},$$

$$D^C = \operatorname{span}\{Z_a\}_{a\in I\cup\bar{I}}, \quad T^CM = \operatorname{span}\{Z_A\}_{A\in I\cup\bar{I}\cup I_0},$$

where $\{e_\alpha;\varphi e_\alpha;\xi,\ \alpha\in I\}$ is an orthonormal φ-bases of TM. The index sets I, $\bar{I}$, I_0 are $\{1,\dots,n\}$, $\{\bar{1},\dots,\bar{n}\}$, $\{0\}$ respectively and $Z_0 = \xi$.

In the following lemmas we shall describe in terms of the essential (i.e. non zero in general) complex components with respect to the comlpex bases $\{Z_A\}_{A\in I\cup\bar{I}\cup I_0}$, the **C**-linear continuations of the fundamental tensors, which will be denoted by the same letters. All the lemmas follow directly from Lemma 1.1. and the definitions of the corresponding tensors.

At first we note that the essential components of:

— the metric g are $g_{\alpha\bar{\beta}} = \overline{g_{\bar{\alpha}\beta}}$, $g_{00} = 1$, and (g^{AB}) denotes the matrics, inverce to (g_{AB}):$g_{\alpha\bar{\beta}}g^{\bar{\beta}\mu} = \delta^{\mu}_{\alpha}$ (Kronecker delta);
— the form η is $\eta_0 = 1$;
— the covariant derivative of $\Phi - F$:

$$F_{ABC} = F(Z_A, Z_B, Z_C) = -F_{ACB} = \overline{F_{\bar{A}\bar{B}\bar{C}}}$$

$$= -F_{A\tilde{B}\tilde{C}} + (\eta \wedge F_{A0*})_{BC} \quad \text{are} \quad F_{A\beta\gamma},\ F_{A\beta 0}.$$

Here $(F_{A0*})_B = F_{A0B}$, $F_{A\tilde{B}\tilde{C}} = F(Z_A, \varphi Z_B, \varphi Z_C)$ and φZ_A, simply denoted as $Z_{\tilde{A}}$, are as follows $\varphi Z_\alpha = iZ_\alpha$, $\varphi Z_{\bar{\beta}} = -iZ_{\bar{\beta}}$, $\varphi Z_0 = 0$.

Lemma 2.1. *The essential components of* $d\eta$:

$$\eta_{AB} = d\eta(Z_A, Z_B) = -\eta_{BA} = \overline{\eta_{\bar{A}\bar{B}}} = \frac{1}{2}(F_{A\tilde{B}0} - F_{B\tilde{A}0}), \quad \textit{are}$$

$$2\eta_{\alpha\bar{\beta}} = -i(F_{\alpha\bar{\beta}0} + F_{\bar{\beta}\alpha 0}), \quad 2\eta_{\alpha\beta} = i(F_{\alpha\beta 0} - F_{\beta\alpha 0}), \quad 2\eta_{\alpha 0} = i\omega_\alpha.$$

Lemma 2.2. *The essential components of* $d\Phi$:

$$\Phi_{ABC} = d\Phi(Z_A, Z_B, Z_C) = \overline{\Phi_{\bar{A}\bar{B}\bar{C}}} = -3 \underset{(A,B,C)}{\mathfrak{S}} F_{ABC}, \quad \textit{are}$$

$$3\Phi_{\alpha\beta\gamma} = -\underset{(\alpha,\beta,\gamma)}{\mathfrak{S}} F_{\alpha\beta\gamma}, \quad 3\Phi_{\bar{\alpha}\beta\gamma} = -F_{\bar{\alpha}\beta\gamma},$$

$$3\Phi_{0\beta\gamma} = 2i\eta_{\beta\gamma} - F_{0\beta\gamma}, \quad 3\Phi_{0\bar{\beta}\gamma} = 2i\eta_{\bar{\beta}\gamma} + F_{\gamma\bar{\beta}0}.$$

We put $[Z_A, Z_B] = C^C_{AB} Z_C$; $C^C_{AB} = -C^C_{BA} = \overline{C^{\bar{C}}_{\bar{A}\bar{B}}}$. Obviously $C^0_{AB} = -2\eta_{AB}$. We denote by Φ^C_{AB} and η^B_A the contractions $g^{SC}\Phi_{ABS}$ and $g^{SB}\eta_{AS}$, respectively.

Lemma 2.3. *The essential components of the Nijenhuis tensor of type* (1,2):

$$N(Z_A, Z_B) = N^C_{AB} Z_C; \quad N^C_{AB} = -N^C_{BA} = \overline{N^{\bar{C}}_{\bar{A}\bar{B}}}$$

and of the corresponding tensor of type (0,3):

$$N_{ABC} = N^S_{AB} g_{SC} = -N_{BAC} = \overline{N_{\bar{A}\bar{B}\bar{C}}}$$

$$= -3\Phi_{\bar{A}BC} + 3\Phi_{\bar{B}AC} + 2\begin{vmatrix} \eta_A & \eta_B \\ \eta_{AC} & \eta_{BC} \end{vmatrix} + 2F_{CB\bar{A}} - 2\eta_{AB}F_{C0\bar{B}}$$

satisfy the following equations:

(a) $$\begin{pmatrix} [Z_\alpha, Z_\beta] \\ [Z_\alpha, Z_{\bar{\beta}}] \\ [Z_\alpha, Z_0] \end{pmatrix} = \begin{pmatrix} C^\sigma_{\alpha\beta} & -\frac{1}{4}N^{\bar{\sigma}}_{\alpha\beta} & -\frac{1}{2}N^0_{\alpha\beta} \\ C^\sigma_{\alpha\bar{\beta}} & C^{\bar{\sigma}}_{\alpha\bar{\beta}} & -2\eta_{\alpha\bar{\beta}} \\ C^\sigma_{\alpha 0} & -\frac{1}{4}N^{\bar{\sigma}}_{\alpha 0} & -N^0_{\alpha 0} \end{pmatrix} \begin{pmatrix} Z_\sigma \\ Z_{\bar{\sigma}} \\ Z_0 \end{pmatrix};$$

(b) $N_{\alpha\beta\gamma} = -6i\Phi_{\alpha\beta\gamma} + 2iF_{\gamma\beta\alpha} = -12i\Phi_{\alpha\beta\gamma} + (N_{\gamma\beta\alpha} - N_{\gamma\alpha\beta})$,
$N_{\alpha\beta 0} = N^0_{\alpha\beta} = 4\eta_{\alpha\beta} = -6i\Phi_{\alpha\beta 0} + 2iF_{0\beta\alpha} = -2C^0_{\alpha\beta}$,
$N_{\alpha 0\gamma} = -3i\Phi_{\alpha 0\gamma} - 2\eta_{\alpha\gamma} + 2iF_{\gamma 0\alpha}$,
$N_{\alpha 00} = N^0_{\alpha 0} = 2\eta_{\alpha 0} = -C^0_{\alpha 0}$.

Lemma 2.4. *The essential components of* $d\Phi$: Φ_{ABC} *and the coefficients* C^C_{AB} *are related by:*

(a) $3i\Phi_{\overline{\alpha}\beta\gamma} = Z_\beta g_{\overline{\alpha}\gamma} - Z_\gamma g_{\overline{\alpha}\beta} - C^\sigma_{\beta\gamma} g_{\sigma\overline{\alpha}} + \begin{vmatrix} C^{\overline{\sigma}}_{\overline{\alpha}\beta} & C^{\overline{\sigma}}_{\alpha\gamma} \\ g_{\beta\overline{\sigma}} & g_{\gamma\overline{\sigma}} \end{vmatrix};$

(b) $3i\Phi_{\alpha\bar{\beta}0} = Z_0 g_{\alpha\bar{\beta}} + C^{\bar{\sigma}}_{\bar{\beta}0} - C^\sigma_{0\alpha} g_{\sigma\bar{\beta}};$

(c) $3i\Phi_{\alpha\beta\gamma} = -\frac{1}{4} \underset{(\alpha,\beta,\gamma)}{\mathfrak{S}} N_{\alpha\beta\gamma};$

(d) $3i\Phi_{\alpha\beta 0} = \frac{1}{2}(N_{\alpha 0\beta} - N_{\beta 0\alpha}).$

Lemma 2.5. *The essential components of* $\mathcal{L}g$:

$$(\mathcal{L}g)_{ABC} = (\mathcal{L}_{Z_A} g)(Z_B, Z_C) = Z_A g_{BC} - C^S_{AB} g_{CS} - C^S_{AC} g_{BS}$$

$$= (\mathcal{L}g)_{ACB} = \overline{(\mathcal{L}g)_{\bar{A}\bar{B}\bar{C}}}, \quad \textit{are}$$

$(\mathcal{L}g)_{\alpha\beta\gamma} = \frac{1}{4}(N_{\alpha\beta\gamma} + N_{\alpha\gamma\beta}),$ $\quad (\mathcal{L}g)_{\alpha\beta 0} = \frac{1}{2}(N_{\alpha\beta 0} + N_{\alpha 0\beta}),$

$(\mathcal{L}g)_{\alpha 00} = 2N_{\alpha 00},$ $\quad (\mathcal{L}g)_{0\alpha\overline{\beta}} = 3i\Phi_{0\alpha\overline{\beta}},$

$(\mathcal{L}g)_{0\alpha\beta} = \frac{1}{2}(N_{0\alpha\beta} + N_{0\beta\alpha}),$ $\quad (\mathcal{L}g)_{0\alpha 0} = N_{0\alpha 0},$

$$(\mathcal{L}g)_{\overline{\alpha}\beta\gamma} = -C^{\overline{\sigma}}_{\overline{\alpha}\beta} g_{\overline{\sigma}\gamma} - C^{\overline{\sigma}}_{\overline{\alpha}\gamma} g_{\beta\overline{\sigma}}, \quad (\mathcal{L}g)_{\alpha\overline{\beta}\gamma} = Z_\alpha\, g_{\overline{\beta}\gamma} - C^{\overline{\sigma}}_{\alpha\overline{\beta}} g_{\overline{\sigma}\gamma} - C^{\sigma}_{\alpha\gamma} g_{\overline{\beta}\sigma},$$

$$(\mathcal{L}g)_{\alpha\bar{\beta}0} = 2\eta_{\alpha\overline{\beta}} - C^{\sigma}_{\alpha 0} g_{\overline{\beta}\sigma}.$$

Lemma 2.6. *The essential components of* $\mathcal{L}\eta$*:*

$$(\mathcal{L}\eta)_{AB} = (\mathcal{L}_{Z_A}\eta)Z_B = -C^0_{AB}$$

are

$$(\mathcal{L}\eta)_{\alpha\beta} = \frac{1}{2} N_{\alpha\beta 0}, \quad (\mathcal{L}\eta)_{\alpha 0} = N_{\alpha 00}, \quad (\mathcal{L}\eta)_{\alpha\bar{\beta}} = 2\eta_{\alpha\bar{\beta}}.$$

Lemma 2.7. *The essential components of* $\mathcal{L}\varphi$*:*

$$(\mathcal{L}\varphi)_{AB} = (\mathcal{L}_{Z_A}\varphi)Z_B = [Z_A, \varphi Z_B] - \varphi[Z_A, Z_B] = \overline{(\mathcal{L}\varphi)_{\bar{A}\bar{B}}}$$

are

$$(\mathcal{L}\varphi)_{\alpha\beta} = -\frac{i}{2}(N^{\bar{\sigma}}_{\alpha\beta} Z_{\bar{\sigma}} + N^0_{\alpha\beta} Z_0), \quad (\mathcal{L}\varphi)_{\alpha\bar{\beta}} = -2i(C^{\sigma}_{\alpha\bar{\beta}} Z_\sigma - \eta_{\alpha\bar{\beta}} Z_0),$$

$$(\mathcal{L}\varphi)_{\alpha 0} = -\frac{i}{2}(N^{\bar{\sigma}}_{\alpha 0} Z_{\bar{\sigma}} + 2C^{\sigma}_{\alpha 0} Z_\sigma), \quad (\mathcal{L}\varphi)_{0\alpha} = i(N^{\bar{\sigma}}_{0\alpha} Z_{\bar{\sigma}} + N^0_{0\alpha} Z_0).$$

Lemmas 2.4, 2.5, 2.7 give the geometric meaning of the coefficients $C^{\sigma}_{\alpha B}$:

Lemma 2.8. *For the coefficients* $C^{\sigma}_{\alpha B}$ *the following expressions are valid:*

$$C^{\sigma}_{\alpha\beta} = -g^{\sigma\bar{\gamma}}[3i\Phi_{\alpha\beta\bar{\gamma}} + (\mathcal{L}g)_{\alpha\beta\bar{\gamma}} - (\mathcal{L}g)_{\beta\alpha\bar{\gamma}}],$$

$$C^{\sigma}_{\alpha\bar{\beta}} = -\frac{i}{2} g^{\sigma\bar{\gamma}} g[(\mathcal{L}\varphi)_{\alpha\bar{\beta}}, Z_{\bar{\gamma}}], \quad C^{\sigma}_{\alpha 0} = i g^{\sigma\bar{\gamma}} g[(\mathcal{L}\varphi)_{\alpha 0}, Z_{\bar{\gamma}}].$$

Lemma 2.9. *The Lee-form* θ *has the following components with respect to the bases* $\{Z^A\}$, $A \in I \cup \bar{I} \cup I_0$*:*

$$\theta_0 = -\frac{1}{2n}\delta\eta = \frac{3i}{4n}\sum_{\alpha=1}^{n} \Phi_{0\alpha\bar{\alpha}}, \quad \theta_\beta = \overline{\theta_{\bar{\beta}}} = \frac{1}{2(n-1)}(\delta\Phi)_\beta$$

$$= \frac{i}{2(n-1)}\left(\frac{3}{2}\sum_{\alpha=1}^{n} \Phi_{\alpha\bar{\alpha}\beta} + \omega_\beta\right).$$

Lemma 2.10. *The essential Cristoffel's symbols for the Levi-Civita connection ∇:*

$$\Gamma^C_{AB} = \Gamma^C_{BA} + C^C_{AB} = \overline{\Gamma^{\bar{C}}_{\bar{A}\bar{B}}}$$

$$= \frac{1}{2}g^{SC}(Z^A g_{BC} + Z^B g_{CA} - Z^C g_{AB} + C^S_{AB}g_{SC} + C^S_{CA}g_{SB} - C^S_{BC}g_{SA}),$$

are

(a) $\Gamma^{\sigma}_{\alpha\beta} = -\frac{3}{2}i\Phi^{\sigma}_{\alpha\beta} + g^{\sigma\bar{\gamma}}(Z_\alpha g_{\beta\bar{\gamma}} + C^{\bar{\sigma}}_{\bar{\gamma}\alpha}g_{\bar{\sigma}\beta}),$ (b) $\Gamma^{\bar{\sigma}}_{\bar{\alpha}\beta} = \frac{3}{2}i\Phi^{\bar{\sigma}}_{\bar{\alpha}\beta},$

(c) $\Gamma^{\bar{\sigma}}_{\alpha\beta} = -\frac{1}{8}N^{\bar{\sigma}}_{\alpha\beta} + \frac{1}{8}g^{\gamma\bar{\sigma}}(N_{\alpha\gamma\beta} + N_{\beta\gamma\alpha}),$ (d) $\Gamma^0_{\bar{\alpha}\beta} = \frac{3}{2}i\Phi^0_{\bar{\alpha}\beta} - \eta_{\bar{\alpha}\beta},$

(e) $\Gamma^0_{\alpha\beta} = -\frac{1}{4}N^0_{\alpha\beta} + \frac{1}{4}(N_{\alpha 0\beta} + N_{\beta 0\alpha}),$ (f) $\Gamma^0_{0\alpha} = -N^0_{0\alpha},$

and as consequences ($\Gamma^{\sigma}_{A0} = -g^{\sigma\bar{\beta}}\Gamma^0_{A\bar{\beta}}$, $F_{A\beta\gamma} = 2i\Gamma^{\bar{\sigma}}_{A\beta}g_{\bar{\sigma}\gamma}$, $F_{A\beta 0} = i\Gamma^0_{A\beta} = -i\Gamma^{\bar{\sigma}}_{A0}g_{\bar{\sigma}\beta}$),

(g) $\Gamma^{\bar{\sigma}}_{\alpha 0} = \frac{1}{4}g^{\bar{\sigma}\beta}(N_{\alpha\beta 0} - N_{\alpha 0\beta} - N_{\beta 0\alpha}),$ (h) $\Gamma^{\sigma}_{\alpha 0} = -\frac{3}{2}i\Phi^{\sigma}_{\alpha 0} + \eta^{\sigma}_{\alpha},$

(i) $\Gamma^{\bar{\sigma}}_{00} = g^{\bar{\sigma}}N^0_{0\alpha}.$

Finally, by using the characterization conditions for everyone of the twelve basic classes, introduced by the projections of F to each of them [2] and the lemmas above, we get the full describtion of each basic class in termes of the **C**-equations for the **C**-essential components of the fundamental tensors.

Theorem 2.1. *Let $M^{2n+1}(\varphi, \xi, \eta, g)$ be an almost contact metric manifold. The characterization conditions M to belong to each of the twelve basic classes, in the following table are given:*

class	characterization C-equations	$\delta\eta$	$\delta\Phi\circ\varphi$	$\delta\Phi(\xi)$
W_1	$F_{00\alpha}=\omega_\alpha=-2i\eta_{\alpha 0}=-iN_{\alpha 00}$ $=i(\mathcal{L}g)_{\alpha 0}$	0	$\omega\circ\varphi$	0
W_2	$F_{\bar\alpha\beta 0}=\frac{f(\xi)}{2n}g_{\bar\alpha\beta}=-i\eta_{\bar\alpha\beta}=F_{\beta\bar\alpha 0}$	0	0	$f(\xi)$
W_3	$F_{\bar\alpha\beta 0}=\frac{f^*(\xi)}{2ni}g_{\bar\alpha\beta}=\frac{1}{2i}(\mathcal{L}_\xi g)_{\bar\alpha\beta}$ $=-\frac{3}{2}\Phi_{\bar\alpha\beta 0}=-F_{\beta\bar\alpha 0}$	$-f^*(\xi)$	0	0
W_4	$F_{\bar\alpha\beta 0}=-i\eta_{\bar\alpha\beta}=F_{\beta\bar\alpha 0}$	0	0	0
W_5	$F_{\bar\alpha\beta 0}=\frac{1}{2i}(\mathcal{L}_\xi g)_{\bar\alpha\beta}=-\frac{3}{2}\Phi_{\bar\alpha\beta 0}=-F_{\beta\bar\alpha 0}$	0	0	0
W_6	$F_{\alpha\beta 0}=\frac{1}{2i}(\mathcal{L}_\xi g)_{\alpha\beta}=\frac{1}{2i}N_{0\alpha\beta}=F_{\beta\alpha 0}$	0	0	0
W_7	$F_{\alpha\beta 0}=-i\eta_{\alpha\beta}=\frac{1}{4i}N_{\alpha\beta 0}=-F_{\beta\alpha 0}$	0	0	0
W_8	$F_{0\alpha\beta}=-3\Phi_{0\alpha\beta}=N_{0\alpha\beta}$	0	0	0
W_9	$F_{\bar\alpha\beta\gamma}=-3\Phi_{\bar\alpha\beta\gamma}=2(\Phi_{\bar\alpha *}\wedge\theta)_{\beta\gamma}$	0	$f\circ\varphi$	0
W_{10}	$F_{\bar\alpha\beta\gamma}=-3\Phi_{\bar\alpha\beta\gamma}$	0	0	0
W_{11}	$F_{\alpha\beta\gamma}=-\Phi_{\alpha\beta\gamma}=\frac{1}{4i}N_{\alpha\beta\gamma}$	0	0	0
W_{12}	$F_{\alpha\beta\gamma}=F_{\gamma\beta\alpha}-F_{\beta\gamma\alpha}=\frac{1}{2i}N_{\gamma\beta\alpha}$	0	0	0

Some special classes of almost contact metric manifolds one can see in [1,3]. We denote, that the classes W_9, W_{10}, W_{11}, W_{12} are the almost contact analogs to the four basic classes of almost Hermitian (AH)-manifolds[4], namely: $W_4 = CK$ (conformally Kahlerian), $W_3 = SH$ (special Hermitian), $W_1 = NK$ (nearly Kahlerian) and $W_2 = AK$ (almost Kahlerian) respectively.

3 Integrability conditions for almost contact manifolds

The four classes of Hermitian manifolds $M^{2n}(J,g)$, characterised by $N = [J,J] \equiv 0$, are the class K of Kahelerian manifolds: $\nabla\Phi \equiv 0$, W_3, W_4 and $W_3 \oplus W_4$. The well known Newlender-Nirenberg's theorem states that the vanishing of the Nijenhuis tensor N is necessary and sufficient condotion an almost complex manifold $M^{2n}(J)$ to be a complex manifold, i.e. to carry local coordinate covering, compatible with the pseudogroup of holomorphic transformations in $\mathbf{R}^{2n} \equiv \mathbf{C}^n : H_{2n} = \{f : \mathbf{R}^{2n}(J_0) \to \mathbf{R}^{2n}(J_0) : f_* \circ J_0 = J_0 \circ f_*\}$, where J_0 is the standard almost complex structure on $\mathbf{R}^{2n} = \{(x^\alpha, y^\alpha = x^{n+\alpha})\}$.

For the commutators of the complex basic vector fields of T_pM^{2n} : $Z_\alpha = e_\alpha - iJe_\alpha$, $Z_{\bar\alpha} = \overline{Z_\alpha}$, the following decomposition is valid

$$\begin{pmatrix} [Z_\alpha, Z_\beta] \\ [Z_\alpha, Z_{\bar\beta}] \end{pmatrix} = \begin{pmatrix} C^\sigma_{\alpha\beta} & -\frac{1}{4}N^{\bar\sigma}_{\alpha\beta} \\ C^\sigma_{\alpha\bar\beta} & C^{\bar\sigma}_{\alpha\bar\beta} \end{pmatrix} \begin{pmatrix} Z_\sigma \\ Z_{\bar\sigma} \end{pmatrix} \tag{3.1}$$

The condition $N \equiv 0$ allows one to introduce local complex coordinate functions. This procedure is well known [5,6] and we'll rewrite them by using (3.1) in a way, appropriate for our considerations.

Corollary 3.1. *Let $M^{2n}(J)$ be* **R***-analical manifold with* **R***-analical almost complex structure J. Then the implication $N \equiv 0 \Longrightarrow C^\sigma_{\alpha\beta} = C^{\bar\sigma}_{\alpha\bar\beta} = -\overline{C^\sigma_{\beta\bar\alpha}} \equiv 0$, is valid, i.e. the matrix in (3.1) vanishes when $N \equiv 0$, in the sence of the existence of local complex coordinate covering $\{(U_\alpha; \{z^\alpha\}),\ \alpha \in I\}$ such that $Z_a = \dfrac{\partial}{\partial z^a}$, $a \in I \cup \bar I$.*

Now the Newlender-Nirenberg's theorem we formulate in the following way:

Theorem 3.1. *Let $M^{2n}(J)$ be almost complex manifold. Then the following conditions are equivalent:*
(i) $N^{\bar\sigma}_{\alpha\beta} = 0$; (ii) *$M^{2n}(J)$ is complex manifold;*
(iii) *J is Lie-parallel on T^{10}_pM : $(\mathcal{L}_U J)V = 0$, U, $V \in T^{10}_pM$.*

Further, we are going to describe the integrability conditions for almost contact manifolds by specifying the results of Yano and Ishihara [7]. It will be considered only **R**-analytical case, i.e. $(2n+1)$-dimensional **R**-analytical manifold with **R**-analytical almost contact structure (φ, ξ, η). We shall use also the decomposition, stated in Lemma 2.3-(a), which is an analog to (3.1).

Theorem 3.2. *Let $M(\varphi, \xi, \eta)$ be almost contact manifold. Then the following conditions are equivalent:*
(i) *the contact distribution D is involutive;* (ii) $N^0_{\alpha\beta} = 0$, $\eta_{\alpha\bar\beta} = 0$;
(iii) *η is Lie-parallel on D : $(\mathcal{L}_U\eta)V \equiv 0$, U, $V \in D$.*

Proof. From Lemmas 2.3 and 2.6 for the essential components of $hd\eta$, we get

$$2\eta_{\alpha\bar\beta} = (\mathcal{L}\eta)_{\alpha\bar\beta}, \quad 4\eta_{\alpha\beta} = (\mathcal{L}\eta)_{\alpha\beta} = N^0_{\alpha\beta}.$$

Now, the equivalence of (ii) and (iii) is clear. The equivalence of (i) and (ii) follows from the well known fact that D is involutive iff $hd\eta \equiv 0$ [3]. □

Definition 3.1 ([7]). The almost contact structure (φ, ξ, η) (resp. the almost contact manifold) is siad to be partially integrable if the contact distribution D is involutive and the restriction $\varphi_{|D}$ to everyone integral manifold $\mathcal{D}$ for D is an integrable almost complex structure.

Theorem 3.3. *Let $M(\varphi, \xi, \eta)$ be almost contact manifold. Then the following conditions are equivalent:*

(i) (φ, ξ, η) *is partially integrable;* (ii) $N^0_{\alpha\beta} = N^{\bar{\sigma}}_{\alpha\beta} = \eta_{\alpha\bar{\beta}} = 0$;

(iii) η *is Lie-parallel on D and φ is Lie-parallel on D^{10} (D^{01})* : $(\mathcal{L}_u\varphi)v \equiv 0$, $u,\ v \in D^{10}_p\ (D^{01}_p)$.

Proof. Let (φ, ξ, η) be partially integrable. Then D is involutive and Theorem 3.2. implies $N^0_{\alpha\beta} = \eta_{\alpha\bar{\beta}} = 0$. The integrability of $\varphi_{|D}$ is equivalent to $[\varphi_{|D}, \varphi_{|D}] = 0$. Now, the equation $[\varphi, \varphi](hx, hy) = N(hx, hy)$, and Theorem 3.1. imply $N^{\bar{\sigma}}_{\alpha\beta} \equiv 0$, which proves the implication (i)$\Rightarrow$(ii). Inversely: (ii)$\Rightarrow$(i) is obviously.

The equivalence of (ii) and (iii) follows from Theorem 3.1, and the decomposition, stated in Lemma 2.7: $(\mathcal{L}\varphi)_{\alpha\beta} = -\frac{i}{2}(N^{\bar{\sigma}}_{\alpha\beta}Z_{\bar{\sigma}} + N^0_{\alpha\beta}Z_0)$. □

From Corollary 3.1 and Theorem 3.3 we get

Corollary 3.2. *Let $M(\varphi, \xi, \eta)$ be partially integrable almost contact manifold. Then M carries local coordinate covering $\{(U; \{x^\alpha, y^\alpha = x^{n+\alpha}, t\})\}$, $\alpha \in I$, such that for any integral manifold $\mathcal{D}$ for D there exists an immersion $i : \mathcal{D} \to M$, so that, for the adapted local coordinate covering for $\mathcal{D}$*

$$\{(U_{|\mathcal{D}}; \{x^\alpha, y^\alpha, t = \text{const}\}),\ \alpha \in I\}$$

$$\xleftrightarrow{1-1} \{(U_{|\mathcal{D}}; \{z^\alpha = x^\alpha + iy^\alpha, t = \text{const}\}),\ \alpha \in I\},$$

it follows $Z_{\alpha|U_{|\mathcal{D}}} \equiv \dfrac{\partial}{\partial z^\alpha}$.

So, the first two rows of the matrix in Lemma 2.3-(a) vanish when $N^0_{\alpha\beta} = N^{\bar{\sigma}}_{\alpha\beta} = \eta_{\alpha\bar{\beta}} = 0$, in the sence of Corollary 3.2.

Remark 1. The existence of the local coordinate covering, described in Corollary 3.2, one can introduced by a way, similar to that of Frölicher [5]. Realy, if $M^{2n+1}(\varphi, \xi, \eta)$ is $\mathbf{R}$-analytical manifold having $\mathbf{R}$-analytical almost contact structure and $\{(U; \{x^a, t\}), a \in I \cup \bar{I}\}$ is local coordinate covering, then:

(1) $\xi \neq 0$, locally can be choosen as $\dfrac{\partial}{\partial t}$ (vector field rectifying theorem);

(2) the components of φ and of η with respect to $\{(U;\{x^a,t\}), a \in I \cup \bar{I}\}$:

$$\varphi(\frac{\partial}{\partial x^a}) = \varphi_a^b \frac{\partial}{\partial x^b} + \varphi_a^0 \frac{\partial}{\partial t}, \quad \eta = dt + \eta_a dx^a, \quad \text{satisfy:}$$

$$\varphi_a^b \varphi_b^c = -\delta_a^c, \quad \varphi_b^0 \varphi_a^b = \eta_a, \quad \varphi_a^0 = -\varphi_a^b \eta_b;$$

(3) the null-space of the set of forms $\mathcal{J} = \{\eta,\, \omega^c = (\varphi_b^c + i\delta_b^c)dx^b\}$ is

$$\mathcal{N}\mathcal{J} = \operatorname{span}\left\{ v_b = (\varphi_b^c - i\delta_b^c)(\frac{\partial}{\partial x^c} - \eta_c \frac{\partial}{\partial t})\right\}, \quad b, c = 1, \ldots, 2n;$$

(4) $\mathcal{J}$ is a differential ideal iff

$$\begin{cases} d\eta(v_b, v_g) = 0 \\ d\omega^c(v_p, v_q) = 0 \end{cases}$$

$$\Longleftrightarrow \begin{cases} (\varphi_b^d - i\delta_b^d)(\varphi_g^a - i\delta_g^a)\left(\dfrac{\partial \eta_a}{\partial x^d} - \dfrac{\partial \eta_d}{\partial x^a} + \dfrac{\partial \eta_d}{\partial t}\eta_a - \dfrac{\partial \eta_a}{\partial t}\eta_d\right) = 0 \\[2ex] (\varphi_p^r - i\delta_p^r)(\varphi_q^b - i\delta_q^b)\left(\dfrac{\partial \varphi_b^c}{\partial x^r} - \dfrac{\partial \varphi_r^c}{\partial x^b} + \dfrac{\partial \varphi_r^c}{\partial t}\eta_b - \dfrac{\partial \varphi_b^c}{\partial t}\eta_r\right) = 0. \end{cases}$$

These are the Frobenius conditions for the totally integrability of the system $\begin{cases} \eta = 0 \\ \omega^c = 0 \end{cases}$ and are equivalent to the vanishing of$[\varphi, \varphi]$ and of $hd\eta$. Here the components of φ and η depend on t. Then there exists smooth transformation

$$\psi : (x^\alpha, x^{n+\alpha}, t) \in U \mapsto (u^\beta(x^\alpha, x^{n+\alpha}, t), v^\beta(x^\alpha, x^{n+\alpha}, t), q(x^\alpha, x^{n+\alpha}, t)) \in V,$$

$\alpha, \beta \in I$, so that, the systems $\begin{cases} dz^\beta = 0, \ \beta \in I \\ dq = 0 \end{cases}$ and $\begin{cases} \eta = 0 \\ \omega^c = 0 \end{cases}$ are equivalent, where $z^\beta = u^\beta + iv^\beta$.

The functions $u^\beta(x^\alpha, x^{n+\alpha}, t)$, $v^\beta(x^\alpha, x^{n+\alpha}, t)$, $q(x^\alpha, x^{n+\alpha}, t)$ are solutions of the system of homogenius linear partial differential equations

$$\begin{cases} dz^\beta(v_p) = 0, \ \beta \in I \\ dq(v_p) = 0 \end{cases} \iff \begin{cases} \varphi_p^r \dfrac{\partial u^\beta}{\partial x^r} - \varphi_p^r \eta_r \dfrac{\partial u^\beta}{\partial t} = -\dfrac{\partial v^\beta}{\partial x^p} + \eta_p \dfrac{\partial v^\beta}{\partial t} \\ \varphi_p^r \dfrac{\partial v^\beta}{\partial x^r} - \varphi_p^r \eta_r \dfrac{\partial v^\beta}{\partial t} = \dfrac{\partial u^\beta}{\partial x^p} - \eta_p \dfrac{\partial u^\beta}{\partial t}, \quad \beta \in I \, . \\ \dfrac{\partial q}{\partial x^p} - \eta_p \dfrac{\partial q}{\partial t} = 0 \end{cases}$$

The local covering $\{(U; \{u^\beta, v^\beta, q\}), \beta \in I\}$ is the required coordinate covering.

Theorem 3.4. *Let $M(\varphi, \xi, \eta)$ be almost contact manifold. Then the following conditions are equivalent:*

(i) $N_{\alpha 0}^{0} = N_{\alpha 0}^{\bar\sigma} = 0$;

(ii) *φ (and consequently η) is Lie-parallel with respect to ξ:* $\mathcal{L}_\xi \varphi \equiv 0 \quad (\mathcal{L}_\xi \eta \equiv 0)$.

Proof. The equivalence follows from the decompositions, in Lemmas 2.6, 2.7.:

$$(\mathcal{L}_\xi \eta) Z_\alpha = -N_{\alpha 0}^{0}, \quad (\mathcal{L}_\xi \varphi) Z_\alpha = i \left(N_{\alpha 0}^{\bar\sigma} Z_{\bar\sigma} + N_{\alpha 0}^{0} Z_0 \right). \qquad \square$$

Theorems 3.2. and 3.4. imply

Theorem 3.5. *Let $M(\varphi, \xi, \eta)$ be almost contact manifold with involutive contact distribution D. The structure tensor field φ (and qonsecuently the form η) is Lie-parallel with respect to ξ iff for arbitrarry integral manifold $\mathcal{D}$ of D immersed in M, the components φ_i^j of φ (η_i of η) with respect to an adapted local coordinate system $(U_{|\mathcal{D}}; \{x^\alpha, y^\alpha = x^{n+\alpha}, t = \text{const}\})$, $\alpha \in I$, don't depend on t.*

Proof. Let $\mathcal{L}_\xi \varphi \equiv 0$ and let $(U_{|\mathcal{D}}; \{x^\alpha, y^\alpha = x^{n+\alpha}, t = \text{const}\})$, $\alpha \in I$, be adapted local coordinate system for immersion $i : \mathcal{D} \to M$ of an integral manifold $\mathcal{D}$ of D in M. Then

$$\xi = \frac{\partial}{\partial t}, \quad (\mathcal{L}_\xi \varphi)(\frac{\partial}{\partial x^i}) = \mathcal{L}_\xi (\varphi_i^j \frac{\partial}{\partial x^j})$$
$$= (\frac{\partial \varphi_i^j}{\partial t}) \frac{\partial}{\partial x^j} = 0 \quad \text{and hence} \quad \frac{\partial \varphi_i^j}{\partial t} = 0.$$

Analogousely from $\mathcal{L}_\xi \varphi \equiv 0 \Rightarrow \mathcal{L}_\xi \eta \equiv 0$, it follows $\dfrac{\partial \eta_i}{\partial t} = 0$. The inverce is clear. □

More over by the conditions of Theorem 3.5 the equality $(\mathcal{L}_{(\partial/\partial x^i)}\varphi)\xi = 0$ with respect to adapted local complex coordinate system is valid too and hence on $(U_{|\mathcal{D}}; \{z^\alpha = x^\alpha + iy^\alpha, t = \text{const}\})$, $\alpha \in I$, it follows $(\mathcal{L}_{Z_\alpha}\varphi)\xi = 0$. By the other hand Lemma 2.7 states the decomposition

$$(\mathcal{L}_{Z_\alpha}\varphi)\xi = (\mathcal{L}\varphi)_{\alpha 0} = -i(\frac{1}{2}N^{\bar\sigma}_{\alpha 0}Z_{\bar\sigma} + C^\sigma_{\alpha 0}Z_\sigma).$$

So we get the following

Corollary 3.3. *Let $M(\varphi, \xi, \eta)$ be almost contact manifold, such that $N^0_{\alpha\beta} = N^0_{\alpha 0} = N^{\bar\sigma}_{\alpha 0} = \eta_{\alpha\bar\beta} = 0$. Then $C^\sigma_{\alpha 0} = 0$, i.e. the matrix in Lemma 2.3-(a) has vanishing last row and last column (in the sence of Theorem 3.5).*

Definition 3.2 ([7]). The structure (φ, ξ, η) (resp. the almost contact manifold) is said to be integrable if (φ, ξ, η) is partially integrable and the components φ^j_i of φ with respect to an adapted local coordinate system $(U_{|\mathcal{D}}; \{x^\alpha, y^\alpha = x^{n+\alpha}, t = \text{const}\})$, $\alpha \in I$, for any integral manifold $\mathcal{D}$ of D, immersed in M, does not depend on t.

Theorem 3.6. *Let $M(\varphi, \xi, \eta)$ be almost contact manifold. Then the following conditions are equivalent:*
(i) (φ, ξ, η) *is integrable;* (ii) $N \equiv 0$, $d\eta \equiv 0$.

Proof. Lemma 2.3 implies that the essential components of N are $N^{\bar\sigma}_{\alpha\beta}$, $N^0_{\alpha\beta}$, $N^{\bar\sigma}_{\alpha 0}$, of $d\eta$ are $\eta_{\alpha\beta}$, $\eta_{\alpha\bar\beta}$, $\eta_{\alpha 0}$, and $N^0_{\alpha\beta} = 4\eta_{\alpha\beta}$, $N^0_{\alpha 0} = \eta_{\alpha 0}$ are valid. Theorems 3.2, 3.3, 3.4 imply the equivalence of (i) and (ii). □

Corollary 3.4. *Let $M(\varphi, \xi, \eta)$ be almost contact manifold, such that $N \equiv 0$, $d\eta \equiv 0$. Then the matrix in Lemma 2.3-(a) is vanishing, i.e. $Z_A = \partial/\partial z^A$, $A \in I \cup \bar I \cup I_0$, $z^0 = t$, $Z_0 = \partial/\partial t = \xi$.*

Finally by using Theorem 2.1 we get

Theorem 3.7. (i) The class of partially integrable almost contact metric manifolds is exatly

$$W_1 \oplus W_3 \oplus W_5 \oplus W_6 \oplus W_8 \oplus W_9 \oplus W_{10};$$

(ii) The class of integrable almost contact metric manifolds is exactly

$$W_3 \oplus W_5 \oplus W_9 \oplus W_{10}.$$

References

1. D. Blair, *Contact manifolds in Riemmannian geometry*, Lecture Notes in Math., vol.509, Springer-Verlag, Berlin, 1976.
2. V. Alexiev and G. Ganchev, *On the classificstion of almost contact metric Manifolds*, Math. and Educ. in Math., Proc. of the XV Spring Conf. of UBM. Sunny Beach, 1986, pp. 155–161.
3. G. Ganchev and V. Alexiev, *On some classes of almost contact metric Manifolds*, Math. and Educ. in Math., Proc. of the XV Spring Conf. of UBM. Sunny Beach, 1986, pp. 186–191.
4. A. Gray, L. Hervella, *The Sixteen Classes of Almost Hermitian Manifolds and Their Linear Invariants*, Ann. di Mat. pura ed appl. **CXXIII**(1980), 35–58.
5. A. Frölicher, *Zur Differentialgeometrie der komplexen Strukturen*, Math. Ann. **129**(1955), 50–95.
6. S. Dimiew, *Functions presque holomorphes, complex analysis*, in Banach cent. publ., vol. 11, Warsaw 1983.
7. K. Yano and S. Ishihara, *On integrability conditions of a structure f, satisfying $f^3 + f = 0$*, Quart. J. Math. **15**(1964), 217–222.

FOUR-DIMENSIONAL SUBMANIFOLDS OF SEVEN-DIMENSIONAL ALMOST CONTACT MANIFOLDS WITH B-METRIC

G.V. NAKOVA

Department of Mathematics, "Vasil Levski" Land Forces Academy, 5000 Veliko Tarnovo, Bulgaria

1 Introduction

A classification of almost contact manifolds with B-metric $(M^{2n+1}, \varphi, \xi, \eta, g)$ with respect to the covariant derivative of the fundamental tensor φ of type (1,1) is given in [7]. Submanifolds of almost contact manifolds with B-metric of codimension 2 in the cases, when ξ is a tangential vector field of the submanifold, ξ belongs to the normal section and ξ is arbitrary are considered in [8] and [9]. An example of a 4-dimensional Lorentzian submanifold of a 5-dimensional almost contact manifold with B-metric is studied in [11]. In this paper we consider classes of 7-dimensional almost contact manifolds with B-metric, constructed as a direct product of 4-dimensional almost complex manifolds with B-metric and 3-dimensional almost contact manifolds with B-metric. Four-dimensional submanifolds of 7-dimensional almost contact manifolds with B-metric are studied and examples of such submanifolds are constructed.

2 Preliminaries

Let (M^{2n}, J, h) be a $2n-$dimensional almost complex manifold with B-metric, i.e. J is a complex structure and h is a metric on M^{2n} such that [5]:

$$J^2X = -X; \qquad h(JX, JY) = -h(X,Y), \tag{1}$$

for all vector fields X, Y on M^{2n}. The tensor field F of type (0,3) on the manifold M^{2n} is defined [5] by $F(X,Y,Z) = h((\nabla_X)Y, Z)$, where ∇ is the Levi-Civita connection of h, X, Y, $Z \in \mathcal{X}M^{2n}$.

The following 1-forms are associated with F:

$$\theta(x) = h^{ij}F(e_i, e_j, x), \quad \theta^*(x) = h^{ij}F(e_i, Je_j, x), \tag{2}$$

where $x \in T_pM^{2n}$, $\{e_i\}$, $(i = 1, 2, \dots, 2n)$, is an arbitrary basis of T_pM^{2n}, (h^{ij}) is the inverse matrix [5] of (h_{ij}).

A classification of the almost complex manifolds with B-metric with respect to the tensor F is given in [5], where are defined three basic classes W_i $(i = 1, 2, 3)$. The class $W_0 : F = 0$ of the Kaehler manifolds with B-metric belongs to each of the basic classes W_i $(i = 1, 2, 3)$.

Let $(M^{2n+1}, \varphi, \xi, \eta, g)$ be a $(2n+1)$-dimensional almost contact manifold with B-metric, i.e. (φ, ξ, η) is an almost contact structure [1] and g is a metric [7] on M^{2n+1}such that:

$$\varphi^2 = -id + \eta \otimes \xi; \quad \eta(\xi) = 1, \tag{3}$$

$$g(\varphi X, \varphi Y) = -g(X, Y) + \eta(X)\eta(Y), \tag{4}$$

where X, Y are arbitrary differentiable vector fields on M^{2n+1}.

Let ∇ be the Levi-Civita connection of the metric g. The tensor field F of type (0,3) on M^{2n+1} is defined by [7]

$$F(X, Y, Z) = g((\nabla_X \varphi)Y, Z), \quad X, Y, Z \in \mathcal{X}M^{2n+1}.$$

The following 1-forms are associated with F:

$$\theta(x) = g^{ij}F(e_i, e_j, x), \quad \theta^*(x) = g^{ij}F(e_i, \varphi e_j, x), \quad \omega(x) = F(\xi, \xi, x), \tag{5}$$

where $x \in T_pM^{2n+1}$, $\{e_i, \xi\}$, $(i = 1, 2, \ldots, 2n)$, is a basis of T_pM^{2n+1} and (g^{ij}) is the inverse matrix [7] of (g_{ij}).

A classification of the almost contact manifolds with B-metric with respect to the tensor F is given in [7], where are defined eleven basic classes $\mathcal{F}_i$ $(i = 1, 2, \ldots, 11)$. The class $\mathcal{F}_0 : F = 0$ belongs to each of the basic classes.

Let R be the curvature tensor field of type (1,3) of the n-dimensional Riemannian manifold (M, g) with Levi-Civita connection ∇ of g, i.e.

$$R(X, Y, Z) = \nabla_X \nabla_Y Z - \nabla_Y \nabla_X Z - \nabla_{[X,Y]} Z, \quad X, Y, Z \in \mathcal{X}M. \tag{6}$$

The corresponding to R tensor field of type (0,4) is given by

$$R(X, Y, Z, W) = g(R(X, Y, Z), W). \tag{7}$$

In [6] the following curvatures for the section $\alpha = \{x, y\}$ in the tangential space of the almost complex manifold (M^{2n}, J, h) with B-metric are defined

$$K(\alpha; p) = K(x, y) = \frac{R(x, y, y, x)}{\pi_1(x, y, y, x)}, \tag{8}$$

for every nondegenerate section α with respect to h. This is the usual Riemannian sectional curvature,

$$\widetilde{K}(\alpha; p) = \widetilde{K}(x, y) = \frac{R(x, y, y, Jx)}{\pi_1(x, y, y, x)}, \tag{9}$$

for every nondegenerate section α with respect to h. $\widetilde{K}(\alpha;p)$ is said to be an *associated sectional curvature*.

A section α is said to be *totally real* if $J\alpha \perp \alpha$.

3 Classes of almost contact manifolds with B-metric, derived from almost complex manifolds with B-metric and almost contact manifolds with B-metric

In this section we shall construct an almost contact manifold with B-metric as a direct product of an almost complex manifold with B-metric and an almost contact manifold with B-metric.

It is known [2] that the product manifold of any two differentiable manifolds M and N of dimensions m and n, respectively, is an $(m+n)$-dimensional differentiable manifold.

Let (M^{2n}, J, h) be a $2n$-dimensional almost complex manifold with B-metric and let $(M'^{2m+1}, \varphi', \xi', \eta', g')$ be a $(2m+1)$−dimensional almost contact manifold with B-metric. A vector field on $\widetilde{M}^{2n+2m+1} = M^{2n} \times M'^{2m+1}$ will be denoted by $\widetilde{X} = (X, X') = X + X'$, where $X \in \mathcal{X}M^{2n}$, $X' \in \mathcal{X}M'^{2m+1}$. It is known [2] that the product manifold of an almost contact metric manifold and an almost Hermitian manifold is an almost contact metric manifold. Analogously as in [2] and [3], an almost contact structure on $\widetilde{M}^{2n+2m+1} = M^{2n} \times M'^{2m+1}$ can be defined by

$$\widetilde{\varphi}(X, X') = (JX, \varphi' X'), \quad \widetilde{\xi} = (0, \xi'), \quad \widetilde{\eta}(X, X') = \eta'(X'). \tag{10}$$

We set

$$\widetilde{g}((X, X'), (Y, Y')) = h(X, Y) + g'(X', Y'), \tag{11}$$

for arbitrary vector fields $(X, X'), (Y, Y') \in \mathcal{X}\widetilde{M}^{2n+2m+1}$.

It is easy to check that so defined structure tensors satisfy the equalities (3) and (4), and consequently the structure $(\widetilde{\varphi}, \widetilde{\xi}, \widetilde{\eta}, \widetilde{g})$ on $\widetilde{M}^{2n+2m+1}$ is an almost contact structure with B-metric, i.e. $(\widetilde{M}^{2n+2m+1}, \widetilde{\varphi}, \widetilde{\xi}, \widetilde{\eta}, \widetilde{g})$ is an almost contact manifold with B-metric.

Let $f \in FM^{2n}$, $f' \in FM'^{2m+1}$, $\widetilde{f} \in F\widetilde{M}^{2n+2m+1}$, $X, Y \in \mathcal{X}M^{2n}$ and $X', Y' \in \mathcal{X}M'^{2m+1}$. For each coordinate neighborhood $(U \times V, x^1, x^2, \dots, x^{2n}, y^1, y^2, \dots, y^{2m+1})$ on $\widetilde{M}^{2n+2m+1}$, where $(U, x^1, x^2, \dots, x^{2n})$ and $(V, y^1, y^2, \dots, y^{2m+1})$ are coordinate neighborhoods of M^{2n} and M'^{2m+1}, respectively, we have

$$f = f(x^1, x^2, \dots, x^{2n}), \qquad f' = f'(y^1, y^2, \dots, y^{2m+1}),$$

$$\widetilde{f} = \widetilde{f}(x^1, x^2, \dots, x^{2n}, y^1, y^2, \dots, y^{2m+1});$$

$$X = X^\alpha \frac{\partial}{\partial x^\alpha}, \ X^\alpha \in FM^{2n},$$

$$Y = Y^\alpha \frac{\partial}{\partial x^\alpha}, \ Y^\alpha \in FM^{2n}, \quad (\alpha = 1, 2, \dots, 2n);$$

$$X' = X'^i \frac{\partial}{\partial y^i}, \quad X'^i \in FM'^{\,2m+1},$$

$$Y' = Y'^i \frac{\partial}{\partial y^i}, \quad Y'^i \in FM'^{2m+1}, \quad (i = 1, 2, \dots, 2m+1).$$

Hence

$$\begin{aligned} Xf' &= (X^\alpha \frac{\partial}{\partial x^\alpha}) f' = X^\alpha (\frac{\partial f'}{\partial x^\alpha}) = 0, \\ X'f &= (X'^i \frac{\partial}{\partial y^i}) f = X'^i (\frac{\partial f}{\partial y^i}) = 0. \end{aligned} \tag{12}$$

Lemma 3.1. *Let* $\widetilde{X} = (X, X')$, $\widetilde{Y} = (Y, Y') \in \mathcal{X}\widetilde{M}^{2n+2m+1}$, $X, Y \in \mathcal{X}M^{2n}$, $X', Y' \in \mathcal{X}M'^{2m+1}$, $\widetilde{f} \in F\widetilde{M}^{2n+2m+1}$. *Then* $[\widetilde{X}, \widetilde{Y}] = [X, Y] + [X', Y']$.

Proof.

$$\begin{aligned} [X, Y']\widetilde{f} &= X(Y'\widetilde{f}) - Y'(X\widetilde{f}) = X(Y'^j \frac{\partial \widetilde{f}}{\partial y^j}) - Y'(X^\alpha \frac{\partial \widetilde{f}}{\partial x^\alpha}) \\ &= X(Y'^j) \frac{\partial \widetilde{f}}{\partial y^j} + Y'^j . X(\frac{\partial \widetilde{f}}{\partial y^j}) - Y'(X^\alpha). \frac{\partial \widetilde{f}}{\partial x^\alpha} - X^\alpha . Y'(\frac{\partial \widetilde{f}}{\partial x^\alpha}). \end{aligned}$$

Taking into account (12) and $\dfrac{\partial^2 \widetilde{f}}{\partial y^j \partial x^\alpha} = \dfrac{\partial^2 \widetilde{f}}{\partial x^\alpha \partial y^j}$ we obtain $[X, Y']\widetilde{f} = 0$. Analogously we verify that $[X', Y]\widetilde{f} = 0$. Then $[\widetilde{X}, \widetilde{Y}] = [X + X', Y + Y'] = [X, Y] + [X, Y'] + [X', Y] + [X', Y'] = [X, Y] + [X', Y']$. □

Let $\widetilde{\nabla}$, ∇ and ∇' be the Levi-Civita connections of the metrics $\widetilde{g}$, h and g' respectively. The following equalities for $\widetilde{\nabla}$, ∇ and ∇' are valid.

$$\begin{aligned} 2\widetilde{g}(\widetilde{\nabla}_{\widetilde{X}} \widetilde{Y}, \widetilde{Z}) =& \widetilde{X}\widetilde{g}(\widetilde{Y}, \widetilde{Z}) + \widetilde{Y}\widetilde{g}(\widetilde{X}, \widetilde{Z}) - \widetilde{Z}\widetilde{g}(\widetilde{X}, \widetilde{Y}) + \widetilde{g}([\widetilde{X}, \widetilde{Y}], \widetilde{Z}) \\ & - \widetilde{g}([\widetilde{X}, \widetilde{Z}], \widetilde{Y}) - \widetilde{g}([\widetilde{Y}, \widetilde{Z}], \widetilde{X}), \quad \widetilde{X}, \widetilde{Y}, \widetilde{Z} \in \mathcal{X}\widetilde{M}^{2n+2m+1}. \end{aligned} \tag{13}$$

$$2h(\nabla_X Y, Z) = Xh(Y,Z) + Yh(X,Z) - Zh(X,Y) + h([X,Y],Z) - h([X,Z],Y) - h([Y,Z],X), \quad X,Y,Z \in \mathcal{X}M^{2n}. \tag{14}$$

$$2g'(\nabla'_{X'}Y', Z') = X'g'(Y',Z') + Y'g'(X',Z') - Z'g'(X',Y') + g'([X',Y'],Z') - g'([X',Z'],Y') - g'([Y',Z'],X'), \quad X',Y',Z' \in \mathcal{X}M'^{2m+1}. \tag{15}$$

According to (11), (12), Lemma 3.1., (14) and (15), from (13) we get

$$\widetilde{g}(\widetilde{\nabla}_{\widetilde{X}}\widetilde{Y}, \widetilde{Z}) = h(\nabla_X Y, Z) + g'(\nabla'_{X'}Y', Z'). \tag{16}$$

Lemma 3.2. *Let F, F' and $\widetilde{F}$ be the tensor fields of type (0,3) on (M^{2n}, J, h), $(M'^{2m+1}, \varphi', \xi', \eta', g')$ and $(\widetilde{M}^{2n+2m+1}, \widetilde{\varphi}, \widetilde{\xi}, \widetilde{\eta}, \widetilde{g})$, respectively, defined by*

$$F(X,Y,Z) = h((\nabla_X J)Y, Z), \quad X,Y,Z \in \mathcal{X}M^{2n},$$
$$F'(X',Y',Z') = g'((\nabla'_{X'}\varphi')Y', Z'), \quad X',Y',Z' \in \mathcal{X}M'^{2m+1},$$
$$\widetilde{F}(\widetilde{X},\widetilde{Y},\widetilde{Z}) = \widetilde{g}((\widetilde{\nabla}_{\widetilde{X}}\widetilde{\varphi})\widetilde{Y}, \widetilde{Z}), \quad \widetilde{X},\widetilde{Y},\widetilde{Z} \in \mathcal{X}\widetilde{M}^{2n+2m+1}.$$

Then

$$\widetilde{F}(\widetilde{X},\widetilde{Y},\widetilde{Z}) = F(X,Y,Z) + F'(X',Y',Z').$$

Proof. Having in mind the definitions of F, F' and $\widetilde{F}$, (10) and (16) we obtain $\widetilde{F}(\widetilde{X},\widetilde{Y},\widetilde{Z}) = F(X,Y,Z) + F'(X',Y',Z')$. □

Lemma 3.3. *For the 1-forms $\widetilde{\theta}$, $\widetilde{\theta}^*$, $\widetilde{\omega}$, associated with $\widetilde{F}$ on $(\widetilde{M}^{2n+2m+1}, \widetilde{\varphi}, \widetilde{\xi}, \widetilde{\eta}, \widetilde{g})$, we have*

$$\widetilde{\theta}(\widetilde{X}) = \theta(X) + \theta'(X'), \quad \widetilde{\theta}^*(\widetilde{X}) = \theta^*(X) + \theta^{*'}(X'), \quad \widetilde{\omega}(\widetilde{X}) = \omega'(X'),$$

where $\widetilde{X} = (X, X') \in \mathcal{X}\widetilde{M}^{2n+2m+1}$, $X \in \mathcal{X}M^{2n}$, $X' \in \mathcal{X}M'^{2m+1}$. The 1-forms θ, θ^* are associated with F on (M^{2n}, J, h) and θ', $\theta^{*'}$, ω' are associated with F' on $(M'^{2m+1}, \varphi', \xi', \eta', g')$.

Proof. Let $\{e_i, Je_i\}$, $(i = 1, 2, \ldots, n)$, be a J-basis of T_pM^{2n}, $p \in M^{2n}$ and let $\{e'_j, \varphi'e'_j, \xi'\}$, $(j = 1, 2, \ldots, m)$ be a φ'-basis of $T_qM'^{2m+1}$, $q \in M'^{2m+1}$, where $h(e_i, e_i) = g'(e'_j, e'_j) = 1$,

$$h(Je_i, Je_i) = g'(\varphi'e'_j, \varphi'e'_j) = -1, \quad (i = 1,2,\ldots,n), (j = 1,2,\ldots,m).$$

Consequently,

$$\{(e_i, 0), (0, e'_j), (Je_i, 0), (0, \varphi'e'_j), (0, \xi')\}, \quad (i = 1,2,\ldots,n, j = 1,2,\ldots,m)$$

is a $\widetilde{\varphi}$-basis of $T_{(p,q)}\widetilde{M}^{2n+2m+1}$. According to (2) and (5) we compute

$$\widetilde{\theta}(\widetilde{X}) = \theta(X) + \theta'(X'), \quad \widetilde{\theta}^*(\widetilde{X}) = \theta^*(X) + \theta^{*'}(X'),$$

$$\widetilde{\omega}(\widetilde{X}) = \widetilde{F}(\widetilde{\xi}, \widetilde{\xi}, \widetilde{X}) = F'(\xi', \xi', X') = \omega'(X'). \qquad \square$$

Further we consider 7-dimensional almost contact manifolds with B-metric, constructed as direct product of a 4-dimensional almost complex manifold with B-metric and a 3-dimensional almost contact manifold with B-metric.

It is known that an almost complex manifold with B-metric (M^4, J, h) cannot belong to the class [10] W_3 and an almost contact manifold with B-metric $(M'^3, \varphi', \xi', \eta', g')$ cannot belong to the classes [7] $\mathcal{F}_2$, $\mathcal{F}_3$, $\mathcal{F}_6$, $\mathcal{F}_7$.

It is clear, when $(M^4, J, h) \in W_0$ and $(M'^3, \varphi', \xi', \eta', g') \in \mathcal{F}_i$ or $(M^4, J, h) \in W_j$ and $(M'^3, \varphi', \xi', \eta', g') \in \mathcal{F}_0$, then $\widetilde{M}^7 = M^4 \times M'^3$ belongs to the class $\mathcal{F}_i$ ($i = 0, 1, 4, 5, 8, 9, 10, 11$) or to the class $\mathcal{F}_j$ ($j = 1, 2$), respectively.

Next, we take into consideration the cases, when $\widetilde{M}^7 = M^4 \times M'^3$, $(M^4, J, h) \in W_i$ ($i = 1, 2$), and $(M'^3, \varphi', \xi', \eta', g') \in \mathcal{F}_j$ ($j = 1, 4, 5, 10, 11$).

In [7] the eleven classes of almost contact manifolds with B-metric are characterized by conditions for the tensor F. Below, we give the characterization conditions with respect to the tensor F for the classes $\mathcal{F}_i \oplus \mathcal{F}_j$, $i \neq j$, ($i = 1, 2$), ($j = 2, 4, 5, 10, 11$).

The class $\mathcal{F}_1 \oplus \mathcal{F}_2$:

$$\begin{gathered} F(\xi, Y, Z) = F(X, \xi, Z) = 0, \\ F(X, Y, \varphi Z) + F(Y, Z, \varphi X) + F(Z, X, \varphi Y) = 0. \end{gathered} \tag{17}$$

The class $\mathcal{F}_1 \oplus \mathcal{F}_4$:

$$\begin{aligned} F(X, Y, Z) =& \frac{1}{2n}(g(X, \varphi Y)\theta(\varphi Z) + g(X, \varphi Z)\theta(\varphi Y) \\ &- g(\varphi X, \varphi Y)\theta(hZ) - g(\varphi X, \varphi Z)\theta(hY)) \\ &- \frac{\theta(\xi)}{2n}(\eta(Y)g(\varphi X, \varphi Z) + \eta(Z)g(\varphi X, \varphi Y)). \end{aligned} \tag{18}$$

The class $\mathcal{F}_1 \oplus \mathcal{F}_5$:

$$\begin{aligned} F(X,Y,Z) =& \frac{1}{2n}(g(X,\varphi Y)\theta(\varphi Z) + g(X,\varphi Z)\theta(\varphi Y) \\ & - g(\varphi X,\varphi Y)\theta(hZ) - g(\varphi X,\varphi Z)\theta(hY)) \\ & - \frac{\theta^*(\xi)}{2n}(\eta(Y)g(\varphi X,Z) + \eta(Z)g(\varphi X,Y)). \end{aligned} \tag{19}$$

The class $\mathcal{F}_1 \oplus \mathcal{F}_{10}$:

$$\begin{aligned} F(X,Y,Z) =& \frac{1}{2n}(g(X,\varphi Y)\theta(\varphi Z) + g(X,\varphi Z)\theta(\varphi Y) \\ & - g(\varphi X,\varphi Y)\theta(hZ) - g(\varphi X,\varphi Z)\theta(hY)) \\ & + \eta(X)F(\xi,\varphi Y,\varphi Z). \end{aligned} \tag{20}$$

The class $\mathcal{F}_1 \oplus \mathcal{F}_{11}$:

$$\begin{aligned} F(X,Y,Z) =& \frac{1}{2n}(g(X,\varphi Y)\theta(\varphi Z) + g(X,\varphi Z)\theta(\varphi Y) \\ & - g(\varphi X,\varphi Y)\theta(hZ) - g(\varphi X,\varphi Z)\theta(hY)) \\ & + \eta(X)(\eta(Y)\omega(Z) + \eta(Z)\omega(Y)). \end{aligned} \tag{21}$$

The class $\mathcal{F}_2 \oplus \mathcal{F}_4$:

$$\begin{aligned} & F(\xi,Y,Z) = 0, \\ & F(X,Y,\varphi Z) + F(Y,Z,\varphi X) + F(Z,X,\varphi Y) \\ & \quad = \frac{\theta(\xi)}{2n}(\eta(Y)g(X,\varphi Z) + \eta(Z)g(X,\varphi Y) + \eta(X)g(Y,\varphi Z)). \end{aligned} \tag{22}$$

The class $\mathcal{F}_2 \oplus \mathcal{F}_5$:

$$\begin{aligned} & F(\xi,Y,Z) = 0, \\ & F(X,Y,\varphi Z) + F(Y,Z,\varphi X) + F(Z,X,\varphi Y) \\ & \quad = -\frac{\theta^*(\xi)}{2n}(\eta(Y)g(\varphi X,\varphi Z) + \eta(Z)g(\varphi X,\varphi Y) + \eta(X)g(\varphi Y,\varphi Z)). \end{aligned} \tag{23}$$

The class $\mathcal{F}_2 \oplus \mathcal{F}_{10}$:

$$\begin{aligned} & F(X,Y,\xi) = 0, \\ & F(X,Y,\varphi Z) + F(Y,Z,\varphi X) + F(Z,X,\varphi Y) \\ & \quad = -\eta(X)F(\xi,\varphi Y,Z) - \eta(Y)F(\xi,\varphi Z,X) - \eta(Z)F(\xi,\varphi X,Y), \\ & \theta = 0. \end{aligned} \tag{24}$$

The class $\mathcal{F}_2 \oplus \mathcal{F}_{11}$:

$$\begin{aligned} &F(X,Y,\varphi Z) + F(Y,Z,\varphi X) + F(Z,X,\varphi Y) \\ &\quad = \eta(X)\eta(Y)\omega(\varphi Z) + \eta(Y)\eta(Z)\omega(\varphi X) + \eta(Z)\eta(X)\omega(\varphi Y), \qquad (25) \\ &\theta = \omega. \end{aligned}$$

Theorem 3.1. *Let (M^4, J, h) be an almost complex manifold with B-metric and let $(M'^3, \varphi', \xi', \eta', g')$ be an almost contact manifold with B-metric and $\widetilde{M}^7 = M^4 \times M'^3$. Then*

1) $M^4 \in W_1$, $M'^3 \in \mathcal{F}_1 \Rightarrow \widetilde{M}^7 \in \mathcal{F}_1$.

2) $M^4 \in W_1$, $M'^3 \in \mathcal{F}_i \Rightarrow \widetilde{M}^7 \in \mathcal{F}_1 \oplus \mathcal{F}_i$, $(i = 4,\ 5,\ 10,\ 11)$.

3) $M^4 \in W_2$, $M'^3 \in \mathcal{F}_i \Rightarrow \widetilde{M}^7 \in \mathcal{F}_2 \oplus \mathcal{F}_i$, $(i = 1,\ 4,\ 5,\ 10, 11)$.

Proof. Using (10), (11), Lemma 3.2, Lemma 3.3, (17)÷(25) and the characterization conditions for the basic classes of almost complex manifolds with B-metric and of almost contact manifolds with B-metric, we establish the true of 1), 2), 3). □

Remark 3.1. Theorem 3.1 is valid and in the cases, when the almost complex manifold with B-metric, belonging to the class W_i, $(i = 1,\ 2)$ is of dimension $2n$ and the almost contact manifold with B-metric, belonging to the class $\mathcal{F}_j$, $(j = 1,\ 4,\ 5,\ 10,\ 11)$ is of dimension $(2m+1)$.

4 Four-dimensional submanifolds of seven-dimensional almost contact manifolds with B-metric

Let $(\overline{M}^7, \varphi, \xi, \eta, g)$ be a 7-dimensional almost contact manifold with B-metric g and let M be a submanifold of codimension 3 of $\overline{M}^7$. We assume that there exists a normal space $\{N, \varphi N, \xi\}$ defined globally over the submanifold M^4such that:

$$g(N,N) = -1, \quad g(N,\varphi N) = 0, \quad \eta(N) = 0. \qquad (26)$$

Now, if $X \in \mathcal{X}M^4$ then $\varphi X \in \mathcal{X}M^4$ and $\varphi^2 X = -X$. We denote the restriction of φ on M^4 by J. From $J^2X = -X$, $X \in \mathcal{X}M^4$ it follows that J is a complex structure on M^4. The restriction of g on M^4 we denote by the same letter. Using (4) and $\eta(X) = 0$ we verify that $g(JX,JY) = -g(X,Y)$ for arbitrary vector fields X, Y on M^4. Hence, the submanifold (M^4, J, g) of $(\overline{M}^7, \varphi, \xi, \eta, g)$ is an almost complex manifold with B-metric.

Denoting by $\overline{\nabla}$ and ∇ the Levi-Civita connections of the metric g on $\overline{M}^7$

and M^4, respectively, the formulas of Gauss and Weingarten are

$$\begin{aligned}\overline{\nabla}_X Y = \nabla_X Y - g(A_N X, Y)N + g(A_{\varphi N} X, Y)\varphi N + g(A_\xi X, Y)\xi, \\ \overline{\nabla}_X \xi = -A_\xi X + \alpha(X)N + \beta(X)\varphi N, \\ \overline{\nabla}_X N = -A_N X + \gamma(X)\varphi N + \alpha(X)\xi, \\ \overline{\nabla}_X \varphi N = -A_{\varphi N} X + \gamma(X)N - \beta(X)\xi,\end{aligned} \tag{27}$$

where X, $Y \in \mathcal{X}M^4$, A_N, $A_{\varphi N}$, A_ξ are the second fundamental tensors with respect to N, φN, and ξ respectively, and α, β, γ are 1-forms on M^4.

Let $\bar{R}$ and R be the curvature tensors of $\overline{M}^7$ and M^4, respectively. Then for the equation of Gauss from (27) and [4] we have

$$\begin{aligned}\bar{R}(x, y, z, w) =& R(x, y, z, w) + \pi_1(A_N x, A_N y, z, w) \\ &- \pi_1(A_{\varphi N} x, A_{\varphi N} y, z, w) - \pi_1(A_\xi x, A_\xi y, z, w),\end{aligned} \tag{28}$$

for arbitrary x, y, z, $w \in T_pM^4$, $p \in M^4$. The tensors π_1, π_2, π_3 are given in [6].

According to (27) and $\varphi \equiv J$ on M^4 we compute

$$\begin{aligned}(\overline{\nabla}_x \varphi)y =& (\nabla_x J)y + (g(A_{\varphi N} x, y) - g(A_N x, Jy))N \\ &+ (g(A_{\varphi N} x, Jy) + g(A_N x, y))\varphi N + g(A_\xi x, Jy)\xi,\end{aligned} \tag{29}$$

x, $y \in T_pM^4$, $p \in M^4$.

Because of (29) we have

$$\bar{F}(x, y, z) = F(x, y, z), \quad x, y, z \in T_pM^4. \tag{30}$$

From the last equality for the restriction of $\bar{\theta}$ on M^4 we obtain

$$\bar{\theta}(x)_{|M^4} = \theta(x), \quad x \in T_pM^4. \tag{31}$$

Theorem 4.1. *Let $(\overline{M}^7, \varphi, \xi, \eta, g)$ be an almost contact manifold with B-metric and let (M^4, J, g) be a submanifold of $\overline{M}^7$, defined by (26). Then*

1) $\overline{M}^7 \in \mathcal{F}_i \Rightarrow M^4 \in W_i$, $(i = 1,\ 2)$;

2) $\overline{M}^7 \in \mathcal{F}_i \Rightarrow M^4 \in W_0$, $(i = 0,\ 3,\ 4,\ 5,\ 6,\ 7,\ 8,\ 9,\ 10, 11)$.

Proof. Having in mind (30), (31), Lemma 2. from [10] and characterization conditions for the basic classes of almost complex manifolds with B-metric and of almost contact manifolds with B-metric we verify immediately that 1) and 2) are valid. □

5 Examples of submanifolds of a seven-dimensional almost contact manifold with B-metric of codimension three

Example 5.1. Let $\overline{M} = (R^7, \varphi, \xi, \eta, g)$ [7]. Identifying the point $p = (u^1, u^2, u^3, v^1, v^2, v^3, t)$ in $\overline{M}$ with its position vector Z, we define a submanifold M by the equalities:

$$g(Z,Z) = a, \quad g(Z, \varphi Z) = b, \quad \eta(Z) = 0, \tag{32}$$

where a and b are constants. M is a 4-dimensional submanifold of $\overline{M}$ and the vector fields Z, φZ, ξ are normal to T_pM^4. For $N_1 = \lambda Z + \mu\varphi Z$, $N_2 = \varphi N_1 = \lambda\varphi Z - \mu Z$, $N_3 = \xi$ by

$$\lambda = -\frac{b}{\sqrt{a+\sqrt{a^2+b^2}}.\sqrt{2(a^2+b^2)}}, \qquad \mu = \sqrt{\frac{a+\sqrt{a^2+b^2}}{2(a^2+b^2)}}$$

we get

$$\begin{aligned} g(N_1,N_1) = -1, \qquad g(N_2,N_2) = g(N_3,N_3) = 1, \\ g(N_1,N_2) = g(N_1,N_3) = g(N_2,N_3) = 0. \end{aligned} \tag{33}$$

From $\xi \perp \frac{\partial Z}{\partial u^i} \Rightarrow \eta(\frac{\partial Z}{\partial u^i}) = 0 \Rightarrow \varphi^2(\frac{\partial Z}{\partial u^i}) = -\frac{\partial Z}{\partial u^i}$, $(i = 1,\ 2,\ 3,\ 4)$, where $\{\frac{\partial Z}{\partial u^i}\}_{i=1}^4$ is a local basis of T_pM^4. Then,

$$\varphi^2 x = -x, \quad x \in T_pM^4. \tag{34}$$

Consequently, φ is a complex structure on T_pM^4 and on the all manifolds M^4 φ is an almost complex structure. We denote the restriction of φ on M^4 by J and the restriction of g on M^4 by the same letter. It is easy to check that $g(Jx, Jy) = -g(x,y)$, x, $y \in T_pM^4$, i.e. the restriction of g on M^4 is a B-metric. We can conclude that the submanifold (M^4, J, g), defined by (32) is an almost complex manifold with B-metric.

Let us denote with $\overline{\nabla}$ and ∇ the Levi-Civita connections of the metric g on $\overline{M}^7$ and M^4, respectively. Since $\overline{\nabla}$ is flat, i.e. $\overline{\nabla}_X Z = X$, Z being the position vector field and X being an arbitrary vector field on M^4, formulas of Gauss and Weingarten are

$$\begin{aligned} \overline{\nabla}_X Y =& \nabla_X Y + (\lambda g(X,Y) + \mu g(X,JY))N_1 \\ &+ (-\lambda g(X,JY) + \mu g(X,Y))N_2, \\ \overline{\nabla}_X\xi = 0,& \quad \overline{\nabla}_X N_1 = \lambda X + \mu JX, \quad \overline{\nabla}_X N_2 = \lambda JX - \mu X, \end{aligned} \tag{35}$$

for X, $Y \in \mathcal{X}M^4$.

Using (34), (35) we calculate $(\nabla_X J)Y = 0$ and hence $F(X,Y,Z) = 0$, where X, Y, $Z \in \mathcal{X}M^4$. Thus, the submanifold (M^4, J, g) of the flat $\mathcal{F}_0$-manifold $\overline{M} = (R^7, \varphi, \xi, \eta, g)$, defined by (32) is an almost complex manifold with B-metric in the class W_0.

Now, from (35), [4] and $\bar{R} = 0$ for the curvature tensor R of the submanifold M^4 we find

$$\begin{aligned} R(x,y,z,w) &= (\mu^2 - \lambda^2)(\pi_1(x,y,z,w) - \pi_2(x,y,z,w)) + 2\lambda\mu\pi_3(x,y,z,w) \\ &= \frac{a}{a^2+b^2}(\pi_1(x,y,z,w) - \pi_2(x,y,z,w)) - \frac{b}{a^2+b^2}\pi_3(x,y,z,w) \end{aligned} \tag{36}$$

for x, y, z, $w \in T_pM^4$.

Proposition 5.1. *Let (M^4, J, g) be a submanifold of the flat $\mathcal{F}_0$-manifold $\overline{M} = (R^7, \varphi, \xi, \eta, g)$, defined by the equalities (32). Then M^4 has constant totally real sectional curvatures $K(\alpha;p) = \nu = \frac{a}{a^2+b^2}$, $\widetilde{K}(\alpha;p) = \widetilde{\nu} = -\frac{b}{a^2+b^2}$.*

Proof. Let $\alpha = \{x,y\}$, x, $y \in T_pM^4$ be a totally real section. From (36) it follows that

$$R(x,y,y,x) = \frac{a}{a^2+b^2}(\pi_1(x,y,y,x) - \pi_2(x,y,y,x)) - \frac{b}{a^2+b^2}\pi_3(x,y,y,x).$$

Then the formulas (8) and (9) imply immediately $K(\alpha;p) = \nu = \frac{a}{a^2+b^2}$ and $\widetilde{K}(\alpha;p) = \widetilde{\nu} = -\frac{b}{a^2+b^2}$. □

Corollary 5.1. *From the Example 5.1. by $a = -1$, $b = 0$ we obtain the submanifold M^4, defined by the equalities:*

$$g(Z,Z) = -1, \quad g(Z,\varphi Z) = 0, \quad \eta(Z) = 0. \tag{37}$$

So defined, the submanifold has constant totally real sectional curvatures $\nu = -1$, $\widetilde{\nu} = 0$.

Remark 5.1. The submanifold, considered in Corollary 5.1. is a sphere of Kotelnikov-Study.

Remark 5.2. It is shown in Theorem 1 [6] that for a Kaehler manifold with B-metric of pointwise constant totally real sectional curvatures $\nu(p)$ and $\widetilde{\nu}(p)$, both functions $\nu(p)$, $\widetilde{\nu}(p)$ are constant by $2n \geq 6$. Example 5.1 is an example of a 4-dimensional Kaehler manifold with B-metric of constant totally real sectional curvatures ν and $\widetilde{\nu}$.

References

1. D. Blair, *Contact manifolds in Riemannian geometry*, Lect. Notes in Math. 7, Springer Verlag, Berlin, 1976, pp. 19–35.
2. S. Kanemaki, *On quasi-Sasakian manifolds*, Lect. Notes in Diff. Geom., Stefan Banach Int. Math. Center, Warsaw, 1979.
3. J. Oubina, *A classification for almost contact structures, II*, Lect. Notes, Santiago de Compostela, Spain, 1979.
4. K. Yano and M. Kon, *CR-Submanifolds of Kaehlerian and Sasakian manifolds*, in Progress in mathematics, Vol. 30, Birkhauser, Boston, Basel, Stuttgart, 1983.
5. G. Ganchev and A. Borisov, *Note an almost complex manifolds with Norden metric*, C. R. Acad. Bulg. Sci. **39**(1986), 31–34.
6. A. Borisov and G. Ganchev, *Curvature properties of Kaehlerian manifolds with B-metric*, in Proc. XIV Spring Conf. UBM, Sunny Beach, 1985, pp. 220–226.
7. G. Ganchev, V. Mihova and K. Gribachev, *Almost contact manifolds with B-metric*, Math. Balkanica. **7**(1993), 262–276.
8. G. Nakova and K. Gribachev, *Submanifolds of some almost contact manifolds with B-metric with codimension two-I*, Math. Balkanica. **11**(1997).
9. ———, *Submanifolds of some almost contact manifolds with B-metric with codimension two-II*, Math. Balkanica. **11**(1997).
10. M. Manev, *Examples of almost contact manifolds with B-metric, derived from almost complex manifolds with B-metric*, Plovdiv University "PaissiiHilendarski", Bulgaria, Scientific works, Vol. 32, Book 3, Mathematics, 1995.
11. G. Nakova and K. Gribachev, *On the Lorentzian submanifolds of an almost contact manifold with B-metric of dimension 5*, to appear.

GEOMETRY OF REAL HYPERSURFACES IN A COMPLEX PROJECTIVE SPACE

SADAHIRO MAEDA
Department of Mathematics, Shimane University, Matsue 690-8504, Japan
e-mail: smaeda@math.shimane-u.ac.jp

In this paper, we survey some chracterizations of homogeneous real hypersurfaces in a complex projective space and pose some open problems on real hypersurfaces.

0 Introduction

Let $P_n(\mathbb{C})$ be an n-dimensional complex projective space with Fubini-Study metric of constant holomorphic sectional curvature 4, and let M be a real hypersurface of $P_n(\mathbb{C})$. Then M has an almost contact metric structure (ϕ, ξ, η, g) induced from the Kaehler structure J of $P_n(\mathbb{C})$. This structure is a useful tool in the study of real hypersurfaces of a Kaehler manifold (cf. [3,4,5,9]). Typical examples of real hypersurfaces in $P_n(\mathbb{C})$ are *homogeneous* real hypersurfaces, which are given as orbits under subgroups of the projective unitary group $PU(n+1)$.

Takagi ([10]) determined all homogeneous real hypersurfaces in $P_n(\mathbb{C})$. By virtue of his work, we find that a homogeneous real hypersurface in $P_n(\mathbb{C})$ is locally congruent to one of the six model spaces of type A_1, A_2, B, C, D and E. They are realized as tubes of constant radius over compact Hermitian symmetric spaces of rank 1 or rank 2 (for details, see Theorem A). A homogeneous real hypersurface of type A_1 is usually called a *geodesic hypersphere*.

In the study of real hypersurfaces in $P_n(\mathbb{C})$, many differential geometers are interested in the following two problems:

(1) Give a characterization of homogeneous real hypersurfaces in $P_n(\mathbb{C})$.

(2) Construct non-homogeneous *nice* real hypersurfaces in $P_n(\mathbb{C})$ and characterize such examples.

In this paper, we concentrate on Problem 1. In submanifold theory it is interesting to know the shape of a Riemannian submanifold by observing the extrinsic shape of geodesics of the submanifold. From this point of view it is known that a hypersurface M^n in $\mathbb{R}^{n+1}$ is locally a standard sphere if and only if all geodesics of M are circles with nonzero curvature in $\mathbb{R}^{n+1}$.

First, we shall provide some characterizations of homogeneous real hypersurfaces in $P_n(\mathbb{C})$ by observing the shape of geodesics on the real hypersurfaces as curves in $P_n(\mathbb{C})$ (see, Theorems 1, 2, 3).

Second, by investigating the holomorphic distribution T^0M which is defined by $(T^0M)_p = \{ X \in T_p(M) \mid X \perp \xi \}$ for $p \in M$, we shall give some characterizations of homogeneous real hypersurfaces in $P_n(\mathbb{C})$ (see, Theorems 4, 5).

In section 3, we review gap theorems for a minimal real hypersurface M with respect to curvatures of M (that is, the scalar curvature, the sectional curvature or the Ricci curvature of M). The main purpose of section 3 is to pose some open problems related to this topic.

1 Preliminaries

Let M be an orientable real hypersurface of $P_n(\mathbb{C})$ and N be a unit normal vector field on M. The Riemannian connections $\tilde{\nabla}$ in $P_n(\mathbb{C})$ and ∇ in M are related by the following formulas for any vector fields X and Y on M:

$$\tilde{\nabla}_X Y = \nabla_X Y + g(AX, Y)N, \tag{1.1}$$

$$\tilde{\nabla}_X N = -AX, \tag{1.2}$$

where g denotes the Riemannian metric of M induced from the Fubini-Study metric G of $P_n(\mathbb{C})$ and A is the shape operator of M in $P_n(\mathbb{C})$. An eigenvector X of the shape operator A is called a *principal curvature vector*. Also an eigenvalue λ of A is called a *principal curvature*. In the following, we denote by V_λ the eigenspace of A associated with eigenvalue λ. It is known that M has an *almost contact metric structure* induced from the Kaehler structure J of $P_n(\mathbb{C})$. Namely, we can define a tensor field ϕ of type $(1,1)$, a vector field ξ and a 1-form η on M by

$$g(\phi X, Y) = G(JX, Y) \quad \textit{and} \quad g(\xi, X) = \eta(X) = G(JX, N).$$

Then we have

$$\phi^2 X = -X + \eta(X)\xi, \; g(\xi,\xi) = 1, \; \phi\xi = 0. \tag{1.3}$$

From (1.1) we find

$$(\nabla_X \phi)Y = \eta(Y)AX - g(AX, Y)\xi, \tag{1.4}$$

$$\nabla_X \xi = \phi AX. \tag{1.5}$$

Let R denote the curvature tensor of M. Then we have the following Gauss and Codazzi equations:

$$\begin{aligned} g(R(X,Y)Z,W) = g(Y,Z)g(X,W) - g(X,Z)g(Y,W) + g(\phi Y,Z)g(\phi X,W) \\ - g(\phi X,Z)g(\phi Y,W) - 2g(\phi X,Y)g(\phi Z,W) \\ + g(AY,Z)g(AX,W) - g(AX,Z)g(AY,W), \end{aligned} \tag{1.6}$$

$$(\nabla_X A)Y - (\nabla_Y A)X = \eta(X)\phi Y - \eta(Y)\phi X - 2g(\phi X,Y)\xi. \tag{1.7}$$

In the following, we use the same terminology and notations as above unless otherwise stated. Now we prepare without proof the following in order to prove Theorem 1.

Theorem A ([10]). *Let M be a homogeneous real hypersurface of $P_n(\mathbb{C})$. Then M is a tube of radius r over one of the following Kaehler submanifolds:*

(A_1) *hyperplane $P_{n-1}(\mathbb{C})$, where $0 < r < \frac{\pi}{2}$,*

(A_2) *totally geodesic $P_k(\mathbb{C})(1 \leqq k \leqq n-2)$, where $0 < r < \frac{\pi}{2}$,*

(B) *complex hyperquadric Q_{n-1}, where $0 < r < \frac{\pi}{4}$,*

(C) *$P_1(\mathbb{C}) \times P_{\frac{n-1}{2}}(\mathbb{C})$, where $0 < r < \frac{\pi}{4}$ and $n(\geqq 5)$ is odd,*

(D) *complex Grassmann $G_{2,5}(\mathbb{C})$, where $0 < r < \frac{\pi}{4}$ and $n = 9$,*

(E) *Hermitian symmetric space $SO(10)/U(5)$, where $0 < r < \frac{\pi}{4}$ and $n = 15$.*

The numbers of distinct principal curvatures of these homogeneous real hypersurfaces are 2, 3, 3, 5, 5, 5, respectively.

Theorem B ([4]). *Let M be a real hypersurface of $P_n(\mathbb{C})$. Then M has constant principal curvatures and ξ is a principal curvature vector if and only if M is locally congruent to a homogeneous real hypersurface.*

Proposition A ([9]). *Assume that ξ is a principal curvature vector and the corresponding principal curvature is α. Then α is locally constant. In addition, $A\phi X = \frac{\alpha\lambda+2}{2\lambda-\alpha}\phi X$ holds for any $X(\perp\xi) \in V_\lambda$.*

Proposition B ([8]). *Let M be a homogeneous real hypersurface given as a tube of radius r $(0 < r < \frac{\pi}{2})$ over a totally geodesic $P_k(\mathbb{C})$ $(1 \leqq k \leqq n-1)$ in $P_n(\mathbb{C})$. Take orthonormal vectors $v_1, v_2, \ldots, v_{2n-2}$ orthogonal to ξ at an arbitrary point p of M in such a way that $v_1, v_2, \ldots, v_{2k}$ (resp. $v_{2k+1},$*

$v_{2k+2}, \ldots, v_{2n-2}$) *are principal curvature vectors with principal curvature* $-\tan r$ (*resp.* $\cot r$). *Then* $v_1, v_2, \ldots, v_{2n-2}$ *satisfy the following:*

(*i*) *All geodesics* γ_i *on* M *with* $\gamma_i(0) = p$ *and* $\dot\gamma_i(0) = v_i$ $(1 \leqq i \leqq 2k)$ *are circles of curvature* $\tan r$ *in* $P_n(\mathbb{C})$.

(*ii*) *All geodesics* γ_i *on* M *with* $\gamma_i(0) = p$ *and* $\dot\gamma_i(0) = v_i$ $(2k+1 \leqq i \leqq 2n-2)$ *are circles of curvature* $\cot r$ *in* $P_n(\mathbb{C})$.

Finally we review the definition of circles in Riemannian geometry. A curve $\gamma = \gamma(s)$ in a Riemannian manifold M parametrized by its arclength s is called a *circle*, if there exists a field of unit vectors $Y = Y(s)$ along the curve which satisfies the differential equations: $\nabla_{\dot\gamma}\dot\gamma = \kappa Y$ and $\nabla_{\dot\gamma} Y = -\kappa\dot\gamma$, where $\kappa(\geqq 0)$ is a constant, which is called the *curvature* of the circle γ and $\nabla_{\dot\gamma}$ denotes the covariant differentiation along γ with respect to the Riemannian connection ∇ of M. As a matter of course, a circle of null curvature is nothing but a geodesic.

2 Characterizations of homogeneous real hypersurfaces

The main theorem in this paper is the following theorem. For the detailed proof, see [2].

Theorem 1. *Let* M *be a connected real hypersurface of* $P_n(\mathbb{C})$. *Then* M *is locally congruent to a homogeneous real hypersurface if and only if there exist such orthonormal vectors* $v_1, v_2, \ldots, v_{2n-2}$ *orthogonal to* ξ *at each point* p *of* M *that all geodesics* $\gamma_i = \gamma_i(s)$ *on* M *with* $\gamma_i(0) = p$ *and* $\dot\gamma_i(0) = v_i$ $(1 \leqq i \leqq 2n-2)$ *are circles in* $P_n(\mathbb{C})$ *with nonzero curvature.*

Sketch of proof. Let M be a homogeneous real hypersurface (of radius r over a certain Kaehler submanifold) in $P_n(\mathbb{C})$. We shall show one by one the fact that, by taking orthonormal vectors $v_1, \ldots, v_{2n-2}$ orthogonal to ξ at an arbitrary point p of M in such a way that $v_1, \ldots, v_{2n-2}$ are principal curvature vectors, all geodesics $\gamma_i = \gamma_i(s)$ on M with $\gamma_i(0) = p$ and $\dot\gamma_i(0) = v_i$ $(1 \leqq i \leqq 2n-2)$ are circles of nonzero curvature.

Let M be of type A_1 or type A_2. See Proposition B.

Now let M be of type B. Then $T_p(M) = V_{r_1} \oplus V_{r_2} \oplus \{\xi\}_{\mathbb{R}}$ at any point p of M, where $r_1 = \frac{1+x}{1-x}$, $r_2 = \frac{x-1}{x+1}$, $\alpha = x - \frac{1}{x}$, $x = \cot r$ $(0 < r < \frac{\pi}{4})$ (for details, see [11]). We remark that $\phi V_{r_1} = V_{r_2}$ (see Proposition A). A direct computation shows that the principal foliation V_{r_1} (resp. V_{r_2}) on M is integrable, and moreover that any leaf of V_{r_1} (resp. V_{r_2}) is a totally geodesic submanifold of M. Hence any leaf L_{r_1} of the principal foliation V_{r_1} is totally umbilic in $P_n(\mathbb{C})$ (see, (1.1)). L_{r_1} is locally congruent to a totally umbilic hypersurface of constant curvature d (with $\sqrt{d-1} = |r_1|$) in $P^n(\mathbb{R})$ which is a totally real totally geodesic submanifold of $P_n(\mathbb{C})$. This shows that every geodesic of L_{r_1} is a circle in $P_n(\mathbb{C})$ of positive curvature $|r_1|$. Hence we can check the case of type B.

Next, let M be one of type C, type D and type E. Let $x = \cot r$ $(0 < r < \frac{\pi}{4})$. Then $T_p(M)$ is decomposed as: $T_p(M) = V_{r_1} \oplus V_{r_2} \oplus V_{r_3} \oplus V_{r_4} \oplus \{\xi\}_{\mathbb{R}}$, where $r_1 = \frac{1+x}{1-x}$, $r_2 = \frac{x-1}{x+1}$, $r_3 = x$, $r_4 = -\frac{1}{x}$ and $\alpha = x - \frac{1}{x}$. Note that $\phi V_{r_1} = V_{r_2}$, $\phi V_{r_3} = V_{r_3}$, $\phi V_{r_4} = V_{r_4}$ (see, Proposition A). By virtue of the discussion in the case of type B we have only to study the foliations V_{r_3} and V_{r_4}. We remark that both of V_{r_3} and V_{r_4} are not integrable. Let $T = V_{r_3} \oplus \{\xi\}_{\mathbb{R}}$. Then a similar computation yields that any leaf L of the distribution T is a totally geodesic submanifold of M. The manifold L is locally congruent to a homogeneous real hypersurface (with unit normal vector $-N$) of type A_1 of radius $(\frac{\pi}{2} - r)$ in $P_{m+1}(\mathbb{C})$ which is a holomorphic totally geodesic submanifold of $P_n(\mathbb{C})$, where $2m = \dim V_{r_3}$ (see, Theorem 1 in [3]). Hence every geodesic orthogonal to ξ of L is a circle in $P_n(\mathbb{C})$ of positive curvature $\cot r$ (see, Proposition B). The same discussion as above tells us that any leaf K of the distribution $V_{r_4} \oplus \{\xi\}_{\mathbb{R}}$ is a totally geodesic submanifold of M. The manifold K is locally congruent to a homogeneous real hypersurface (with unit normal vector N) of type A_1 of radius r in $P_{k+1}(\mathbb{C})$ which is a holomorphic totally geodesic submanifold of $P_n(\mathbb{C})$, where $2k = \dim V_{r_4}$. Then every geodesic orthogonal to ξ of K is a circle in $P_n(\mathbb{C})$ of positive curvature $\tan r$ (see, Proposition B). Thus we can check the cases of type C, type D and type E.

Conversely, let M be a real hypersurface satisfying the hypothesis of our Theorem. Let $\gamma_i = \gamma_i(s)$ $(1 \leqq i \leqq 2n-2)$ be geodesics of M with $\gamma_i(0) = p$ and $\dot{\gamma}_i(0) = v_i$. Then they satisfy

$$\widetilde{\nabla}^2_{\dot{\gamma}_i} \dot{\gamma}_i = -k_i^2 \dot{\gamma}_i \tag{2.1}$$

for some positive constants k_i. On the other hand, from (1.1) and (1.2) we have

$$\widetilde{\nabla}^2_{\dot{\gamma}_i} \dot{\gamma}_i = g((\nabla_{\dot{\gamma}_i} A)\dot{\gamma}_i, \dot{\gamma}_i)N - g(A\dot{\gamma}_i, \dot{\gamma}_i)A\dot{\gamma}_i. \tag{2.2}$$

Comparing the tangential components of (2.1) and (2.2), we obtain

$$g(A\dot{\gamma}_i, \dot{\gamma}_i)A\dot{\gamma}_i = k_i^2 \dot{\gamma}_i$$

so that at $s = 0$

$$g(Av_i, v_i)Av_i = k_i^2 v_i.$$

Hence

$$Av_i = k_i v_i \qquad or \qquad Av_i = -k_i v_i \quad (1 \leqq i \leqq 2n-2). \tag{2.3}$$

Equation (2.3) implies that ξ is a principal curvature vector.

We consider the open dense subset $\mathcal{U} = \{ p \in M \mid$ the multiplicity of each principal curvature of M in $P_n(\mathbb{C})$ is constant on some neighborhood $\mathcal{V}_p$ of $p \}$ of M. We remark that all principal curvatures are differentiable on $\mathcal{U}$ and in a sufficiently small neighborhood of any point p in $\mathcal{U}$ the principal curvature vectors can be chosen to be smooth.

In the following, we shall study on a fixed neighborhood $\mathcal{V}_p$. We divide the discussion into the following two cases at the fixed point $p(\in \mathcal{U})$:

(I) For any i $(1 \leqq i \leqq 2n-2)$, $\alpha \neq k_i$ and $\alpha \neq -k_i$.

(II) For some i $(1 \leqq i \leqq 2n-2)$, $\alpha = k_i$ or $\alpha = -k_i$.

(I) The tangent space $T_p(M)$ is decomposed as;

$$\begin{aligned} T_p(M) =& \{\xi\}_{\mathbb{R}} \oplus \{v \in T_p(M) \mid Av = -k_{i_1} v\} \oplus \{v \in T_p(M) \mid Av = k_{i_1} v\} \oplus \cdots \\ & \oplus \{v \in T_p(M) \mid Av = -k_{i_g} v\} \oplus \{v \in T_p(M) \mid Av = k_{i_g} v\}, \end{aligned}$$

where $0 < k_{i_1} < k_{i_2} < \cdots < k_{i_g}$ and g is the number of distinct positive k_j $(j = 1, \ldots, 2n-2)$. Hence our discussion guarantees that every k_{i_j} is differentiable on $\mathcal{V}_p$. Next, we shall show the constancy of k_{i_j}. It suffices to check the case that $Av_{i_j} = k_{i_j} v_{i_j}$. By hypothesis $v_{i_j} k_{i_j} = 0$. For any v_l $(1 \leqq l \neq i_j \leqq 2n-2)$, since A is symmetric, we see

$$g((\nabla_{v_{i_j}} A) v_l, v_{i_j}) = g(v_l, (\nabla_{v_{i_j}} A) v_{i_j}) \tag{2.4}$$

It follows from Equation (2.4) that $v_l k_{i_j} = 0$. Thus we can see that $X k_{i_j} = 0$ for any $X (\perp \xi) \in T_p(M)$. Next, we shall check $\xi k_{i_j} = 0$. It follows from (1.5) and Proposition A that

$$(\nabla_\xi A) V_{i_j} - (\nabla_{V_{i_j}} A)\xi = (\xi k_{i_j}) V_{i_j} + (k_{i_j} I - A) \nabla_\xi V_{i_j} - k_{i_j} \left(\alpha - \frac{\alpha k_{i_j} + 2}{2k_{i_j} - \alpha} \right) \phi V_{i_j}.$$

This, together with (1.7), implies $\xi k_{i_j} = 0$. Therefore we can see that the differential dk_{i_j} of k_{i_j} vanishes at the point p, which tells us that every $k_{i_j} (> 0)$ is constant on $\mathcal{V}_p$, since p is an arbitrary point of $\mathcal{V}_p$.

(II) On the fixed neighborhood $\mathcal{V}_p$ the multiplicity of each principal curvature of M in $P_n(\mathbb{C})$ is constant. Note that α is constant (cf. Proposition A). These, combined with the fact that every principal curvature (function) is continuous on M, yield that $\alpha = k_i$ or $\alpha = -k_i$ on $\mathcal{V}_p$, so that k_i is constant on

$\mathcal{V}_p$. The discussion in the case (I) shows the constancy of the other distinct k_j.

Thus we find that every principal curvature is locally constant on the open dense subset $\mathcal{U}$ of the connected real hypersurface M. Therefore the continuity of all principal curvatures on M gives the conclusion (see, Theorem B). We finally note that the case (II) does not happen at any point of M. □

In the hypothesis of Theorem 1 we do not need to suppose that we take the vectors $\{v_1, \ldots, v_{2n-2}\}$ as *a local field of orthonormal frames* in M. However, for each homogeneous real hypersurface M, we can take a local field of orthonormal frames in M satisfying the hypothesis of our Theorem.

It is well-known that there does not exist a real hypersurface all of whose geodesics are circles in $P_n(\mathbb{C})$. Every circle in Theorem 1 is a simple closed curve which lies on some totally real totally geodesic $P^2(\mathbb{R})$ in $P_n(\mathbb{C})$. We note that for any homogeneous real hypersurface M, at each point p of M the geodesic $\gamma = \gamma(s)$ with $\gamma(0) = p$ and $\dot{\gamma}(0) = \xi$ is also a circle in $P_n(\mathbb{C})$ which is a simple closed curve lying on some holomorphic totally geodesic $P_1(\mathbb{C})$ in $P_n(\mathbb{C})$. All circles in $P_n(\mathbb{C})$ are simple curves. However, a circle in $P_n(\mathbb{C})$ is not necessarily closed (cf. [1]).

Due to the discussion in the proof of Theorem 1 we get the following fact:

(i) Every homogeneous real hypersurface except for type A_1 and type A_2 in $P_n(\mathbb{C})$ admits some real projective space (of the standard metric) as a totally geodesic submanifold.

(ii) Every homogeneous real hypersurface except for type B in $P_n(\mathbb{C})$ admits a geodesic hypersphere (in some $P_k(\mathbb{C})$ which is a holomorphic totally geodesic submanifold of $P_n(\mathbb{C})$) as a totally geodesic submanifold.

As immediate consequences of Theorem 1 we obtain the following:

Theorem 2. *Let M be a real hypersurface of $P_n(\mathbb{C})$. Then M is locally congruent to a geodesic hypersphere of radius r $(0 < r < \frac{\pi}{2})$ if and only if there exist such orthonormal vectors $v_1, v_2, \ldots, v_{2n-2}$ orthogonal to ξ at each point p of M that all geodesics of M through p in the direction $v_i + v_j$ $(1 \leqq i \leqq j \leqq 2n-2)$ are circles with nonzero curvature.*

Theorem 3. *Let M be a real hypersurface of $P_n(\mathbb{C})$. Then M is locally congruent to either a geodesic hypersphere of radius r $(0 < r < \frac{\pi}{2})$ or a homogeneous real hypersurface of type A_2 with radius $\frac{\pi}{4}$ if and only if there exist such orthonormal vectors $v_1, v_2, \ldots, v_{2n-2}$ orthogonal to ξ at each* *point p of M that all geodesics* $\gamma_i = \gamma_i(s)$ *on* M *with* $\gamma_i(0) = p$ *and* $\dot{\gamma}_i(0) = v_i$ $(1 \leqq i \leqq 2n-2)$ *are circles with the same nonzero curvature.*

By virtue of the discussion in the proof of Theorem 1 we establish the following:

Theorem 4. *Let* M *be a real hypersurface of* $P_n(\mathbb{C})$, $n \geqq 3$. *Then* M *is locally congruent to a homogeneous real hypersurface of type* B *if and only if the holomorphic distribution* $T^0M = \{X \in TM \mid \eta(X) = 0\}$ *satisfies the following three conditions:*

(1) T^0M *is decomposed as the direct sum of principal foliations* V_{λ_i}*'s of* M *in* $P_n(\mathbb{C})$ *with* $\dim V_{\lambda_i} \geqq 2$.

(2) *Every principal foliation* V_{λ_i} *in condition* (1) *is integrable.*

(3) *Every leaf of any principal foliation* V_{λ_i} *in condition* (1) *is a totally geodesic submanifold of the real hypersurface* M.

Sketch of proof. Theorem 1 tells us that the manifold M satisfying the hypothesis of Theorem 4 is homogeneous in $P_n(\mathbb{C})$. Hence we get the conclusion. □

Theorem 5. *Let* M *be a real hypersurface of* $P_n(\mathbb{C})$. *Then* M *is locally congruent to a homogeneous real hypersurface of type* A_1 *or type* A_2 *if and only if the holomorphic distribution* T^0M *satisfies the following two conditions:*

(1) T^0M *is decomposed as the direct sum of principal foliations* V_{λ_i}*'s of* M *in* $P_n(\mathbb{C})$.

(2) *For each principal foliation* V_{λ_i} *in condition* (1), *the distribution* $V_{\lambda_i} \oplus \{\xi\}_{\mathbb{R}}$ *is integrable.*

Sketch of proof. First, from condition (1) we know that ξ is principal. By condition (2), for any $X(\in V_{\lambda_i})$, we have $\nabla_X\xi - \nabla_\xi X \in V_{\lambda_i} \oplus \{\xi\}_{\mathbb{R}}$ $(i = 1, \ldots, d)$. We note that $\nabla_X\xi - \nabla_\xi X$ is perpendicular to ξ for any $X(\in V_{\lambda_i})$, so that

$$A(\nabla_X\xi - \nabla_\xi X) = \lambda_i(\nabla_X\xi - \nabla_\xi X) \qquad \textit{for any } X \in V_{\lambda_i}.$$

This, together with (1.4) and Proposition A, shows

$$(A - \lambda_i I)\nabla_\xi X = \lambda_i\left(\frac{\alpha\lambda_i + 2}{2\lambda_i - \alpha} - \lambda_i\right)\phi X. \tag{2.5}$$

It follows from (1.4), (2.5) and Proposition A that

$$(\nabla_X A)\xi - (\nabla_\xi A)X = \lambda_i(\alpha - \lambda_i)\phi X - (\xi\lambda_i)X.$$

On the other hand, the Codazzi equation (1.7) implies

$$(\nabla_X A)\xi - (\nabla_\xi A)X = -\phi X.$$

Hence, the principal curvature λ_i is a solution for the quadratic equation: $\lambda_i^2 - \alpha\lambda_i - 1 = 0$. This yields that $\phi V_{\lambda_i} = V_{\lambda_i}$ $(i = 1, \ldots, d)$, so that M is of type A_1 or type A_2 (cf. [9]). As a matter course these two homogeneous real hypersurfaces satisfy the hypothesis of Theorem 5. □

We finally remark that if we omit the condition (3) of Theorem 4, Theorem 4 is not true. There exist non-homogeneous real hypersurfaces in $P_n(\mathbb{C})$ satisfying the conditions (1) and (2) of Theorem 4.

3 Pinching problems for minimal real hypersurfaces

In this section, we present all results without proof. Let M be a minimal real hypersurface of $P_n(\mathbb{C})$ of constant holomorphic sectional curvature 4. We denote by S the Ricci tensor of type (1,1) of M. Then Equation (1.6) shows

$$SX = (2n+1)X - 3\eta(X)\xi - A^2X. \tag{3.1}$$

Let ρ be the scalar curvature of M. Then (3.1) implies that $\rho = 4(n^2-1) - \operatorname{trace} A^2$, so that $\rho \leqq 4(n^2-1)$ holds for any minimal real hypersurface M. The following is well-known.

Theorem C ([6]). *Let M be a compact orientable minimal real hypersurface of $P_n(\mathbb{C})$. Suppose that the scalar curvature ρ of M satisfies $\rho \geqq 2(2n^2-n-1)$. Then $\rho = 2(2n^2-n-1)$ and M is congruent to one of minimal homogeneous real hypersurfaces of type A_1 and type A_2.*

We are interested in the scalar curvature of the other homogeneous minimal real hypersurfaces. By a direct computation we have

Proposition 1. *Let M be a minimal homogeneous real hypersurface of type B, C, D or E in $P_n(\mathbb{C})$. Then the scalar curvature ρ of M satisfies $\rho = 2(2n^2-3n-1)$.*

Proposition 1 tells us that we cannot distinguish minimal homogeneous real hypersurfaces of type B, C, D and E in terms of scalar curvatures ρ.

Motivated by Theorem C and Proposition 1, we pose the following:

Problem 1. *Let M be a minimal non-homogeneous real hypersurface of $P_n(\mathbb{C})$. Does there exist M (even in local) with the scalar curvature ρ satisfying the following inequalities?*

$$2(2n^2-3n-1) \leqq \rho \leqq 2(2n^2-n-1).$$

Remark 1. A direct calculation asserts that each homogeneous real hypersurface M (which is a tube of radius r) is minimal in the following cases:

(A_1) $\cot r = \sqrt{2n-1}$,

(A_2) $\cot r = \sqrt{\frac{2k+1}{2n-2k-1}}$,

(B) $\cot r = \sqrt{n} + \sqrt{n-1}$,

(C) $\cot r = \frac{\sqrt{n}+\sqrt{2}}{\sqrt{n-2}}$,

(D) $\cot r = \sqrt{5}$,

(E) $\cot r = \frac{\sqrt{15}+\sqrt{6}}{3}$.

Next, we consider the problem on sectional curvatures. It is known that

Theorem D ([5]). *Let M be a compact orientable minimal real hypersurface of $P_n(\mathbb{C})$. Suppose that the sectional curvature K satisfies $K \geqq \frac{1}{2n-1}$. Then M is congruent to the minimal geodesic hypersphere.*

Remark 2. The sectional curvature K of the minimal geodesic hypersphere satisfies $\frac{1}{2n-1} \leqq K \leqq \frac{8n-3}{2n-1}$.

In relation to Theorem D, we pose the following:

Problem 2. *Let M be a compact orientable minimal real hypersurface of $P_n(\mathbb{C})$. If the sectional curvature K of M satisfies $K \geqq 0$, is M congruent to one of minimal homogeneous real hypersurfaces of type A_1 and type A_2?*

Finally, we study the pinching problem on Ricci curvatures. Equation (3.1) shows that every Ricci curvature of any minimal real hypersurface M is not greater than $2n+1$. We call $g(SX, X)$ the Ricci curvature of M with respect to the unit vector X.

We here calculate the Ricci curvature of any minimal homogeneous real hypersurface.

Proposition 2. *Let M be a minimal homogeneous real hypersurface in $P_n(\mathbb{C})$. We denote by R the Ricci curvature of M. Then R satisfies the following inequalities:*

(A_1) $\frac{2n-2}{2n-1} \leqq R \leqq \frac{4n^2-2}{2n-1}$,

(A_2) $2n - \frac{2n-2k-1}{2k+1} - \frac{2k+1}{2n-2k-1} \leqq R \leqq 2n+1 - \frac{2n-2k-1}{2k+1}$,

(B) $-2n+2 \leqq R \leqq 2n+1 - \frac{\sqrt{n}-1}{\sqrt{n}+1}$,

(C) $n+2-\sqrt{n(n-2)} \leqq R \leqq n+2+\sqrt{n(n-2)}$,

(D) $\frac{31-3\sqrt{5}}{2} \leqq R \leqq \frac{31+3\sqrt{5}}{2}$,

(E) $27-\sqrt{15} \leqq R \leqq 27+\sqrt{15}$.

Proposition 2 yields the following

Problem 3. *Let M be a compact orientable minimal real hypersurface of $P_n(\mathbb{C})$. If every Ricci curvature of M is not less than $2n-2$, is M congruent to the minimal homogeneous real hypersurface of type A_2 with $2k = n-1$?*

An easy computation tells us the following local statement.

Proposition C ([7]). *Let M be a minimal real hypersurface of $P_n(\mathbb{C})$, $n \geqq 3$. Suppose that the following inequalities hold $2n-2 \leqq$ (every Ricci curvature of M) $\leqq 2n$. Then M is locally congruent to the minimal homogeneous real hypersurface of type A_2 with $2k = n-1$.*

Remark 3. From the viewpoint of Problem (1) it is important to compute various geometric invariants of homogeneous real hypersurfaces. For instance, we shall consider the following invariant:

Let M be a real hypersurface of $P_n(\mathbb{C})$. Then by an easy computation we find that ξ is principal if and only if $\nabla_\xi \phi = 0$. In general, for any real hypersurface M the following relation holds;

$$||\nabla\phi||^2 = 2\text{trace } A^2 - 2||A\xi||^2 \geqq 0.$$

We shall compute $||\nabla\phi||^2$ for each homogeneous minimal real hypersurface.

Proposition 3. *Let M be a minimal homogeneous real hypersurface of $P_n(\mathbb{C})$. Then $||\nabla\phi||^2$ of M is described as follows:*

(A_1) $\frac{4(n-1)}{2n-1}$,

(A_2) $4\left\{\frac{(n-1-k)(2k+1)}{2n-1-2k}+\frac{k(2n-1-2k)}{2k+1}\right\}$,

(B) $4(n+1)$,

(C) $12n-4-\frac{16}{n-2}$,

(D) $\frac{488}{5}(=97.6)$,

(E) $\frac{512}{3}(=170.6\cdots)$.

Proposition 3 shows that $\|\nabla\phi\|^2$ takes the minimum value in case of type A_1 and takes the maximum value in case of type C.

References

1. T. Adachi, S. Maeda and S. Udagawa, *Circles in a complex projective space*, Osaka J. Math. **32**(1995), 709–719.
2. T. Adachi, M. Kimura and S. Maeda, *A characterization of all homogeneous real hypersurfaces in a complex projective space by observing the extrinsic shape of geodesics*, to appear in Archiv der Math.
3. T. Cecil and P. Ryan, *Focal sets and real hypersurfaces in complex projective space*, Trans. Amer. Math. Soc. **269**(1982), 481–499.
4. M. Kimura, *Real hypersurfaces and complex submanifolds in complex projective space*, Trans. Amer. Math. Soc. **296**(1986), 137–149.
5. M. Kon, *Real minimal hypersurfaces in a complex projective space*, Proc. Amer. Math. Soc. **79**(1980), 285–288.
6. H.B. Lawson, *Rigidity theorems in rank-*1 *symmetric spaces*, J. Diff. Geom. **4**(1970), 349–357.
7. S. Maeda, *Real hypersurfaces of a complex projective space II*, Bull. Australian Math. Soc. **30**(1984), 123–127.
8. S. Maeda and K. Ogiue, *Characterizations of geodesic hyperspheres in a complex projective space by observing the extrinsic shape of geodesics*, Math. Z. **225**(1997), 537–542.
9. Y. Maeda, *On real hypersurfaces of a complex projective space*, J. Math. Soc. Japan **28**(1976), 529–540.
10. R. Takagi, *On homogeneous real hypersurfaces in a complex projective space*, Osaka J. Math. **10**(1973), 495–506.
11. R. Takagi, *Real hypersurfaces in a complex projective space with constant principal curvatures I, II*, J. Math. Soc. Japan **27**(1975), 43–53, 507–516.

LENGTH SPECTRUM OF CIRCLES AND KÄHLER MAGNETIC FIELDS ON COMPLEX SPACE FORMS

TOSHIAKI ADACHI
Department of Mathematics, Nagoya Institute of Technology
Nagoya 466-8555, Japan
E-mail: adachi@math.kyy.nitech.ac.jp

We study how lengths of closed circles are distributed on the real line for a complex projective space and for a complex hyperbolic space. Our study shows lengths of closed circles on these spaces are related to classical number theory.

1 Introduction

This note sketches recent progress [A1], [A5], [AM2], [AMU1] in studying geometry of circles on complex space forms. Let $(M, \langle\ ,\ \rangle)$ be a Riemannian manifold with a Riemannian metric $\langle\ ,\ \rangle$. A smooth curve γ on M is said to be a *circle* of *geodesic curvature* κ_γ if it is parametrized by its arc-length and satisfies

$$\nabla_{\dot\gamma}\nabla_{\dot\gamma}\dot\gamma = -\kappa_\gamma^2\dot\gamma. \tag{C}$$

Here κ_γ is a non-negative constant. The equation (C) is equivalent to the following equations of Frenet-Serret type:

$$\begin{cases} \nabla_X X = & \kappa_\gamma Y \\ \nabla_X Y = -\kappa_\gamma X, & \end{cases}$$

where $X = \dot\gamma$ and Y is a unit vector field along γ. Since a circle γ is parametrized by its arc-length, it satisfies $\|\nabla_{\dot\gamma}\dot\gamma\| = \kappa_\gamma$. Hence a curve is a circle of null geodesic curvature if and only if it is a geodesic. For given a positive constant κ and an orthonormal pair of tangent vectors $u, v \in T_pM$ at an arbitrary point p, we have a unique circle γ with $\gamma(0) = p$, $\dot\gamma(0) = u$, $(\nabla_{\dot\gamma}\dot\gamma)(0) = \kappa v$. When M is complete, each circle $\gamma(t)$ is defined for $-\infty < t < \infty$.

On a Euclidean space $\mathbb{R}^n$ a circle of geodesic curvature κ is a circle of radius $1/\kappa$ in the sense of Euclidean geometry, and on a standard sphere, it is a small circle. We call two smooth curves γ_1, γ_2 on M are *congruent* if there exist an isometry φ of M and a constant t_0 with $\varphi \circ \gamma_1(t) = \gamma_2(t + t_0)$ for every t. On a real space form, that is a standard sphere, a Euclidean space and a hyperbolic space, two circles are congruent if and only if they have the same geodesic curvatures. But the situation is not the same on non-flat complex space forms, which are a complex projective space and a complex hyperbolic space. For a circle γ of positive geodesic curvature on a Kähler manifold (M, J) with a complex structure J, we define its *complex torsion* τ_γ by

$$\tau_\gamma = \frac{1}{\|\nabla_{\dot\gamma}\dot\gamma\|}\langle\dot\gamma, J\nabla_{\dot\gamma}\dot\gamma\rangle = \langle X, JY\rangle,$$

which is constant along γ. In their paper [MO], Maeda and Ohnita show that on a non-flat complex space form, two circles of positive geodesic curvature are congruent if and only if they have the same geodesic curvatures and the same complex torsions. On the other hand, even on complex space forms geodesics are congruent each other. Thus it is natural to consider that some properties of the Kähler structure reflect on properties of circles. We call a circle with complex torsion ±1 *holomorphic* and a circle with null complex torsion *totally real*.

2 Background

Some circles on a Kähler manifold M can be interpreted in terms of physics. In view of static magnetic fields on a Euclidean 3-space, we call a closed 2-form a *magnetic field*. For a magnetic field $\mathbb{B}$ we denote by $\Omega_{\mathbb{B}} : TM \longrightarrow TM$ the skew symmetric operator on the tangent bundle of M defined by

$$\mathbb{B}(u, v) = \langle u, \Omega_{\mathbb{B}}(v)\rangle \qquad \text{for } u,\, v \in TM.$$

We call a smooth curve γ a *trajectory* for $\mathbb{B}$ if it satisfies

$$\nabla_{\dot\gamma}\dot\gamma = \Omega_{\mathbb{B}}(\dot\gamma). \tag{T}$$

One can easily see that each trajectory has constant speed. We call a trajectory *normal* if it has unit speed. When $\mathbb{B}$ has a global vector potential $\mathbb{A}$, that is a 1-form with $d\mathbb{A} = \mathbb{B}$, the equation (T) is the Eular-Lagrange equation for the following energy functional for smooth paths c:

$$E_{\mathbb{A}}(c) = \int_0^1 \{\frac{1}{2}\|\dot c(t)\|^2 + \mathbb{A}(\dot c(t))\}dt.$$

Needless to say, even if $\mathbb{B}$ does not have global vector potentials the equation (T) makes sense. The most elementary magnetic field is a magnetic field with constant strength, which means that it has a parallel skew symmetric operator: $\nabla\Omega_{\mathbb{B}} = 0$. One of important examples of such magnetic fields is a constant multiple of the Kähler form on a Kähler manifold. For a constant k we put $\mathbb{B}_k = k \times$ (Kähler form). Since the skew symmetric operator $\Omega_{\mathbb{B}_k}$

turns to a constant multiple of the complex structure J (i.e. $\Omega_{\mathbb{B}_k} = kJ$), we find a trajectory for $\mathbb{B}_k$ satisfies $\nabla_{\dot\gamma}\dot\gamma = kJ\dot\gamma$. Hence it is a holomorphic circle if it is normal.

For more treatment of trajectories for magnetic fields, see for example [Go], [S] and [A4].

3 Circles on complex space forms

In this section we shall see some fundamental properties of circles on non-flat complex space forms. On a (real) hyperbolic space $H^n(-c)$ of constant curvature $-c$, a circle γ is bounded if and only if its geodesic curvature κ_γ is greater than $\sqrt{c}$. When κ_γ is not greater than $\sqrt{c}$, it is unbounded in both directions (i.e. $\gamma([0,\infty))$ and $\gamma((-\infty,0])$ are unbounded sets), and has limit points in the ideal boundary ∂H^n as a Hadamard manifold:

$$\gamma(\infty) = \lim_{t\to\infty}\gamma(t),\ \gamma(-\infty) = \lim_{t\to-\infty}\gamma(t) \in \partial H^n.$$

In particular, when $\kappa_\gamma = \sqrt{c}$ it is horocyclic. Here we call a circle γ *horocyclic* if it satisfies the following conditions.

i) It is unbounded in both directions and has a single limit point; $\gamma(\infty) = \gamma(-\infty)$.

ii) When it crosses to a geodesic ρ with $\rho(\infty) = \gamma(\infty)$, it crosses orthogonally.

Similar properties hold for circles on a complex hyperbolic space.

Proposition 1 ([AM1]). *On a complex hyperbolic space $\mathbb{C}H^n(-c)$ of constant holomorphic sectional curvature $-c$, there exists a monotone increasing function $\kappa : [0,1] \longrightarrow \mathbb{R}$ with the following properties.*

0) *$\kappa(0) = \sqrt{c}/2$ and $\kappa(1) = \sqrt{c}$.*

1) *A circle γ is bounded if and only if $\kappa_\gamma > \kappa(|\tau_\gamma|)$.*

2) *If a circle γ satisfies $\kappa_\gamma \leqq \kappa(|\tau_\gamma|)$, it has limit points $\gamma(\infty)$, $\gamma(-\infty)$ in the ideal boundary $\partial\mathbb{C}H^n$.*

3) *When $\kappa_\gamma < \kappa(|\tau_\gamma|)$, these two limit points do not coincide,*

4) *When $\kappa_\gamma = \kappa(|\tau_\gamma|)$, it is horocyclic.*

Every circle lies on a totally geodesic complex hyperbolic plane. In particular, every holomorphic circle lies on a totally geodesic complex hyperbolic line, and every totally real circle lies on a totally geodesic real hyperbolic plane.

We shall now concern ourselves on the point whether circles are closed or not. We call a smooth curve γ *closed* if there is a constant ω with $\gamma(t+\omega) = \gamma(t)$ for all t. The minimum positive ω with this property is called the *length* or the *prime period* of a closed curve γ, and is denoted by $length(\gamma)$. A smooth curve which is not closed is said to be open. For convenience we set $length(\gamma) = \infty$ in this case. On a Euclidean space $\mathbb{R}^n$ and on a standard sphere $S^n(c)$ of constant curvature c, all circles are closed. When their geodesic curvature are κ, their lengths are $2\pi/\kappa$ and $2\pi/\sqrt{\kappa^2+c}$, respectively. On a complex projective space $\mathbb{C}P^n(c)$ of constant holomorphic sectional curvature c, all holomorphic circles and totally real circles are clearly closed, and their lengths are $2\pi/\sqrt{\kappa^2+c}$ and $4\pi/\sqrt{4\kappa^2+c}$ respectively when their geodesic curvatures are κ. But circles with complex torsions $0 < |\tau| < 1$ are a bit complicated.

Proposition 2 ([AMU1], [AM1]). *On a complex projective space $\mathbb{C}P^n$, for each κ (> 0), there exist infinitely many closed circles and infinitely many open circles of geodesic curvature κ.*

For a complex hyperbolic space $\mathbb{C}H^n(-c)$, the same thing is true for $\kappa > \sqrt{c}/2$.

On a complex projective space $\mathbb{C}P^n(c)$, a circle γ whose complex torsion $\tau_\gamma \neq 0, \pm 1$ is closed if and only if one of (hence all) ratios of three solutions a, b, d ($a < b < d$) for the following cubic equation is rational:

$$c\lambda^3 - (4\kappa_\gamma^2 + c)\lambda + 2\sqrt{c}\kappa_\gamma\tau_\gamma = 0.$$

When γ is closed its length is

$$length(\gamma) = \frac{4\pi}{\sqrt{c}} \cdot \text{ L.C.M.}\left\{\frac{1}{b-a}, \frac{1}{d-a}\right\},$$

where the least common multiple L.C.M.(α, β) for positive constants α, β denotes the minimum value of the set $\{\alpha m \mid m = 1, 2, \cdots\} \cap \{\beta m \mid m = 1, 2, \cdots\}$. One can see their feature through an isometric embedding of a flat torus. In his paper [N], Naitoh constructs an isometric embedding ϕ of a flat torus into a complex projective space $\mathbb{C}P^n(c)$ such that every geodesic ρ on this torus is mapped to a circle $\phi \circ \rho$ of geodesic curvature $\sqrt{2c}/4$. We also find all circles on complex space forms do not have multiple points.

For results of circles in connection with submanifold theory see [AKM] and [AMO]. For circles on other symmetric space, see [MT], [A3] and [AMU2].

4 Length spectrum of circles

We now mention on length spectrum of circles. On real space forms, two circles are congruent if and only if they have the same geodesic curvatures, hence are congruent if and only if they have the same length. We here pose a problem: Is it true that two closed circles on a non-flat complex space form are congruent if they have the same lengths and the same geodesic curvatures? Recall that circles on a non-flat complex space form are congruent if and only if they have the same geodesic curvatures and the same absolute values of complex torsions. So our problem is equivalent to the problem that lengths of circles indicate their complex torsions or not. The answer is no.

Theorem 1 ([AM2], [A5]). *On a non-flat complex space form, for each* κ (>0), *we can not distinguish congruence classes of closed circles of geodesic curvature* κ *only by their length.*

In order to explain a bit more precisely, we here introduce some notations. Let $Cir(M)$ denote the set of all congruence classes of circles on M. We define the *length spectrum* of circles as the map

$$\mathcal{L} : Cir(M) \ni [\gamma] \longmapsto length(\gamma) \in \mathbb{R} \cup \{\infty\}.$$

We denote by $Cir_\kappa(M)$ the set of all congruence classes of circles of geodesic curvature κ, and define $\mathcal{L}_\kappa$ as the restriction of $\mathcal{L}$ on this set. We also call the images of $\mathcal{L}$ and $\mathcal{L}_\kappa$ on the real line

$$LSpec(M) = \mathcal{L}(Cir(M)) \cap \mathbb{R}, \quad LSpec_\kappa(M) = \mathcal{L}(Cir_\kappa(M)) \cap \mathbb{R}$$

the length spectrum of circles and the length spectrum of circles of geodesic curvature κ respectively. For example, the length spectrum of circles on real space forms are as follows;

$$LSpec_\kappa(S^n(c)) = \left\{ \frac{2\pi}{\sqrt{\kappa^2 + c}} \right\}, \quad LSpec(S^n(c)) = \left(0, \frac{2\pi}{\sqrt{c}}\right],$$

$$LSpec_\kappa(H^n(-c)) = \left\{ \frac{2\pi}{\sqrt{\kappa^2 - c}} \right\}, \text{ if } \kappa > \sqrt{c}, \quad LSpec(H^n(-c)) = (0, \infty).$$

For non-flat complex space forms, although the structure of the length spectrum of circles of geodesic curvature κ is much complicated, it essentially does not depend on κ. On a complex projective space $\mathbb{C}P^n(c)$, for each κ there exist a constant C_κ and a injection

$$\Phi_\kappa : Cir_\kappa(\mathbb{C}P^n(c)) \setminus \{[\gamma_{\kappa,1}]\} \longrightarrow Cir_{\frac{\sqrt{2c}}{4}}(\mathbb{C}P^n(c)) \setminus \{[\gamma_{\frac{\sqrt{2c}}{4},1}]\}$$

with $\mathcal{L}_\kappa = C_\kappa \cdot \mathcal{L}_{\sqrt{2c}/4} \circ \Phi_\kappa$. Here $[\gamma_{\kappa,1}]$ denotes the congruence class of circles of geodesic curvature κ and complex torsion 1. For a complex hyperbolic space $\mathbb{C}H^n(-c)$, we denote by $\mathcal{M}_\kappa$ the set of all congruence classes of bounded circles of geodesic curvature κ. Trivially $\mathcal{M}_\kappa = Cir_\kappa(\mathbb{C}H^n(-c))$ for $\kappa > \sqrt{c}$. For each κ $(> \sqrt{c}/2)$ there exist a constant C'_κ and a injection

$$\Phi'_\kappa : \mathcal{M}_\kappa \longrightarrow \mathcal{M}_{\sqrt{c}} = Cir_{\sqrt{c}}(\mathbb{C}H^n(-c)) \setminus \{[\gamma_{\sqrt{c},1}]\}$$

(which is bijective for $\sqrt{c}/2 < \kappa \le \sqrt{c}$) with $\mathcal{L}_\kappa = C'_\kappa \cdot \mathcal{L}_{\sqrt{c}} \circ \Phi'_\kappa$. We also have a bijection

$$\Psi : Cir_{\sqrt{c}}(\mathbb{C}H^n(-c)) \setminus \{[\gamma_{\sqrt{c},1}]\} \longrightarrow Cir_{\frac{\sqrt{2c}}{4}}(\mathbb{C}P^n(c)) \setminus \{[\gamma_{\frac{\sqrt{2c}}{4},1}]\}$$

with $\mathcal{L}_{\sqrt{c}} = \frac{1}{\sqrt{2}} \cdot \mathcal{L}_{\sqrt{2c}/4} \circ \Psi$. In view of Nitoh's embedding we find that

$$\begin{aligned} LSpec_{\sqrt{2c}/4}(\mathbb{C}P^n(c)) = & \left\{ \frac{8\pi}{3\sqrt{2c}} \right\} \bigcup \left\{ \frac{8\pi}{\sqrt{6c}} \right\} \\ & \bigcup \left\{ \frac{8\pi}{3\sqrt{2c}} \sqrt{3p^2+q^2} \mid p > q, pq \text{ is even}, \right. \\ & \qquad \left. p \text{ and } q \text{ are mutually prime positive numbers} \right\} \\ & \bigcup \left\{ \frac{4\pi}{3\sqrt{2c}} \sqrt{3p^2+q^2} \mid p > q, pq \text{ is odd}, \right. \\ & \qquad \left. p \text{ and } q \text{ are mutually prime positive numbers} \right\} \end{aligned}$$

With the aid of those bijections we obtain our results on length spectrum of circles by use of classical number theory. For positive λ we call the cardinality $m_{\mathcal{L}}$ of the set $\mathcal{L}^{-1}(\lambda)$ the *multiplicity* of $\mathcal{L}$ at λ. When it is equal to 1, we call a spectrum λ *simple*. We here summarize some of our results on tables (see [AM2], [A5] for more detail).

Length spectrum of circles of geodesic curvature κ

	$\mathbb{C}H^n(-c)$	$\mathbb{C}P^n(c)$
$LSpec_\kappa$	$\emptyset$ if $0 \le \kappa \le \frac{\sqrt{c}}{2}$ discrete, unbounded, if $\kappa > \frac{\sqrt{c}}{2}$	discrete, unbounded
multiplicity m_κ	$\limsup_{\lambda\to\infty} m_\kappa(\lambda) = \infty$, if $\kappa > \frac{\sqrt{c}}{2}$	$\limsup_{\lambda\to\infty} m_\kappa(\lambda) = \infty$
1st length spectrum	$\frac{4\pi}{\sqrt{4\kappa^2-c}}$, simple, if $\kappa > \frac{\sqrt{c}}{2}$ (length spectrum of totally real circles)	$\frac{2\pi}{\sqrt{\kappa^2+c}}$, simple (length spectrum of holomorphic circles)

Full length spectrum of circles

	$\mathbb{C}H^n(-c)$	$\mathbb{C}P^n(c)$
LSpec	$(0,\infty)$	$(0,\infty)$
multiplicity $m_{\mathcal{L}}$	finite at each λ monotone increasing polynomial growth left continuous $m_{\mathcal{L}}(\lambda) \geq 2$ $m_{\mathcal{L}}(\lambda) = 2$ if and only if $0 < \lambda \leq \frac{8\pi}{3\sqrt{c}}$	finite at each λ $m_{\mathcal{L}}(\lambda) \geq 1$ $m_{\mathcal{L}}(\lambda) = 1$ if and only if $\frac{2}{\sqrt{c}}\pi < \lambda \leq \frac{4}{3}\sqrt{\frac{5}{c}}\pi$

5 Unbounded trajectories for Kähler magnetic fields

In this section we shall mention on unbounded circles with complex torsion ± 1. As we see in section 2, they are normal trajectories for some Kähler magnetic field.

For trajectories for magnetic fields we can define some notions along the same lines as for geodesics. Let $\mathbb{B}$ be a magnetic field on M. For a univ tangent vector $u \in UM$, we denote by γ_u the normal trajectory for $\mathbb{B}$ with the initial vector $\dot{\gamma}_u(0) = u$. The *magnetic flow* $\mathbb{B}\varphi_t : UM \longrightarrow UM$ on the unit tangent bundle is defined by

$$\mathbb{B}\varphi_t(u) = \dot{\gamma}_u(t).$$

The *magnetic exponential map* $\mathbb{B}exp_p : T_pM \longrightarrow M$ is given by

$$\mathbb{B}\mathrm{exp}_p(v) = \begin{cases} p, & \text{if } v = 0 \\ \dot{\gamma}_{v/\|v\|}(\|v\|), & \text{otherwise.} \end{cases}$$

For Kähler magnetic flows on a complex projective space, they are conjugate each other in the strong sense. This means that there exist a diffeomorphism $g_{k,\ell} : UM \longrightarrow UM$ and a constant λ with

$$g_{k,\ell}^{-1} \circ \mathbb{B}_\ell\varphi_t \circ g_{k,\ell} = \mathbb{B}_k\varphi_{\lambda t}.$$

They are rotation flows. For Kähler magnetic flows on a complex hyperbolic space we have the following.

Theorem 2 ([A1]). *On a complex hyperbolic space $\mathbb{C}H^n(-c)$ of constant holomorphic sectional curvature $-c$, we can classify Kähler magnetic flows into three conjugacy classes in the strong sense:*

i) *rotation flows: when* $|k| > \sqrt{c}$,

ii) *horocyclic flows: when* $k = \pm\sqrt{c}$,

iii) *hyperbolic flows: when* $|k| < \sqrt{c}$.

In particular, when $|k| < \sqrt{c}$ *Kähler magnetic flow* $\mathbb{B}_k\varphi_t$ *is conjugate to the geodesic flow* $\varphi_t = \mathbb{B}_0\varphi_t$: *That is, there exists a diffeomorphism* $g_k : U\mathbb{C}H^n(-c) \longrightarrow U\mathbb{C}H^n(-c)$ *with*

$$g_k^{-1} \circ \mathbb{B}_k\varphi_t \circ g_k = \varphi_{\sqrt{c-k^2}t/\sqrt{c}}.$$

This property is inherited to quotient manifolds of complex hyperbolic space.

Thus trajectories for Kähler magnetic fields $\mathbb{B}_k$ with $|k| < \sqrt{c}$ on a complex hyperbolic space $\mathbb{C}H^n(-c)$ have similar properties as of geodesics. When $|k| < \sqrt{c}$, for any distinct points $p, q \in \mathbb{C}H^n(-c)$ there exists a unique normal trajectory segment for $\mathbb{B}_k$ from p to q. Its length ℓ satisfies

$$\sqrt{c}\sinh\left(\frac{1}{2}\sqrt{c-\kappa^2}\ell\right) = \sqrt{c-\kappa^2}\sinh\left(\frac{1}{2}\sqrt{c}d(p,q)\right), \tag{R}$$

where $d(p,q)$ denotes the distance between p and q. On a Hadamard manifold every axial isometry φ has an axis, a unique geodesic translated by φ. The same thing holds for trajectory for Kähler magnetic fields on a complex hyperbolic space.

Theorem 3. *For every isometry* $\varphi \in Iso(\mathbb{C}H^n(-c)) \setminus \{Id\}$ *on a complex hyperbolic space* $\mathbb{C}H^n(-c)$, *there exists a unique normal trajectory* γ *for a Kähler magnetic field* $\mathbb{B}_k$ ($|k| < \sqrt{c}$) *which is translated by* φ. *This means that there is a positive* $\omega_{\varphi,k}$ *with* $\varphi \circ \gamma(t) = \gamma(t + \omega_{\varphi,k})$ *for every* t.

It is interesting to study geometric characterization of the constant $\omega_{\varphi,k}$. For $\omega_\varphi = \omega_{\varphi,0}$, it is known that

$$\omega_\varphi = \min\{\, d(p,\varphi(p)) \mid p \in \mathbb{C}H^n(-c)\,\},$$

and if $\omega_\varphi = d(p_0, \varphi(p_0))$ then the axis for φ is the geodesic joining p_0 and $\varphi(p_0)$. Clearly by Theorem 2 we have $\omega_{\varphi,k} = \sqrt{c/(c-k^2)}\,\omega_\varphi$, but in view of (R) it is not easy to get such characterization The constant $\omega_{\varphi,k}$ also deeply related with lengths of closed trajectories on a quotient manifold $M = \mathbb{C}H^n/\Gamma$ of a complex hyperbolic space. As a direct consequence of Theorem 2 we have the following result.

Corollary . *Let* $M = \mathbb{C}H^n(-c)/\Gamma$ *be a compact quotient of a complex hyperbolic space* $\mathbb{C}H^n(-c)$. *When* $|k| < \sqrt{c}$, *the asymptotic behaviour of the* *number of closed normal trajectories for $\mathbb{B}_k$ with respect to their prime periods is described with the topological entropy h_k of the Kähler magnetic flow $\mathbb{B}_k\varphi_t$ on UM as follows:*

$$\sharp\{\,\gamma \mid \textit{normal trajectory for } \mathbb{B}_k \textit{ with length}(\gamma) < \lambda\,\} \sim \frac{1}{h_k\lambda}\exp(h_k\lambda).$$

Here, for two positive function $f,\ g : [0,\infty) \to [0,\infty)$ we denote by $f \sim g$ if they satisfy $\lim_{\lambda\to\infty} f(\lambda)/g(\lambda) = 1$.

The topological entropy $h_k = n\sqrt{c-k^2}$ of a Kähler magnetic flow $\mathbb{B}_k\varphi_t$ on a complex hyperbolic space $\mathbb{C}H^n(-c)$ is interpreted as the volume growth of magnetic balls $\mathbb{B}_k B_r(\cdot)$:

$$h_k = \lim_{r\longrightarrow\infty} \frac{1}{r}\log Vol\left(\mathbb{B}_k B_r(\cdot)\right),$$

where

$$\mathbb{B}_k B_r(p) = \{\mathbb{B}_k \exp_p(tu) \mid u \in U_pM, 0 \le t < r\}.$$

In the study of distribution of prime periods of hyperbolic flows, zeta functions play an important role. The dynamical zeta function for a Kähler magnetic flow $\mathbb{B}_k\varphi_t$ on a manifold $M = \mathbb{C}H^n(-c)/\Gamma$ is defined by

$$\zeta_{M,k}(s) = \prod_{\gamma}\Big\{1 - \exp\big(-s\cdot length(\gamma)\big)\Big\}^{-1},$$

where γ runs over all (prime) closed normal trajectories for $\mathbb{B}_k$ on M. By Theorem 2, we see that the zeta function $\zeta_{M,k}$ for a Kähler magnetic flow and the zeta function $\zeta_M = \zeta_{M,0}$ for the geodesic flow have the following relation:

$$\zeta_{M,k}(s) = \zeta_M\left(\sqrt{\frac{c}{c-k^2}}\,s\right).$$

Therefore we find that

1) it is absolutely convergent and holomorphic on the domain $\Re(s) > h_k$,
2) it has a nowhere vanishing meromorphic extension to the hole plane,
3) it has a simple pole at $s = h_k$,

(see [Ga], and also for example [PP], [AS]). For the zeta function of Selberg type for a Käler magnetic flow $\mathbb{B}_k\varphi_t$ which is defined as

$$Z_{M,k}(s) = \prod_{\ell=0}^{\infty}\prod_{\gamma}\Big\{1 - \exp\big(-(s+\ell)\cdot length(\gamma)\big)\Big\},$$

we still have some work to do (cf. [H], [P]).

Acknowledgment

This research is partially supported by Grant-in-Aid for Encouragement of Young Scientists (No. 09740055), The Ministry of Education, Science and Culture.

References

[A1] T. Adachi, *Kähler magnetic flows on a manifold of constant holomorphic sectional curvature*, Tokyo J. Math. **18**(1995), 473–483.

[A2] ———, *Circles on a quaternionic space form*, J. Math. Soc. Japan **48**(1996), 205–227.

[A3] ———, *Curvature bound and trajectories for magnetic fields on a Hadamard surface*, Tsukuba J. Math. **20**(1996), 225–230.

[A4] ———, *A comparison theorem for magnetic Jacobi fields*, Proc. Edinburgh Math. Soc. **40**(1997), 293–308.

[A5] ———, *Length spectrum of circles on a complex hyperbolic space*, to appear in Nagoya. Math. J. **153**(1999).

[AKM] T. Adachi, M. Kimura and S. Maeda, *A characterization of all homogeneous real hypersurfaces in a complex projective space by observing the extrinsic shape of geodesics*, to appear in Archiv der Math.

[AM1] T. Adachi and S. Maeda, *Global behaviours of circles in a complex hyperbolic spaces*, Tsukuba J. Math. **21**(1997), 29–42.

[AM2] ———, *Length spectrum of circles in a complex projective spaces*, Osaka Math. J. **35**(1998), 553–565.

[AMO] T. Adachi, S. Maeda and K. Ogiue, *Extrinsic shape of circles and standard imbedding of projective spaces*, Manuscripta math. **93**(1997), 267–272.

[AMU1] T. Adachi, S. Maeda and S. Udagawa, *Circles in a complex projective spaces*, Osaka J. Math. **32**(1995), 709–719.

[AMU2] ———, *Simpleness and closedness of circles in compact Hermitian symmetric spaces*, preprint.

[AS] T. Adachi and T. Sunada, *Twisted Perron-Frobenius theorem and L-functions*, J. Func. Anal. **71**(1987), 1–46.

[C] A. Comtet, *On the Landau levels on the hyperbolic plane*, Ann. of Phys. **173**(1987), 185–209.

[Ga] R. Gangolli, *Zeta functions of Selberg's type for compact space forms of symmetric spaces of rank one*, Ill. J. Math. **21**(1977), 1–42.

[Go] N. Gouda, *Magnetic flows of Anosov type*, Tôhoku Math. J. **49**(1997), 165–183.

[H] D. Hejal, *The Selberg trace formula for $PSL(2, \mathbb{R})$ I, II*, Lecture Notes in Math. 548(1976), 1001(1983).

[KS] A. Katsuda and T. Sunada, *Closed orbits in homology classes*, Inst. Hautes Études Sci. Publ. Math. **71**(1990), 5–32.

[MO] S. Maeda and Y. Ohnita, *Herical geodesic immersion into complex space forms*, Geom. Dedicata **30**(1989), 93–114.

[MT] K. Mashimo and K. Tojo, *Circles in Riemannian symmetric spaces*, preprint.

[N] H. Naitoh, *Isotropic submanifolds with parallel second fundamental form in $P^m(\mathbb{C})$*, Osaka J. Math. **18**(1981), 427–464.

[NY] K. Nomizu and K. Yano, *On circles and spheres in Riemannian geometry*, Math. Ann. **210**(1974), 163–170.

[P] S.J. Patterson, *The Laplace operator on a Riemann surface*, Compo. Math. **31**(1975), 83–107.

[PP] W. Parry and M. Pollicott, *An analogue of the prime number theorem for closed orbits of Axiom A flows*, Ann. of Math. **118**(1983), 573–591.

[S] T. Sunada, *Magnetic flows on a Riemann surface*, Proc. KAIST Math. Workshop 8(1993) *Analysis and geometry*, 93–108.

QUATERNIONIC-TYPE STRUCTURES AND ALMOST HYPERBOLIC PSEUDODISTANCES

PIERRE DOLBEAULT

Université Pierre et Marie Curie (Paris VI), Mathématiques, Case 247; 4, place Jussieu, F-75252 Paris, Cédex 05, France

JERZY KALINA

Institute of Mathematics, Technical University of Łódź, Al. Politechniki 11, PL-93-590 Łódź, Poland

JULIAN LAWRYNOWICZ

Institute of Mathematics, Polish Academy of Sciences, Łódź Branch, Narutowicza 56, PL-90-136 Łódź, Poland

Chair of Solid State Physics, University of Łódź, Pomorska 149/153, PL-90-236 Łódź, Poland

In 1989 two of us (P.D. and J.Ł.) had introduced a Dirichlet integral-type biholomorphic-invariant pseudodistance connected with bordered holomorphic chains whose regular part was treated as a Riemann surface [6]. The condition for a complex manifold that the pseudodistance on it was a distance defined a class of hyperbolic-like manifolds which had an important property of extendability of holomorphic mappings, analogous to the hyperbolic manifolds, Stein spaces, and complex spaces with a Stein covering. Further results in this direction were published by G. Boryczka and L.M. Tovar in 1996 [2] and by us in 1998 [5]. The present paper introduces an extension of the staff to hyperkählerian holomorphic chains related to a two-dimensional sphere in $\mathbb{R}^3$ and complex submanifolds of $\mathbb{C}^2$ and $\mathbb{C}^4$ with six standard complex structures related to Hermitian Hurwitz pairs, corresponding to the Kałuża-Klein, Penrose, and exciton theories (J.Ł. and O. Suzuki 1998 [1]). This is done by means of the atomization method of J.Ł. and O. Suzuki.

1 Introduction

The present paper provides an extension of our previous work [5] to hyperkählerian holomorphic chains related to a two-dimensional sphere in $\mathbb{R}^3$ and complex structures.

Research of the second and third author partially supported by the State Committee for Scientific Research (KBN) grant PB 2 P03A 016 10. Section 2 of the paper has been included in the lecture by J.Ł. during the IV International Symposium on Complex Structures and Vector Fields at St. Constantine near Varna on September 4, 1998.

The importance of the subject is motivated by a number of results by A. Andreotti, W. Stoll and K. Kobayashi, referred to in [6], and by [12]. The authors believe that this introduction to a new approach will open new possibilities in continuing those lines, in particular in the aspect of interrelations between the complex dynamics and hyperbolic geometry.

2 Hyperkählerian manifolds and the atomization theorem

Suppose that (X, g) is a hyperkählerian manifold of real dimension $4n$, $n = 1, 2, \ldots$. This means that there exist on X two complex structures I and J such that g is a Kähler metric for each of them and $IJ = -JI$ (cf. e.g. [1], Proposition 14.10). Then g is still a Kähler metric for the complex structure $K = IJ$. Moreover, if $\underline{\sigma} = (x, y, z) \in S^2 \subset \mathbb{R}^3$, where S^2 is a two-dimensional sphere, then

$$J_{\underline{\sigma}} := \underline{\sigma} \cdot (I, J, K) = xI + yJ + zK \tag{1}$$

is again a complex structure of X and g is a Kähler metric for it.

We are specially interested in complex structures on isometric embeddings ι:

$$\mathbb{C}^2 \simeq \mathbb{R}^4 \xrightarrow{\iota} G(2,4),\ \mathbb{R}^4 \ni \mathbf{x} \overset{\iota}{\longmapsto} \sum_{\alpha=1}^{3} x^\alpha S_\alpha + x^4 I_4,$$

resp. (2)

$$\mathbb{C}^4 \simeq \mathbb{R}^8 \xrightarrow{\iota} G(8,16),\ \mathbb{R}^8 \ni \mathbf{x} \overset{\iota}{\longmapsto} \sum_{\alpha=1}^{7} x^\alpha S_\alpha + x^8 I_8,$$

so that they are real parts of holomorphic mappings. Here I_n is the $n \times n$ unit matrix and S_α, I_4 resp. S_α, I_7 are generators of the central real Clifford algebra of order $(0,4)$ or $(2,2)$ resp. $(0,8)$, $(4,4)$, $(2,6)$ or $(6,2)$. This means that the generators S_α can be chosen as follows [9].

Case $(0,4)$: $S_\alpha = \sigma_\alpha$, $\alpha = 1, 2, 3$, where σ_α are the familiar *Pauli matrices*:

$$\sigma_1 = \begin{pmatrix} 0 & 1 \\ 1 & 0 \end{pmatrix}, \quad \sigma_2 = \begin{pmatrix} 0 & -i \\ i & 0 \end{pmatrix}, \quad \sigma_3 = \begin{pmatrix} 1 & 0 \\ 0 & -1 \end{pmatrix}.$$

The corresponding space-time element $ds^2 = c^2dt^2 - dx^2 - dy^2 - dz^2 - d\tau^2$ plays the crucial role in the Kałuża-Klein theory.

Case $(2,2)$: $S_1 = i\sigma_1, S_2 = i\sigma_2, S_3 = \sigma_3$.
The corresponding space-time element $ds^2 = c^2dt^2 - dx^2 - dy^2 - dz^2 + d\tau^2$ plays the crucial role in the Penrose theory.

The following cases, apart from their mathematical beauty, are important in the two-body theory with constraints, e.g, in a study of excitons. Let

$$\operatorname{diag}(A,B) := \begin{pmatrix} A & 0 \\ 0 & B \end{pmatrix} \quad \text{etc. and} \quad \operatorname{diag}^*(A,B) := \begin{pmatrix} 0 & B \\ A & 0 \end{pmatrix} \text{ etc.,}$$

where A, B, etc. denote square matrices.

Case $(0,8)$: $S_\alpha = \operatorname{diag}(\sigma_\alpha, -\sigma_\alpha, -\sigma_\alpha, \sigma_\alpha)$, $\alpha = 1, 2, 3$,

$$S_4 = \begin{pmatrix} 0 & I_2 & 0 & 0 \\ I_2 & 0 & 0 & 0 \\ 0 & 0 & 0 & -I_2 \\ 0 & 0 & -I_2 & 0 \end{pmatrix}, \quad S_5 = \begin{pmatrix} 0 & iI_2 & 0 & 0 \\ -iI_2 & 0 & 0 & 0 \\ 0 & 0 & 0 & -iI_2 \\ 0 & 0 & iI_2 & 0 \end{pmatrix},$$

$$S_6 = \begin{pmatrix} 0 & I_4 \\ I_4 & 0 \end{pmatrix}, \qquad S_7 = \begin{pmatrix} 0 & -iI_4 \\ -iI_4 & 0 \end{pmatrix}.$$

Case $(4,4)$: $S_\alpha = \operatorname{diag}(\sigma_\alpha, \sigma_\alpha, -\sigma_\alpha, -\sigma_\alpha)$, $\alpha = 1, 2, 3$,

$$\begin{aligned} S_4 &= \operatorname{diag}^*(-I_2, I_2, -I_2, I_2), \\ S_5 &= \operatorname{diag}^*(iI_2, iI_2, iI_2, iI_2), \end{aligned} \quad S_6 = \begin{pmatrix} 0 & I_4 \\ -I_4 & 0 \end{pmatrix}, \quad S_7 = \begin{pmatrix} 0 & 0 & iI_2 & 0 \\ 0 & 0 & 0 & -iI_2 \\ iI_2 & 0 & 0 & 0 \\ 0 & -iI_2 & 0 & 0 \end{pmatrix}.$$

Case $(2,6)$: $S_\alpha = \operatorname{diag}(\sigma_\alpha, -\sigma_\alpha, \sigma_\alpha, -\sigma_\alpha)$, $\alpha = 1, 2, 3$,

$$S_4,\ S_5 \text{ as in the case } (0,8),\ S_6 = \begin{pmatrix} 0 & iI_4 \\ iI_4 & 0 \end{pmatrix}, \quad S_7 = \begin{pmatrix} 0 & I_4 \\ -I_4 & 0 \end{pmatrix}.$$

Case $(6,2)$: $S_\alpha = \operatorname{diag}^*(i\sigma_\alpha, i\sigma_\alpha, i\sigma_\alpha, i\sigma_\alpha)$, $\alpha = 1, 2, 3$,

$$\begin{aligned} S_4 &= \operatorname{diag}^*(-I_2, I_2, -I_2, I_2), \\ S_7 &= \operatorname{diag}(\sigma_3, -\sigma_3, -\sigma_3, \sigma_3), \end{aligned} \quad S_5 \text{ resp. } S_6 \text{ as } S_6 \text{ resp. } S_7 \text{ in the case } (4,4).$$

Lawrynowicz and Suzuki [10] have proved

Theorem 1. *In each of the cases* $(0,4)$, $(0,2)$, *resp.* $(0,8)$, $(4,4)$, $(2,6)$, $(6,2)$ *there exists a complex structure* $I = I[\iota]$ *on the coresponding holomorphic embedding* (2), *with the property that the embedding is the real part of a holomorphic mapping.*

The proof is effective: the embeddings in question are constructed explicitely with the help of an original method which may be called the *atomization method.* Let

$u = x^4 + ix^3, v = x^2 + ix^1$ in the case of (0,4),

$u = x^4 + ix^3, v = x^1 + ix^2$ in the case of (2,2),

$u = x^3 + ix^8, v = x^1 + ix^2, w = x^4 + ix^5, t = x^6 + ix^7$ in the case of (0,8),

$u = x^3 + ix^8, v = x^1 + ix^2, w = x^6 + ix^7, t = x^4 + ix^5$ in the case of (4,4),

$u = x^3 + ix^8, v = x^1 + ix^2, w = x^7 + ix^6, t = x^4 + ix^5$ in the case of (2,6),

$u = x^4 + ix^3, v = x^1 + ix^2, w = x^5 + ix^6, t = x^7 + ix^8$ in the case of (6,2).

Now we introduce eleven 2×2 complex matrices which we call *atoms*:

$$A_1 = \begin{pmatrix} u & v \\ -\bar{v} & \bar{u} \end{pmatrix}, \quad A_2 = \begin{pmatrix} u & v \\ \bar{v} & \bar{u} \end{pmatrix}, \quad A_3 = \begin{pmatrix} \bar{u} & \bar{v} \\ v & -u \end{pmatrix}, \quad A_4 = \begin{pmatrix} -u & -\bar{v} \\ -v & \bar{u} \end{pmatrix},$$

$$A_{11} = \begin{pmatrix} \bar{u} & -v \\ \bar{v} & u \end{pmatrix}, \quad A_5 = \begin{pmatrix} w & 0 \\ 0 & w \end{pmatrix}, \quad A_6 = \begin{pmatrix} \bar{w} & 0 \\ 0 & \bar{w} \end{pmatrix},$$

$$A_7 = \begin{pmatrix} t & 0 \\ 0 & t \end{pmatrix}, \quad A_8 = \begin{pmatrix} \bar{t} & 0 \\ 0 & \bar{t} \end{pmatrix}, \quad A_9 = \begin{pmatrix} t & 0 \\ 0 & -\bar{t} \end{pmatrix}, \quad A_{10} = \begin{pmatrix} \bar{t} & 0 \\ 0 & -t \end{pmatrix}.$$

The atomization method allows us to find the following explicit formulae for the embeddings in question, where $\zeta = (u, v) \in \mathbb{C}^2$ and $z = (u, v, w, t) \in \mathbb{C}^4$:

$$\iota(\zeta) = A_1 \text{ for } (0,4),$$

$$\iota(\zeta) = A_2 \text{ for } (2,2),$$

$$\iota(z) = \begin{pmatrix} A_3 & A_5 & A_7 & 0 \\ A_6 & A_4 & 0 & A_7 \\ A_8 & 0 & A_4 & -A_5 \\ 0 & A_8 & -A_6 & A_3 \end{pmatrix} \text{ for } (0,8),$$

$$\iota(z) = \begin{pmatrix} A_3 & 0 & A_5 & A_7 \\ 0 & A_3 & -A_8 & A_6 \\ -A_6 & A_7 & A_4 & 0 \\ -A_8 & -A_5 & 0 & A_4 \end{pmatrix} \text{ for } (4,4),$$

$$\iota(z) = \begin{pmatrix} A_3 & A_7 & A_5 & 0 \\ A_8 & A_4 & 0 & A_5 \\ A_6 & 0 & A_3 & -A_7 \\ 0 & -A_6 & -A_8 & A_4 \end{pmatrix} \text{ for } (2,6),$$

$$\iota(z) = \begin{pmatrix} A_9 & 0 & A_5 & A_{11} \\ 0 & -A_{10} & A_4 & A_6 \\ -A_6 & A_1 & -A_{10} & 0 \\ A_{11} & -A_5 & 0 & A_9 \end{pmatrix} \text{ for } (6,2).$$

3 An anologue of the hyperbolic pseudodistance related to hyperkählerian holomorphic chains

Consider the standard measure $d\tau$ on S^2, and denote by psh (X, σ) the class of all functions on X, $n = 1$ or 2, C^2-plurisubharmonic there with respect to $J_{\underline{\sigma}}$ given by (1) *with I described in Theorem* 1. Denote by $\mathrm{ph}(X, \underline{\sigma})$ the analogous class of all pluriharmonic functions. Let psh $X := \bigcap_{\underline{\sigma}} \mathrm{psh}(X, \underline{\sigma})$ and ph $X := \bigcap_{\underline{\sigma}} \mathrm{ph}(X, \underline{\sigma})$. For any $\mathcal{J}_{\underline{\sigma}}$ there is a decomposition $d = \delta_{\underline{\sigma}} + \bar{\delta}_{\underline{\sigma}}$ and we can consider the conjugate operator $d^c_{\underline{\sigma}} := i(\bar{\delta}_{\underline{\sigma}} - \delta_{\underline{\sigma}})$. For a *submanifold X* of $\mathbb{C}^{2n}$, $n = 1$ or 2, we shall consider, subsequenly, two cases: psh X contains noncostant functions and psh X contains only constant functions.

3.1. psh X *contains nonconstant functions.* Assume that X is (k, m)-hyperbolic for $k = 1$ or 2, and a fixed $m \geq n$, in the sense of Eisenman-Kobayashi [7, 8, 11]. Set $\alpha = 1 - m/n$. For a fixed elementary chain Γ_j [4], let

$$\mu^{\alpha}_{\Gamma_j}(\underline{\sigma})[u] := \inf_{\Gamma'_j \subset \Gamma_j} \{\mu_1(\gamma'_j, f_j^{-1})_m / \mu_2(\Gamma'_j, f_j^{-1})_m | \int_{S^2} (\int_{\Gamma'_j} du \wedge d^c_{\underline{\sigma}} u) d\underline{\sigma}|\}, \tag{3}$$

where $\mu = \mu_1$ and $\mu = \mu_2$ are the intermediate one- and two-dimensional measures [8], u belongs to an admissible family $F[\mathcal{U}] = \mathrm{adm}(X, \mathcal{U})$ of pluriharmonic functions, defined in the usual way [6] for a given finite open covering $\mathcal{U}$ of X and the infimum in (3) is taken over all compact connected C^1-cycles of dimension one within Γ_j [5].

3.2. psh X *contains only contant functions.* Then we can define the corresponding families $\mathrm{Psh}(X, I)$, $\mathrm{Psh}(X, J)$, $\mathrm{Psh}(X, K)$, decompositions $\partial = \partial_I + \partial_I$ etc., and the conjugate operators $d^c_I := i(\overline{\partial}_I - \overline{\partial}_I)$ etc. Here $\mathrm{Psh}(X, I)$ etc. stands for $\mathrm{psh}(X, (1, 0, 0))$ etc. and ∂_I etc. for $\partial_{(1,0,)}$ etc. Then we can modify definition (1) as follows:

$$\mu^{\alpha}_{\Gamma_j}[u] := \inf_{\Gamma'_j \subset \Gamma_j} \{[\mu_1(\gamma'_j, f_j^{-1})_m / \mu_2(\Gamma'_j, f_j^{-1})_m] | \int_{\Gamma'_j} du \wedge d^c_I u|\}, \quad u \in F[\mathcal{U}, I]$$

and, analogously, for J and K.

(4)

3.3. We have

Lemma 1. *The expressions* (3) *and* (4) *are well defined.*

Thus with any bordered holomorphic chain passing through points z_0, z of X, such that $\mu_1(\gamma'_j, f_j^{-1})_m$ is uniformly bounded in Γ, we may associate the expression

$$\mu_\Gamma^\alpha(z_0, z; \underline{\sigma})[u] := \sum_j \mu_{\Gamma_j}^\alpha(z_{j-1}, z_j)[u]$$

or, alternatively, the expressions $\mu_\Gamma^\alpha(z_0, z; I)[u]$ etc. Using this expressions, we set

$$\mu_X^\alpha(z_0, z; \underline{\sigma})[u, \mathcal{U}] := \inf\{\mu_\Gamma^\alpha(z_0, z; \underline{\sigma})[u] : \Gamma \text{ passing through } z_0, z\}$$

or, alternatively, we define $\mu_X^\alpha(z_0, z; I)(u, \mathcal{U})$ etc. coresspondingly.

Finally, we define an *almost hyperbolic quaternionic-type pseudodistance*:

$$\rho_X^\alpha(z_0, z; \underline{\sigma})[\mathcal{U}] := \sup\{\mu_X^\alpha(z_0, z; \underline{\sigma})[u, \mathcal{U}] : \ u \in F[\mathcal{U}]\} \tag{5}$$

or, alternatively,

$$\rho_X^\alpha(z_0, z)[\mathcal{U}] := \max\{\rho_X^\alpha(z_0, z; I)[\mathcal{U}], \rho_X^\alpha(z_0, z; J)[\mathcal{U}], \rho_X^\alpha(z_0, z; K)[\mathcal{U}]\}, \tag{6}$$

where

$$\rho_X^\alpha(z_0, z; I)[\mathcal{U}] := \sup\{\mu_X^\alpha(z_0, z; I)[u, \mathcal{U}] : \ u \in F[\mathcal{U}, I]\}$$

and, analogously, for J and K.

We have to prove that (5) and (6) are finite and that they are pseudodistances indeed. Since $J_{\underline{\sigma}}$ is of type described in Theorem 1, we have the following counterparts of Lemmas 2 and 3, and Remark 1 in [5].

Lemma 2. *Let z_0, z_1 and z_2 be points on a (k, m)-hyperbolic n-dimensional complex submanifold X of $\mathbb{C}^{2n}$, $n = 1$ or 2, $k = 1$ or 2, and a fixed $m \geq n$. Set $\alpha = 1 - n/m$. Then, for any locally finite open covering $\mathcal{U}$ of X, we have*

$$\rho_X^\alpha(z_0, z_2; \underline{\sigma})[\mathcal{U}] \leq \rho_X^\alpha(z_0, z_1; \underline{\sigma})[\mathcal{U}] + \rho_X^\alpha(z_1, z_2; \underline{\sigma})[\mathcal{U}]$$

or, alternatively,

$$\rho_X^\alpha(z_0, z_2)[\mathcal{U}] \leq \rho_X^\alpha(z_0, z_1)[\mathcal{U}] + \rho_X^\alpha(z_1, z_2)[\mathcal{U}],$$

provided that we can introduce on X a structure $J_{\underline{\sigma}}$ given in (1) and described there, where the structure $I = I[\iota]$ has to be one of those whose existence is assured by Theorem 1 (they are constructed explicitly in the proof of that theorem).

Lemma 3. *Let z_0, z be points on a (k, m)-hyperbolic manifold X satisfying the hypotheses of Lemma 2. Then, for any locally finite open covering $\mathcal{U}$ of X, we have $\rho_X^\alpha(z_0, z; \underline{\sigma})[\mathcal{U}] < +\infty$ or, alternatively, $\rho_X^\alpha(z_0, z)[\mathcal{U}] < +\infty$.*

Lemma 4. *Under the hypotheses of Lemma 2,* $\rho^{\alpha}_{X}(z, z_0; \underline{\sigma})[\mathcal{U}] \geq 0$ *and*

$$\rho^{\alpha}_{X}(z, z_0; \underline{\sigma})[\mathcal{U}] = \rho^{\alpha}_{X}(z_0, z; \underline{\sigma})[\mathcal{U}]$$

or, alternatively, $\rho^{\alpha}_{X}(z_0, z)[\mathcal{U}] \geq 0$ *and* $\rho^{\alpha}_{X}(z_0, z)[\mathcal{U}] = \rho^{\alpha}_{X}(z, z_0)[\mathcal{U}]$. *If* $z = z_0$, *then the length of* $\phi^{-1}[\gamma']$ *can be as small as we desire, so* $\rho^{\alpha}_{X}(z_0, z; \underline{\sigma})[\mathcal{U}] = 0$ *or, alternatively,* $\rho^{\alpha}_{X}(z_0, z)[\mathcal{U}] = 0$.

From Lemmas 3 and 4 we infer

Theorem 2. *Let* X *be a* (k, m)*-hyperbolic* n*-dimensional complex submanifold of* $\mathbb{C}^{2n}$, $n = 1$ *or* 2, $k = 1$ *or* 2, *and a fixed* $m \geq n$. *Set* $\alpha = 1 - n/m$. *Moreover, suppose that we can introduce on* X *a structure* $J_{\underline{\sigma}}$ *given in* (1) *and described there, where the structure* $I = I[\iota]$ *has to be one of those whose existence is assured by Theorem 1 (they are constructed explicitly in the proof of that theorem). Then, for any locally finite open covering* $\mathcal{U}$ *of* X, *the corresponding expression* ρ^{α}_{X} *given by* (5) — *or, alternatively, by* (6) — *is a continuous pseudodistance.*

By Theorem 2, we trivially get (for the proof, cf. [2]):

Corollary 1. *Let* $(X, \mathcal{U})$ *and* $(Y, \mathcal{V})$ *be two* (k, m)*-hyperbolic* n*-dimensional complex submanifolds of* $\mathbb{C}^{2n}$, $n = 1$ *or* 2, *with* k, m, α *as in Theorem 2, locally finite open coverings* $\mathcal{U}$ *and* $\mathcal{V}$, *and admissible families* $F[\mathcal{U}]$ *and* $F[\mathcal{V}]$ *of pluriharmonic functions. Let* $f : X \to Y$ *be a biholomorphic mapping such that* $f[\mathcal{U}] = \mathcal{V}$. *Moreover, suppose that we can introduce on* X *a structure* $J_{\underline{\sigma}}$ *as in Theorem 2 and, analogously (in the case 3.1), the structure* $J_{f_{*}\underline{\sigma}}$ *on* Y. *Then*

$$\rho^{\alpha}_{X}(z_0, z; \underline{\sigma})[\mathcal{U}] = \rho^{\alpha}_{Y}(f(z_0), f(z); f_{*}\underline{\sigma})[\mathcal{V}] \quad \text{for} \quad z_0, z \in X$$

or, alternatively

$$\rho^{\alpha}_{X}(z_0, z)[\mathcal{U}] = \rho^{\alpha}_{Y}(f(z_0), f(z))[\mathcal{V}] \quad \text{for} \quad z_0, z \in X.$$

Theorem 2 and Corollary 1 motivate the following definition. Let X be a (k, m)-hyperbolic n-dimensional complex submanifold of $\mathbb{C}^{2n}$, $n = 1$ or 2, $k = 1$ or 2, and a fixed $m \geq n$. Set $\alpha = 1 - n/m$. Moreover, suppose that we can introduce on X a structure $J_{\underline{\sigma}}$ given in (1) and described there, where the structure $I = I[\iota]$ has to be one of those whose existence is assured by Theorem 1 (they are constructed explicity in the proof of that theorem). If, for a locally finite open covering $\mathcal{U}$ of X, $\rho^{\alpha}_{X}(\ ,\ ; \underline{\sigma})[\mathcal{U}]$ is a distance for some $\underline{\sigma}$, i.e. $\rho^{\alpha}_{X}(z_0, z; \underline{\sigma})[\mathcal{U}] > 0$ for $z_0 \neq 0$ and some $\underline{\sigma}$ (or, alternatively, $\rho^{\alpha}_{X}(\ ,\ )[\mathcal{U}]$

is a distance, i.e. $\rho_X^\alpha(z_0, z)[\mathcal{U}] > 0$ for $z_0 \neq z$), then X is called an $(\alpha, \mathcal{U})$-*almost hyperbolic quaternionic-type manifold.* An $(\alpha, \mathcal{U})$-almost hyperbolic quaternionic-type manifold is said to be complete if it is complete with respect to $\rho_X^\alpha(\ ,\ ;\underline{\sigma})[\mathcal{U}]$ for sòme $\underline{\sigma}$ or, alternatively, with respect to $\rho_X^\alpha(\ ,\ )[\mathcal{U}]$. Almost hyperbolic manifolds in the sense of [5] are — in general — not almost hyperbolic quaternionic-type manifolds.

4 The expression $\rho_X^\alpha(z_0, z)$ as an almost hyperbolic quaternionic-type pseudodistance

We start with proving

Theorem 3. *Let $(X, \mathcal{U})$ and $(Y, \mathcal{V})$ be two (k, m)-hyperbolic n-dimensional complex submanifolds of $\mathbb{C}^{2n}$, $n = 1$ or 2, with k, m, α as in Theorem 2, locally finite open coverings $\mathcal{U}$ and $\mathcal{V}$, and admissible families $F[\mathcal{U}]$ and $F[\mathcal{V}]$ of pluriharmonic functions. Let $f : X \longrightarrow Y$ be a proper holomorphic mapping such that $f^{-1}[\mathcal{V}] \subset \mathcal{U}$. Moreover, suppose that we can introduce on X a structure $J_{\underline{\sigma}}$ as in Theorem 2 and, analogously (in the case 3.1), the structure $J_{f_*\underline{\sigma}}$ on Y. Then*

$$\rho_X^\alpha(z_0, z; \underline{\sigma})[\mathcal{U}] \geq \rho_Y^\alpha(f(z_0), f(z), f_*\underline{\sigma})[\mathcal{V}] \quad \text{for} \quad z_0, z \in X \tag{7}$$

or, alternatively

$$\rho_X^\alpha(z_0, z)[\mathcal{U}] \geq \rho_Y^\alpha(f(z_0), f(z))[\mathcal{V}] \quad \text{for} \quad z_0, z \in X \tag{8}$$

Proof. Given $u \in F[\mathcal{U}]$, we have $u \circ f \in F[\mathcal{U}]$. For each elementary chain Γ_j either $f(\Gamma_j)$ is one point or, since the image of any elementary chain passing through points z_0, z of X, is a bordered holomorphic chain passing through the points $f(z_0)$, $f(z)$ of $Y = f[X]$ (see [2], Lemma 1), $f[\Gamma_j]$ is a one-dimensional complex variety and the restriction $f|\Gamma_j : \Gamma_j \longrightarrow f[\Gamma_j]$ is a finite ramified covering. By the definition of ρ_X^α and the above observation, taking into account the suprema over $V = F[\mathcal{V}]$ and $U = F[\mathcal{U}]$, we arrive at (7) or, alternatively, at (8), as desired. □

Besides, arguing as in the proof of Proposition 4 in [6], we get

Theorem 4. *Let $(X, \mathcal{U})$ and $(Y, \mathcal{V})$ be two (l, m)-hyperbolic m-dimensional manifolds as in Theorem 3 such that Y is a finite-sheeted covering manifold of X with covering projection $\pi : Y \to X$, every $u \in \mathcal{U}$ is well covered by π, and $\mathcal{V} = \pi^{-1}[\mathcal{U}]$. Let z_0, $z \in X$ and $s_0, s \in Y$ so that $\pi(s_0) = z_0$, $\pi(s) = z$,* *and (in the case 3.1)* $\underline{\sigma} = \pi_* \underline{\sigma}'$. *Then*

$$\rho_X^\alpha(z_0, z; \underline{\sigma})[\mathcal{U}] = \min\{\rho_Y^\alpha(s_0, s; \underline{\sigma}')[\mathcal{V}] : \ s \in Y, \ \pi(s_0) = z_0 \ \textit{and} \ \pi(s) = z\}$$

or, alternatively,

$$\rho_X^\alpha(z_0, z)[\mathcal{U}] = \min\{\rho_Y^\alpha(s_0, s)[\mathcal{V}] : \ s \in Y, \ \pi(s_0) = z_0 \ \textit{and} \ \pi(s) = z\}.$$

Since a (k, m)-hyperbolic n-dimensional complex manifold $(X, \mathcal{U})$ with a locally finite open covering $\mathcal{U}$ induces a locally finite open covering $\mathcal{U}'$ on any submanifold X' of X, we also have:

Corollary 2. *An n-dimensional complex submanifold $(X,\ \mathcal{U})$ of a complete $(\alpha, \mathcal{U})$-almost hyperbolic quaternionic-type manifold is an $(\alpha, \mathcal{U} \cap X')$-almost hyperbolic quaternionic-type manifold. If, in addition, X' is closed, it is also complete.*

Corollary 3. *The submanifold $X' = \{z \in X : f(z) = 0\}$ of an $(\alpha, \mathcal{U})$-almost hyperbolic quaternionic-type manifold X, where f is a holomorphic function on X, is an $(\alpha, \mathcal{U} \times X')$-almost hyperbolic quaternionic-type manifold.*

Proof. It is sufficient to observe that the embedding of X' into X is proper holomorphic since X' is a closed submanifold of X, and to apply Proposition 1 of [2] in its modified version corresponding to (k, m)-hyperbolic manifolds. □

The next step is to prove the following analogue of Theorem 4.10 in [8]:

Corollary 4. *Let X be an $(\alpha, \mathcal{U})$-almost hyperbolic quaternionic-type manifold and f a holomorphic function of X. Then the open submanifold $X' = \{z \in X : f(z) \neq 0\}$ of X is the $(\alpha, \mathcal{U} \cap X')$-almost hyperbolic manifold.*

Proof. It is easy to observe that X' is an open complex submanifold of X and thus, it is holomorphically embedded by a holomorphic inclusion mapping $i_X : X' \to X$. □

References

1. A.L. Besse, *Einstein Manifolds*, (Ergebnisse der Math. und ihrer Grenzgebiete (3) 10), Springer-Verlag, Berlin 1987, xii + 510 pp.
2. G. Boryczka and L.M. Tovar, *Hyperbolic-like manifolds. Geometrical properties and holomorphic mappings*, in: Generalizations of Complex

Analysis and Their Applications in Physics. Ed. by J. Ławrynowicz (Banach Center Publications 37), PWN — Polish Scientific Publishers, Warsaw, 1996, pp. 53–66.

3. S.S. Chern, H.I. Levine and L. Nirenberg, *Intrinsic norms on a complex manifold*, in: Global Analysis, Papers in honor of K. Kodaira, ed. by D.C. Spencer and S. Iynaga, Univ. of Tokyo Press and Princeton Univ. Press, Tokyo 1969, pp. 119–139; reprinted in S.S. Chern: Selected Papers, Springer-Verlag, New York-Heidelberg-Berlin 1978, pp. 371–391.
4. P. Dolbeault, *Sur les chaines maximalement complexes au bord donné*, Proc. Sympos. Pure Math. **44**(1986), 171–205.
5. ———, J. Kalina and J. Ławrynowicz, *Introduction to almost hyperbolic pseudodistances via intermediate dimensional-invariant measures*, Univ. Mariae Curie-Skłodowska Sect. A **52**(1998), 9–14.
6. ——— and J. Ławrynowicz, *Holomorphic chains and extendability of holomorphic mappings*, in: Deformations of Mathematical Structures. Complex Analysis with Physical Applications. Selected papers from the Seminar on Deformations, Łódź-Lublin 1985/87. Edited by J. Ławrynowicz, Kluwer Academic Publishers, Dordrecht–Boston–London 1989, pp. 191–204.
7. E. Eisenman (now Pelles), *Intrinsic Measures on Complex Manifolds and Holomorphic Mappings*, Mem. Amer. Math. Soc. **96**(1970), ii + 80 pp.
8. S. Kobayashi, *Hyperbolic Manifolds and Holomorphic Mappings* (Pure and Appl. Math. 2), Marcel Dekker, New York 1970, x + 148 pp.
9. J. Ławrynowicz and O. Suzuki, *An approach to the* 5-, 9-, *and* 13-*dimensional complex dynamics II*, Bull. Soc. Sci. Lettres Łódź 48 Sér. Rech. Déform. **26**(1998), 23–48.
10. ———, *An introduction to pseudotwistors*, to appear.
11. D.A. Pelles (formerly Eisenman), *Holomorphic maps which preserve intrinsic measure*, Amer. J. Math. **97**(1975), 1–15.
12. T. Takakura, *Degeneration of Riemann surfaces and intermediate polarization of the moduli space of flat connections*, Invent. Math. **123**(1996), 53–66.

TOPOLOGY AND CLASSICAL GAUGE THEORY

GREGORY L. NABER
California State University, Department of Mathematics
Chico, CA 95929–0525, USA

This sequence of five lectures will attempt to survey the mutual interaction of topology and gauge theory. A general mathematical framework for classical gauge theories will be introduced and examples will be described which illustrate such notions as "topological charge" of gauge and Higgs fields. Donaldson's theorem on the intersection form of a smooth 4-manifold will be used to illustrate the impact had on low dimensional topology by the differential equations arising in gauge theory. The lectures will conclude with a brief overview of the Seiberg-Witten equations and some results on their solutions on flat space.

1 Framework for classical gauge theories

The emphasis of these lectures will be on concrete examples that illuminate the interaction between topology and classical gauge field theory. In order to have a context within which to place these examples, however, we will begin with an explicit enumeration of the basic mathematical ingredients required to describe, at the classical level, the interaction of a particle with a gauge field.

1. A smooth, oriented, (semi-) Riemannian manifold X.

   Generally, this will be space ($\mathbb{R}^3$), a spacetime (e.g., $\mathbb{R}^{1,3}$), a Euclidean (Wick rotated) version of a spacetime (e.g., $\mathbb{R}^4$), a compactification of one of these (e.g., $S^4 = \mathbb{R}^4 \cup \{\infty\}$), an open submanifold of one of these (e.g., $\mathbb{R}^3 - \{(0,0,0)\}$), or some homotopy equivalent (e.g., $S^2 \cong \mathbb{R}^3 - \{(0,0,0)\}$). The particles and fields "live" in X.

2. A finite dimensional vector space $\mathcal{V}$ equipped with an inner product $\langle\ ,\ \rangle$.

   The particles have wavefunctions that take values in $\mathcal{V}$. The choice of $\mathcal{V}$ is dictated by the internal structure of the particle (e.g., phase, spin, isospin, etc.) and so $\mathcal{V}$ is called the *internal space*. Typical examples are $\mathbb{C}$ (spin 0 charged particles), $\mathbb{C}^4$ (Dirac electrons), $\mathbb{C}^8$ (nucleons), or the Lie algebra $\mathcal{G}$ of some Lie group G (Higgs fields). From the inner product $\langle\ ,\ \rangle$ one computes squared norms of $\mathcal{V}$-valued objects and, from these, formulates the action principles that govern the dynamics (see 8 below).

3. A matrix Lie group G and a representation $\rho : G \to GL(\mathcal{V})$ of G

on $\mathcal{V}$ that is orthogonal with respect to the inner product $\langle\ ,\ \rangle$ on $\mathcal{V}$ ($\langle \rho(g)(v), \rho(g)(w) \rangle = \langle v, w \rangle$ for all $g \in G$ and all $v,\ w \in \mathcal{V}$).

G will generally be one of the classical groups (e.g., $U(1)$, $SU(2)$, $SO(4)$, $SL(2,\mathbb{C})$, etc.) or a product of these. In general, G describes a symmetry of the physical system under consideration, while ρ describes the particular type of invariance that a particle's wavefunction exhibits under this symmetry. More specifically, the Lie group G plays the following dual role:

(a) The inner product $\langle\ ,\ \rangle$ on $\mathcal{V}$ determines a class of orthonormal bases, or frames, in $\mathcal{V}$ and these are related by the elements of G, i.e., if P is the collection of all such frames, then there is a (right) action of G on P which sends any frame $p \in P$ to a new frame $p \cdot g \in P$. By fixing (arbitrarily) some frame at the outset one can therefore identify the elements of G with the frames.

(b) G also acts on $\mathcal{V}$ (on the left) via the representation ρ ($v \to (\rho(g))(v) = g \cdot v$) and so acts on the wavefunction at each point. If $\psi(p)$ is a value of the wavefunction, described relative to the frame $p \in P$, then

$$\psi(p \cdot g) = g^{-1} \cdot \psi(p)$$

is its description relative to the new frame $p \cdot g$.

The right action of G on P transforms frames in the internal space and the left action of G on $\mathcal{V}$ describes the corresponding transformation law for the wavefunction.

4. A smooth principal G-bundle $G \hookrightarrow P \xrightarrow{\mathcal{P}} X$ over X.

Typical examples are trivial (product) bundles

$$(\text{e.g., } SU(2) \hookrightarrow \mathbb{R}^4 \times SU(2) \xrightarrow{\mathcal{P}} \mathbb{R}^4)$$

and Hopf bundles

$$(\text{e.g., } U(1) \hookrightarrow S^3 \xrightarrow{\mathcal{P}} \mathbb{CP}^1, \text{ and } SU(2) \hookrightarrow S^7 \xrightarrow{\mathcal{P}} \mathbb{HP}^1).$$

At each $x \in X$ the fiber $\mathcal{P}^{-1}(x)$ is a copy of G, thought of as the set of all frames in the internal space $\mathcal{V}$ at x. A local cross-section $s : V \to P$ (V open in X and $\mathcal{P} \circ s = \mathrm{id}_V$) is a smooth selection of a frame at each point of V relative to which wavefunctions can be described on V. Such a cross-section is called a local *gauge*.

5. A connection ω on $G \hookrightarrow P \xrightarrow{\mathcal{P}} X$ with curvature Ω.

   In classical electrodynamics an electromagnetic field is generally modeled by a 2-form $\boldsymbol{F}$ defined on space or spacetime (i.e., on X). The corresponding potential is a 1-form $\boldsymbol{A}$ with $\boldsymbol{F} = d\boldsymbol{A}$. $\boldsymbol{F}$ is globally defined on X, but, in general, potentials are only locally defined so that a complete description of $\boldsymbol{F}$ will require several potentials with domains that cover X. In non-Abelian gauge theories even the field strengths are, in general, only locally defined on X. By virtue of the manner in which these locally defined forms on X are related on the intersections of their domains one can piece them together into globally defined forms on the bundle space P. These are the connection ω and its curvature Ω. Given ω and Ω one retrieves the physical potentials and fields by choosing a local gauge (cross-section) $s : V \to P$ and pulling back to X: $\mathcal{A} = s^*\omega$ is the *local gauge potential* and $\mathcal{F} = s^*\Omega$ is the *local field strength* (both in gauge s). Another local gauge $s' : V' \to P$ with $V \cap V' \neq \emptyset$ will be related to s on $V \cap V'$ by $s' = s \cdot g$, where $g : V \cap V' \to G$ and $\cdot$ is the right action of G on P. One generally writes s^g rather than s' to explicitly display this transition function g. The corresponding potential and field strength are written $\mathcal{A}^g$ and $\mathcal{F}^g$, respectively, and their relationship to $\mathcal{A}$ and $\mathcal{F}$ are given by

$$\mathcal{A}^g = g^{-1}\mathcal{A}g + g^{-1}dg \tag{1}$$

   and

$$\mathcal{F}^g = g^{-1}\mathcal{F}g \tag{2}$$

   on $V \cap V'$. A local section $s : V \to P$ is a choice of internal coordinates at each $x \in V$. An internal coordinate transformation to a new section $s^g : V' \to P$ is accomplished by a map $g : V \cap V' \to G$ and called a *local gauge transformation*. An analogue of the relativity principle, called the *gauge principle*, holds that such gauge transformations alter only the appearance, and not the physics, of the situation, i.e., $\mathcal{A}^g$ and $\mathcal{F}^g$ represent the same potential and field strength as $\mathcal{A}$ and $\mathcal{F}$, only in different internal coordinates.

6. A global cross-section ϕ of the vector bundle $P \times_\rho \mathcal{V}$ associated to $G \hookrightarrow P \xrightarrow{\mathcal{P}} X$ by the representation ρ (or, equivalently, a $\mathcal{V}$-valued map $\psi : P \to \mathcal{V}$ on P that is equivariant, i.e., $\psi(p \cdot g) = g^{-1} \cdot \psi(p)$ for all $g \in G$ and $p \in P$).

   Particles coupled to (i.e., experiencing the effects of) the gauge field determined by ω have locally defined wavefunctions ϕ taking values in $\mathcal{V}$

that are obtained by solving field equations (see 8 below) involving the local potentials $\mathcal{A}$. A change of gauge ($s \to s^g = s \cdot g$) changes the wavefunction by the representation ρ so these local wavefunctions piece together into a globally defined object called a *matter field* that can be described in either of the two equivalent ways referred to in 6.

7. A non-negative, smooth, real-valued function $U : \mathcal{V} \to \mathbb{R}$ that is invariant under the action of G on $\mathcal{V}$, i.e., $U(g \cdot v) = U(v)$ for all $g \in G$ and $v \in \mathcal{V}$.

   U is to be regarded as a potential function with $U \circ \psi$ describing the self-interaction energy of the matter field ψ. Typically, this will depend only on $||\psi||^2 = \langle \psi, \psi \rangle$, e.g., $\frac{1}{2}m||\psi||^2$ or $\frac{\lambda}{8}(||\psi||^2 - 1)^2$, where m and λ are non-negative constants.

8. An action (energy) functional $A(\omega, \phi)$, the stationary points of which are the physically significant field configurations (ω, ϕ).

   Typically, this functional has the general form

$$A(\omega, \phi) = c \int_X (||\boldsymbol{F}_\omega||^2 + c_1||d^\omega \phi||^2 + c_2 U \circ \phi) d\mathrm{vol}$$

   where c is a normalizing constant, c_1 and c_2 are coupling constants, $\boldsymbol{F}_\omega$ is a globally defined 2-form on X with values in the adjoint bundle $\mathrm{ad}\, P$ which locally pulls back to the gauge field strengths $\mathcal{F}$, $d^\omega \phi$ is the covariant exterior derivative of ϕ and the norms arise from the metric on X, the inner product on $\mathcal{V}$ and the Killing form on the Lie algebra $\mathcal{G}$ of G. The Euler-Lagrange equations for the action $A(\omega, \phi)$ are the appropriate field equations of the gauge theory (e.g., Maxwell's equations, Yang-Mills equations, Yang-Mills-Higgs equations). In the special case in which $c_1 = c_2 = 0$ (so that, effectively, there are no matter fields present and $A(\omega, \phi)$ depends only on ω) the action is called the *Yang-Mills action* and written

$$\mathcal{YM}(\omega) = c \int_X ||\boldsymbol{F}_\omega||^2 d\mathrm{vol}.$$

   The corresponding Euler-Lagrange equations can be written

$$d^\omega {}^*\boldsymbol{F}_\omega = 0,$$

   where ${}^*\boldsymbol{F}_\omega$ is the Hodge dual of $\boldsymbol{F}_\omega$. These then are the *Yang-Mills equation*. Any solution also satisfies a purely geometrical constraint known as the *Bianchi identity*:

$$d^\omega \boldsymbol{F}_\omega = 0.$$

2 The prototypical example: Dirac monopoles

In order to understand the physical and historical origins of this model for classical gauge theories we will describe the first, and simplest example. Begin with the source free Maxwell equations, which we write as

$$\begin{aligned} \vec{\nabla}\cdot(\vec{E}+i\,\vec{B}) &= 0 \\ \frac{\partial}{\partial t}(\vec{E}+i\,\vec{B})+i\vec{\nabla}\times(\vec{E}+i\,\vec{B}) &= \vec{0} \end{aligned} \tag{3}$$

These equations have many symmetries: they are Lorentz invariant, gauge invariant and conformally invariant. In addition, they possess what we will call a $U(1)$-duality symmetry: if $\vec{E}+i\,\vec{B}$ is a solution and φ is a real number, then $e^{i\varphi}(\vec{E}+i\,\vec{B})$ is also a solution. When $\varphi=\frac{\pi}{2}$ this reduces to the familiar fact that the replacements $\vec{B}\to\vec{E}$ and $\vec{E}\to-\vec{B}$ carry solutions onto solutions. This last symmetry is lost, of course, if one includes electric charges and currents, but can be restored with the inclusion of (hypothetical) magnetic charges and currents. A magnetic charge (or monopole) is a point object assumed to create the "magnetic analogue" of a Coulomb field. Specifically, if the object is located at the origin in $\mathbb{R}^3$ and (ρ,φ,θ) represent standard spherical coordinates, then a monopole field is given by

$$\vec{E}=\vec{0},\quad \vec{B}=\frac{n/2}{\rho^2}\hat{e}_\rho \quad \text{on} \quad \mathbb{R}^3-\{(0,0,0)\}, \tag{4}$$

where $\hat{e}_\rho$ is the outward unit radial field.

Remark. We have written the strength (magnetic charge) of the monopole as $n/2$ ($n\in\mathbb{Z}$) in compliance with the Dirac Quantization Condition (Appendix I).

One can write the field strength 2-form $\boldsymbol{F}$ for the monopole field in the usual way. In Cartesian coordinates the result is

$$\boldsymbol{F}=\frac{n/2}{\rho^3}(x\,dy\wedge dz-y\,dx\wedge dz+z\,dx\wedge dy), \tag{5}$$

while, in spherical coordinates, $\boldsymbol{F}$ becomes

$$\boldsymbol{F}=\frac{n}{2}\sin\varphi\,d\varphi\wedge d\theta \tag{6}$$

(note that this last expression is independent of ρ and so may be thought of as a 2-form on S^2). It is easy to see that $\boldsymbol{F}$ is not exact on $\mathbb{R}^3-\{(0,0,0)\}$, i.e., there does not exist a 1-form $\boldsymbol{A}$ on $\mathbb{R}^3-\{(0,0,0)\}$ for which $\boldsymbol{F}=d\boldsymbol{A}$

everywhere ($\int_{S^2} \boldsymbol{F} = 2n\pi$, but the existence of such an $\boldsymbol{A}$ would, by Stokes' Theorem, give a value of 0 for the integral). Dirac was well aware of this and also knew (although, perhaps not in these terms) that if one deletes from $\mathbb{R}^3$ not only the location of the monopole, but also some ray extending from it to infinity (a so-called *Dirac string*) then the result is a manifold whose 2nd de Rham cohomology is trivial

$$H^2_{\text{de Rham}}(\mathbb{R}^3 - \text{Dirac string}) = 0$$

so that every 2-form on it (e.g., $\boldsymbol{F}$) is exact. Taking the string to be the non-negative z-axis one can explicitly write down a 1-form $\boldsymbol{A}_S$ on $\mathbb{R}^3 - \{(0,0,z) : z \geq 0\}$ with $d\boldsymbol{A}_S = \boldsymbol{F}$ there.

$$\boldsymbol{A}_S = \frac{n/2}{\rho(\rho - z)}(y\,dx - x\,dy) = -\frac{n}{2}(1 + \cos\varphi)d\theta \tag{7}$$

Similarly, on $\mathbb{R}^3 - \{(0,0,z) : z \leq 0\}$,

$$\boldsymbol{A}_N = \frac{n/2}{\rho(\rho + z)}(y\,dx - x\,dy) = \frac{n}{2}(1 - \cos\varphi)d\theta \tag{8}$$

satisfies $d\boldsymbol{A}_N = \boldsymbol{F}$. Together the domains of $\boldsymbol{A}_S$ and $\boldsymbol{A}_N$ cover all of $\mathbb{R}^3 - \{(0,0,0)\}$. Thus, although $\boldsymbol{F}$ has no globally defined potential on $\mathbb{R}^3 - \{(0,0,0)\}$, it does have two locally defined potentials whose domains exhaust the domain of F.

One of the most remarkable things about the 1-forms $\frac{n}{2}(\pm 1 - \cos\varphi)d\theta$ is that, historically, they arose, quite independently of their role in physics as the locally defined potentials for a Dirac monopole field, in the study of the geometry of principal $U(1)$-bundles over S^2. To see how this comes about it will be convenient to regard $\boldsymbol{A}_S$ and $\boldsymbol{A}_N$ as 1-forms on S^2 and introduce a factor of $-i$ to obtain forms that take values in the pure imaginary complex numbers Im$\mathbb{C}$. Thus, we let $U_S = S^2 - \{(0,0,1)\}$ and $U_N = S^2 - \{(0,0,-1)\}$ and define

$$\mathcal{A}_S = -i\,\boldsymbol{A}_S = \frac{n}{2}i\,(1 + \cos\varphi)d\theta \qquad \text{on} \quad U_S \tag{9}$$

and

$$\mathcal{A}_N = -i\,\boldsymbol{A}_N = -\frac{n}{2}i\,(1 - \cos\varphi)d\theta \qquad \text{on} \quad U_N. \tag{10}$$

Notice that, on $U_S \cap U_N$, $\mathcal{A}_N - \mathcal{A}_S = -ni\,d\theta$ which, for reasons that will become clear shortly, we write as

$$\mathcal{A}_N = e^{ni\theta}\mathcal{A}_S e^{-ni\theta} + e^{ni\theta}de^{-ni\theta}. \tag{11}$$

The simplest (nontrivial) examples of principal $U(1)$-bundles over S^2 are obtained from the complex Hopf bundle $U(1) \hookrightarrow S^3 \xrightarrow{\mathcal{P}} \mathbb{CP}^1$. Here we regard S^3 as the submanifold of $\mathbb{C}^2$ consisting of those (z^1, z^2) with $|z^1|^2 + |z^2|^2 = 1$. The right action of $U(1)$ on S^3 is given by $p \cdot g = (z^1, z^2) \cdot g = (z^1 g, z^2 g)$ for all $p = (z^1, z^2) \in S^3$ and all $g \in U(1)$. The orbits are copies of $U(1) \cong S^1$ and $\mathcal{P}$ is the quotient map that identifies each such circle to a point in the projective space $\mathbb{CP}^1$. The map $[z^1, z^2] \to (2\text{Re}(\bar{z}^1 z^2), 2\text{Im}(\bar{z}^1 z^2), |z^1|^2 - |z^2|^2)$ is a diffeomorphism of $\mathbb{CP}^1$ onto S^2 and composing with $\mathcal{P}$ gives a principal $U(1)$-bundle over S^2 denoted

$$U(1) \hookrightarrow S^3 \xrightarrow{\mathcal{P}_1} S^2.$$

Another (inequivalent) bundle is obtained by composing $\mathcal{P}$ with the diffeomorphism $[z^1, z^2] \to (2\text{Re}(\bar{z}^1 z^2), -2\text{Im}(\bar{z}^1 z^2), \ |z^1|^2 - |z^2|^2)$ of $\mathbb{CP}^1$ onto S^2 and we denote this

$$U(1) \hookrightarrow S^3 \xrightarrow{\mathcal{P}_{-1}} S^2.$$

Any principal bundle over S^2 can be trivialized over U_N and U_S (since these are contractible) and so has a single transition function $g_{SN} : U_N \cap U_S \to U(1)$ that characterizes the bundle up to equivalence. For $U(1) \hookrightarrow S^3 \xrightarrow{\mathcal{P}_1} S^2$ and $U(1) \hookrightarrow S^3 \xrightarrow{\mathcal{P}_{-1}} S^2$ these are given by $g_{SN}(\varphi, \theta) = e^{-\theta i}$ and $g_{SN}(\varphi, \theta) = e^{\theta i}$, respectively. Similarly, one can define, for any integer n, a principal $U(1)$-bundle

$$U(1) \hookrightarrow P_n \xrightarrow{\mathcal{P}_n} S^2$$

with transition function $g_{SN}(\varphi, \theta) = e^{-n\theta i}$. For $n = 0$ this is the trivial bundle $U(1) \hookrightarrow S^2 \times U(1) \xrightarrow{\mathcal{P}_0} S^2$. For $n \neq 0, 1, -1$, P_n is the quotient of S^3 by a $\mathbb{Z}_n$-action (i.e., a lens space) and $\mathcal{P}_n$ is the map induced on P_n by $\mathcal{P}_1$ if $n > 1$ and by $\mathcal{P}_{-1}$ if $n < -1$. The details are not important at the moment. What is important is that the collection of $U(1)$-bundles over S^2 that we have just enumerated exhausts all of the equivalence classes of such bundles. Thus, any $U(1)$-bundle over S^2 is characterized up to equivalence by an integer n.

Any principal bundle $G \hookrightarrow P \xrightarrow{\mathcal{P}} X$ admits (in general, many) connections. These are 1-forms on the bundle space P with values in the Lie algebra $\mathcal{G}$ of G. Each of the $U(1)$-bundles $U(1) \hookrightarrow P_n \xrightarrow{\mathcal{P}_n} S^2$, $n \in \mathbb{Z}$, has defined on it a natural connection ω_n arising from the Im$\mathbb{C}$-valued 1-form $in\text{Im}(\bar{z}^1\, dz^1 + \bar{z}^2\, dz^2)$ on $\mathbb{C}^2$ (recall that the Lie algebra of $U(1)$ can be identified with Im $\mathbb{C}$). Trivializing the bundles over U_N and U_S and denoting the corresponding cross-sections by s_N and s_S one finds that the pullbacks $s_N^* \omega_n$

and $s_S^*\omega_n$ are nothing other than the Dirac monopole potentials $\mathcal{A}_N$ and $\mathcal{A}_S$, respectively.

$$
\begin{aligned}
s_S^*\omega_n &= \frac{n}{2}i(1+\cos\varphi)d\theta, \\
s_N^*\omega_n &= -\frac{n}{2}i(1-\cos\varphi)d\theta.
\end{aligned}
\tag{12}
$$

Now, a connection is completely characterized by its family of pullbacks by the cross-sections corresponding to a trivializing cover. Conversely, given a principal bundle $G \hookrightarrow P \xrightarrow{\mathcal{P}} X$, a trivializing cover $\{U_j\}_{j\in J}$ with transition functions $g_{ij} : U_j \cap U_i \to G$ and $\mathcal{G}$-valued 1-forms $\mathcal{A}_j$ on U_j for each $j \in J$ with the property that

$$
\mathcal{A}_j = g_{ij}^{-1}\mathcal{A}_i g_{ij} + g_{ij}^{-1}dg_{ij} \tag{13}
$$

on $U_j \cap U_i$ whenever this intersection is nonempty, there exists a unique connection ω on P such that $\mathcal{A}_j = s_j^*\omega$, $j \in J$, where s_j is the cross-section corresponding to the trivialization on U_j. Since (11) is just (13) for the transition function of $U(1) \hookrightarrow P_n \xrightarrow{\mathcal{P}_n} S^2$, we find that the Dirac monopole of strength $\frac{n}{2}$ is uniquely represented by the connection ω_n on P_n.

These observations were intended to motivate those items in our General Framework that suggest modeling physical fields by connections on principal bundles, but they also lead us to our first encounter with topology. We have alluded to the fact that any principal $U(1)$-bundle over S^2 is equivalent to some $U(1) \hookrightarrow P_n \xrightarrow{\mathcal{P}_n} S^2$ so that such bundles are in one-to-one correspondence with the integers. The integer labeling the bundle is twice the strength of the Dirac monopole that lives on the bundle. There is, however, another way to view this integer that is purely topological. Any principal bundle $G \hookrightarrow P \xrightarrow{\mathcal{P}} X$ has associated with it a characteristic de Rham cohomology class $c_1(P)$ called its 1st Chern class (Appendix II). This is a cohomology class of real-valued 2-forms on X that can be computed from *any* connection on the bundle and whose integral over any closed, oriented 2-dimensional submanifold of X is necessarily an integer. Principal $U(1)$-bundles over S^2 are characterized by their 1st Chern classes and, indeed, by their integrals

$$
c_1(P)[S^2] = \int_{S^2} c_1(P)
$$

over S^2 (the so-called Chern number of the bundle). For the bundles $U(1) \hookrightarrow P_n \xrightarrow{\mathcal{P}_n} S^2$ one finds that $c_1(P_n)[S^2] = n$. Thus, the magnetic charge of a Dirac monopole is directly encoded in the topology of the bundle on which the monopole lives as a connection. By contrast, the existence of a globally defined potential for the (electric) Coulomb field has the consequence that any such field is modeled by a connection on a trivial bundle (1st Chern class zero) so that, in particular, the electric charge is not encoded in the topology of the bundle. Magnetic charge is "topological", but electric charge is not.

3 Spin zero electrodynamics

In the hope that the example of the Dirac monopole has now clarified the rationale behind modeling gauge fields with connections on principal bundles, we move on to the remaining items in the General Framework, which we illustrate with the simplest example available. These are dynamical in nature, involving the interaction of a particle with a gauge field, and so the appropriate choice for the base manifold X is now a spacetime. We will focus attention on the simplest situation in which X is an open submanifold of Minkowski spacetime $\mathbb{R}^{1,3}$ (think of $\mathbb{R}^{1,3} - X$ as the worldlines of the particles creating the electromagnetic field, which have been carved out of Minkowski spacetime to leave X). An electromagnetic field is then modeled by a connection on a principal $U(1)$-bundle over X and we wish to describe the interaction of a charged particle with this field. The simplest case is that of a charged spin zero particle since the wavefunction will then have only one complex component. We therefore take the vector space $\mathcal{V}$ to be $\mathbb{C}$ (thought of as a 2-dimensional real vector space) and write the standard inner product on $\mathbb{C}$ as

$$\langle z_1, z_2 \rangle = \frac{1}{2}(z_1 \bar{z}_2 + \bar{z}_1 z_2). \tag{14}$$

For each integer q we define a representation $\mathcal{P}_q : U(1) \to GL(\mathbb{C})$ by

$$(\rho_q(g))(z) = g \cdot z = g^q z, \tag{15}$$

where we identify an element g of $U(1)$ with its sole entry (a complex number of modulus 1). These are, in fact, all of the irreducible representations of $U(1)$ on $\mathbb{C}$ and each is clearly orthogonal with respect to the standard inner product $\langle \ , \ \rangle$ on $\mathbb{C}$.

Now, let $U(1) \hookrightarrow P \xrightarrow{\mathcal{P}} X$ be a principal $U(1)$-bundle over X and ω a connection on it. Since $U(1)$ is Abelian, all brackets in the Lie algebra $u(1) = \operatorname{Im}\mathbb{C}$ are zero so the curvature Ω is given by $\Omega = d\omega$. A matter field can then be identified with a $\mathbb{C}$-valued map $\psi : P \to \mathbb{C}$ on P that is equivariant, i.e., satisfies

$$\psi(p \cdot g) = g^{-1} \cdot \psi(p) = g^{-q}\psi(p) \tag{16}$$

for all $p \in P$ and $g \in U(1)$. As a potential function $U : \mathbb{C} \to \mathbb{R}$ we take

$$U(z) = \frac{1}{2}m\langle z, z\rangle = \frac{1}{2}m|z|^2, \tag{17}$$

where $m \geq 0$ is a constant. Since $\langle\ ,\ \rangle$ is orthogonal with respect to ρ_q, U is invariant under the action of $U(1)$ on $\mathbb{C}$.

To describe the action functional we introduce a bit of notation. If $s : V \to P$ is a local cross-section of $U(1) \hookrightarrow P \xrightarrow{\mathcal{P}} X$ we write the electromagnetic gauge potential $\mathcal{A} = s^*\omega$ and field strength $\mathcal{F} = s^*\Omega$ as

$$\mathcal{A} = s^*\omega = -i\boldsymbol{A}$$

and

$$\mathcal{F} = s^*\Omega = d\mathcal{A} = -i\, d\boldsymbol{A} = -i\boldsymbol{F},$$

where $\boldsymbol{A}$ and $\boldsymbol{F}$ are real-valued forms on V. For any other cross-section $s^g : V' \to P$ with $V \cap V' \neq \emptyset$,

$$\mathcal{A}^g = g^{-1}\mathcal{A}g + g^{-1}dg = \mathcal{A} + g^{-1}dg$$

and

$$\mathcal{F}^g = g^{-1}\mathcal{F}g = \mathcal{F}$$

on $V \cap V'$ because $U(1)$ is Abelian. In particular, there is a globally defined field strength $\mathcal{F}$ on X. Relative to standard coordinates x^0, x^1, x^2, x^3 on $\mathbb{R}^{1,3}$ we write

$$\mathcal{A} = \mathcal{A}_\alpha\, dx^\alpha = -iA_\alpha\, dx^\alpha$$

and

$$\mathcal{F} = \frac{1}{2}\mathcal{F}_{\alpha\beta}\, dx^\alpha \wedge dx^\beta = -\frac{1}{2}iF_{\alpha\beta}\, dx^\alpha \wedge dx^\beta,$$

where

$$\mathcal{F}_{\alpha\beta} = \partial_\alpha\mathcal{A}_\beta - \partial_\beta\mathcal{A}_\alpha = -i(\partial_\alpha A_\beta - \partial_\beta A_\alpha) = -i\, F_{\alpha\beta}$$

($\partial_\alpha = \partial/\partial x^\alpha$, $\alpha = 0,\ 1,\ 2,\ 3$). The connection ω determines a covariant exterior derivative $d^\omega\psi$ of ψ ($d\psi$ acting on the horizontal part of each tangent vector) whose pullback by s is given in standard coordinates by

$$(\partial_\alpha - iqA_\alpha)\psi\, dx^\alpha.$$

The invariant inner product $\langle\ ,\ \rangle$ on $\mathbb{C}$ together with the Minkowski inner product on X combine to give a natural gauge invariant and Lorentz invariant norm for $d^\omega\psi$:

$$||d^\omega\psi||^2 = (\partial_\alpha\psi - iqA_\alpha\psi)(\partial^\alpha\bar{\psi} + iqA^\alpha\bar{\psi}), \tag{18}$$

where $A^\alpha = \eta^{\alpha\beta}A_\beta$ and $\partial^\alpha = \eta^{\alpha\beta}\partial_\beta$, $\alpha = 0, 1, 2, 3$ (η being the matrix diag$(1, -1, -1, -1)$ of the Minkowski inner product). With this and an appropriate choice of normalizing and coupling constants, we obtain the action functional

$$\begin{aligned} A(\omega, \psi) = \int_X \Big[& -\frac{1}{4}F_{\alpha\beta}F^{\alpha\beta} + \frac{1}{2}(\partial_\alpha\psi - iqA_\alpha\psi)(\partial^\alpha\bar{\psi} + iqA^\alpha\bar{\psi}) \\ & + \frac{1}{2}m|\psi|^2 \Big] dx^0 \wedge dx^1 \wedge dx^2 \wedge dx^3, \end{aligned} \tag{19}$$

where $F^{\alpha\beta} = \eta^{\alpha a}\eta^{\beta b}F_{ab}$, $\alpha, \beta = 0, 1, 2, 3$. The corresponding Euler-Lagrange equations are

$$(\partial_\alpha - iqA_\alpha)(\partial^\alpha - iqA^\alpha)\psi + m^2\psi = 0, \tag{20}$$

$$d^*\boldsymbol{F} = 0, \tag{21}$$

where * denotes the Minkowski spacetime Hodge star operator. Since $-i\boldsymbol{F}$ is the pullback of a curvature form it satisfies the Bianchi identity which gives

$$d\boldsymbol{F} = 0 \tag{22}$$

as well. The last two equations are equivalent to Maxwell's (source free) equations, while (20) is the Klein-Gordon equation for a scalar field of mass m and charge q coupled to the electromagnetic field determined by the local gauge potential $\mathcal{A} = -iA_\alpha dx^\alpha$.

We will describe in detail another example of the General Framework ($SU(2)$-Yang-Mills-Higgs Theory on $\mathbb{R}^3$) somewhat later. Generalizing the discussion above to accomodate spin one-half particles involves certain technical complications that we do not wish to become involved in at the moment. Basically, these arise from the fact that, according to Dirac, a charged, massive, spin one-half particle corresponds to a (reducible) representation of $SL(2,\mathbb{C})$ on $\mathbb{C}^4$ and therefore lives in a spinor bundle over spacetime rather than a complex line bundle. Nevertheless, the electromagnetic field to which it is coupled is a connection on a $U(1)$-bundle. The formalism of the General Framework requires that these two be spliced together into a single $SL(2,\mathbb{C}) \times U(1)$-bundle on which both may be thought to live.

4 Instantons and Donaldson's theorem

Dirac magnetic monopoles have potentials that are simply pullbacks to S^2 by standard cross-sections of the natural connection on a principal $U(1)$-bundle over S^2. In particular, the monopoles of lowest strength arise from the natural connection on the complex Hopf bundle. There is a quaternionic analogue of the complex Hopf bundle which also admits a natural connection and this also yields something interesting when pulled back by standard cross-sections.

Regard S^7 as the submanifold of $\mathbb{H}^2$ consisting of those (q^1, q^2) with $|q^1|^2 + |q^2|^2 = 1$. Denote by $Sp(1)$ the group of quaternions of norm 1 and define a right action of $Sp(1)$ on S^7 by $p \cdot g = (q^1, q^2) \cdot g = (q^1 g, q^2 g)$ for all $p = (q^1, q^2) \in S^7$ and all $g \in Sp(1)$. The orbits are copies of $Sp(1) \cong S^3$ and we let $\mathcal{P} : S^7 \to \mathbb{HP}^1$ be the quotient map that identifies each such S^3 to a point in the projective space $\mathbb{HP}^1 \cong S^4$. Then

$$Sp(1) \hookrightarrow S^7 \xrightarrow{\mathcal{P}} \mathbb{HP}^1$$

is the quaternionic Hopf bundle. Its natural connection ω is the restriction to S^7 of the $\operatorname{Im}\mathbb{H}$-valued 1-form $\operatorname{Im}(\bar{q}^1 dq^1 + \bar{q}^2 dq^2)$ on $\mathbb{H}^2$ (recall that the Lie algebra $sp(1)$ of $Sp(1)$ is just $\operatorname{Im}\mathbb{H}$). If $U_k = \{ [q^1, q^2] \in \mathbb{HP}^1 : q^k \neq 0 \}$ for $k = 1, 2$, then $\{ U_1, U_2 \}$ is a trivializing cover for the bundle and the corresponding cross-sections $s_k : U_k \to S^7$ are given by

$$s_1([q^1, q^2]) = (|q^1|, q^2 (q^1)^{-1} |q^1|)$$

and

$$s_2([q^1, q^2]) = (q^1 (q^2)^{-1} |q^2|, |q^2|).$$

Pulling the connection ω back by these cross-sections (and writing both pullbacks in coordinates defined by the chart $[q^1, q^2] \to q = q^2 (q^1)^{-1}$ on $U_1 \cap U_2$) gives

$$\mathcal{A}_1 = s_1^* \omega = \operatorname{Im}\left(\frac{\bar{q}}{1 + |q|^2} \, dq \right) \tag{23}$$

and

$$\mathcal{A}_2 = s_2^* \omega = \operatorname{Im}\left(\frac{q}{|q|^2 (1 + |q|^2)} \, d\bar{q} \right). \tag{24}$$

The connection ω is uniquely determined by either one of these pullbacks (given, say, $s_1^*\omega$ on U_1, the transformation law $s_2^*\omega = g_{12}^{-1}(s_1^*\omega) g_{12} + g_{12}^{-1} dg_{12}$ determines $s_2^*\omega$ on $U_1 \cap U_2$, which misses only one point of U_2 and continuity determines $s_2^*\omega$ at this point). Thus, we will restrict our attention to $\mathcal{A}_1$.

Ignoring, for a moment, its origin, one can regard $\mathcal{A}_1 = \operatorname{Im}\left(\frac{\bar{q}}{1+|q|^2} dq \right)$ simply as an $\operatorname{Im}\mathbb{H}$-valued 1-form on $\mathbb{R}^4 = \mathbb{H}$. So regarded, it is precisely the famous *BPST instanton* solution to the Euclidean Yang-Mills equations discovered in [5]. The corresponding field strength $\mathcal{F}_1 = d\mathcal{A}_1 + \mathcal{A}_1 \wedge \mathcal{A}_1$ is easily computed and found to be

$$\mathcal{F}_1 = d\mathcal{A}_1 + \mathcal{A}_1 \wedge \mathcal{A}_1 = \frac{1}{(1+|q|^2)^2} d\bar{q} \wedge dq. \tag{25}$$

This is anti-self-dual (ASD) in the sense that ${}^*\mathcal{F}_1 = -\mathcal{F}_1$, where * is the Hodge dual corresponding to the Euclidean metric and usual orientation on $\mathbb{R}^4$ (because of the Bianchi Identity any ASD potential necessarily satisfies the Yang-Mills equations). Furthermore, it has Yang-Mills action

$$\mathcal{YM}(\mathcal{A}_1) = \int_{\mathbb{R}^4} -\text{trace}(\mathcal{F}_1 \wedge {}^*\mathcal{F}_1) = 8\pi^2, \tag{26}$$

which is, in particular, finite.

Remark. Wedge products of $sp(1)$-valued differential forms are most conveniently computed by identifying $Sp(1)$ with $SU(2)$. Then $sp(1)$ is identified with $su(2)$ and any form with values in $su(2)$ can be viewed as a matrix of complex-valued forms. Wedge products are then computed by forming matrix products with the entries multiplied by the ordinary wedge product of complex forms.

Such finite action ASD potentials on $\mathbb{R}^4$ are called *instantons*. One can produce more examples of such objects in the following way: Let $f : S^7 \to S^7$ be a conformal diffeomorphism that respects the right action of $Sp(1)$ on S^7 (i.e., $f(p \cdot g) = f(p) \cdot g$ for all $p \in S^7$ and $g \in Sp(1)$). Then, if ω is the natural connection on $Sp(1) \hookrightarrow S^7 \to \mathbb{HP}^1$, $f^*\omega$ is also a connection on this bundle and its pullback to $\mathbb{R}^4$ via the cross-section s_1 is an instanton on $\mathbb{R}^4$. One can, in this way, construct, for each $n \in \mathbb{H}$ and each $\lambda > 0$, the *generic BPST instanton*

$$\mathcal{A}_{\lambda,n} = \text{Im}\left(\frac{\bar{q} - \bar{n}}{\lambda^2 + |q-n|^2} dq\right) \tag{27}$$

with *center* n and *scale* λ (so $\mathcal{A}_1 = \mathcal{A}_{1,0}$). Computing the corresponding field strengths gives

$$\mathcal{F}_{\lambda,n} = d\mathcal{A}_{\lambda,n} + \mathcal{A}_{\lambda,n} \wedge \mathcal{A}_{\lambda,n} = \frac{\lambda^2}{(\lambda^2 + |q-n|^2)^2} d\bar{q} \wedge dq. \tag{28}$$

The Yang-Mills action is given by

$$\mathcal{YM}(\mathcal{A}_{\lambda,n}) = \int_{\mathbb{R}^4} -\text{trace}(\mathcal{F}_{\lambda,n} \wedge {}^*\mathcal{F}_{\lambda,n}) = 8\pi^2. \tag{29}$$

We need to understand the reason that all of these instantons on $\mathbb{R}^4$ have the same Yang-Mills action (total field strength). Note that any $\mathrm{Im}\,\mathbb{H}$-valued 1-form on $\mathbb{R}^4$ can be regarded as the gauge potential for a connection on the trivial $\mathrm{Im}\,\mathbb{H}$-bundle over $\mathbb{R}^4$, but that the $\mathcal{A}_{\lambda,n}$ are more. They are pullbacks to $\mathbb{R}^4$ of a connection on the nontrivial Hopf bundle over $\mathbb{HP}^1 \cong S^4$. Turning matters about, one might say that the connection on the trivial bundle over $\mathbb{R}^4$ "extends to S^4" in the sense that $S^4 \cong \mathbb{R}^4 \cup \{\infty\}$ is the 1-point compactification of $\mathbb{R}^4$ and, due to its asympototic behavior as $|x| \to \infty$ in $\mathbb{R}^4$, the connection extends to the point at infinity. Note, however, that the extension process involves not only the connection, but the bundle as well. A remarkable theorem of Karen Uhlenbeck [18] asserts that the same sort of thing can be done for any finite action Yang-Mills potential on $\mathbb{R}^4$. In particular, if $\mathcal{A}$ is any ASD $\mathrm{Im}\,\mathbb{H}$-valued 1-form on $\mathbb{R}^4$ with $\mathcal{YM}(\mathcal{A}) < \infty$, then there exists a unique $Sp(1)$-bundle $\mathcal{P} : P \to S^4$ over S^4 and a connection ω on P whose pullback by some cross-section s is $\mathcal{A}$ ($s^*\omega = \mathcal{A}$). Furthermore, the bundle to which $\mathcal{A}$ "extends" is uniquely determined by $\mathcal{YM}(\mathcal{A})$ in the following way. Just as $U(1)$-bundles over S^2 are determined up to equivalence by their 1st Chern class, so principal $Sp(1)$-bundles over S^4 are characterized up to equivalence by their 2nd Chern class $c_2(P)$ (Appendix II). Indeed, two such bundles are equivalent if and only if they have the same 2nd Chern number

$$c_2(P)[S^4] = \int_{S^4} c_2(P)$$

(always an integer). This integral can be computed from any connection on P. Moreover, since there is an orientation preserving conformal diffeomorphism from S^4 minus a point onto $\mathbb{R}^4$, $c_2(P)[S^4]$ can be computed by integrating pullbacks over $\mathbb{R}^4$. Choosing a connection ω on P with an ASD pullback $\mathcal{A}$ to $\mathbb{R}^4$ one finds that

$$c_2(P)[S^4] = \frac{1}{8\pi^2}\mathcal{YM}(\mathcal{A}). \tag{30}$$

Thus, the Yang-Mills action for a finite action ASD potential $\mathcal{A}$ on $\mathbb{R}^4$ is directly encoded in the topology of the bundle over S^4 to which $\mathcal{A}$ extends by Uhlenbeck's Theorem. Notice, however, that the value of $\mathcal{YM}(\mathcal{A})$ is entirely determined by the asymptotic behavior of the field strength $\mathcal{F}$ on $\mathbb{R}^4$ so it is this physical characteristic of the gauge field that is represented by the Chern number. Physicists call $-c_2(P)[S^4]$ the *instanton number*, or *topological charge*, of $\mathcal{A}$. The reason that all of the BPST instantons $\mathcal{A}_{\lambda,n}$ have the same Yang-Mills action is now clear: they all extend to (come from) the same $Sp(1)$-bundle over S^4.

We have seen that, for finite action ASD potentials on $\mathbb{R}^4$, the total field strength is a topological invariant, not unlike the magnetic charge of a Dirac monopole. Being an integer the topological charge cannot be altered through a continuous variation of the field; it is conserved, but for purely topological reasons. We wish to argue now that the asymptotics of a finite action Yang-Mills field are always encoded as an integer in the topology of the bundle to which it extends, even for potentials that are not ASD.

Observe that $\mathcal{YM}(\mathcal{A}) < \infty$ implies that $\|\mathcal{F}\|^2 = -2\text{trace}(\mathcal{F} \wedge {}^*\mathcal{F})$ must approach zero sufficiently fast as $|x| \to \infty$ in $\mathbb{R}^4$. This, together with $\mathcal{F} = d\mathcal{A} + \mathcal{A} \wedge \mathcal{A}$, would seem to require a similar decay for the components of $\mathcal{A}$ and their first derivatives. However, due to the gauge freedom available in the choice of $\mathcal{A}$, this is not the case. All that is necessary is that there exist some local gauge transformation g, defined for sufficiently large $|x|$, such that the potential $\mathcal{A}^g$ in this new gauge has components that decay sufficiently fast (g need only be defined for large $|x|$ because an integral over any compact set in $\mathbb{R}^4$ is necessarily finite). If such a g exists and S^3_R is a 3-sphere in $\mathbb{R}^4$ of sufficiently large radius R that it is contained in the domain of g, then

$$g|S^3_R : S^3_R \longrightarrow Sp(1)$$

is a map from S^3 to S^3 and so determines an element of the homotopy group $\pi_3(S^3)$. But $\pi_3(S^3) \cong \mathbb{Z}$ and an isomorphism is provided by the Brouwer degree deg. Thus, the asymptotic behavior of $\mathcal{F}$ determines g, which determines $[g|S^3_R] \in \pi_3(S^3)$, which gives an integer $\deg(g|S^3_R)$ (the restriction of g to two such spheres are clearly homotopic and therefore have the same degree). The various possible asymptotic behaviors for Yang-Mills field strengths $\mathcal{F}$ on $\mathbb{R}^4$ fall into "homotopy classes", each labeled by an integer. These integers also label the corresponding $Sp(1)$-bundles over S^4 to which the gauge potentials extend.

Remark. Briefly, the reason for this is as follows. $S^4 = \mathbb{R}^4 \cup \{\infty\}$ consists of two copies of the closed 4-dimensional disc (upper and lower hemispheres) glued together along the equator which is a copy of S^3 and which we can take to be S^3_R. The restriction of any bundle over S^4 to either of these discs is trivial because the disc is contractible. This provides a trivializing cover of S^4 consisting of just two trivializations and hence essentially one transition function g. This one transition function determines the bundle up to equivalence and its restriction to the equator is a map from S^3 to $Sp(1) \cong S^3$. Furthermore, any map from S^3 to S^3 can be regarded as such a restriction and homotopic maps correspond to equivalent bundles. In particular, this is true of the restricted gauge transformation $g|S^3_R$ so the integer $\deg(g|S^3_R)$

uniquely determines an $Sp(1)$-bundle over S^4.

Let $G \hookrightarrow P \xrightarrow{\mathcal{P}} X$ be an arbitrary principal G-bundle. A *global gauge transformation* is an automorphism of the bundle, i.e., a diffeomorphism $f : P \to P$ of P onto itself that satisfies $f(p \cdot g) = f(p) \cdot g$ for all $p \in P$ and $g \in G$ and for which the induced map on X is the identity:

$$\begin{array}{ccc} P & \xrightarrow{f} & P \\ \downarrow & & \downarrow \\ X & \xrightarrow[\mathrm{id}_X]{} & X \end{array}$$

Two connections ω and ω' on P are said to be *gauge equivalent* if there exists a global gauge transformation f such that $\omega' = f^*\omega$. The set of all gauge equivalence classes of connections on P is called the *moduli space of connections* on P. These are the configuration spaces of quantum field theory and their structure is, in general, quite complicated. We consider a special case.

Let $Sp(1) \hookrightarrow S^7 \to \mathbb{HP}^1$ be the quaternionic Hopf bundle. As was the case for the complex Hopf bundle there are two natural ways to identify the projective space $\mathbb{HP}^1$ with S^4 and these give inequivalent bundles with Chern numbers ± 1. We will consider the bundle $Sp(1) \hookrightarrow S^7 \to S^4$ with Chern number 1 (i.e., instanton number -1). Any connection on this bundle is uniquely determined by its gauge potential under the cross-section corresponding to stereographic projection from the north pole of S^4. This stereographic projection is an orientation preserving diffeomorphism and so preserves the Hodge dual. We will say that a connection ω on $Sp(1) \hookrightarrow S^7 \to S^4$ is anti-self-dual (ASD) if the corresponding gauge potential $\mathcal{A}$ on $\mathbb{R}^4$ is ASD. The set of ASD connections is invariant under global gauge transformations so we may consider the set $\mathcal{M}$ of gauge equivalence classes of ASD connections on $Sp(1) \hookrightarrow S^7 \to S^4$. This then is the same as the set of gauge equivalence classes of ASD potentials $\mathcal{A}$ on $\mathbb{R}^4$ with $\mathcal{YM}(\mathcal{A}) = 8\pi^2$.

Each BPST potential $\mathcal{A}_{\lambda,n}$ determines a point $[\mathcal{A}_{\lambda,n}]$ in $\mathcal{M}$ and it is easy to see that $[\mathcal{A}_{\lambda,n}] = [\mathcal{A}_{\lambda',n'}]$ if and only if $\lambda' = \lambda$ and $n' = n$. Remarkably, Atiyah, Hitchin and Singer [4] have proved that *every* element of $\mathcal{M}$ is represented by some $\mathcal{A}_{\lambda,n}$. Thus, the map

$$(\lambda, n) \in (0, \infty) \times \mathbb{H} \longrightarrow [\mathcal{A}_{\lambda,n}] \in \mathcal{M}$$

is a bijection. This picture of $\mathcal{M}$ as the half-space $(0, \infty) \times \mathbb{H}$ in $\mathbb{R}^5$ is remarkably simple and pleasing, but there is another which is even more instructive.

This arises from the fact that there is an orientation preserving, conformal diffeomorphism of $(0,\infty) \times \mathbb{H}$ onto the open 5-dimensional ball B^5 in $\mathbb{R}^5$. Indeed, one can introduce "spherical coordinates" on $\mathcal{M}$ that yield a picture of $\mathcal{M}$ as B^5 with $\mathcal{A}_{1,0}$ at its center. Proceeding radially outward from $[\mathcal{A}_{1,0}]$ toward a point on $\partial B^5 = S^4$ one encounters potentials all of which have the same center n, but which become more and more concentrated, i.e., for which the scale $\lambda \to 0$. A particularly pleasing aspect of this picture is that the base manifold S^4 of the underlying bundle appears naturally as the boundary of the moduli space $\mathcal{M}$ in some compactification of $\mathcal{M}$ (the closed 5-dimensional ball $\bar{B}^5 = B^5 \cup S^4$). One sees clearly how the topologies of the underlying 4-manifold S^4 and the moduli space $\mathcal{M}$ are tied together. We now very briefly describe an amazing generalization of this scenario due to Simon Donaldson [7].

We let X denote a compact, oriented, simply connected, smooth 4-manifold. $H_2(X;\mathbb{Z})$ will denote its second homology group (with integer coefficients) and

$$Q_X : H_2(X;\mathbb{Z}) \times H_2(X;\mathbb{Z}) \longrightarrow \mathbb{Z}$$

its intersection form.

Remark. $H_2(X;\mathbb{Z})$ is a finitely generated, free Abelian group and each of its elements can be identified with a certain equivalence class of smoothly embedded, oriented, closed surfaces (2-manifolds with boundary) $\sum$ in X. Very roughly, the definition of Q_X goes as follows: For $\alpha_1, \alpha_2 \in H_2(X;\mathbb{Z})$ one can select surfaces $\sum_1, \sum_2$ representing them that intersect transversally (i.e., at each intersection point the tangent spaces to $\sum_1$ and $\sum_2$ span the tangent space to X). An intersection point p is assigned the value 1 if an oriented basis for $T_p(\sum_1)$ together with an oriented basis for $T_p(\sum_2)$ gives an oriented basis for $T_p(X)$; otherwise it is assigned the value -1. Then $Q_X(\alpha_1, \alpha_2)$ is the (necessarily finite) sum of these values over all the intersection points. Q_X is a symmetric, bilinear form and is, moreover, unimodular, i.e., if $\alpha_1, \ldots, \alpha_t$ is a basis for $H_2(X;\mathbb{Z})$ over $\mathbb{Z}$, then the matrix $(Q_X(\alpha_i, \alpha_j))$ has determinant ± 1. Here are a few examples:

X	$H_2(X;\mathbb{Z})$	Q_X
S^4	0	$\emptyset$
$\mathbb{CP}^2$	$\mathbb{Z}$	(1)
$\overline{\mathbb{CP}^2}$	$\mathbb{Z}$	(-1)
$S^2 \times S^2$	$\mathbb{Z} \oplus \mathbb{Z}$	$\begin{pmatrix} 0 & 1 \\ 1 & 0 \end{pmatrix}$

The intersection form can also be defined in the category of topological manifolds, but we will not enter into this here.

It has been known for some time that the intersection form is a basic invariant for compact 4-manifolds. In 1949, Whitehead proved that two compact, simply connected 4-manifolds X_1 and X_2 have the same homotopy type if and only if their intersection forms are equivalent (i.e., there exist bases for $H_2(X_1;\mathbb{Z})$ and $H_2(X_2;\mathbb{Z})$ relative to which Q_{X_1} and Q_{X_2} have the same matrix). In 1982, Freedman showed that *every* unimodular, symmetric, integer bilinear form is the intersection form of at least one (and at most two) compact, simply connected *topological* 4-manifold(s). In particular, this is true of the vast, impenetrable maze of positive definite forms (when the rank is 40 there are at least 10^{51} equivalence classes of definite forms). Donaldson has shown that the differential topologist need not venture into this maze because only *one* positive definite, unimodular, symmetric integer bilinear form can arise as the intersection form of a compact, simply connected *smooth* 4-manifold.

Donaldson's 1983 Theorem. *Let X be a compact, oriented, simply connected, smooth 4-manifold with positive definite intersection form Q_X. Then Q_X is standard, i.e., there is a basis for $H_2(X;\mathbb{Z})$ over $\mathbb{Z}$ relative to which the matrix of Q_X is the identity matrix.*

Donaldson's Theorem is remarkable, but still more remarkable is its proof, which is a byproduct of the analysis of an instanton moduli space for X. We (very, very) briefly sketch the idea. Consider the bundle over X analogous to the Hopf bundle over S^4, i.e., the $Sp(1)$-bundle

$$Sp(1) \hookrightarrow P \xrightarrow{\mathcal{P}} X$$

over X with Chern number 1. Next, choose a Riemannian metric $\boldsymbol{g}$ on X. Both the bundle and the metric are to be regarded as auxiliary structures to facilitate the study of X. From $\boldsymbol{g}$ and the given orientation for X one obtains a Hodge star operator and thereby a notion of ASD connection on P. Indeed, a deep result of Taubes [17] asserts that, for manifolds satisfying the hypotheses we have assumed of X, the bundle $Sp(1) \hookrightarrow P \xrightarrow{\mathcal{P}} X$ actually admits ASD connections. Thus, we may introduce the moduli space $\mathcal{M}(\boldsymbol{g})$ of ASD connections on P. This moduli space *does* depend on the choice of $\boldsymbol{g}$ and, for a randomly chosen Riemannian metric, little can be said about its structure. One can show, however, that, for some choice of $\boldsymbol{g}$ (indeed, for "almost every" choice of $\boldsymbol{g}$), the moduli space $\mathcal{M}$ has all of the following properties.

(i) If m denotes half the number of homology classes $\alpha \in H_2(X;\mathbb{Z})$ for which $Q_X(\alpha,\alpha) = 1$, then there exist points $p_1, \ldots, p_m \in \mathcal{M}$ such that $\mathcal{M} - \{p_1, \ldots, p_m\}$ is a smooth, orientable 5-manifold.

(ii) Each p_i, $i = 1, \ldots, m$, has a neighborhood in $\mathcal{M}$ that is homeomorphic to a cone over $\mathbb{CP}^2$ with p_i at the vertex (the cone over $\mathbb{CP}^2$ is the quotient of $\mathbb{CP}^2 \times [0,1]$ obtained by identifying all points of the form $(p, 1)$).

(iii) There is a compact set $K \subseteq \mathcal{M}$ such that $\mathcal{M} - K$ is a submanifold of $\mathcal{M} - \{p_1, \ldots, p_m\}$ diffeomorphic to $X \times (0,1)$ (see Fig. 1.)

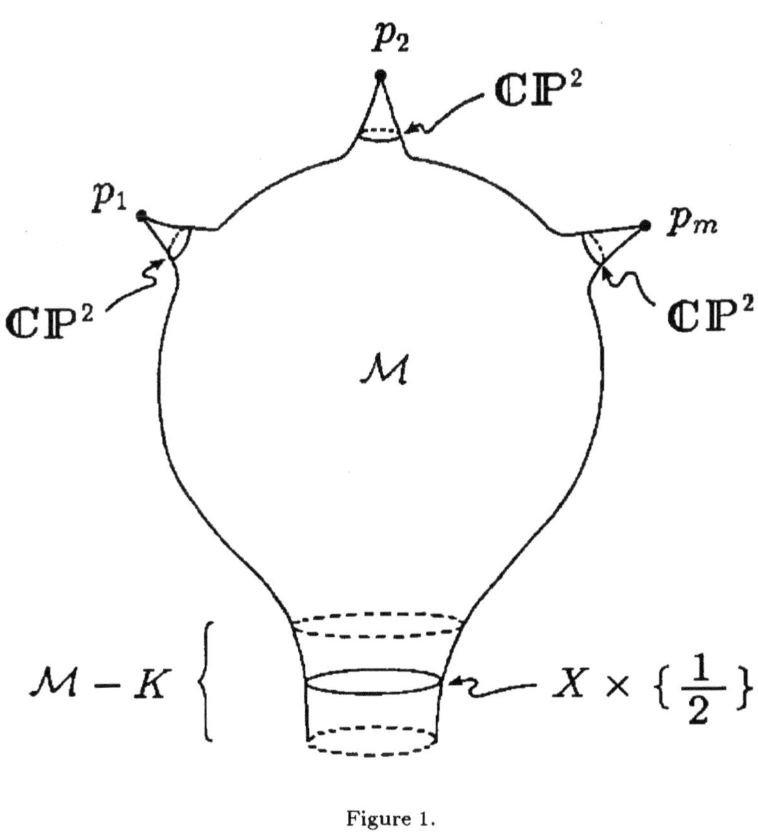

Figure 1.

Now we build a new space $\mathcal{M}_0$ from $\mathcal{M}$ by cutting off the top half of each cone and the bottom half of the cylinder $X \times (0,1)$ (see Fig. 2.).

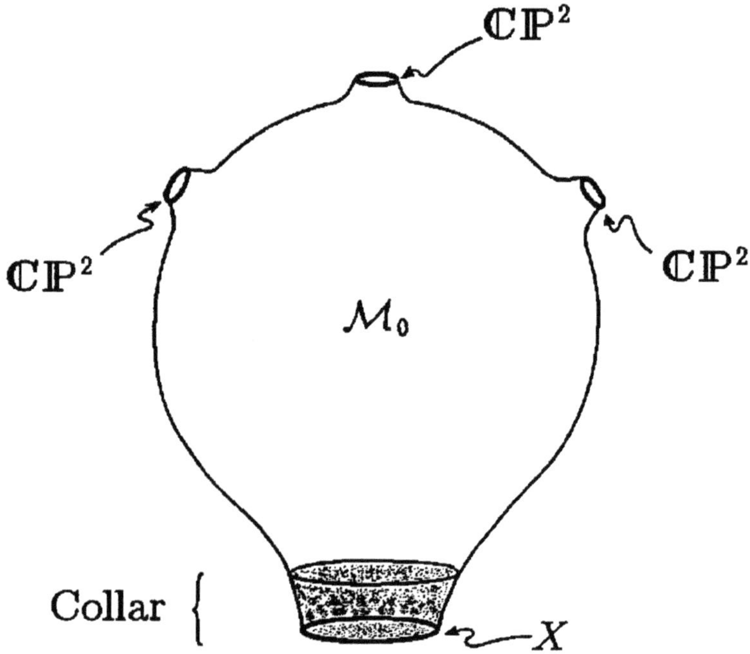

Figure 2.

$\mathcal{M}_0$ is compact (because K is compact). It is also a manifold with boundary whose boundary consists of the disjoint union of a copy of X and m copies of $\mathbb{CP}^2$.

Now, in general, if X_1 and X_2 are two n-manifolds and if there exists an $(n+1)$-manifold M with boundary whose boundary is a disjoint union of X_1 and X_2, then M is called a *cobordism* between X_1 and X_2. X_1 and X_2 are then said to be *cobordant*. Thus, $\mathcal{M}_0$ is a cobordism between X and a disjoint union of $\mathbb{CP}^2$'s. As it happens, the signature of the intersection form of a 4-manifold is a cobordism invariant. This fact, together with the positive definiteness of Q_X, the known intersection form of $\bigsqcup \mathbb{CP}^2$ and a bit of integer linear algebra suffice to produce a basis for $H_2(X;\mathbb{Z})$ relative to which the matrix for Q_X is the identity.

Donaldson's 1983 Theorem was the first gauge-theoretic assault on problems in the topology of smooth 4-manifolds. Subsequent developments in what came to be known as Donaldson theory yielded spectacular results, but at a cost in labor that seemed to grow exponentially with each new success. A breakthrough occurred in 1993 when Kronheimer and Mrowka isolated the (apparently large) class of 4-manifolds of "simple type" and proved that, for these at least, Donaldson theory had some realistic chance of becoming effectively computable. At precisely this moment, however, fate (or rather Ed Witten) intervened and the subject of smooth 4-manifold topology took an entirely new turn.

Remark. Very briefly, this is the story. In 1988, Witten [19] showed that computing the Donaldson polynomial invariants of a 4-manifold X was equivalent to computing correlation functions in a certain "$N = 2$ supersymmetric Yang-Mills theory" formulated on X. The Donaldson polynomials are diffeomorphism invariants and therefore independent of the Riemannian metric on X. Witten studied a 1-parameter family $\{g_t : t > 0\}$ of such metrics in which the parameter t played the role of a coupling constant. The Donaldson polynomials are visible when t is small (i.e., in the weak coupling, or ultraviolet, regime). Generally, computations in quantum field theory are intractable in the strong coupling, or infrared, regime (i.e., when t is large), but, at least in principle, one now has available an alternative view of the Donaldson invariants. In 1994, Seiberg and Witten [16] exploited a "duality" of the field theory in question not unlike the electric/magnetic duality of electromagnetism (supplemented by the addition of magnetic monopoles) and the fact that certain parts of their action functional were holomorphic in the coupling constants, to obtain exact analytic results in the infrared regime and therefore a new way of computing Donaldson invariants. In either regime these invariants are obtained from a moduli space of "classical solutions". In the infrared, it is the moduli space of the "Seiberg-Witten monopole equations" and these are vastly simpler (primarily because the solutions are $U(1)$, i.e., Abelian, connections). We will have more to say about the Seiberg-Witten equations later.

5 $SU(2)$-Yang-Mills-Higgs theory on $\mathbb{R}^3$

Here we provide another illustration of the General Framework described in Section 1 and indicate how the associated matter fields (called Higgs fields in this case) fall into topological types with charges that are conserved, not because of any symmetry of the action, but for purely topological reasons. The underlying base manifold is $X = \mathbb{R}^3$ with its usual Riemannian metric $\boldsymbol{g}$ and orientation μ and with standard coordinates x^1, x^2, x^3. The Lie group G is $SU(2)$ (which is isomorphic to $Sp(1)$, of course, but here the matrix picture is more convenient and traditional). The internal space $\mathcal{V}$ is the Lie algebra $su(2)$ of $SU(2)$ with the positive definite inner product $\langle\ ,\ \rangle$ defined by

$$\langle A, B\rangle = -2\mathrm{trace}(AB) \tag{31}$$

for all A, $B \in su(2)$ ($-\mathrm{trace}(AB)$ is the Killing form on $su(2)$ and the 2 is for convenience). The representation $\rho : G \to GL(\mathcal{V})$ is the adjoint representation $ad : SU(2) \to GL(su(2))$ defined by

$$(ad(g))(A) = ad_g(A) = gAg^{-1} \tag{32}$$

for all $g \in SU(2)$ and all $A \in su(2)$. This is clearly orthogonal with respect to the inner product $\langle\ ,\ \rangle$ defined by (31). Since $\mathbb{R}^3$ is contractible, any principal bundle over it is necessarily trivial and we will trivialize at the outset and take as our $SU(2)$-bundle over $\mathbb{R}^3$ the product bundle

$$SU(2) \hookrightarrow \mathbb{R}^3 \times SU(2) \xrightarrow{\mathcal{P}} \mathbb{R}^3,$$

where the right action of $SU(2)$ on $\mathbb{R}^3 \times SU(2)$ is given by $p \cdot g = (x, h) \cdot g = (x, hg)$. Thus, we have a natural global cross-section $s : \mathbb{R}^3 \to \mathbb{R}^3 \times SU(2)$ defined by $s(x) = (x, \mathbf{1})$. Any other global cross-section $s^g : \mathbb{R} \to \mathbb{R}^3 \times SU(2)$ then has the form $s^g(x) = (x, g(x))$ for some smooth map $g : \mathbb{R}^3 \to SU(2)$.

A connection on $SU(2) \hookrightarrow \mathbb{R}^3 \times SU(2) \xrightarrow{\mathcal{P}} \mathbb{R}^3$ is an $su(2)$-valued 1-form on $\mathbb{R}^3 \times SU(2)$. An $su(2)$-valued form is most conveniently thought of as a matrix of complex-valued forms in the following way: Take as a basis for $su(2)$ the matrices

$$T_1 = -\frac{1}{2}i\sigma_1 \qquad T_2 = -\frac{1}{2}i\sigma_2 \qquad T_3 = -\frac{1}{2}i\sigma_3 \tag{33}$$

where

$$\sigma_1 = \begin{pmatrix} 0 & 1 \\ 1 & 0 \end{pmatrix} \qquad \sigma_2 = \begin{pmatrix} 0 & -i \\ i & 0 \end{pmatrix} \qquad \sigma_3 = \begin{pmatrix} 1 & 0 \\ 0 & -1 \end{pmatrix}$$

are the Pauli spin matrices (the i produces elements of $su(2)$, the $\frac{1}{2}$ makes $\{T_1, T_2, T_3\}$ orthonormal with respect to $\langle\ ,\ \rangle$ and the minus sign ensures that the resulting identification of $su(2)$ with $\mathbb{R}^3$ via $x^1T_1 + x^2T_2 + x^3T_3 \to (x^1, x^2, x^3)$ is orientation preserving). Then any $su(2)$-valued form $\boldsymbol{\alpha}$ on $\mathbb{R}^3$ can be written

$$\boldsymbol{\alpha} = \boldsymbol{\alpha}^a T_a = -\frac{1}{2}\begin{pmatrix} \boldsymbol{\alpha}^3 i & \boldsymbol{\alpha}^2 + \boldsymbol{\alpha}^1 i \\ -\boldsymbol{\alpha}^2 + \boldsymbol{\alpha}^1 i & -\boldsymbol{\alpha}^3 i \end{pmatrix} \tag{34}$$

where each $\boldsymbol{\alpha}^a$ is a real-valued form. One can then write (if $\boldsymbol{\alpha}$ is a k-form)

$$\boldsymbol{\alpha}^a = \sum_{i_1<\cdots<i_k} \alpha^a_{i_1\cdots i_k} dx^{i_1} \wedge \cdots \wedge dx^{i_k} = \frac{1}{k!}\alpha^a_{i_1\cdots i_k} dx^{i_1} \wedge \cdots \wedge dx^{i_k}$$

for $a = 1, 2, 3$ and then

$$\boldsymbol{\alpha} = \boldsymbol{\alpha}^a T_a = \frac{1}{k!}\alpha_{i_1\cdots i_k} dx^{i_1} \wedge \cdots \wedge dx^{i_k} \tag{35}$$

where

$$\alpha_{i_1\cdots i_k} = \alpha^a_{i_1\cdots i_k} T_a = -\frac{1}{2}\begin{pmatrix} \alpha^3_{i_1\cdots i_k} i & \alpha^2_{i_1\cdots i_k} + \alpha^1_{i_1\cdots i_k} i \\ -\alpha^2_{i_1\cdots i_k} + \alpha^1_{i_1\cdots i_k} i & -\alpha^3_{i_1\cdots i_k} i \end{pmatrix}.$$

Exterior derivatives, Hodge duals, etc. of such forms are computed entrywise in (34). Wedge products are computed by writing both forms as in (34) and forming the matrix product, with entries multiplied by the ordinary wedge.

Now, a connection on $SU(2) \hookrightarrow \mathbb{R}^3 \times SU(2) \xrightarrow{\mathcal{P}} \mathbb{R}^3$ is an $su(2)$-valued 1-form $\boldsymbol{\omega}$ on $\mathbb{R}^3 \times SU(2)$. Since the bundle is trivial, such a connection is uniquely determined by the gauge potential $\mathcal{A} = s^*\omega$, which is an $su(2)$-valued 1-form on $\mathbb{R}^3$. Moreover, *any* $su(2)$-valued 1-form on $\mathbb{R}^3$ is the pullback by s of a unique connection on the trivial bundle so we may restrict attention entirely to globally defined gauge potentials $\mathcal{A}$ on $\mathbb{R}^3$. A gauge transformation $g : \mathbb{R}^3 \to SU(2)$ gives a new section $s^g : \mathbb{R}^3 \to \mathbb{R}^3 \times SU(2)$, and a new potential

$$\mathcal{A}^g = g^{-1}\mathcal{A}g + g^{-1}dg,$$

where dg is the entrywise exterior derivative of $g : \mathbb{R}^3 \to SU(2)$. The curvature $\Omega = d^\omega\omega = d\omega + \omega \wedge \omega$ is likewise determined by the field strength $\mathcal{F} = s^*\Omega = d\mathcal{A} + \mathcal{A} \wedge \mathcal{A}$. A gauge transformation gives

$$\mathcal{F}^g = g^{-1}\mathcal{F}g.$$

When, in our General Framework, the internal space $\mathcal{V}$ is the Lie algebra $\mathcal{G}$ of the group G and the representation $\rho : G \to GL(\mathcal{V})$ is the adjoint representation, then a corresponding matter field is called a *Higgs field*. Again because of the triviality of the bundle in our present circumstances, one can identify a Higgs field with its pullback by a global cross-section, i.e., with a map $\psi : \mathbb{R}^3 \to su(2)$ which, under a gauge transformation $g : \mathbb{R}^3 \to SU(2)$, transforms as

$$\psi^g = g^{-1}\psi g$$

(this is equivariance). A common choice for the Higgs self-interaction potential $U : su(2) \to \mathbb{R}$ required in item 7 of the General Framework is the so-called *Georgi-Glashow potential*

$$U(A) = \frac{\lambda}{8}(\langle A, A\rangle - 1)^2, \tag{36}$$

where $\lambda \geq 0$ is a constant. With this we can now record the *Yang-Mills-Higgs action functional*

$$\begin{aligned} A(\mathcal{A}, \psi) &= \frac{1}{2}\int_{\mathbb{R}^3} (\|\mathcal{F}\|^2 + \|d^{\mathcal{A}}\psi\|^2 + \frac{\lambda}{4}(\| \psi \|^2 - 1)^2) dx^1 \wedge dx^2 \wedge dx^3 \\ &= \int_{\mathbb{R}^3} (-\text{trace } (\mathcal{F} \wedge {}^*\mathcal{F}) - \text{trace } (d^{\mathcal{A}}\psi \wedge {}^* d^{\mathcal{A}}\psi)) - \frac{\lambda}{8} {}^*(\| \psi \|^2 - 1)^2), \end{aligned} \tag{37}$$

where $d^{\mathcal{A}}\psi = d\psi + [\mathcal{A}, \psi] = d\psi + \mathcal{A} \wedge \psi - \psi \wedge \mathcal{A}$.

This then is the basic model for Yang-Mills-Higgs theory on $\mathbb{R}^3$. Generally, one is interested in finite action, stationary configurations $(\mathcal{A}, \psi)$ for the action (36). The requirement that $A(\mathcal{A}, \psi) < \infty$ implies that, as $|x| \to \infty$ in $\mathbb{R}^3$, $\|\mathcal{F}\| \to 0$, $\|d^{\mathcal{A}}\psi\| \to 0$ and, at least if $\lambda \neq 0$, $\|\psi\| \to 1$. One can show that, even when $\lambda = 0$, one loses nothing by restricting attention to those configurations $(\mathcal{A}, \psi)$ for which $\|\psi\| \to 1$ uniformly as $|x| \to \infty$ in $\mathbb{R}^3$. Somewhat more precisely, we have the following (see [12]): For any finite action critical point for $A(\mathcal{A}, \psi)$ with $\lambda = 0$ there exists a constant $c \geq 0$ such that $\|\psi\| \to c$ uniformly as $|x| \to \infty$. If $c = 0$, then $(\mathcal{A}, \psi)$ is trivial. If $c \neq 0$, then one can rescale to obtain a new configuration $(\mathcal{A}', \psi')$ given by $(\mathcal{A}'(x), \psi'(x)) = (c^{-1}\mathcal{A}(c^{-1}x), c^{-1}\psi(c^{-1}x))$. Then $(\mathcal{A}', \psi')$ is also a finite action critical point of (37) with $\lambda = 0$ and satisfies $\|\psi'\| \to 1$ uniformly as $|x| \to \infty$ in $\mathbb{R}^3$.

What we intend to do then may seem rather strange. Having gone to all the trouble of introducing the rather special Georgi-Glashow potential (36) we now propose to take $\lambda = 0$ and retain only a "virtual" self-interaction in the form of the boundary condition $\|\psi\| \to 1$ as $|x| \to \infty$ in $\mathbb{R}^3$.

Remark. Before abandoning the self-interaction term, however, we make one observation concerning the action (37) with $\lambda \neq 0$. Notice that $(\mathcal{A}, \psi)$ will achieve its absolute minimum value of zero whenever $\mathcal{A}$ is taken to be zero and ψ is a constant in $su(2)$ with $\|\psi\| = 1$. These absolute minima are the ground states of our system and the selection of any one such is said to *break the symmetry* from $SU(2)$ to $U(1)$. The reason is that a gauge transformation $g : \mathbb{R}^3 \to SU(2)$ transforms ψ according to $\psi \to g^{-1}\psi g$ and this will leave the ground state invariant if and only if g maps into the isotropy subgroup of ψ in $SU(2)$, which is easily seen to be a copy of $U(1)$.

This said, we now take as our action

$$A(\mathcal{A},\psi) = \frac{1}{2}\int_{\mathbb{R}^3} (\|\mathcal{F}\|^2 + \|d^{\mathcal{A}}\psi\|^2) dx^1 \wedge dx^2 \wedge dx^3$$
$$= \int_{\mathbb{R}^3} (-\text{trace } (\mathcal{F} \wedge {}^*\mathcal{F}) - \text{trace } (d^{\mathcal{A}}\psi \wedge {}^*d^{\mathcal{A}}\psi)) \tag{38}$$

and seek stationary points in the configuration space

$$\mathcal{C} = \left\{ (\mathcal{A},\psi) : A(\mathcal{A},\psi) < \infty \ \lim_{R\to\infty} \sup_{|x|\geq R} |1 - \|\psi\|| = 0 \right\}. \tag{39}$$

The Euler-Lagrange equations for (38), called the *Yang-Mills-Higgs equations*, are

$$\begin{cases} {}^*d^{\mathcal{A}*}\mathcal{F} = [d^{\mathcal{A}}\psi, \psi] \\ {}^*d^{\mathcal{A}*}d^{\mathcal{A}}\psi = 0 \end{cases}. \tag{40}$$

As was the case for the pure Yang-Mills equations on $\mathbb{R}^4$, however, there is a simpler set of first order equations (analogous to the ASD equations on $\mathbb{R}^4$) whose solutions represent the absolute minima of (38) and we will deal exclusively with these. Noting that

$$\|\mathcal{F}\|^2 + \|d^{\mathcal{A}}\psi\|^2 = \left\|\mathcal{F} \mp {}^*d^{\mathcal{A}}\psi\right\|^2 \pm 2\langle \mathcal{F}, {}^*d^{\mathcal{A}}\psi\rangle \tag{41}$$

we conclude that the action (38) will achieve an absolute minimum when

$$\mathcal{F} = \pm{}^* d^{\mathcal{A}}\psi. \tag{42}$$

These are called the *(Bogomolny) monopole equations*.

Remark. The appellation "monopole" derives from a certain exact solution to these equations discovered by t'Hooft, Polyakov, Prasad and Sommerfield. In spherical co-ordinates on $\mathbb{R}^3$ this solution (with the minus sign in (42)) is given by

$$\mathcal{A} = -\frac{1}{2}\begin{pmatrix} \mathcal{A}^3 i & \mathcal{A}^2 + \mathcal{A}^1 i \\ -\mathcal{A}^2 + \mathcal{A}^1 i & -\mathcal{A}^3 i \end{pmatrix}$$
$$\psi = -\frac{1}{2}\begin{pmatrix} \psi^3 i & \psi^2 + \psi^1 i \\ -\psi^2 + \psi^1 i & -\psi^3 i \end{pmatrix} \tag{43}$$

where

$$
\begin{aligned}
\mathcal{A}^1 &= -\frac{\rho}{\sinh\rho}(\sin\theta d\varphi + \cos\theta\sin\varphi d\theta)\\
\mathcal{A}^2 &= \frac{\rho}{\sinh\rho}(\cos\theta d\varphi - \sin\theta\sin\varphi d\theta)\\
\mathcal{A}^3 &= -(1-\cos\varphi)d\theta\\
\psi^1 &= \psi^2 = 0\\
\psi^3 &= \coth\rho - \frac{1}{\rho}.
\end{aligned}
\tag{44}
$$

Notice that, despite appearances to the contrary, this is a globally defined, smooth object on all of $\mathbb{R}^3$ (the component functions are actually real analytic everywhere, even at $\rho = 0$). Furthermore, when viewed from a distance (i.e., as $\rho \to \infty$), the Higgs field approaches the constant value T_3 (since $\coth\rho - \frac{1}{\rho} \to 1$), the first two components of $\mathcal{A}$ approach zero (since $\frac{\rho}{\sinh\rho} \to 0$) and $\mathcal{A}^3$, which is independent of ρ, remains fixed at $-(1-\cos\varphi)d\theta$. Thus, for large ρ, $\mathcal{A}$ is effectively

$$
\begin{pmatrix}
\frac{1}{2}i(1-\cos\varphi)d\theta & 0\\
0 & -\frac{1}{2}i(1-\cos\varphi)d\theta
\end{pmatrix}
$$

which should be compared with (10). The most interesting thing about the appearance of the Dirac monopole here is that it was entirely voluntary. In classical electromagnetic theory, magnetic monopoles are, but certainly need not be, inserted by hand, whereas, in $SU(2)$ Yang-Mills-Higgs theory, they appear of their own accord. We remark in passing that, choosing the plus sign in (42), the same procedure that leads to (44) gives the same result, but with ψ replaced by $-\psi$. Henceforth we will restrict our attention to the equation $\mathcal{F} = -d^{\mathcal{A}}\psi$.

It will be convenient to have in hand a number of alternative formulas for the Yang-Mills-Higgs action (38) for a configuration $(\mathcal{A}, \psi)$ that satisfies the monopole equations. The first is a simple consequence of (41) and basic properties of the Hodge dual.

$$
A(\mathcal{A},\psi) = \int_{\mathbb{R}^3} \mathrm{Tr}(\mathcal{F}\wedge d^{\mathcal{A}}\psi) \qquad \text{(monopoles)} \tag{45}
$$

Here we use Tr for 2 trace. Computing this integral for the t'Hooft-Polyakov-Prasad-Sommerfield monopole gives a value of 4π. We normalize the action by 4π and define the *monopole number* $N(\mathcal{A}, \psi)$ by

$$N(\mathcal{A}, \psi) = \frac{1}{4\pi} \int_{\mathbb{R}^3} \mathrm{Tr}(\mathcal{F} \wedge d^{\mathcal{A}} \psi). \tag{46}$$

Remark. Shortly we will describe another way of computing $N(\mathcal{A}, \psi)$ that will make it clear that it is, in fact, an integer. Indeed, it can be shown that the right-hand side of (46) is an integer for any $(\mathcal{A}, \psi)$ in the configuration space $\mathcal{C}$ defined by (39), whether or not it satisfies the monopole equations. In fact, $\mathcal{C}$ has a natural topology whose connected components consist of those configurations $(\mathcal{A}, \psi)$ for which $N(\mathcal{A}, \psi)$ takes some fixed integer value. See [11] for these results.

The next integral formula for the monopole number is obtained by writing $\mathbb{R}^3$ as $\lim_{R\to\infty} \int_{|x|\le R}$, noting that $\mathrm{Tr}(\mathcal{F} \wedge d^{\mathcal{A}} \psi) = d(\mathrm{Tr}(\psi\mathcal{F}))$ and applying Stokes' Theorem to each $\int_{|x|\le R}$. The result is

$$N(\mathcal{A}, \psi) = \lim_{R\to\infty} \frac{1}{4\pi} \int_{S_R^2} \mathrm{Tr}(\psi\mathcal{F}), \tag{47}$$

where $S_R^2 = \{ x \in \mathbb{R}^3 : |x| = R \}$. Since $\|\psi\| \to 1$ as $|x| \to \infty$ there exists an R_0 such that $|x| > R_0$ implies $\|\psi(x)\| > \frac{1}{2}$. For $|x| > R_0$ we may define

$$\hat{\psi}(x) = \|\psi(x)\|^{-1} \psi(x)$$

and, for any $R > R_0$,

$$\hat{\psi}_R = \hat{\psi} | S_R^2.$$

One can show [12] that, in (47), ψ can be replaced by $\hat{\psi}$ so

$$\begin{aligned} N(\mathcal{A}, \psi) &= \lim_{R\to\infty} \frac{1}{4\pi} \int_{S_R^2} \mathrm{Tr}(\hat{\psi}\mathcal{F}) \\ &= \lim_{R\to\infty} \frac{1}{4\pi} \int_{S_R^2} \|\psi\|^{-1} \mathrm{Tr}(\psi\mathcal{F}). \end{aligned} \tag{48}$$

The reason for preferring $\hat{\psi}$ and $\hat{\psi}_R$ can be seen as follows: Each $\hat{\psi}_R$ can be regarded as a map from S^2 to S^2 and so determines an element $[\hat{\psi}_R]$ of the homotopy group $\pi_2(S^2)$. Since $\hat{\psi}$ is smooth for $R > R_0$, $\hat{\psi}_R$ varies smoothly with $R > R_0$ so this homotopy class is actually independent of $R > R_0$ and one may simply denote it $[\hat{\psi}]$. One can show that $[\hat{\psi}]$ is gauge invariant (i.e., $\|\psi^g\|^{-1}\psi^g$ is homotopic to $\|\psi\|^{-1}\psi$ on $|x| > R_0$ for any gauge transformation g) and so one would expect that $[\hat{\psi}]$ has some physical significance. Indeed,

$\pi_2(S^2) \cong \mathbb{Z}$ with an isomorphism provided by the Brouwer degree (winding number) and one can prove that, for any $R > R_0$,

$$N(\mathcal{A}, \psi) = \deg(\hat{\psi}_R) = \deg(\|\psi\|^{-1}\psi|S_R^2). \tag{49}$$

There is an explicit formula for calculating these Brouwer degrees as integrals:

$$N(\mathcal{A}, \psi) = -\frac{1}{4\pi}\int_{S_R^2} \mathrm{Tr}(\hat{\psi}\, d\hat{\psi} \wedge d\hat{\psi}) \qquad (R > R_0). \tag{50}$$

From (49) it is clear that the monopole number is, in fact, an integer and so, in particular, is stable under continuous perturbations of the fields. Both (49) and (50) express the monopole number of the configuration entirely in terms of the Higgs field ψ and its behavior on large spheres, but this, of course, does not mean that $N(\mathcal{A}, \psi)$ does not depend on $\mathcal{A}$ since ψ and $\mathcal{A}$ are coupled in the monopole equations (42). From our point of view the most significant aspect of these formulas is their topological nature. Just as instantons on $\mathbb{R}^4$ are characterized by a topological invariant (an element of $\pi_3(S^3)$, or, equivalently, a Chern class) so monopoles on $\mathbb{R}^3$ fall into distinct topological types according to the homotopy class of the Higgs field on large spheres. One thinks of the monopole number as another instance of a topological charge.

Remark. There is yet another way to see the topological nature of this magnetic charge/monopole number. Briefly, one can associate with a monopole $(\mathcal{A}, \psi)$ and any S_R^2 $(R > R_0)$ a principal $U(1)$-bundle over S_R^2 that is a reduction of the trivial bundle $SU(2) \hookrightarrow S_R^2 \times SU(2) \to S_R^2$ to $U(1)$ and show that $N(\mathcal{A}, \psi)$ is the 1st Chern class of this bundle (the $U(1)$-bundle is obtained by breaking the $SU(2)$ symmetry to $U(1)$ through the selection of some "ground state" ψ_0 in the image of $\hat{\psi}_R : S_R^2 \to S^2_{su(2)}$).

We conclude this section with a few observations on the physical origins and significance of these $SU(2)$ Yang-Mills-Higgs monopoles. For this we momentarily return to the (anti-) self-dual equations on $\mathbb{R}^4$ (Section 4). Finite action solutions to these equations are instantons, but we now abandon this requirement and seek instead solutions that are *static*, i.e., independent of the standard coordinate x^4 (a static solution cannot have finite action on $\mathbb{R}^4$ unless it is trivial). If $\tilde{\mathcal{A}} = \tilde{\mathcal{A}}_1 dx^1 + \tilde{\mathcal{A}}_2 dx^2 + \tilde{\mathcal{A}}_3 dx^3 + \tilde{\mathcal{A}}_4 dx^4$ is an $su(2)$-valued 1-form on $\mathbb{R}^4$ with each $\tilde{\mathcal{A}}_\alpha$ independent of x^4, then it is a simple matter to write out the (A)SD equations (${}^*\tilde{\mathcal{F}} = \pm\tilde{\mathcal{F}}$) and find that they reduce to

$$\tilde{\mathcal{F}}_{ij} = \pm\sum_{k=1}^{3} \epsilon_{ijk}(\partial_k \tilde{\mathcal{A}}_4 + [\tilde{\mathcal{A}}_k, \tilde{\mathcal{A}}_4]), \qquad i,j = 1,2,3. \tag{51}$$

Now, fix (arbitrarily) some value x_0^4 of x^4 and consider the submanifold $\mathbb{R}^3 \times \{x_0^4\}$ of $\mathbb{R}^4$ (henceforth written simply $\mathbb{R}^3$). For $i = 1, 2, 3$, we let $\mathcal{A}_i = \tilde{\mathcal{A}}_i|\mathbb{R}^3$ and then define $\mathcal{A} = \mathcal{A}_1 dx^1 + \mathcal{A}_2 dx^2 + \mathcal{A}_3 dx^3$ on $\mathbb{R}^3$. One shows that $\mathcal{A}$ is a gauge potential on $\mathbb{R}^3$ and the corresponding field strength $\mathcal{F} = \frac{1}{2}\mathcal{F}_{ij} dx^i \wedge dx^j$ has components that are the restrictions of $\tilde{\mathcal{F}}_{ij}$ to $\mathbb{R}^3$. Note that $\tilde{\mathcal{A}}_4$ does not enter into either the potential or the field strength on $\mathbb{R}^3$. However, if we define $\psi : \mathbb{R}^3 \to su(2)$ by

$$\psi = \tilde{\mathcal{A}}_4|\mathbb{R}^3 \tag{52}$$

then the static, (A)SD equations (51) on $\mathbb{R}^4$ become, when restricted to $\mathbb{R}^3$,

$$\mathcal{F}_{ij} = \pm \sum_{k=1}^{3} \epsilon_{ijk}(\partial_k \psi + [\mathcal{A}_k, \psi]), \qquad i, j = 1, 2, 3, \tag{53}$$

and this is simply the coordinate form of the Bogomolny monopole equations (42). Thus, we may think of Bogomolony monopoles as static solutions to the (A)SD equations on $\mathbb{R}^4$.

Remark. The process by which we arrived at this interpretation is the simplest instance of a phenomenon known as *dimensional reduction*.

This view of Bogomolny monopoles provides some hint as to the "source" of their magnetic charge. Recall that any solution to the (A)SD equations on $\mathbb{R}^4$, including those that are static, also satisfies the full Yang-Mills equations $d^{\mathcal{A}*}\mathcal{F} = 0$ and the Bianchi identity $d^{\mathcal{A}}\mathcal{F} = 0$ and these are regarded as a generalization of Maxwell's equations. The static version of Maxwell's equations that contains both electric and magnetic charge densities (ρ_e and ρ_m, respectively) is (in appropriate units) $d^*\boldsymbol{F} = 0$, $d\boldsymbol{F} = {}^*\rho_m$, and $\nabla^2\psi = \rho_e$, where $\boldsymbol{F}$ is the magnetic field 2-form and ψ is the electric potential. Noting that $d^{\mathcal{A}}\mathcal{F} = 0$ is equivalent to

$$d\mathcal{F} = -[\mathcal{A}, \mathcal{F}]$$

one can view the commutator term as playing the role of a magnetic charge density. The role of the Higgs field ψ (or, more to the point, the boundary condition $\|\psi\| \to 1$ as $|x| \to \infty$) is to break the symmetry at large distances from $SU(2)$ down to $U(1)$.

6 Seiberg-Witten monopoles on $\mathbb{R}^4$

We have already alluded (at the end of Section 4) to the revolution that took place in low dimensional topology in 1994 with the appearance of the

Seiberg-Witten equations. In this final section we will provide a brief overview of what these equations say and how they are used to study 4-manifolds. We conclude by proving a theorem of Witten which asserts that there are no finite energy solutions to these equations on flat $\mathbb{R}^4$ and contrasting this result with some recently discovered solutions on Minkowski space that contain a Dirac monopole potential.

We begin with some Lie groups. Recall that $SU(2)$ can be identified with Spin(3), the double cover of $SO(3)$. Since $SO(4) \cong SO(3) \oplus SO(3)$ one finds that the double cover Spin(4) of $SO(4)$ is just $SU(2) \times SU(2)$. Our interest here is in the double cover of $SO(4) \times U(1)$ which one can show is the group

$$\text{Spin}^{\text{c}}(4) \cong SU(2) \times SU(2) \times U(1)/\pm 1, \tag{54}$$

where ± 1 acts on all factors by multiplication.

Remark. The reason we are interested in $\text{Spin}^{\text{c}}(4)$ and the fact that it is given by (54) are seen most clearly when viewed from the perspective of Clifford algebras. Briefly, the situation is as follows: The Clifford algebra $Cl(4)$ of $\mathbb{R}^4$ (with its usual positive definite inner product) can be identified with the algebra of all 2×2 quaternionic matrices (or the algebra of 4×4 complex matrices obtained by identifying a quaternion $q = \alpha + \beta j$, α, $\beta \in \mathbb{C}$, with the 2×2 complex matrix

$$\begin{pmatrix} \alpha & \beta \\ -\overline{\beta} & \overline{\alpha} \end{pmatrix}$$

in the usual way). Identifying $\mathbb{R}^4$ with $\begin{pmatrix} 0 & q \\ -\overline{q} & 0 \end{pmatrix} \in Cl(4)$ one has $\mathbb{R}^4 \subseteq Cl(4)$. Now, Spin(4) appears naturally as a subset of $Cl(4)$. The complexified Clifford algebra $Cl(4) \otimes \mathbb{C}$ contains a natural complex analogue of Spin(4). This is $\text{Spin}^{\text{c}}(4)$ and it turns out to be the double cover of $SO(4) \times U(1)$.

While on the subject of Clifford algebras we remark also that $Cl(4) \otimes \mathbb{C}$ is actually the complex algebra of *all* 4×4 complex matrices and so can be identified with the algebra $\text{End}_{\mathbb{C}}(\mathbb{C}^4)$ of all endomorphisms of the complex vector space $\mathbb{C}^4$. Since

$$\mathbb{R}^4 \hookrightarrow Cl(4) \hookrightarrow Cl(4) \otimes \mathbb{C} \cong \text{End}_{\mathbb{C}}(\mathbb{C}^4)$$

we find that $\mathbb{R}^4$ acts on $\mathbb{C}^4$ by regarding each of its elements as an endomorphism. This action is called Clifford multiplication and has the property that it interchanges the two subspaces

$$W_+ = \left\{ \begin{pmatrix} z_1 \\ z_2 \\ 0 \\ 0 \end{pmatrix} : z_1, z_2 \in \mathbb{C} \right\}$$

and

$$W_- = \left\{ \begin{pmatrix} 0 \\ 0 \\ z_3 \\ z_4 \end{pmatrix} : z_3, z_4 \in \mathbb{C} \right\}$$

of $\mathbb{C}^4$.

From (54) it follows that there are natural homomorphisms of $\mathrm{Spin}^c(4)$ onto $SO(4) \cong SU(2) \times SU(2)/\pm 1$, $U(1) \cong U(1)/\pm 1$ and $U(2) \cong SU(2) \times U(1)/\pm 1$ obtained by "forgetting" one or more of the factors. Specifically, if we denote by $[u, v, a]$ an element (equivalence class) in $\mathrm{Spin}^c(4)$, then one defines maps

$$\begin{aligned} \pi &: \mathrm{Spin}^c(4) \longrightarrow SO(4) \\ \lambda &: \mathrm{Spin}^c(4) \longrightarrow U(1) \\ s_\pm &: \mathrm{Spin}^c(4) \longrightarrow U(2) \end{aligned}$$

by $\pi([u,v,a]) = [u,v]$, $\lambda([u,v,a]) = [a]$, $s_+([u,v,a]) = [v,a]$ and $s_-([u,v,a]) = [u,a]$. These maps, together with the natural actions of $SO(4)$ on $\mathbb{R}^4$, $U(1)$ on $\mathbb{C}$ and $U(2)$ on $\mathbb{C}^2$, give rise to representations of $\mathrm{Spin}^c(4)$ on $\mathbb{R}^4$, $\mathbb{C}$ and $\mathbb{C}^2$ which we denote also π, λ, and $s_\pm$. One should think of the representation s_+ as acting on the copy W^+ of $\mathbb{C}^2$ in $\mathbb{C}^4$ and s_- as acting on W^-.

Now let X be a smooth, oriented 4-manifold (for topological applications compactness and simple connectivity are generally assumed as well). Choose a Riemannian metric for X and let $SO(4) \hookrightarrow F(X) \xrightarrow{\mathcal{P}_F} X$ denote the oriented orthonormal frame bundle. A spin structure for X is a lift of the frame bundle to a principal Spin(4)-bundle over X and these exist if and only if the second Stiefel-Whitney class $w_2(X)$ of X is trivial. We are interested in the complex analogue of a spin structure. A *Spin^c-structure* for X consists of a principal $\mathrm{Spin}^c(4)$-bundle

$$\mathrm{Spin}^c(4) \hookrightarrow S^c(X) \xrightarrow{\mathcal{P}_c} X$$

together with a map $\Pi : S^c(X) \to F(X)$ satisfying

$$\mathcal{P}_F \circ \Pi = \mathcal{P}_C \tag{55}$$

and

$$\Pi(p \cdot g) = \Pi(p) \cdot \pi(g) \tag{56}$$

for all $p \in S^c(X)$ and all $g \in \mathrm{Spin}^{\mathbb{C}}(4)$. One can show that $\mathrm{Spin}^{\mathbb{C}}$-structures exist for every oriented 4-manifold and the first step in writing down the Seiberg-Witten equations on X is to choose one. Given such a structure the representations $\lambda : \mathrm{Spin}^{\mathbb{C}}(4) \to GL(\mathbb{C})$ and $s_\pm : \mathrm{Spin}^{\mathbb{C}}(4) \to GL(\mathbb{C}^2)$ determine associated vector bundles

$$L = S^{\mathbb{C}}(X) \times_\lambda \mathbb{C}$$

and

$$S^{\mathbb{C}}_\pm(X) = S^{\mathbb{C}}(X) \times_{s_\pm} \mathbb{C}^2.$$

L is called the *determinant line bundle* of the $\mathrm{Spin}^{\mathbb{C}}$-structure and $S^{\mathbb{C}}_\pm(X)$ are the *positive* and *negative spinor bundles*. We denote by L° the principal $U(1)$-bundle associated with L and refer to the cross-sections of $S^{\mathbb{C}}_\pm(X)$ as *positive* and *negative spinor fields* on X.

The data for the Seiberg-Witten equations consists of a $U(1)$-connection A on L° (the gauge field) and a positive spinor field ψ (the matter field). The equations themselves read

$$\not{D}_A \psi = 0 \tag{57}$$

$$F_A^+ = \sigma^+((\psi \otimes \psi^*)_0). \tag{58}$$

We will briefly describe the meaning of the symbols that appear in these equations and then write them out explicitly in local orthonormal coordinates. In (57), $\not{D}_A$ is called the Dirac operator associated with A and arises in the following way: Since $\mathrm{Spin}^{\mathbb{C}}(4)$ is the double cover of $SO(4) \times U(1)$, $S^{\mathbb{C}}(X)$ is the double cover of the fiber product $SO(4) \times U(1) \hookrightarrow F(X) \times L^\circ \to X$. The Levi-Civita connection on $F(X)$ together with the connection A on L° determine a connection on $F(X) \times L^\circ$ which therefore lifts to a connection on the double cover $S^{\mathbb{C}}(X)$. This connection then determines a covariant derivative

$$\nabla_A : \Gamma(S^{\mathbb{C}}_+(X)) \longrightarrow \Gamma(S^{\mathbb{C}}_+(X) \otimes T^*(X))$$

on the associated vector bundle $S^{\mathbb{C}}_+(X)$. Composing with Clifford multiplication (which interchanges $S^{\mathbb{C}}_+(X)$ and $S^{\mathbb{C}}_-(X)$) gives the Dirac operator

$$\not{D}_A : \Gamma(S^{\mathbb{C}}_+(X)) \longrightarrow \Gamma(S^{\mathbb{C}}_-(X)).$$

Remark. In a bit more detail, here's what "composing with Clifford multiplication" means. Choose a local orthonormal frame field $\{e_1, e_2, e_3, e_4\}$ for X. For any $\psi \in \Gamma(S^{\mathbb{C}}_+(X))$, $\nabla_A\psi$ is a 1-form with values in $S^{\mathbb{C}}_+(X)$ so each $\nabla_A\psi(e_i)$ is a section of $S^{\mathbb{C}}_+(X)$. Then e_i acts on $\nabla_A\psi(e_i)$ by Clifford multiplication at each point to give a section $e_i \cdot \nabla_A\psi(e_i)$ of $S^{\mathbb{C}}_-(X)$. The Dirac operator is defined by

$$\not{D}_A\psi = \sum_{i=1}^{4} e_i \cdot \nabla_A\psi(e_i). \tag{59}$$

The first of the Seiberg-Witten equations (57) then simply requires that the positive spinor field ψ satisfy the "Dirac equation".

Regarding (58) we recall that A is a connection on a $U(1)$-bundle so its curvature is given by $F_A = dA$. The self-dual part of F_A is $F_A^+ = \frac{1}{2}(F_A + {}^*F_A)$, where * is the Hodge dual on X. $\psi \otimes \psi^*$ is the endomorphism of $S^{\mathbb{C}}_+(X)$ determined by ψ and its Hermitian conjugate ψ^*. Writing $\psi = \begin{pmatrix}\psi_1\\ \psi_2\end{pmatrix}$ it is given by

$$\psi \otimes \psi^* = \begin{pmatrix}\psi_1\\ \psi_2\end{pmatrix}\begin{pmatrix}\bar{\psi}_1 & \bar{\psi}_2\end{pmatrix} = \begin{pmatrix}|\psi_1|^2 & \psi_1\bar{\psi}_2\\ \psi_2\bar{\psi}_1 & |\psi_2|^2\end{pmatrix}.$$

The tracefree part of $\psi \otimes \psi^*$ is

$$(\psi \otimes \psi^*)_0 = \begin{pmatrix}\frac{1}{2}(|\psi_1|^2 - |\psi_2|^2) & \psi_1\bar{\psi}_2\\ \psi_2\bar{\psi}_1 & \frac{1}{2}(|\psi_2|^2 - |\psi_1|^2)\end{pmatrix}. \tag{60}$$

The σ^+ in (58) is an ismorphism

$$\sigma^+ : \mathrm{End}_0(S^{\mathbb{C}}_+(X)) \longrightarrow \Lambda^2_+(X) \otimes \mathbb{C}$$

from the tracefree endomorphisms of $S^{\mathbb{C}}_+(X)$ to the complex-valued self-dual 2-forms on X. This isomorphism arises naturally in the context of Clifford algebras, but we will content ourselves with writing out explicitly its value on an endomorphism of the type $(\psi \otimes \psi^*)_0$. Relative to local orthonormal coordinates $\{x^0, x^1, x^2, x^3\}$, a basis for the real self-dual 2-forms on X is $\{dx^0 \wedge dx^1 + dx^2 \wedge dx^3,\ dx^0 \wedge dx^2 + dx^3 \wedge dx^1,\ dx^0 \wedge dx^3 + dx^1 \wedge dx^2\}$. With $(\psi \otimes \psi^*)_0$

written as in (60),

$$
\begin{aligned}
\sigma^+((\psi\otimes\psi^*)_0) = \frac{1}{4}i\big[&-(|\psi_1|^2-|\psi_2|^2)(dx^0\wedge dx^1+dx^2\wedge dx^3)\\
&+2\mathrm{Im}(\psi_1\bar{\psi}_2)(dx^0\wedge dx^2+dx^3\wedge dx^1)\\
&-2\mathrm{Re}(\psi_1\bar{\psi}_2)(dx^0\wedge dx^3+dx^1\wedge dx^2)\big].
\end{aligned}
\tag{61}
$$

Now, if we write $F_A = \sum_{\alpha<\beta} F_{\alpha\beta}dx^\alpha\wedge dx^\beta$, then the components of F_A^+ relative to this same basis are

$$
\begin{aligned}
F^+ =&\frac{1}{2}(F_{01}+F_{23})(dx^0\wedge dx^1+dx^2\wedge dx^3)\\
&+\frac{1}{2}(F_{02}+F_{31})(dx^0\wedge dx^2+dx^3\wedge dx^1)\\
&+\frac{1}{2}(F_{03}+F_{12})(dx^0\wedge dx^3+dx^1\wedge dx^2).
\end{aligned}
\tag{62}
$$

Thus, the second Seiberg-Witten equation (58) requires that

$$
\begin{aligned}
F_{01}+F_{23} &= -\frac{1}{2}i(|\psi_1|^2-|\psi_2|^2)\\
F_{02}+F_{31} &= -i\mathrm{Im}(\bar{\psi}_1\psi_2)\\
F_{03}+F_{12} &= -i\mathrm{Re}(\bar{\psi}_1\psi_2).
\end{aligned}
\tag{63}
$$

Finally, if one writes $A = A_\alpha dx^\alpha$, then $F_{ij} = \partial_i A_j - \partial_j A_i$ and the two Seiberg-Witten equations take on the following explicit local form.

$$
\begin{pmatrix} -(\partial_0+A_0)+i(\partial_1+A_1) & (\partial_2+A_2)+i(\partial_3+A_3)\\ -(\partial_2+A_2)+i(\partial_3+A_3) & -(\partial_0+A_0)-i(\partial_1+A_1)\end{pmatrix}\begin{pmatrix}\psi_1\\ \psi_2\end{pmatrix} = \begin{pmatrix}0\\0\end{pmatrix}
\tag{64}
$$

$$
\begin{aligned}
(\partial_0A_1-\partial_1A_0)+(\partial_2A_3-\partial_3A_2) &= -\frac{1}{2}i(|\psi_1|^2-|\psi_2|^2)\\
(\partial_0A_2-\partial_2A_0)+(\partial_3A_1-\partial_1A_3) &= -i\mathrm{Im}(\bar{\psi}_1\psi_2)\\
(\partial_0A_3-\partial_3A_0)+(\partial_1A_2-\partial_2A_1) &= -i\mathrm{Re}(\bar{\psi}_1\psi_2).
\end{aligned}
\tag{65}
$$

One can show (although we will not do so here) that solutions to the Seiberg-Witten equations are also solutions the Euler-Lagrange equations for the energy functional

$$
E(A,\psi) = \int_X (\|\nabla_A\psi\|^2 + \frac{s}{4}\|\psi\|^2 + \frac{1}{4}\|\psi\|^4 + \|F_A\|^2)d\mathrm{vol}
\tag{66}
$$

where s is the scalar curvature of X. Witten has shown that if X admits a Riemannian metric for which $s \geq 0$ on X, then every solution (A,ψ) to the Seiberg-Witten equations has $\psi \equiv 0$ (for any choice of the Spin$^{\mathbb{C}}$-structure).

He showed also that if $X = \mathbb{R}^4$ (standard metric and orientation), then $E(A,\psi) < \infty$ implies $\psi \equiv 0$ and $F_A \equiv 0$. We will prove this last result shortly.

Applications of the Seiberg-Witten equations to topology proceed in a manner entirely analogous to that in Donaldson theory. The group $C^\infty(X, S^1)$ acts on a solution (A,ψ) in the following way: For each $u \in C^\infty(X, S^1)$ we define

$$u \cdot (A,\psi) = (u^{-1}Au + u^{-1}du, u^{-1}\psi) = (A + u^{-1}du, u^{-1}\psi).$$

Then (A,ψ) satisfies the Seiberg-Witten equations if and only if $u \cdot (A,\psi)$ does so and we say that two solutions are gauge equivalent if and only if they differ by the action of some $u \in C^\infty(X, S^1)$ (this is equivalent to saying that they differ by an automorphism of $\mathrm{Spin}^{\mathrm{c}}(4) \hookrightarrow S^{\mathrm{c}}(X) \to X$ that covers the identity on $F(X)$). One thereby obtains a moduli space which, under certain topological restrictions on X ($b_2^+(X) > 0$ and $c_1(L^\circ) \neq 0$ rationally) and for generic choice of the Riemannian metric, has the structure of a smooth, oriented, *compact* manifold. The structure of the moduli space contains much information about the topology of X (Witten has conjectured that it contains all of the information contained in the Donaldson invariants). The great advantage of the Seiberg-Witten moduli space is its compactness which eliminates most of the technical difficulties associated with the analysis of the anti-self-dual moduli spaces in Donaldson theory. See [13] and [15] for introductions to Seiberg-Witten theory.

Our final objective in this section is the proof of Witten's theorem on finite energy solutions to the Seiberg-Witten equations on $\mathbb{R}^4$. We will actually prove a bit more and will then briefly describe some interesting (infinite energy) solutions on flat space.

We will denote the standard coordinates on $\mathbb{R}^4$ by x^0, x^1, x^2, x^3 and use ∂_i for $\partial/\partial x^i$, $i = 0, 1, 2, 3$. Since $\mathbb{R}^4$ is contractible all of the relevant bundles are trivial and, in particular, there is an essentially unique $\mathrm{Spin}^{\mathrm{c}}$-structure. Furthermore, we can identify the spinor field and the connection with their pullbacks by the canonical global cross-sections. We will use the same notation for these and write

$$\psi = \begin{pmatrix} \psi_1 \\ \psi_2 \end{pmatrix} : \mathbb{R}^4 \longrightarrow \mathbb{C}^2$$

and

$$A = A_\alpha dx^\alpha,$$

where

$$A_\alpha : \mathbb{R}^4 \longrightarrow \mathrm{Im}\mathbb{C}, \qquad \alpha = 0, 1, 2, 3.$$

Thus,

$$F = F_A = dA = \sum_{\alpha<\beta} F_{\alpha\beta} dx^\alpha \wedge dx^\beta = \sum_{\alpha<\beta} (\partial_\alpha A_\beta - \partial_\beta A_\beta) dx^\alpha \wedge dx^\beta.$$

The associated covariant derivative is given by

$$\nabla\psi = \nabla_A\psi = (\nabla_\alpha\psi)dx^\alpha = (\partial_\alpha\psi + A_\alpha\psi)dx^\alpha.$$

The Dirac operator is then

$$\not{D}_A\psi = -\nabla_0\psi + I\nabla_1\psi + J\nabla_2\psi + K\nabla_3\psi,$$

where

$$I = \begin{pmatrix} i & 0 \\ 0 & -i \end{pmatrix}, \quad J = \begin{pmatrix} 0 & 1 \\ -1 & 0 \end{pmatrix}, \quad K = \begin{pmatrix} 0 & i \\ i & 0 \end{pmatrix}.$$

Finally, the adjoint

$$\not{D}_A^* : \Gamma(S_-^{\mathbb{C}}(X)) \longrightarrow \Gamma(S_+^{\mathbb{C}}(X))$$

of the Dirac operator is

$$\not{D}_A^*\phi = \nabla_0\phi + I\nabla_1\phi + J\nabla_2\phi + K\nabla_3\phi.$$

We record the adjoint only because, at one point in the proof, we will need to appeal to a special case of the famous *Weitzenböck formula* from differential geometry:

$$\not{D}_A^*\not{D}_A\psi + \sum_{\alpha=0}^{3} \nabla_\alpha\nabla_\alpha\psi = (F_{01} + F_{23})I\psi + (F_{02} + F_{31})J\psi + (F_{03} + F_{12})K\psi.$$

(This can be verified by direct calculation in the present case.) The energy functional now assumes the form

$$E(A,\psi) = \int_{\mathbb{R}^4} \left(\sum_{\alpha=0}^{3} \|\nabla_\alpha\psi\|^2 + \frac{1}{4}\|\psi\|^4 + \sum_{\alpha<\beta} |F_{\alpha\beta}|^2 \right) d\text{vol},$$

where $\| \ \|^2$ indicates the squared norm on $\mathbb{C}^2$ arising from the Hermitian inner product $\langle z, w\rangle = \bar{z}_1 w_1 + \bar{z}_2 w_2$ and $| \ |^2$ is the squared modulus on Im$\mathbb{C}$. In this notation the Seiberg-Witten equations take the form

$$\nabla_0\psi = I\nabla_1\psi + J\nabla_2\psi + K\nabla_3\psi \qquad (\not{D}_A\psi = 0) \tag{67}$$

$$F_{01} + F_{23} = -\frac{1}{2}\psi^* I\psi \tag{68}$$

$$F_{02} + F_{31} = -\frac{1}{2}\psi^* J\psi \tag{69}$$

$$F_{03} + F_{12} = -\frac{1}{2}\psi^* K\psi. \tag{70}$$

Theorem (Witten). *Let A be an $\mathrm{Im}\mathbb{C}$-valued 1-form on $\mathbb{R}^4$ and ψ a smooth map from $\mathbb{R}^4$ to $\mathbb{C}^2$. Suppose (A,ψ) satisfies the Seiberg-Witten equations (67)-(70). Then*
(a) $\psi \in L^2(\mathbb{R}^4) \Longrightarrow \psi \equiv 0$
(b) $E(A,\psi) < \infty \Longrightarrow \psi \equiv 0$ *and* $F_A \equiv 0$.

Proof. Let $\Delta = -\sum_{\alpha=0}^{3} \frac{\partial^2}{(\partial x^\alpha)^2}$ be the usual Laplacian on $\mathbb{R}^4$. Simple calculations show that

$$\Delta\|\psi\|^2 = -2\sum_{\alpha=0}^{3} \frac{\partial}{\partial x^\alpha}\mathrm{Re}\langle\psi, \nabla_\alpha\psi\rangle \tag{71}$$

and

$$\frac{\partial}{\partial x^\alpha}\mathrm{Re}\langle\psi, \nabla_\alpha\psi\rangle = \|\nabla_\alpha\psi\|^2 + \mathrm{Re}\langle\psi, \nabla_\alpha\nabla_\alpha\psi\rangle. \tag{72}$$

Now, by (67), $\not{D}_A\psi = 0$ so the Weitzenböck formula gives

$$\sum_{\alpha=0}^{3} \nabla_\alpha\nabla_\alpha\psi = (F_{01}+F_{23})I\psi + (F_{02}+F_{31})J\psi + (F_{03}+F_{12})K\psi. \tag{73}$$

Substituting (72) into (71) and using (73) yields

$$\begin{aligned}\Delta\|\psi\|^2 = &-2\sum_{\alpha=0}^{3}\|\nabla_\alpha\psi\|^2 - 2\mathrm{Re}\langle\psi, (F_{01}+F_{23})I\psi\rangle \\ &- 2\mathrm{Re}\langle\psi, (F_{02}+F_{31})J\psi\rangle - 2\mathrm{Re}\langle\psi, (F_{03}+F_{12})K\psi\rangle.\end{aligned} \tag{74}$$

Now, by (68),

$$\langle\psi, (F_{01}+F_{23})I\psi\rangle = \left\langle\psi, \left(-\frac{1}{2}\psi^* I\psi\right) I\psi\right\rangle = \frac{1}{2}|\psi^* I\psi|^2.$$

Similarly, (69) and (70) give

$$\langle\psi, (F_{02}+F_{31})J\psi\rangle = \frac{1}{2}|\psi^* J\psi|^2$$

and

$$\langle\psi, (F_{03}+F_{12})K\psi\rangle = \frac{1}{2}|\psi^* K\psi|^2.$$

With these, (74) becomes

$$\Delta\|\psi\|^2 = -2\sum_{\alpha=0}^{3}\|\nabla_\alpha\psi\|^2 - |\psi^* I\psi|^2 - |\psi^* J\psi|^2 - |\psi^* K\psi|^2 \tag{75}$$

and we can now prove part (a) of the Theorem. Indeed, (75) implies that $\Delta||\psi||^2 \leq 0$ on all of $\mathbb{R}^4$ so the real-valued function $x \to ||\psi(x)||^2$ is subharmonic on $\mathbb{R}^4$. Consequently, it satisfies a Mean Value Inequality on $\mathbb{R}^4$. Specifically, for any $r > 0$ and any $x \in \mathbb{R}^4$, we have

$$||\psi(x)||^2 \leq \frac{2}{\pi^2 r^4} \int_{B_r(x)} ||\psi||^2 d\text{vol}, \tag{76}$$

where $B_r(x)$ is the closed ball of radius r about x in $\mathbb{R}^4$. Assuming $\psi \in L^2(\mathbb{R}^4)$,

$$\int_{B_r(x)} ||\psi||^2 d\text{vol} \leq \int_{\mathbb{R}^4} ||\psi||^2 d\text{vol} < \infty.$$

Thus, if

$$c = \frac{2}{\pi^2} \int_{\mathbb{R}^4} ||\psi||^2 d\text{vol},$$

we find that, for each $x \in \mathbb{R}^4$,

$$||\psi(x)||^2 \leq \frac{2}{r^4}$$

for every $r > 0$. It follows that $\psi(x) = 0$ for every x.

The proof of (b) is similar, but requires a few preliminary calculations. The first of these is routine and yields

$$|\psi^* I \psi|^2 + |\psi^* J \psi|^2 + |\psi^* K \psi|^2 = ||\psi||^4. \tag{77}$$

Next we appeal to an identity from vector calculus ($\Delta(fg) = f\Delta g - 2\nabla f \cdot \nabla g + g\Delta f$) to obtain

$$\Delta||\psi||^4 = -2(\nabla||\psi||^2) \cdot (\nabla||\psi||^2) + 2||\psi||^2 \Delta||\psi||^2. \tag{78}$$

Now, substituting (77) into (75) gives

$$\Delta||\psi||^2 = -2 \sum_{\alpha=0}^{3} ||\nabla_\alpha \psi||^2 - ||\psi||^4$$

so (78) becomes

$$\Delta||\psi||^4 = -2(\nabla||\psi||^2) \cdot (\nabla||\psi||^2) - 4||\psi||^2 \sum_{\alpha=0}^{3} ||\nabla_\alpha \psi||^2 - 2||\psi||^6. \tag{79}$$

We conclude that $\Delta||\psi||^4 \leq 0$ on all of $\mathbb{R}^4$ so $x \to ||\psi(x)||^4$ is subharmonic on $\mathbb{R}^4$. Thus, for every $x \in \mathbb{R}^4$ and every $r > 0$,

$$||\psi(x)||^4 \leq \frac{2}{\pi^2 r^4} \int_{B_r(x)} ||\psi||^4 d\text{vol}.$$

The assumption in part (b) is that $E(A,\psi) < 0$ and this implies, in particular, that

$$\int_{\mathbb{R}^4} \|\psi\|^4 \, d\mathrm{vol} < \infty$$

so once again we conclude that $\psi(x) = 0$ for every x.

To complete the proof we must show, still assuming $E(A,\psi) < \infty$, that $F_A \equiv 0$ as well. Now, $\psi \equiv 0$ implies (by (68)–(70)) that $F_A^+ = 0$ so F_A is anti-self-dual, i.e.,

$$F_A = dA = -{}^*dA. \tag{80}$$

From this it follows that F_A is a harmonic 2-form, i.e., that if Δ is the Laplacian on forms, then

$$\begin{aligned}\Delta F_A &= -(d^*d^* + {}^*d^*d)F_A \\ &= -(d^*d^* + {}^*d^*d)(dA) \\ &= -d^*d^*(dA) \\ &= -d^*d^*({}^*dA) \\ &= d^*d(dA) \\ &= 0.\end{aligned}$$

Since

$$\Delta F_A = -\sum_{\alpha<\beta}\left(\sum_{\gamma=0}^{3} \frac{\partial^2 F_{\alpha\beta}}{(\partial x^\gamma)^2}\right) dx^\alpha \wedge dx^\beta$$

we find that each $F_{\alpha\beta}$ satisfies

$$\Delta F_{\alpha\beta} = 0, \qquad \alpha,\beta = 0,1,2,3. \tag{81}$$

Now, since $F_{\alpha\beta} : \mathbb{R}^4 \to \mathrm{Im}\mathbb{C}$, $|F_{\alpha\beta}|^2 = -F_{\alpha\beta}^2$ so

$$\begin{aligned}\Delta\Big(\sum_{\alpha<\beta} |F_{\alpha\beta}|^2\Big) &= -\sum_{\alpha<\beta} \Delta F_{\alpha\beta}^2 \\ &= -\sum_{\alpha<\beta} (F_{\alpha\beta}\Delta F_{\alpha\beta} - 2(\nabla F_{\alpha\beta})\cdot(\nabla F_{\alpha\beta}) + F_{\alpha\beta}\Delta F_{\alpha\beta}) \\ &= 2\sum_{\alpha<\beta} (\nabla F_{\alpha\beta})\cdot(\nabla F_{\alpha\beta}) \qquad \text{(by (81))} \\ &= 2\sum_{\alpha<\beta}\left(\sum_{\gamma=0}^{3}\left(\frac{\partial F_{\alpha\beta}}{\partial x^\gamma}\right)^2\right)\end{aligned}$$

$$= -2\sum_{\alpha<\beta}\left(\sum_{\gamma=0}^{3}\left(\frac{\partial F_{\alpha\beta}}{\partial x^\gamma}\right)\left(-\frac{\partial F_{\alpha\beta}}{\partial x^\gamma}\right)\right)$$

$$= -2\sum_{\alpha<\beta}\left(\sum_{\gamma=0}^{3}\left|\frac{\partial F_{\alpha\beta}}{\partial x^\gamma}\right|^2\right) \leq 0.$$

Thus, $\sum_{\alpha<\beta}|F_{\alpha\beta}|^2$ is subharmonic on $\mathbb{R}^4$ so, as before, $E(A,\psi) < \infty$ implies $\sum_{\alpha<\beta}|F_{\alpha\beta}|^2 \equiv 0$. Thus, $F_{\alpha\beta} \equiv 0$ for all α, $\beta = 0, 1, 2, 3$, i.e., $F_A \equiv 0$. □

One might think that, at least from the point of view of physics, Witten's Theorem would put to rest any thought of looking for solutions to the Seiberg-Witten equations on flat space. As it happens, however, this is not the case. There are variants of the Seiberg-Witten equations for Minkowski space and for other gauge groups (e.g., $SU(2)$) and recently solutions have been found to these equations for which the potential is nothing other than a Dirac monopole (see [9] and [8]). Perhaps the moral to this entire story is that, if magnetic monopoles do not exist in the "real world", it is almost certainly an oversight on the part of the gods, for they exist everywhere in mathematics.

Appendix I

The Dirac quantization condition

Consider a magnetic monopole of strength g located at the origin with local potentials given by $\boldsymbol{A}_S = -g(1+\cos\varphi)d\theta$ on $\mathbb{R}^3 - \{(0,0,z) : z \geq 0\}$ and $\boldsymbol{A}_N = g(1-\cos\varphi)d\theta$ on $\mathbb{R}^3 - \{(0,0,z) : z \leq 0\}$. On the intersection of these domains,

$$\boldsymbol{A}_N = \boldsymbol{A}_S + d(2g\theta). \tag{82}$$

Now consider a charge q in the field of the monopole. Each potential $\boldsymbol{A}_N$ and $\boldsymbol{A}_S$ gives rise (via the Schroedinger equation) to an expression for the wavefunction of q. Denote these ψ_N and ψ_S, respectively. On the intersection of the domains of $\boldsymbol{A}_N$ and $\boldsymbol{A}_S$, (82) implies

$$\psi_N = e^{i(2qg\theta)}\psi_S.$$

For each fixed t, the circle $\rho = 1$, $\varphi = \frac{\pi}{2}$ is in the intersection of these domains. ψ_N and ψ_S are single-valued so, on such a circle, $\theta \to \theta + 2\pi$ must leave ψ_N and ψ_S unchanged. But this implies

$$e^{i(2qg(\theta+2\pi))} = e^{i(2qg\theta)}$$

so

$$e^{i(4\pi qg)} = 1.$$

This can occur only if $4\pi qg = 2n\pi$ for some $n \in \mathbb{Z}$, i.e.,

$$qg = \frac{n}{2}.$$

Thus, for $n \neq 0$,

$$g = n\left(\frac{1}{2q}\right) \quad \text{and} \quad q = n\left(\frac{1}{2g}\right).$$

Appendix II

1st and 2nd Chern classes

We consider a smooth principal G-bundle $G \hookrightarrow P \xrightarrow{\mathcal{P}} X$ with G a matrix Lie group. Let ω be a connection on P with curvature Ω and let $s : V \to P$ and $s^g : V' \to P$ be two local cross-sections with $V \cap V' \neq \emptyset$ and related on $V \cap V'$ by $s = s \cdot g$, where $g : V \cap V' \to G$ is the transition function. The local field strengths $\mathcal{F} = s^*\Omega$ and $\mathcal{F}^g = (s^g)^*\Omega$ are related on $V \cap V'$ by $\mathcal{F}^g = g^{-1}\mathcal{F}g$ and so, in particular, generally do not agree on the intersection of their domains. Thus, the local field strengths do not piece together to give a globally defined 2-form on X. However, since two similar matrices have the same trace, $\text{tr}(\mathcal{F}^g) = \text{tr}(\mathcal{F})$ so these do piece together into a globally defined, complex-valued 2-form on X. One can show that $\frac{i}{2\pi}\text{tr}(\mathcal{F})$ is a real-valued, closed 2-form on X which therefore determines a de Rham cohomology class $\frac{i}{2\pi}[\text{tr}(\mathcal{F})] \in H^2_{\text{de Rham}}(X)$. Furthermore, this cohomology class does not depend on the initial choice of the connection ω from which it arose. It is a characteristic of the bundle $G \hookrightarrow P \xrightarrow{\mathcal{P}} X$ itself and is called the 1st *Chern class* of the bundle:

$$c_1(P) = \frac{i}{2\pi}[\text{tr}(\mathcal{F})] \in H^2_{\text{de Rham}}(X).$$

One can show that $c_1(P)$ is an integral cohomology class in the sense that its integral over any compact, connected, oriented 2-dimensional submanifold of X is an integer. Moreover, principal $U(1)$-bundles over S^2 are characterized up to equivalence by their 1st *Chern number*

$$c_1(P)[S^2] = \int_{S^2} c_1(P).$$

Similarly, $\text{tr}(\mathcal{F}) \wedge \text{tr}(\mathcal{F}) - \text{tr}(\mathcal{F} \wedge \mathcal{F})$ is a well-defined, real-valued, closed 4-form on X and the cohomology class $c_2(P) = -\frac{1}{8\pi^2}[\text{tr}(\mathcal{F}) \wedge \text{tr}(\mathcal{F}) - \text{tr}(\mathcal{F} \wedge \mathcal{F})]$, called the 2nd *Chern class* of the bundle, is independent of the connection on P from which it is computed. Principal $SU(2)$-bundles over S^4 (which have $c_1(P) = 0$) are equivalent if and only if they have the same 2nd *Chern number* $c_2(P)[S^4] = \int_{S^4} c_2(P) \in \mathbb{Z}$.

References

1. J. Arafune, P.G.O. Freund and C.J. Goebel, *Topology of Higgs fields*, J. Math. Phys. **16**(1975), 433–437.
2. M.F. Atiyah, *Geometry of Yang-Mills Fields*, Lezioni Fermiane, Accademia Nazionale dei Lincei Scuola Normale Superione, Piza, 1979.
3. M.F. Atiyah and N. Hitchin, *The Geometry and Dynamics of Magnetic Monopoles*, Princeton Univ. Press, Princeton N. J., 1988.
4. M.F. Atiyah, N.J. Hitchin and I.M. Singer, *Self-duality in four dimensional Riemannian geometry*, Proc. Roy. Soc. Lond. **A362**(1978), 425–461.
5. A. Belavin, A. Polyakov, A. Schwartz and Y. Tyupkin, *Pseudoparticle solutions of the Yang-Mills equations*, Phys. Lett. **59B**(1975), 85–87.
6. P.A.M. Dirac, *Quantised singularities in the electromagnetic field*, Proc. Roy. Soc. **A133**(1931), 60–72.
7. S.K. Donaldson, *An application of gauge theory to four dimensional topology*, J. Diff. Geom. **18**(1983), 279–315.
8. T. Dereli and M. Tekmen, *Wu-Yang monopoles and non-Abelian Seiberg-Witten equations*, e-print hep-th/9806031.
9. P. Freund, *Dirac monopoles and the Seiberg-Witten monopole equations*, J. Math. Phys. **36**(1995), 2673–2674.
10. D.S. Freed and K.K. Uhlenbeck, *Instantons and Four-Manifolds*, MSRI Publications, Springer-Verlag, New York, Berlin, 1984.
11. D.J. Groisser, *SU(2) Yang-Mills theory on* $\mathbf{R}^3$, Ph.D. Thesis, Harvard University, 1983.
12. A.M. Jaffe and C.H. Taubes, *Vortices and Monopoles: Structure of Static Gauge Theories*, Birkhäuser, Boston, 1980.
13. J.W. Morgan, *The Seiberg-Witten Equations and Applications to the Topology of Smooth Four-Manifolds*, Princeton Univ. Press, Princeton N.J., 1996.
14. G.L. Naber, *Topology, Geometry and Gauge Fields: Foundations*, TAM 25, Springer, New York, 1997.
15. D. Salamon, *Spin Geometry and Seiberg-Witten Invariants*, (in preparation).
16. N. Seiberg and E. Witten, *Monopoles, duality and chiral symmetry break-*

ing in $N = 2$ supersymmetric QCD, Nucl. Phys. **B431**(1994), 581–640.

17. C. Taubes, *Self-dual connections on non-self-dual four-manifolds*, J. Diff. Geom. **17**(1982), 139–170.
18. K. Uhlenbeck, *Removable singularities in Yang-Mills fields*, Comm. Math. Phys. **83**(1982), 11–29.
19. E. Witten, *Topological quantum field theory*, Comm. Math. Phys **117**(1988), 353–386.
20. ———, *Monopoles and 4-manifolds*, Math. Res. Letters **1**(1994), 764–796.
21. C.N. Yang and R.L. Mills, *Conservation of isotopic spin and isotopic gauge invariance*, Phys. Rev. **96**(1954), 191–195.

COMPLEX STRUCTURE AND DIRAC THEORY

I.B. PESTOV

Bogoliubov Laboratory of Theoretical Physics,
Joint Institute for Nuclear Research,
141980, Dubna, Russia

In the present article the role of the complex structure in the electrodynamics is investigated. The connection between the complex structure and the theory of charged particles with spin one half is established.

1 First-Order Equations

Let $\Lambda_j = (\Lambda^i_{jk})$ be components of a given affine connection Λ. Then components of any other connection Γ can be represented in the form $\Gamma^i_{jk} = \Lambda^i_{jk} + H^i_{jk}$, where $H_j = (H^i_{jk})$ are components of a tensor field of the type (1,2). We suppose that under gauge transformations the tensor field H is transformed by the following law

$$\bar{H}_i = SH_iS^{-1} + S\nabla_iS^{-1}, \tag{1}$$

where $S = (S^i_j)$ is a nondegenerate tensor field of the type (1,1), $S^{-1} = T^i_j$, $S^i_jT^j_k = \delta^i_k$, ∇_i is a covariant derivative with respect to the connection Λ.

For tensor field $F = (F^i_j)$ of the type (1,1), that will play a key role in what follows, we assume that $\bar{F} = SF$ and

$$D_iF = \nabla_iF + H_iF. \tag{2}$$

Let $\bar{D}_iF = \nabla_iF + \bar{H}_iF$, then from (1) and (2) it follows that

$$\bar{D}_i\bar{F} = S(D_iF), \tag{3}$$

where $\bar{F} = SF$. Thus, F and D_iF are transformed by the same law. We put

$$D_iF = \nabla_iF - FH_i, \tag{4}$$

if $\bar{F} = FS^{-1}$.

Let $C_i = (C^j_{ik})$ be components of a tensor field of the type (1,2). Consider the equation

$$C^iD_iF = mF, \tag{5}$$

where m is a scalar, D_i is a gauge derivative (4), $C^i = g^{ij}C_j$. According to (4) and (5),

$$C^i\bar{D}_i\bar{F} = C^i(D_iF)S^{-1} = m\bar{F}.$$

Hence equation(5) is gauge invariant. If the field F is transformed as $\bar{F} = SF$ then the equation

$$(D_i F)C^i = mF,$$

is also gauge invariant, which can easily be verified with the use of (3). Thus, it is shown that there exist nontrivial gauge-invariant tensor equations of the first order for tensor fields of the type (1,1).

The requirement of gauge invariance does not here impose any constraints on the tensor field C, which allows us to specialize it through its association with the problem of finding real tensor representations of Clifford algebras. The so-called spinor representations of Clifford algebras are well known. The Dirac wave function is the carrier space of a complex, faithful and irreducible representation of the Clifford algebra [1].

The real tensor field C of the type (1,2) will be called the Clifford tensor if its components $C_i = (C^j_{ik})$ satisfy the equation

$$C^k_{im}C^m_{jl} + C^k_{jm}C^m_{il} = 2g_{ij}\delta^k_l,$$

or in the matrix notation, $C_iC_j + C_jC_i = 2g_{ij}E$. We will demonstrate that the class of Clifford tensors is not empty. At every point p of space-time the Clifford tensor should define a real irreducible representation of a Clifford algebra $Cl(g)$. Then, based on the theory of representations of Clifford algebras [2] we conclude that the Clifford tensor can exist only on a manifold M of even dimension, $\dim M = n = 2m$, as all tensor indices run over n values. Next we find that m should obey the equation $2m = 2^m$ that has two solutions, $m = 1$, $m = 2$. Consequently, the Clifford tensor can exist only on manifolds of dimension $n = 2$ or $n = 4$. The case $n = 2$ will be considered separately. Therefore, in what follows $\dim M = 4$ throughout. To satisfy the reality condition, it is necessary to require that the quadratic form $g(X,X) = g_{ij}V^iV^j$ be of the signature $(-,+,+,+)$ or $(-,-,+,+)$. We will dwell upon the first of them for the reason to be explained in the next section.

The literature on physics deals mainly with the signature $(+,-,-,-)$, therefore, we will write the relation defining the Clifford tensor in the form

$$C_iC_j + C_jC_i = -2g_{ij}E. \tag{6}$$

If

$$g_{ij} = \eta_{ab}e^a_ie^b_j, \quad g^{ij} = \eta^{ab}e^i_ae^j_b, \quad e^i_ae^a_j = \delta^i_j,$$

then for the Clifford tensor in the basis $E_a = e^i_a\partial_i$ we have $C^j_{ik} = C^b_{ad}e^d_ke^j_be^a_i$, $\{C_a, C_b\} = -2\eta_{ab}E$, where $\eta_{ab} = (+1,-1,-1,-1)$, $C_a = (C^b_{ad})$. We will solve this equation defining the components of the Clifford tensor in the basis E_a explicitly:

$$C_0 = \begin{pmatrix} 0 & 0 & 0 & -1 \\ 0 & 0 & -1 & 0 \\ 0 & 1 & 0 & 0 \\ 1 & 0 & 0 & 0 \end{pmatrix}, \quad C_1 = \begin{pmatrix} -1 & 0 & 0 & 0 \\ 0 & -1 & 0 & 0 \\ 0 & 0 & 1 & 0 \\ 0 & 0 & 0 & 1 \end{pmatrix},$$

$$C_2 = \begin{pmatrix} 0 & 0 & 0 & 1 \\ 0 & 0 & -1 & 0 \\ 0 & -1 & 0 & 0 \\ 1 & 0 & 0 & 0 \end{pmatrix}, \quad C_3 = \begin{pmatrix} 0 & 0 & 1 & 0 \\ 0 & 0 & 0 & 1 \\ 1 & 0 & 0 & 0 \\ 0 & 1 & 0 & 0 \end{pmatrix}.$$

Hence, the Clifford tensors are proved to exist on a 4-manifold admitting a global field of orthonormal tetrad. The study of a general case of the 4-manifold with the metric of signature $(+,-,-,-)$ is beyond the scope of the present paper. We put $\Lambda^i_{jk} = e^i_a \partial_j e^a_k$. In this case we have

$$\nabla_i C^k_{jl} = 0 \tag{7}$$

that will be useful in what follows.

Now let us explain the choice of the metric signature. Tensor fields of the type $(1, p+1)$, $p = 0, 1, 2, 3, 4$,

$$(C_{i_1 \cdots i_p}{}^k{}_l) = C_{i_1 \cdots i_p} = \frac{1}{p!} \delta^{j_1 \cdots j_p}_{i_1 \cdots i_p} C_{j_1} \cdots C_{j_p},$$

where $\delta^{j_1 \cdots j_p}_{i_1 \cdots i_p}$ is the generalized Kronecker delta, will be called the Clifford basis that contains a unit field $E = (\delta^k_i)$, $p = 0$. For $p = 2$ we have $C_{i_1 i_2}{}^k{}_l = \frac{1}{2}(C^k_{i_1 m} C^m_{i_2 l} - C^k_{i_2 m} C^m_{i_1 l})$ and so on. Let e_{ijkl} be components of a completely antisymmetric Levi-Civita tensor field with $e_{0123} = \varepsilon\sqrt{|g|}$, where ε is a chosen orientation of M, i.e. an even scalar is such that $\varepsilon^2 = 1$ [3]. We introduce the field J of the type (1,1) with the equation

$$J = \frac{1}{4!} e^{ijkl} C_{ijkl}, \tag{8}$$

where $e^{ijkl} = g^{im} g^{jn} g^{kp} g^{lq} e_{mnpq}$. Then $J^2 = -E$ only when the metric has the signature $(+,-,-,-)$. As it is known, the almost complex structure on a real differentiable manifold M is a tensor field J of type (1,1) such that $J^2 = -E$ [4]. Consequently, the field we have introduced (8) is the almost complex structure in space-time. Thus, the choice of correct signature of the metric is connected with a definite property of space-time. In what follows it will be shown what are the laws predetermined by this property of space-time. Here we will only consider the general characteristics of the almost complex structure.

Let space-time be a real differentiable manifold M of dimension $n = 4$. The system of complex local coordinates on M will be defined as a topological mapping of an open region V of the manifold M onto an open region of the number complex space C^2. This mapping makes every point p from V correspond to two complex numbers $z^k = x^k + iy^k$, $k = 1, 2$. The manifold is said to admit the complex analytic structure provided there exists a set of systems of complex local coordinates V_a satisfying the following conditions: the union of all V_a coincides with M, $M = \cup V_a$; in the intersection of two domains $V_a \cap V_b$, the complex local coordinates z^k_a and z^k_b, are holomorphic functions of each other, $z^k_a = z^k_a(z^k_b)$ and $z^k_b = z^k_b(z^k_a)$ (i.e. z^k_a are functions of z^k_b and not of their conjugates $\bar{z}^k_b$).

There is an important correspondence between the almost complex structure J on M and complex analytic structure [4]. Let

$$N^i_{jk} = J^l_j \partial_l J^i_k - J^l_k \partial_l J^i_j - J^i_l \partial_j J^l_k + J^i_l \partial_k J^l_j$$

be components of the Nijenhuis tensor of the almost complex structure J on M. If $N^i_{jk} = 0$, then M admits the complex analytic structure, and vice versa, every complex analytic structure on M induces the almost complex structure J whose torsion tensor equals zero. So, intuitively it is clear that the almost complex structure should have a direct relationship to the complex fields and electromagnetic interactions.

2 Dirac-like equation

Here we will study a Dirac-like equation

$$C^i D_i F = mF, \tag{9}$$

where C is the Clifford tensor (6) and F is a real field of the type (1,1). As equation (9) is gauge-invariant, it remains to be seen how this equation can be derived from the variational principle. To this end, we first consider some properties of the Clifford basis following from its definition and equations (6). We have the relations

$$C_j C_{i_1 \cdots i_p} = C_{j i_1 \cdots i_p} + \frac{1}{p!}(p g_{j k_1} C_{k_2 \cdots k_p} \delta^{k_1 \cdots k_p}_{i_1 \cdots i_p}), \tag{10}$$

$p = 0, 1, 2, 3, 4$. From (6), (10) it follows that a field F may be represented as an expansion on Clifford basis

$$F = \sum_{p=1}^{4} \frac{1}{p!} f_{i_1 \cdots i_p} C^{i_1 \cdots i_p},$$

where

$$f_{i_1 \cdots i_p} = \frac{1}{4}(-1)^{\frac{p(p+1)}{2}} \mathrm{Tr}\,(C_{i_1 \cdots i_p} F).$$

A completely skew-symmetrical covariant tensor field of the type $(0, p)$ is called the p-vector, whereas the 0-vector is simply a scalar field; and the 1-vector, a covector field. So, it may be said that components of p-vectors are coordinates of the field F in the Clifford basis. Then it follows that in the given case the fields of the type (1,1) can be made to correspond to an inhomogeneous differential form ω. The formal expression $\omega_p = \frac{1}{p!} f_{i_1 \cdots i_p} dx^{i_1} \wedge \cdots \wedge dx^{i_p}$ is called the differential form of the degree p, then $\omega = \sum_{p=0}^{4} \omega_p$ [5]. Since in the general case ω has 2^n components and F has n^2 components, we find that under the condition $n^2 = 2^n$ we get again $n = 2$ and $n = 4$. We define the operation of conjugation so that the conjugate field $\hat{F}$ obey the equation

$$\mathrm{Tr}\,(F\hat{F}) = \sum_{p=0}^{4} \frac{1}{p!} f_{i_1 \cdots i_p} f^{i_1 \cdots i_p}.$$

Because $\hat{C}_i = -C_i$, then $\hat{F} = \sum_{p=0}^{4} \frac{1}{p!}(-1)^{\frac{p(p+1)}{2}} f_{i_1 \cdots i_p} C^{i_1 \cdots i_p}$. We also note the relations

$$\widehat{F_1 F_2} = \widehat{F_2}\widehat{F_1}, \quad \mathrm{Tr}\, F = \mathrm{Tr}\,(\widehat{F}), \quad J = \hat{J}, \quad \hat{\hat{F}} = F.$$

The scalar product $\mathrm{Tr}\,(F\widehat{F})$ is invariant under the gauge transformations obeying the equation

$$S\widehat{S} = E. \tag{11}$$

From the (11) it follows that the gauge field H_i must satisfy the equation

$$H_i + \widehat{H_i} = 0. \tag{12}$$

From the (9),(12) it follows that $\widehat{F}$ obey the equation

$$(\nabla_i \widehat{F})C^i + H_i \widehat{F} C^i = -m\widehat{F}.$$

Equation (9) follows from the variational principle with the Lagrangian

$$L_F = \frac{1}{2}\mathrm{Tr}\,(\widehat{F} C^i D_i F) - \frac{m}{2}\mathrm{Tr}\,(\widehat{F} F)$$

under the condition that the torsion covector of the ground connection Λ is zero. In the general case the Euler-Lagrange equation for L_F is as follows

$$C^i(D_i - \frac{1}{2}T_i)F = mF, \tag{13}$$

where $T_i = T^k_{ki}$, $T^i_{jk} = \Lambda^i_{jk} - \Lambda^i_{kj}$. The Lagrangian of interacting fields

$$L = \frac{1}{2}\mathrm{Tr}\,(\widehat{F}C^i D_i F) - \frac{m}{2}\mathrm{Tr}\,(\widehat{F}F) + \frac{1}{4}\mathrm{Tr}\,(B_{ij}B^{ij}), \tag{14}$$

where $B_{ij} = \partial_i\Gamma_j - \partial_j\Gamma_i + [\Gamma_i, \Gamma_j]$ is a strength tensor of the gauge field H^i_{jk}, is invariant under the gauge transformations of type (11). It is interesting that the energy-momentum canonical tensor of the Lagrangian (14) is not symmetrical because

$$W^i_j = \frac{1}{2}\mathrm{Tr}\,(\widehat{F}C^i D_j F) + \mathrm{Tr}\,(B^{il}B_{jl}) - L\delta^i_j$$

but it is gauge invariant.

3 Almost Complex Structure and Dirac Theory

Almost complex structure being an essential property of a real space-time manifold defines interactions of electromagnetic type. This notion is specified by the following natural conditions:

i. The almost complex structure J is an element of the gauge group and generates its two subgroups $G(J)$ and $U(1)$. The subgroup $G(J)$ is composed of the elements of the gauge group obeying the equation

$$SJS^{-1} = J,$$

whereas the second subgroup $U(1)$ consists of elements of the form $Q = \exp(\frac{1}{2}\alpha J) = \cos\frac{\alpha}{2} + J\sin\frac{\alpha}{2}$, where α is a scalar. The reasons for half angles will be clear from further consideration.

ii.The Lagrangian describing interactions of electromagnetic type should be invariant under transformations of the gauge groups $G(J)$ and $U(1)$.

iii. The strength tensor B_{ij} should also be invariant under transformations of the gauge group $U(1)$.

Recall that the correct signature of the metric was found on the basis of the almost complex structure (8). Therefore, it seems of interest to find the corresponding interactions of the electromagnetic type. From (8) it follows that J is not an element of the gauge group G_c that is defined (11). As

$JJ_0 + J_0J = 0$, $\widehat{J} = J$, then $\widehat{J}J_0J = J_0$, where $J_0 = e_0^iC_i$, $J_0^2 = -E$. Hence, we conclude that J is an element of the gauge group given by the equation

$$\widehat{S}J_0S = J_0. \tag{15}$$

From equation (15) and $[J, S] = 0$ it follows that the gauge field H_i should be subjected to the constraints $H_iJ_0 + J_0H_i = 0$ and $H_iJ - JH_i = 0$. We will write the general solution of these equations as follows

$$H_i = \frac{1}{2}A_iJ + \frac{1}{4}(Y_{ijkl}e_0^l)C^{jk}, \tag{16}$$

where A_i is the 4-vector potential and Y_{ijkl} are components of the tensor field skew-symmetrical in the last three indices. Now we shall construct a gauge invariant Lagrangian of interacting fields. The scalar $\mathrm{Tr}\,(FJ_0\widehat{H})$ is invariant under gauge transformations satisfying equation (15). As $\mathrm{Tr}\,(\widehat{F}) = \mathrm{Tr}\,(F)$ and $\widehat{J_0} = -J_0$, then $\mathrm{Tr}\,(FJ_0\widehat{H}) = -\mathrm{Tr}\,(HJ_0\widehat{F})$. From this it follows that $\mathrm{Tr}\,(FJ_0\widehat{F}) = 0$. So, in the case of real fields F the Lagrangian will be massless. Owing to this interesting property we should immediately go over to complex fields $\Psi = (\Psi_j^i)$, $\Psi = \Psi_1 + i\Psi_2$ of the type (1,1). Setting $\Psi_+ = \frac{1}{2}\Psi(E - iJ)$ and $\Psi_- = \frac{1}{2}\Psi(E + iJ)$, let us now consider an expression $\varphi = \mathrm{Tr}\,(\Psi_+J_0\overset{*}{\Psi}_+)$, where $\bar{\Psi} = \Psi_1 - i\Psi_2$ and $\overset{*}{\Psi}_+ = \widehat{\bar{\Psi}}_+$. Performing complex conjugation we obtain $\bar{\varphi} = -\varphi$. Consequently, the invariant $i\mathrm{Tr}\,(\Psi_+J_0\overset{*}{\Psi}_+)$ is a real scalar.

As a result, we arrive at a gauge invariant Lagrangian of the form

$$L = \frac{i}{2}\mathrm{Tr}\,(C^iD_i\Psi_+J_0\overset{*}{\Psi}_+) - \frac{im}{2}\mathrm{Tr}\,(\Psi_+J_0\overset{*}{\Psi}_+) + c.c. + \frac{1}{4}\mathrm{Tr}\,(B_{ij}B^{ij}), \tag{17}$$

where B_{ij} are components of the strength tensor of the connection $\Gamma_i = \Lambda_i + H_i$ and Λ is the integrable connection. From (16) it follows that the Lagrangian (17) can be rewritten as

$$L = L_\Psi - A_iJ^i - \frac{1}{4}F_{ij}F^{ij} - \frac{1}{2}Z_{ijk}J^{ijk} - \frac{1}{8}H_{ijkl}H^{ijkl}, \tag{18}$$

where $Z_{ijk} = Y_{ijkl}e_0^l$;

$$H_{ijkl} = \nabla_iZ_{jkl} + Z_{ikm}Z_{jl}^m - (i \leftrightarrow j) + T_{ij}^mZ_{mkl},$$

$$F_{ij} = \partial_iA_j - \partial_jA_i,$$

$$L_\Psi = \frac{i}{2}\mathrm{Tr}\,(C^i\nabla_i\Psi_+J_0\overset{*}{\Psi}_+) - \frac{i}{2}\mathrm{Tr}\,(\Psi_+J_0\nabla_i\overset{*}{\Psi}_+C^i) - im\mathrm{Tr}\,(\Psi_+J_0\overset{*}{\Psi}_+),$$

$$J^i = \frac{1}{2}\mathrm{Tr}\,(C^i\Psi_+J_0\overset{*}{\Psi}_+), J^{ikl} = \frac{1}{2}\mathrm{Tr}\,(C^i\Psi_+C^{kl}J_0\overset{*}{\Psi}_+).$$

The proof that the 00-component of the field $J^{ij} = \frac{1}{2}\mathrm{Tr}\,(C^i\Psi_+ C^j \overset{*}{\Psi}_+)$ being positive definite can be made by expressing the component J_{ij} through the coordinate of the field Ψ in the Clifford basis. The variational principle produces for Ψ_+ the following equation

$$C^i(\nabla_i - \frac{1}{2}T_i - iA_i)\Psi_+ - C^i\Psi_+ H_i = m\Psi_+, \tag{19}$$

where T_i is the torsion covector of the background connection Λ, and

$$H_i = \frac{1}{4}(Z_{ijk})C^{jk}.$$

Now we set Z_{ijk} in (19) to be zero and study the equation

$$C^i(\nabla_i - \frac{1}{2}T_i - iA_i)\Psi_+ = m\Psi_+. \tag{20}$$

Consider the fields $Q_a = e^i_a C_i, a = 0,1,2,3$, of the type (1,1). From (6) we get $Q_a Q_b + Q_b Q_a = -2\eta_{ab}E$. Since $JQ_a + Q_a J = 0$ and, according to (7) $\nabla_i Q_a = 0$, the operator $\Sigma = iQ_1Q_2$ commutes with J and acts in the space of solutions to equation (20). Let us now solve the equation $\Psi_+\Sigma = \Psi_+$. The field Ψ_+ in the Clifford basis can be written in the form

$$\Psi_+ = \phi E + \phi_i C^i + \frac{1}{2}\phi_{ij}C^{ij} + \frac{i}{3!}(e_{ijkl}\phi^i)C^{jkl} - i\phi J \tag{21}$$

provided that the bivector components satisfy the equation $\phi_{ij} = \frac{i}{2}e_{ijkl}\phi^{kl}$. Setting $\phi_i = z_a e^a_i$, and $\phi_{ij} = z_{ab}e^a_i e^b_j$, we find that the equation $\Psi_+\Sigma = \Psi_+$ is satisfied if $z_0 = z_3$, $z_1 = -iz_2$ and

$$z_{03} = \phi, \quad z_{01} = -iz_{02}, \quad z_{12} = -iz_{03}, \quad z_{23} = -iz_{01}, \quad z_{31} = -iz_{02}.$$

Therefore, putting

$$\begin{aligned}\phi_i &= \psi_1 l_i + \psi_2 m_i, \phi = \psi_4, \phi_{ij}\\ &= \psi_3(l_i n_j - l_j n_i - m_i\bar{m}_j + m_j\bar{m}_i) + \psi_4(l_i m_j - l_j m_i),\end{aligned} \tag{22}$$

where

$$l_i = e^0_i + e^3_i, \quad n_i = e^0_i - e^3_i, \quad m_i = e^1_i + ie^2_i, \quad \bar{m}_i = e^1_i - ie^2_i$$

and ψ_1, ψ_2, ψ_3, ψ_4 are arbitrary complex functions, we derive a general solution of the equation $\Psi_+\Sigma = \Psi_+$. It is worth-while to notice that the known basis $l_i, n_i, m_i, \bar{m}_i$ [6] arises here in algebraic context naturally. Inserting (22) into (21) and (21) into (20) we derive for ψ_1, ψ_2, ψ_3, ψ_4 the following system of equations

$$
\begin{aligned}
(-P_0 - P_3)\psi_3 + (P_1 + iP_2)\psi_4 &= m\psi_1, \\
(-P_1 + iP_2)\psi_3 + (P_0 - P_3)\psi_4 &= m\psi_2, \\
(P_0 - P_3)\psi_1 + (-P_1 - iP_2)\psi_2 &= m\psi_3, \\
(P_1 - iP_2)\psi_1 + (-P_0 - P_3)\psi_2 &= m\psi_4,
\end{aligned}
$$

where

$$P_a = e_a^i(\partial_i - \frac{1}{2}T_i - iA_i) = E_a - \frac{1}{2}T_a - iA_a, \quad a = 0, 1, 2, 3.$$

If we introduce the matrices

$$\gamma^0 = \begin{pmatrix} 0 & 0 & -1 & 0 \\ 0 & 0 & 0 & 1 \\ 1 & 0 & 0 & 0 \\ 0 & -1 & 0 & 0 \end{pmatrix}, \quad \gamma^1 = \begin{pmatrix} 0 & 0 & 0 & 1 \\ 0 & 0 & -1 & 0 \\ 0 & -1 & 0 & 0 \\ 1 & 0 & 0 & 0 \end{pmatrix},$$

$$\gamma^2 = \begin{pmatrix} 0 & 0 & 0 & i \\ 0 & 0 & i & 0 \\ 0 & -i & 0 & 0 \\ -i & 0 & 0 & 0 \end{pmatrix}, \gamma^3 = \begin{pmatrix} 0 & 0 & -1 & 0 \\ 0 & 0 & 0 & -1 \\ -1 & 0 & 0 & 0 \\ 0 & -1 & 0 & 0 \end{pmatrix},$$

then the system of equations for ψ_1, ψ_2, ψ_3, ψ_4 can be rewritten in a symmetrical matrix form

$$\gamma^a P_a \psi = m\psi, \tag{23}$$

where ψ is a matrix column. The matrices γ^a obey the standard relations

$$\gamma^a\gamma^b + \gamma^b\gamma^a = -2\eta^{ab}.$$

Thus, equation (23) is nothing else than the Dirac equation. The Dirac Lagrangian follows from the Lagrangian (17) with the use of the above described substitutions. Introducing the projection operator $P = \frac{1}{4}(E - iJ)(E + iQ_1Q_2)$ we can write the field of the type (1,1) equivalent to the Dirac wave function ψ in the form

$$\Psi_D = \Psi P = \frac{1}{4}\Psi(E - iJ)(E + iQ_1Q_2). \tag{24}$$

So, the connection with the Dirac theory may be considered to be established.

And finally, note that if we go over from the Lagrangian (17) to the Dirac Lagrangian, the almost complex structure J_0 transforms into the matrix γ^0 that camouflages the vector E_0. More details on this subject can be found in [7].

Thus, it is shown that there are interesting and deep connections between the theory of generalized spaces, theory of complex structures on a manifolds and the theory of charged particles with spin one half. Of course, much has to be done in this direction, but the problem is worth efforts as many interesting applications become possible.

References

1. E. Cartan, *The theory of Spinors*, Herman, Paris, 1966.
2. C. Chevalley, *The Algebraic Theory of Spinors*, Columbia University Press, 1954.
3. G. De Rham, *Varetes differentiable: forms harmoniques*, Hermann, Paris, 1955.
4. E.J. Flaherty, *Hermitian and Kählerian Geometry in Relativity*, Springer, Berlin, 1976.
5. T. Eguchi, P.B. Gilkey and A.J. Hansen, Physics Reports **66**(1980), 213.
6. E. Newman and R. Penrose, Journal of Mathematical Physics **3**(1962), 566.
7. I. Pestov, Hadronic Journal Supplement **8**(1993), 99.

HOW TO DESCRIBE PHOTONS AS (3+1)-SOLITONS?

S. DONEV[1] AND D. TRIFONOV[2]
Bulgarian Academy of Sciences
Institute for Nuclear Research and Nuclear Energy
Blvd. Tzarigradsko Chaussee 72
1784 Sofia - Bulgaria
E-mail: [1] *sdonev@inrne.bas.bg;* [2] *dtrif@inrne.bas.bg*

This paper aims to present the pure field part of the newly developed nonlinear Extended Electrodynamics [1-3] innon-relativistic terms, i.e. in terms of the electric and magnetic vector fields $(\mathbf{E}, \mathbf{B})$, and to give explicitly those strictly localized $(3+1)$-soliton solutions of the new equations which have the integral properties of photons. The set of solutions to the new equations contains all solutions to Maxwell's equations as a subclass. The new solutions, called nonlinear, are identified through the non-zero values of two appropriately defined vector fields $\vec{\mathcal{F}}$ and $\vec{\mathcal{M}}$. The important characteristics scale factor, amplitude function, and phase function of a nonlinear solution are defined in a coordinate free way and effectively used. The intrinsic angular momentum (spin) is also defined. A limited super position principle (interference of nonlinear solutions), yielding the well known classical coherence conditions, is found to exist.

1 Introduction

The 19th century physics, due mainly to Faraday and Maxwell, created the theoretical concept of *electromagnetic field*, i.e. extended (in fact, infinite) object, having dynamical structure. The concepts of *flux of a vector field through a 2-dimensional surface* and *circulation of a vector field along a closed curve* were coined and used extensively. Maxwell's equations in their integral form establish where the time-changes of the fluxes of the electric and magnetic fields go to, or come from, in both cases of a closed 2-surface and a 2-surface with a boundary. We note that these fluxes are specific to the continuous character of the physical object under consideration and it is important also to note that Maxwell's field equations have not the sense of direct energy-momentum balance relations as the Newton's law $\dot{\mathbf{p}} = \mathbf{F}$ has. Nevertheless, they are consistent with energy-momentum conservation, as is well known, the corresponding local energy-momentum quantities are quadratic functions of the electric and magnetic vectors.

Although very useful for considerations in finite regions with boundary conditions, the pure field Maxwell's equations have time-dependent vacuum solutions (in the whole space) that give an inadequate description of the real fields. As a rule, if these solutions are time-stable, they occupy the whole

3-space or an infinite subregion of it, and they do not go to zero at infinity, hence, they carry *infinite* energy and momentum. As an example we recall the plane wave solution, given by the electric and magnetic fields of the form

$$\mathbf{E} = \{u(ct+\varepsilon z),\, p(ct+\varepsilon z),\, 0\}\,;$$
$$\mathbf{B} = \{\varepsilon p(ct+\varepsilon z),\, -\varepsilon u(ct+\varepsilon z),\, 0\},\ \ \varepsilon = \pm 1\ ,$$

where u and p are arbitrary differentiable functions. Even if u and p are soliton-like with respect to the coordinate z, they do not depend on the other two spatial coordinates (x, y). Hence, the solution occupies the whole $\mathcal{R}^3$, or its infinite subregion, and clearly it carries infinite integral energy (we use Gauss units)

$$W = \frac{1}{4\pi}\int_{\mathcal{R}^3}(u^2+p^2)dx\,dy\,dz = \infty.$$

In particular, the popular harmonic plane wave

$$u = U_o\cos(\omega t \pm k_z.z),\ \ p = P_o\sin(\omega t \pm k_z.z)\ ,$$
$$k_z^2 = \omega^2,\quad U_o = \mathrm{const}\ ,\quad P_o = \mathrm{const}\ ,$$

clearly occupies the whole 3-space, carries infinite energy

$$W = \frac{1}{4\pi}\int_{\mathcal{R}^3}(U_o + P_o)dx\,dy\,dz = \infty$$

and, therefore, could hardly be a model of a really created field.

On the other hand, according to Cauchy's theorem for the wave equation (which is necessarily satisfied by the components of **E** and **B** in the pure field case), every finite (and smooth enough) initial field configuration is strongly time-unstable [4]: the initial condition blows up radially and goes to infinity (see the next section). Hence, Maxwell's equations *cannot describe finite and time-stable localized fields.* The contradictions between theory and experiment that became clear at the end of the last century were a challenge to theoretical physics. Planck and Einstein created the notion of *elementary field quanta*, named later by Lewis [5] the *photon.* The concept of *photon* proved to be very seminal and has been widely used in the 20th century physics. However, even now, after almost a century, we still do not have a complete and satisfactory self-consistent theory of single photons. It worths recalling at this place the Einstein's remarks of unsatisfaction concerning the linear character of Maxwell theory which makes it not able to describe the microstructure of radiation[6]. Along this line we may also note here some other results and opinions [7,8].

In this paper we consider *single photons* as $(3+1)$-dimensional *extended finite objects*, moving as a whole in a consistent *translational- rotational* manner with the speed of light. Their integral characteristics like *frequency, period* and *spin* are considered to be intrinsically related to their *dynamical structure*, and the most important integral quantity characterizing the single photon seems to be its *intrinsic action*, this being put equal to Planck's constant h. The point-like view on photons can never give a satisfactory explanation of where the characteristic *frequency* of a *free* photon comes from. *Extended Electrodynamics* (EED) [1-3] was built mainly to meet this view, i.e. to give a consistent field description of single photons, and this paper gives a completely non-relativistic approach to the pure field part of EED. This gives new insights into the problem.

Our assumption that single photons are objects of a finite nature has as its mathematical representation the soliton concept. In accordance with this concept the components of $\mathbf{E}$ and $\mathbf{B}$ at every moment t have to be smooth functions and different from zero only inside a finite 3-dimensional region $\Omega_t \subset \mathcal{R}^3$. Moreover, a periodic process of constant frequency has to go inside Ω consistently with the straight line motion as a whole. Hence, we have to be ready to meet all difficulties coming from the unavoidable requirements for nonlinear equations and working with finite functions with respect to the spatial variables (x, y, z). This goes along with Einstein's view that "the whole theory must be based on partial differential equations and their singularity-free solutions" [9].

2 The new equations

We recall first Maxwell's equations in the pure field case:

$$\operatorname{rot}\mathbf{B} - \frac{\partial \mathbf{E}}{\partial \xi} = 0, \quad \operatorname{div}\mathbf{E} = 0, \tag{1}$$

$$\operatorname{rot}\mathbf{E} + \frac{\partial \mathbf{B}}{\partial \xi} = 0, \quad \operatorname{div}\mathbf{B} = 0, \tag{2}$$

where ξ denotes the product ct, c is the velocity of light and t is the time variable. From these equations we get the well known Poynting relation

$$\frac{\partial}{\partial \xi}\frac{\mathbf{E}^2 + \mathbf{B}^2}{8\pi} = -\frac{1}{4\pi}\operatorname{div}(\mathbf{E}\times\mathbf{B}). \tag{3}$$

We recall now why Maxwell's equations (1)-(2) have no (3+1) soliton-like solutions. As we know from textbooks on Classical Electrodynamics (CED), (e.g. [10]) from (1) and (2) it follows that every component U of $\mathbf{E}$ and $\mathbf{B}$ *necessarily* satisfies the wave equation

$$U_{tt} - c^2\left[U_{xx} + U_{yy} + U_{zz}\right] = 0. \tag{4}$$

We are interested in the behaviour of U at $t > 0$, if at $t = 0$ the function U satisfies the initial conditions

$$U|_{t=0} = f(x, y, z) \qquad \frac{\partial U}{\partial t}|_{t=0} = F(x, y, z),$$

where the functions $f(x, y, z)$ and $F(x, y, z)$ are finite, i.e. they are different from zero in some finite region $\Omega_o \subset \mathcal{R}^3$ and have no singularities. Besides, we assume also that f is continuously differentiable up to third order, and F is continuously differentiable up to the second order. Under these conditions Poisson proved (about 1818) that a unique solution $U(x, y, z; t)$ of the wave equation is defined, and it is expressed by the initial conditions f and F through the following formula (a good explanation is given in [4]):

$$U(x, y, z, t) = \frac{1}{4\pi c}\left\{\frac{\partial}{\partial t}\left[\int_{S_{ct}} \frac{f(P)}{r} d\sigma_r\right] + \int_{S_{ct}} \frac{F(P)}{r} d\sigma_r\right\}, \qquad (5)$$

where P is a point on the sphere S of radius $r = ct$, centred at the point (x, y, z), and $d\sigma_r$ is the surface element on $S_{r=ct}$.

The above formula (5) shows the following. In order to get the solution at the moment $t > 0$ at the point (x, y, z), being at an arbitrary position with respect to the region Ω_o, where the initial condition, defined by the two functions f and F, is concentrated, we have to integrate f/r and F/r over a sphere $S_{r=ct}$, centred at (x, y, z) and having a radius of $r = ct$, and then to form the expression (5). Clearly, the solution will be different from zero only if the sphere $S_{r=ct}$ crosses the region Ω_o at the moment $t > 0$. Consequently, if $r_1 = ct_1$ is the shortest distance from (x, y, z) to Ω_o, and $r_2 = ct_2$ is the longest distance from (x, y, z) to Ω_o, then the solution at (x, y, z) will be different from zero only inside the interval (t_1, t_2).

From another point of view this means the following. The initially concentrated in the region Ω_o perturbation begins to expand radially, it comes to the point (x, y, z) at the moment $t_1 > 0$, makes it "vibrate" (i.e. our devices show the availability of a field) during the time interval $\Delta t = t_2 - t_1$, after this the point goes back to its initial state and our devices find no more the field. Through every point out of Ω_o there will pass a wave, and its forefront reaches the point (x, y, z) at the moment t_1 while its backfront leaves the same point at the moment t_2. Roughly speaking, the initial condition "blows up" radially and goes to infinity with the velocity of light.

This rigorous mathematical result shows that *every* finite nonstatic solution of Maxwell's equations in vacuum is time-unstable, so these equations *have no* finite and smooth enough, i.e. nonsingular, time-dependent solutions, which could be used as models of real photons, as viewed by us.

Hence, if we want to describe 3-dimensional time-dependent soliton-like electromagnetic formations (or configurations) it is *necessary* to leave off Maxwell's equations and to look for new equations for $\mathbf{E}$ and $\mathbf{B}$.

On the other hand we know that Maxwell's theory is widely used in almost all natural sciences and electrical engineering, so, it does not seem reasonable to leave it off entirely and to look for a completely new theory. Moreover, in all energy computations for finite volumes, it gives very good results. This suggests to look for some extension of the theory, i.e. to extend in a nonlinear way the equations, keeping all solutions to Maxwell's equations as solutions to the new equations and keeping the energy-momentum relations of Maxwell's theory as relations of the new theory. In the same time we must incorporate new solutions with corresponding to our purpose properties.

The road we are going to follow in searching for the appropriate nonlinearization of (1)-(2) is suggested mainly by two ideas: the idea of *local energy-momentum conservation*, and the idea of *invariance* of Maxwell's equations (1)-(2) with respect to the transformation

$$(\mathbf{E},\mathbf{B}) \rightarrow (-\mathbf{B},\mathbf{E}). \tag{6}$$

We begin by recalling Newton's law $\dot{\mathbf{p}} = \mathbf{F}$: its true sense is *local momentum balance*, i.e. the momentum gained by the particle is lost by the external field. So, if the particle is absent then the field will not lose its momentum and we get the equation $\mathbf{F} = 0$. What we have to do now is to express this $\mathbf{F}$ in terms of the corresponding field functions. In other words, we have to determine in terms of the field functions and their derivatives how much momentum the field is potentially able to transfer to another physical system in case of its presence and ability to interact with the field.

In order to do this in the case of electrodynamics we recall that if the other system consists of charged particles the corresponding force is the well known Lorentz' force, acting on a particle of electric charge e: $\mathbf{F} = e\mathbf{E} + \frac{e}{c}(\mathbf{v}\times\mathbf{B})$. In case of a continuous distribution of particles with charge density of ρ the Lorentz' force is

$$\mathbf{F} = \rho\mathbf{E} + \frac{\rho}{c}(\mathbf{v}\times\mathbf{B}).$$

The corresponding Maxwell's equations with non-zero charge distribution ρ and current $\mathbf{j} = \rho\mathbf{v}$ in this case read:

$$\operatorname{rot}\mathbf{B} - \frac{\partial\mathbf{E}}{\partial\xi} = \frac{4\pi}{c}\mathbf{j}, \quad \operatorname{div}\mathbf{E} = 4\pi\rho.$$

In order to make consistent these last equations with the Lorentz' force in the continuous charge distribution case we substitute $\mathbf{j}$ and ρ from these Maxwell's equations into the above given Lorentz' force. Then we put $\mathbf{F} = 0$. Finally, we forget about the character of the "other system", and interpret $\mathbf{F}$ as the momentum which the field is potentially able to transfer to any physical system that is able to absorb it. In some sense, this defines one of the field's momentum exchanging abilities.

In this way we obtain our first extended equation:

$$\left(\operatorname{rot}\mathbf{B} - \frac{\partial \mathbf{E}}{\partial \xi}\right) \times \mathbf{B} + \mathbf{E}\operatorname{div}\mathbf{E} = 0. \tag{7}$$

This vector equation (7) extends Maxwell's equations (1) in the sense that (7) implies no more (1), i.e. (7) may have solutions which do not satisfy (1) and these new solutions, at least some of them, are considered as *admissible.* The physical sense of (7) is quite clear: *no field momentum is lost in this way.* The nonlinearity of (7) is also obvious.

Now we look for another momentum exchanging ability of the field, different from the first one, and directed in general to new physical systems. In order to come to such an expression we make use of the above mentioned $(\mathbf{E}, \mathbf{B}) \rightarrow (-\mathbf{B}, \mathbf{E})$ invariance of the pure field Maxwell's equations. This invariance is valid also for the energy and momentum (i.e. Poynting's vector) densities:

$$w = \frac{\mathbf{E}^2 + \mathbf{B}^2}{8\pi} \rightarrow \frac{\mathbf{B}^2 + \mathbf{E}^2}{8\pi}, \quad \vec{S} = \frac{c}{4\pi}\mathbf{E} \times \mathbf{B} \rightarrow -\frac{c}{4\pi}\mathbf{B} \times \mathbf{E} = \frac{c}{4\pi}\mathbf{E} \times \mathbf{B}.$$

Moreover, the basic energy-momentum balance relation of Poynting (3), is also invariant with respect to this transformation (6). This suggests that, transforming (7), i.e. replacing in (7) $\mathbf{E}$ by $-\mathbf{B}$ and $\mathbf{B}$ by $\mathbf{E}$, we should obtain also a true and valuable relation, since (7) describes now intra-field local momentum balance. In this way we obtain our second extended vector equation:

$$\left(\operatorname{rot}\mathbf{E} + \frac{\partial \mathbf{B}}{\partial \xi}\right) \times \mathbf{E} + \mathbf{B}\operatorname{div}\mathbf{B} = 0. \tag{8}$$

Hence, (8) defines explicitly another momentum exchanging ability of the field. Note that (8) is obtained from (7) in the same way as (2) is obtained from (1).

We complete this process of extension of Maxwell's equations by adding two new invariant with respect to the same transformation (6) equations, which also have the physical sense of intra-field local energy-momentum balance:

$$\left(\operatorname{rot}\mathbf{E}+\frac{\partial\mathbf{B}}{\partial\xi}\right)\times\mathbf{B}+\left(\operatorname{rot}\mathbf{B}-\frac{\partial\mathbf{E}}{\partial\xi}\right)\times\mathbf{E}-\mathbf{E}\operatorname{div}\mathbf{B}-\mathbf{B}\operatorname{div}\mathbf{E}=0, \tag{9}$$

$$\mathbf{B}.\left(\operatorname{rot}\mathbf{B}-\frac{\partial\mathbf{E}}{\partial\xi}\right)-\mathbf{E}.\left(\operatorname{rot}\mathbf{E}+\frac{\partial\mathbf{B}}{\partial\xi}\right)=0. \tag{10}$$

Remark. In the relativistic formulation of EED [3] the two relations (9) and (10) have a natural interpretation of energy-momentum transfers between F and $*F$, where F is the conventional electromagnetic field tensor constructed by means of $\mathbf{E}$ and $\mathbf{B}$, and $*F$ is its dual (constructed similarly by means of $-\mathbf{B}$ and $\mathbf{E}$). This interpretation says that these transfers are mutual: $F \rightleftarrows *F$, and always in equal quantities. In other words, the relativistic formalism considers the couples $(\mathbf{E},\mathbf{B})$ and $(-\mathbf{B},\mathbf{E})$, or F and $*F$, as two components of a new more general mathematical object [2,3].

Note that under the transformation (6) equations (7) and (8) transform into each other, while equations (9) and (10) are kept the same (up to a sign of the left-hand side).

Equations (7)-(10) constitute our new system of equations for the electromagnetic field in vacuum. Obviously, they do not introduce new parameters, they are non-linear and all solutions to Maxwell's equations are solutions to the new equations. We are going now to study those new solutions to (7)-(10), which satisfy the conditions

$$\operatorname{rot}\mathbf{E}+\frac{\partial\mathbf{B}}{\partial\xi}\neq 0,\quad \operatorname{rot}\mathbf{B}-\frac{\partial\mathbf{E}}{\partial\xi}\neq 0,\quad \operatorname{div}\mathbf{E}\neq 0,\quad \operatorname{div}\mathbf{B}\neq 0. \tag{11}$$

For further convenience, all solutions to (7)-(10), which satisfy (11), will be called *nonlinear*.

3 Properties of the nonlinear solutions

The first two, almost obvious, properties of the nonlinear solutions follow directly from (7) and (8) and are given by the relations

$$\mathbf{E}.\mathbf{B}=0, \tag{12}$$

$$\mathbf{B}.\left(\operatorname{rot}\mathbf{E}+\frac{\partial\mathbf{B}}{\partial\xi}\right)=0,\quad \mathbf{E}.\left(\operatorname{rot}\mathbf{B}-\frac{\partial\mathbf{E}}{\partial\xi}\right)=0. \tag{13}$$

Relation (12) says that the electric and magnetic vectors of every nonlinear solution are *necessarily orthogonal* to each other at every point, the algebraic property that Maxwell's (linear) plane wave solution has. From (13) it follows that the Poynting's relation (3) is true for all nonlinear solutions, and this justifies the usage of all energy-momentum quantities and relations from Maxwell's theory in the set of nonlinear solutions of the new equations. We can consider the left-hand sides of relations (13) as the energy quantities which the field may transfer to some other physical object. We note also the obvious invariance of (12)-(13) with respect to transformation (6).

We are going to show now that all nonlinear solutions satisfy the relation

$$\mathbf{E}^2 = \mathbf{B}^2. \tag{14}$$

In order to prove (14) let's multiply (8) on the left by $\mathbf{B}$. We obtain

$$\mathbf{B}.\left\{\left(\operatorname{rot}\mathbf{E} + \frac{\partial \mathbf{B}}{\partial \xi}\right) \times \mathbf{E}\right\} + \mathbf{B}^2 \operatorname{div}\mathbf{B} = 0. \tag{$*$}$$

Now, multiplying (9) on the left by $\mathbf{E}$ and having in view (12), we obtain

$$\mathbf{E}.\left\{\left(\operatorname{rot}\mathbf{E} + \frac{\partial \mathbf{B}}{\partial \xi}\right) \times \mathbf{B}\right\} - \mathbf{E}^2 \operatorname{div}\mathbf{B} = 0.$$

This last relation is equivalent to

$$-\mathbf{B}.\left\{\left(\operatorname{rot}\mathbf{E} + \frac{\partial \mathbf{B}}{\partial \xi}\right) \times \mathbf{E}\right\} - \mathbf{E}^2 \operatorname{div}\mathbf{B} = 0. \tag{$**$}$$

Now, summing up ($*$) and ($**$), in view of $div\mathbf{B} \neq 0$, we obtain (14).

Relation (14) is also true for the linear plane electromagnetic wave. It requires for all nonlinear solutions a permanent equality of the energy densities carried by the electric and magnetic fields, although permanent mutual energy-momentum flows run between the two fields, which means that these two flows are always in equal quantities.

We introduce now the following two vector fields:

$$\vec{\mathcal{F}} = \operatorname{rot}\mathbf{E} + \frac{\partial \mathbf{B}}{\partial \xi} + \frac{\mathbf{E}\times\mathbf{B}}{|\mathbf{E}\times\mathbf{B}|}\operatorname{div}\mathbf{B}, \tag{15}$$

$$\vec{\mathcal{M}} = \operatorname{rot}\mathbf{B} - \frac{\partial \mathbf{E}}{\partial \xi} - \frac{\mathbf{E}\times\mathbf{B}}{|\mathbf{E}\times\mathbf{B}|}\operatorname{div}\mathbf{E}. \tag{16}$$

It is obvious that on the solutions of Maxwell's equations $\vec{\mathcal{F}}$ and $\vec{\mathcal{M}}$ are equal to zero. Note also that under the transformation (6) we get $\vec{\mathcal{F}} \to -\vec{\mathcal{M}}$ and $\vec{\mathcal{M}} \to \vec{\mathcal{F}}$. We shall show now that on the non-zero nonlinear solutions of our equations (7)-(10) $\vec{\mathcal{F}}$ and $\vec{\mathcal{M}}$ are never zero. Indeed, consider the products $\vec{\mathcal{F}} \times \mathbf{E}$ and $\vec{\mathcal{M}} \times \mathbf{B}$. Since

$$(\mathbf{E}\times\mathbf{B})\times\mathbf{E} = -\mathbf{E}\times(\mathbf{E}\times\mathbf{B}) = -[\mathbf{E}(\mathbf{E}.\mathbf{B}) - \mathbf{B}(\mathbf{E}.\mathbf{E})] = \mathbf{B}(\mathbf{E}^2)$$

and $|\mathbf{E}\times\mathbf{B}| = \mathbf{E}^2 = \mathbf{B}^2$, we obtain

$$\vec{\mathcal{F}}\times\mathbf{E} = \left(\operatorname{rot}\mathbf{E} + \frac{\partial\mathbf{B}}{\partial\xi}\right)\times\mathbf{E} + \mathbf{B}\operatorname{div}\mathbf{B} = 0.$$

In the same way we get $\vec{\mathcal{M}}\times\mathbf{B} = 0$. In other words, on the nonlinear solutions we obtain that $\vec{\mathcal{F}}$ is co-linear to $\mathbf{E}$ and $\vec{\mathcal{M}}$ is co-linear to $\mathbf{B}$. Hence, we can write the relations

$$\vec{\mathcal{F}} = f_1.\mathbf{E},\quad \vec{\mathcal{M}} = f_2.\mathbf{B}, \tag{17}$$

where f_1 and f_2 are two functions, and further we consider the interesting cases $f_1 \neq 0, \infty$, $f_2 \neq 0, \infty$. Note that the physical dimension of f_1 and f_2 is the reciprocal to the dimension of coordinates, i.e. $[f_1] = [f_2] = [\text{length}]^{-1}$.

We shall prove now that $f_1 = f_2$. In fact, making use of the same formula for the double vector product, used above, we easily obtain

$$\begin{aligned}&\vec{\mathcal{F}}\times\mathbf{B} + \vec{\mathcal{M}}\times\mathbf{E}\\ &= \left(\operatorname{rot}\mathbf{E} + \frac{\partial\mathbf{B}}{\partial\xi}\right)\times\mathbf{B} + \left(\operatorname{rot}\mathbf{B} - \frac{\partial\mathbf{E}}{\partial\xi}\right)\times\mathbf{E} - \mathbf{E}\operatorname{div}\mathbf{B} - \mathbf{B}\operatorname{div}\mathbf{E} = 0.\end{aligned}$$

Therefore,

$$\vec{\mathcal{F}}\times\mathbf{B} + \vec{\mathcal{M}}\times\mathbf{E} = f_1\mathbf{E}\times\mathbf{B} + f_2\mathbf{B}\times\mathbf{E} = (f_1 - f_2)\mathbf{E}\times\mathbf{B} = 0.$$

The assertion follows. Now from (14) and (17) it follows also the relation $|\vec{\mathcal{F}}| = |\vec{\mathcal{M}}|$.

Definition 1. The quantity

$$L(\mathbf{E},\mathbf{B}) = \frac{1}{|f_1|} = \frac{1}{|f_2|} = \frac{|\mathbf{E}|}{|\vec{\mathcal{F}}|} = \frac{|\mathbf{B}|}{|\vec{\mathcal{M}}|}, \tag{18}$$

will be called *the scale factor* for the nonlinear solution $(\mathbf{E},\mathbf{B})$.

Obviously, $L(\mathbf{E},\mathbf{B}) = L(-\mathbf{B},\mathbf{E})$, and for all non-zero nonlinear solutions $0 < L < \infty$, while for the linear solutions $L \to \infty$.

4 Photon-like solutions

As we mentioned earlier, we consider photons as finite nonsingular objects moving as a whole along *straight lines* in the 3-space with the velocity of light. The direction of motion is assumed to be that of $(\mathbf{E}\times\mathbf{B})$. This means that the integral curves of $(\mathbf{E}\times\mathbf{B})$ have to be straight lines. If we choose the coordinate system (x, y, z) so that this direction of motion to coincide with the direction of the coordinate line z, in this coordinate system the vector fields $\mathbf{E}$ and $\mathbf{B}$ will have non-zero components only along x and y:

$$\mathbf{E} = (u, p, 0); \quad \mathbf{B} = (m, n, 0),$$

so, $(\mathbf{E} \times \mathbf{B}) = (0,\ 0,\ un - pm)$. Now from $\mathbf{E}.\mathbf{B} = 0$ and $\mathbf{E}^2 = \mathbf{B}^2$ it follows $m = \varepsilon p$ and $n = -\varepsilon u$, $\varepsilon = \pm 1$. Hence,

$$\mathbf{E} = (u, p, 0); \quad \mathbf{B} = (\varepsilon, -\varepsilon u, 0), \quad \mathbf{E} \times \mathbf{B} = [0, 0, -\varepsilon(u^2 + p^2)],$$

and we have to determine just the two functions u and p.

Let's substitute these $\mathbf{E}$ and $\mathbf{B}$ into the left hand sides of equations (7)-(10). We obtain:

$$\left(\operatorname{rot}\mathbf{E} + \frac{\partial \mathbf{B}}{\partial \xi}\right) \times \mathbf{E} + \mathbf{B}\operatorname{div}\mathbf{B} = [0, 0, \varepsilon p(p_\xi - \varepsilon p_z) + \varepsilon u(u_\xi - \varepsilon u_z)];$$

$$\left(\operatorname{rot}\mathbf{B} - \frac{\partial \mathbf{E}}{\partial \xi}\right) \times \mathbf{B} + \mathbf{E}\operatorname{div}\mathbf{E} = [0, 0, \varepsilon u(u_\xi - \varepsilon u_z) + \varepsilon p(p_\xi - \varepsilon p_z)];$$

$$\left(\operatorname{rot}\mathbf{E} + \frac{\partial \mathbf{B}}{\partial \xi}\right) \times \mathbf{B} = [\varepsilon u(p_x - u_y), \varepsilon p(p_x - u_y), -u(p_\xi - \varepsilon p_z) + p(u_\xi - \varepsilon u_z)];$$

$$\left(\operatorname{rot}\mathbf{B} - \frac{\partial \mathbf{E}}{\partial \xi}\right) \times \mathbf{E} = [\varepsilon p(u_x + p_y), \varepsilon u(u_x + p_y), -p(u_\xi - \varepsilon u_z) + u(p_\xi - \varepsilon p_z)];$$

$$-\mathbf{E}\operatorname{div}\mathbf{B} = [-\varepsilon u(p_x - u_y), -\varepsilon p(p_x - u_y), 0];$$

$$-\mathbf{B}\operatorname{div}\mathbf{E} = [-\varepsilon p(u_x + p_y), \varepsilon u(u_x + p_y), 0];$$

$$-\mathbf{E}.\left(\operatorname{rot}\mathbf{E} + \frac{\partial \mathbf{B}}{\partial \xi}\right) = -\varepsilon u(p_\xi - \varepsilon p_z) + \varepsilon p(u_\xi - \varepsilon u_z);$$

$$\mathbf{B}.\left(\operatorname{rot}\mathbf{B} - \frac{\partial \mathbf{E}}{\partial \xi}\right) = -\varepsilon p(u_\xi - \varepsilon u_z) + \varepsilon u(p_\xi - \varepsilon p_z),$$

where the indices of u and p mean the corresponding derivatives.

It is seen that equations (9)-(10) are satisfied identically, and equations (7)-(8) reduce to only one equation, namely

$$u(u_\xi - \varepsilon u_z) + p(p_\xi - \varepsilon p_z) = \frac{1}{2}\left[(u^2 + p^2)_\xi - \varepsilon(u^2 + p^2)_z\right] = 0. \qquad (19)$$

The solution to this equation is

$$u^2 + p^2 = \phi^2(x, y, \xi + \varepsilon z), \qquad (20)$$

where ϕ is an arbitrary differentiable function of its arguments. This relation means that the energy density

$$\frac{1}{8\pi}(\mathbf{E}^2 + \mathbf{B}^2) = \frac{1}{4\pi}(u^2 + p^2)$$

is a running wave along the coordinate z. Hence, our equations determine the field components u and p up to a bounded function $\varphi(x,y,z,\xi)$, $|\varphi|\leq 1$:

$$u=\phi.\varphi, \qquad p=\pm\phi.\sqrt{1-\varphi^2}. \tag{21}$$

Reversely,

$$\phi=\pm\sqrt{u^2+p^2}, \qquad \varphi=\frac{u}{\pm\sqrt{u^2+p^2}}. \tag{22}$$

The above relations show that instead of u and p we can work with ϕ and φ. Equations (7)-(10) require only ϕ to be running wave along z (in this coordinate system), and φ to be bounded function. In all other respects these two smooth functions are *arbitrary.* Hence, they may be chosen *finite* with respect to the spatial coordinates (x,y,z). Hence, *the nonlinear equations* (7)-(10) *allow* $(3+1)$ *soliton-like solutions.* Note that, since φ is bounded, it is sufficient to choose just ϕ to be spatially finite.

We are going to show now that the two functions ϕ and φ have a certain invariant sense and can be introduced in a coordinate free way.

First, let's denote by α the invariant

$$\alpha=\frac{1}{\sqrt{\frac{\mathbf{E}^2+\mathbf{B}^2}{2}}}.$$

Since $\mathbf{E}.\mathbf{B}=0$ we can consider the orthonormal frame

$$\chi=\left(\alpha\mathbf{E},-\alpha\varepsilon\mathbf{B},-\alpha^2\varepsilon\mathbf{E}\times\mathbf{B}\right).$$

This frame is defined at every point where the field is different from zero. At every point we have also the canonical frame of unit vectors

$$\chi^o=[\mathbf{e_x}=(1,0,0),\mathbf{e_y}=(0,1,0),\mathbf{e_z}=(0,0,1)].$$

We represent now the frame vectors of χ through the frame vectors of χ^o, and obtain the matrix

$$\mathcal{A}=\left\|\begin{matrix}\frac{u}{\sqrt{u^2+p^2}} & \frac{-p}{\sqrt{u^2+p^2}} & 0\\ \frac{p}{\sqrt{u^2+p^2}} & \frac{u}{\sqrt{u^2+p^2}} & 0\\ 0 & 0 & 1\end{matrix}\right\|.$$

This matrix has three invariants: $I_1=tr\mathcal{A}$, I_2 =*the sum of all principal minors of second order*, $I_3=det\mathcal{A}$. We find

$$I_1=I_2=\frac{2u}{\sqrt{u^2+p^2}}+1; \qquad I_3=\det\mathcal{A}=1.$$

Clearly, $\frac{1}{2}(I_1 - 1) \leq 1$.

Hence, we can define ϕ and φ through these invariants:

$$\phi = \pm\sqrt{\alpha^{-2} I_3(\mathcal{A})}, \quad \varphi = \frac{1}{2}\left(\operatorname{tr}\mathcal{A} - 1\right). \tag{23}$$

Definition 2. The functions ϕ and φ, defined by (23) will be called *amplitude function* and *phase function* of the corresponding nonlinear solution, respectively. The function *arccos*(φ) will be called *phase* of the solution.

For $\vec{\mathcal{F}}$ and $\vec{\mathcal{M}}$ we obtain

$$\vec{\mathcal{F}} = \left[\varepsilon(p_\xi - \varepsilon p_z), -\varepsilon(u_\xi - \varepsilon u_z), 0\right],$$
$$\vec{\mathcal{M}} = \left[-(u_\xi - \varepsilon u_z), -(p_\xi - \varepsilon p_z), 0\right].$$

We shall express the scale factor L through ϕ and φ. We obtain

$$|\vec{\mathcal{F}}| = |\vec{\mathcal{M}}| = \frac{|\phi||\varphi_\xi - \varepsilon\varphi_z|}{\sqrt{1-\varphi^2}}.$$

Therefore, since $|\mathbf{E}| = |\mathbf{B}| = |\phi|$, the scale factor L is obtained as function of φ and its first derivatives only,

$$L = \frac{|\mathbf{E}|}{|\vec{\mathcal{F}}|} = \frac{|\mathbf{B}|}{|\vec{\mathcal{M}}|} = \frac{\sqrt{1-\varphi^2}}{|\varphi_\xi - \varepsilon\varphi_z|}. \tag{24}$$

Now we shall separate a subclass of nonlinear solutions, called *almost photon-like*, through the following conditions:

$$u\frac{\partial\varphi}{\partial x} + p\frac{\partial\varphi}{\partial y} = 0, \quad p\frac{\partial\varphi}{\partial x} - u\frac{\partial\varphi}{\partial y} = 0, \quad (u^2+p^2)\frac{\partial\varphi}{\partial z} = 0; \quad \frac{\partial L}{\partial\xi} = 0. \tag{25}$$

The *invariant sense* of the first three equations of (25) is that *the phase function φ is a first integral of the vector fields* $\mathbf{E}$, $\mathbf{B}$, $\mathbf{E}\times\mathbf{B}$. From the third equation is clearly seen that in this coordinate system φ does not depend on z. The first two equations of (25), considered as an algebraic linear homogeneous system with respect to the two derivatives, yield $\varphi_x = \varphi_y = 0$ because the corresponding determinant is always nonzero: $u^2 + p^2 \neq 0$. Hence, φ may depend only on ξ. In view of (24) and $\varphi_z = 0$ the fourth equation $\frac{\partial L}{\partial\xi} = 0$ means $L =$constant. Hence, relation (24) turns into equation for φ:

$$L = \frac{\sqrt{1-\varphi^2}}{|\varphi_\xi|} \quad \rightarrow \quad \frac{\partial\varphi}{\partial\xi} = \mp\frac{1}{L}\sqrt{1-\varphi^2}. \tag{26}$$

The obvious solution to this equation is

$$\varphi = \cos\left(\kappa\frac{\xi}{L} + \beta_o\right) = \cos\left(\kappa\frac{c}{L}t + \beta_o\right), \tag{27}$$

where $\kappa = \pm 1$, $\beta_o =$const. Since φ is a periodic function with respect to t, then c/L has the physical interpretation of *frequency*, and this frequency has nothing to do with the concept of frequency in classical electrodynamics since it is defined by L, and L is not defined in Maxwell's theory.

It is clearly seen the consistent translational-rotational behaviour of the solution obtained: the electric and magnetic vectors

$$\mathbf{E} = \left[\phi\cos\left(\kappa\frac{t}{T} + \beta_o\right), \phi\sin\left(\kappa\frac{t}{T} + \beta_o\right), 0\right],$$

$$\mathbf{B} = \left[\varepsilon\phi\sin\left(\kappa\frac{t}{T} + \beta_o\right), -\varepsilon\phi\cos\left(\kappa\frac{t}{T} + \beta_o\right), 0\right]$$

run along z: $\phi = \phi(x, y, \xi \pm z)$, and rotate (left or right: $\kappa = \pm 1$) with the frequency $\nu = (c/L) = 1/T$.

In order to separate the photon-like solutions we recall that the photon's characteristic quantity is its integral intrinsic angular momentum, or spin, being equal to the Planck's constant h. Namely h represents quantitatively in a unified manner the rotational and translational aspects of its dynamical nature: for all photons the product WT has the same value h, although W and T may be different for the different photons. That's why every photon should be able to determine its own scale factor $L = const$ in order to have a *cosine* periodic phase function and to obey the Planck's law: $h = WT = WL/c$. The photon's intrinsic periodic process demonstrates itself in our approach through the (left or right) rotation of the couple $(\mathbf{E}, \mathbf{B})$. Since these two vectors are orthogonal to each other and with equal modules: $|\mathbf{E}| = |\mathbf{B}|$, the basic local quantity appears to be the area of the square defined by $\mathbf{E}$ and $\mathbf{B}$ at every point, and this area is equal to $|\mathbf{E} \times \mathbf{B}|$. During one period T this square performs a full turning around and this gives the local action $|\mathbf{E}\times\mathbf{B}|.T$. In order to obtain the integral action of the solution we have to sum up all these local actions.

The above described idea is easily represented mathematically. In fact, let u, p and $u^2 + p^2$ be spatially finite functions. Then the integral energy

$$W = \frac{1}{4\pi}\int_{\mathcal{R}^3}(u^2 + p^2)dx\,dy\,dz < \infty$$

is finite. For every *almost photon-like* solution we define the *local spin vector*

S by

$$\mathbf{S} = L^2 \frac{\vec{\mathcal{F}} \times \vec{\mathcal{M}}}{4\pi} = \frac{\mathbf{E} \times \mathbf{B}}{4\pi}, \quad L = \text{const}, \quad 0 < L < \infty. \tag{28}$$

Now, the *integral intrinsic action*, or *integral spin* $\mathcal{S}$ of the solution, is defined by

$$\mathcal{S} = \int_{[0,T]} \int_{\mathcal{R}^3} |\mathbf{S}| dx\, dy\, dz\, dt. \tag{29}$$

We obtain

$$\mathcal{S} = WT. \tag{30}$$

We note once again that this approach works because $W = const < \infty$, $0 < L = const < \infty$ and the solution is soliton-like, i.e. it is finite, it has periodic dynamical structure and is time-stable. Clearly, no solution of Maxwell's equations (1)-(2) in the whole space has all these properties.

Definition 3. A nonlinear solution will be called *photon-like* if it is spatially finite, if it satisfies conditions (25) and if its integral spin $\mathcal{S}$ is equal to the Planck constant h: $\mathcal{S} = h$.

Finally, we consider briefly the problem of *interference* of photon-like solutions: if we have two photon-like solutions

$$\begin{aligned}
\mathbf{E}_1 &= \left[\phi_1(x, y, \xi + \varepsilon_1 z) \cos\left(\kappa_1 \frac{\xi}{L_1} + \beta_1\right),\right. \\
&\qquad \left.\phi_1(x, y, \xi + \varepsilon_1 z) \sin\left(\kappa_1 \frac{\xi}{L_1} + \beta_1\right), 0\right], \\
\mathbf{B}_1 &= \left[\varepsilon_1 \phi_1(x, y, \xi + \varepsilon_1 z) \sin\left(\kappa_1 \frac{\xi}{L_1} + \beta_1\right)\right. \\
&\qquad \left.-\varepsilon_1 \phi_1(x, y, \xi + \varepsilon_1 z) \cos\left(\kappa_1 \frac{\xi}{L_1} + \beta_1\right), 0\right], \\
\mathbf{E}_2 &= \left[\phi_2(x, y, \xi + \varepsilon_2 z) \cos\left(\kappa_2 \frac{\xi}{L_2} + \beta_2\right),\right. \\
&\qquad \left.\phi_2(x, y, \xi + \varepsilon_2 z) \sin\left(\kappa_2 \frac{\xi}{L_2} + \beta_2\right), 0\right],
\end{aligned}$$

$$\mathbf{B}_2 = \left[\varepsilon_2\phi_2(x,y,\xi+\varepsilon_2 z)\sin\left(\kappa_2\frac{\xi}{L_2}+\beta_2\right)\right.$$
$$\left. -\varepsilon_2\phi_2(x,y,\xi+\varepsilon_2 z)\cos\left(\kappa_2\frac{\xi}{L_2}+\beta_2\right),0\right],$$

we ask: under what conditions their sum $(\mathbf{E}_1+\mathbf{E}_2, \mathbf{B}_1+\mathbf{B}_2)$ will be again a nonlinear solution? Having done the corresponding elementary computations, we come to the following important conclusion: if

$$\varepsilon_1 = \varepsilon_2, \qquad \kappa_1 = \kappa_2, \qquad L_1 = L_2 \tag{31}$$

then the sum $(\mathbf{E}_1+\mathbf{E}_2, \mathbf{B}_1+\mathbf{B}_2)$ is again a nonlinear solution.

These relations (31) coincide with the well known from CED *coherence* conditions. They say that the two photon-like solutions will interfere, i.e. their sum will be again a solution, if:

1. They propagate along the same direction: $\varepsilon_1 = \varepsilon_2$,
2. They have the same polarization: $\kappa_1 = \kappa_2$,
3. They have the same frequency: $\nu_1 = \nu_2$, i.e. the same scale factors.

Recall that in CED these coherence conditions are experimentally found, they do not follow from the theory as necessary conditions, because CED is a linear theory and the sum of *any* two or more solutions is again a solution. For example, the sum "plane wave + a spherically symmetric field" is again a solution but no interference features are available.

In EED, which is a non-linear theory and there is no superposition principle in general, the interference of photon-like solutions is a *very special case* and it is a remarkable result that the experimentally found coherence conditions (31) appear from the nonlinear equations as *necessary* conditions, otherwise the sum *will not* be a solution.

The computation shows that the energy density w of the sum-solution is given by

$$w = \frac{1}{4\pi}\left[(\phi_1)^2 + (\phi_2)^2 + 2\phi_1\phi_2\cos(\beta_1-\beta_2)\right], \tag{32}$$

and this relation (32) allows to talk about interference instead if superposition: in our approach the superposition of photon-like solutions leads always to interference.

Of course, from the non-linear point of view, these interference phenomena are of some interest only for those soliton-like solutions which at a given moment occupy intersecting regions, otherwise the interference term $2\phi_1\phi_2\cos(\beta_1-\beta_2)$ in (32) is equal to zero. Since the two summonds follow the same direction of motion as a whole, then the sum-solution will be a time-stable solution, but it *will not be* photon-like one.

Acknowledgments

We would like to acknowledge the substantial support of the Bulgarian National Fund "Science Research" through Project $\phi - 718$ in doing this research.

References

1. S. Donev and M. Tashkova, Proc. R. Soc. of Lond. **A 443**(1993), 301.
2. ———, Proc. R. Soc. of Lond. **A 450**(1995), 281.
3. ———, Annales de la Fondation Louis de Broglie **23**(1998), No. 2, 3, 4.
4. S.J. Farlow, Partial Differential Equations for Scientists and Engineers, John Wiley & Sons, Inc., 1982.
5. G.N. Lewis, Nature **118**(1926), 874.
6. A. Einstein, Sobranie Nauchnih Trudov, vols.2, 3, (Nauka, Moskva, 1966) (in Russian).
7. M. Planck, J. Franklin Institute, 1927 (July), p.13.
8. J.J. Thomson, Philos. Mag. Ser. **6**(1924), 48, 737.
9. ———, Philos. Mag. Ser. **6**(1925), 50, 1181.
10. ———, Nature **137**(1936), 23.
11. N. Rashevsky, Philos. Mag. Ser. **7**(1927), 4, 459.
12. W. Honig, Found. Phys. **4**(1974), 367.
13. G. Hunter and R. Wadlinger Phys. Essays **2**(1989), 158.
14. A. Einstein, J. Franklin Institute **221**(1936), 349–382.
15. J.D. Jackson, *Classical electrodynamics*, John Wiley and Sons Inc., New York-London, 1962.

BOUNDARY SURFACE PROPERTIES AND COMMUTATIVITY OF SELF-ADJOINT EXTENSIONS OF CERTAIN OPERATORS OF MATHEMATICAL PHYSICS

TODOR T. TODOROV

Institute of Nuclear Research and Nuclear Energy,
Tzarigradsko choussée 72,
1784 Sofia Bulgaria

Simultaneous measurability of quantum mechanical observables defined as self-adjoint extensions of higher-order ellyptic differential expressions is shown to depend on the geometric properties of the space domain of the quantum system. This gives a further set of physical examples of the first mathematical observation due to Nelson [1] on conditions of strong commutativity of selfadjoint extensions of algebraically commuting on common dense domain essentially selfadjoint operators.

In quantum theory the compatibility, i.e. simultaneous measurability of two quantum mechanical observables A, B represented by (essentially) selfadjoint operators in a Hilbert space $\mathbf{H}$ is expressed mathematically as commutativity of the respective projection valued spectral measures $P_A(\Delta)$, $P_B(\Delta)$, called strong commutativity, i.e.

$$P_A(\Delta)P_B(\Delta) = P_B(\Delta)P_A(\Delta)$$

for all $\Delta \in B$, the set of all Borel subsets on the real line.

Frequently the restriction $\hat{A}_0$ of an observable $\hat{A}$ on a set D_0 of smooth functions is given by the quantization rule as a differential operator. For instance to the components p_1, p_2, p_3 of the momentum observable there correspond respectively the following differential operators:

$$-i\hbar\partial/\partial x_1, \quad -i\hbar\partial/\partial x_2, \quad -i\hbar\partial/\partial x_3.$$

Hence the expression of the quantum version of the classical Hamiltonian operator

$$H^{(2)} = \frac{p_1^2}{2m_1} + \frac{p_2^2}{2m_2} + \frac{p_3^2}{2m_3}$$

of a quasiparticle is

$$\hat{H}_0^2 = -i\hbar^2 \left(\frac{1}{m_1}\frac{\partial^2}{\partial x_1^2} + \frac{1}{m_2}\frac{\partial^2}{\partial x_2^2} + \frac{1}{m_3}\frac{\partial^2}{\partial x_3^2} \right) \tag{1}$$

Here m_1, m_2, m_3 are the masses along the respective coordinate axes x, y, z which are generally unequal as it is the case e.g. of an electron moving in a crystal medium. If the Hamiltonian H is a function of p_1, p_2, p_3 such that its Taylor series exists, then H can be approximated by a finite number of members from this series. After quantization this results in the appearance of partial derivatives in the quantized version $\hat{H}_0^{2R}$ of H with higher derivative of order say $2R$, R-integer which can be written shortly in the following general form:

$$\hat{H}_0^{2R} = \sum_{|I|\leq 2R} \alpha_i D^I,$$

where D denotes a general derivative of order $|I|$.

The observables can then be considered as selfadjoint extensions of the corresponding differential expressions,e.g. $\hat{H}^{2R} \supset \hat{H}_0^{2R}$, $R = 1, 2, \ldots$

Furthermore, as it was first observed by Nelson [1] two selfadjoint operators A, B commuting algebraically i.e. $ABf = BAf$ on a dense domain D_0 may not commute strongly.

In the following we formulate (cf. also [2]) a theorem concerning some geometric properties of the domain boundaries which are necessary for strong commutativity of two self-adjoint extensions $\hat{A}$, $\hat{B}$ of the elliptic partial differential expressions A, B of an arbitrary high (even) order. For a similar theorem but with $R = 1$ cf. [3].

Let Ω be a bounded domain in $\mathbf{R}^n$ with piecewise smooth boundary $\partial\Omega$ and ω be a piece of $\partial\Omega$. Denote the partial derivative $\partial^{|I|}/\partial x_{i_1}, x_{i_2} \ldots x_{i_q}$ by $\partial_x^{|I|}$, $I = \{i_1, i_2, \ldots, i_q\}$, $|I| = q$. If R is an integer and $C^\infty(\bar{\Omega})$ is the set of infinitely differentiable functions on $\bar{\Omega}$ define

$$\begin{aligned} D^\infty(\bar{\Omega}) = \{\, v_k \in C^\infty(\bar{\Omega}) \,:\, v_k(x) &= \partial_x v_k(x) = \partial_x^2 v_k(x) \\ &= \ldots \partial_x^{R-1} v_k(x) = 0,\ x \in \partial\bar{\Omega} \,\} \end{aligned}$$

Let $v_k \in D^\infty(\bar{\Omega})$ and A, B are formally self-adjoint differential expressions with constant coefficients α_i, β_i, a_I, b_I:

$$Av_k = \sum_{i=1}^n \alpha_i \partial^{2R} v_k(x)/\partial x_i^{2R} + \sum_{|I|\leq 2R-1} a_I \partial_x^I v_k(x), \tag{2}$$

$$Bv_k = \sum_{i=1}^n \beta_i \partial^{2R} v_k(x)/\partial x_i^{2R} + \sum_{|I|\leq 2R-1} b_I \partial_x^I v_k(x).$$

Denote by $x = (x_1, x_2, \ldots, x_n)$ a point in $\mathbf{R}^n$. Consider a decomposition of the set of all coordinates of x into subsets with void intersections:

$$x_{i_1}, x_{i_2}, \ldots, x_{i_\pi}; x_{j_1}, x_{j_2}, \ldots, x_{j_\psi}; x_{k_1}, x_{k_2}, \ldots, x_{k_\sigma}.$$

Consider the respective decomposition of the set of ratios $r_i = \alpha_i/\beta_i$, $i = 1, 2, \ldots, n$

$$r_{i_1}, r_{i_2}, \ldots, r_{i_\pi}; r_{j_1}, r_{j_2}, \ldots, r_{j_\psi}; r_{k_1}, r_{k_2}, \ldots, r_{k_\sigma}.$$

Consider the following conditions [Bou], [Coe] and [Dom]:

[Bou]:The piece $\omega \in \partial\Omega$ is supposed to be given by the equation $f(x) = 0$ which is solvable with respect to some x_i and $f \in C^\infty(\bar{\Omega})$.

[Coe]: For the subsets of coefficient ratios one requires the following relations:

$$r_{i_1} = r_{i_2} = \cdots = r_{i_\pi}; r_{j_1} = r_{j_2} = \cdots = r_{j_\psi}; r_{k_1} = r_{k_2} = \cdots = r_{k_\sigma},$$
$$r_{i_1} \neq r_{j_1} \neq r_{k_1}$$

[Dom]:The respective self-adjoint extensions $\hat{A}$, $\hat{B}$ of A, B exist in $L_2(\Omega)$ and satisfy the following relations:

$$D(\hat{A}) = D(\hat{B})$$
$$v(x) \in D(\hat{A}) \bigcap C^\infty(\Omega) \quad \text{implies} \quad v(x) = 0 \text{ if } x \in \partial\Omega.$$

With the above assumptions and notations one has (cf. [2] also for the case of nonconstant coefficients) the following

Theorem. *Let [Bou], [Coe] and [Dom] are verified and let $\hat{A}$, $\hat{B}$ commute strongly. Then on ω the equation $f(x) = 0$ of the boundary $\partial\Omega$ depends only on one coordinate subset from the corresponding decomposition of the coordinates $x_1, x_2, \ldots, x_n$ into subsets.*

This Theorem has the following application. In the outlined above quantum mechanical frames, consider two particles moving in a crystal medium in a volume Ω in $\mathbf{R}^3$. The particles acquire in the general case different masses m_1, m_2, m_3 while having nonzero projections of their velocities along the three coordinate axes x_1, x_2, x_3. After quantization the Hamiltonian has the general form given by (1) (resp. by (2) if $R \geq 1$). Then the Theorem says that the simultaneous measurability (i.e. compatibility) of the values of the energy observables depends on the geometric properties of the surface of the crystal medium. The same applies to a single particle if one considers two different observables, say its Hamiltonian and some function (even power) of its momentum operator.

The above discussion provides further (cf. [4]) physical examples along the lines of the first Nelson [1] mathematical observation of the existence of two algebraically commuting on common dense domain essentially selfadjoint operators which do not necessarily have strongly commuting selfadjoint extensions.

An extension of the discussion on compatibility of observables to the case of two maximal symmetric but not necessarily selfadjoint operators A, B is contained in [5], [6], [7]. There the strong commutativity is defined in terms of the isometry semigroups generated respectively by A, B and positive operator valued measures are used instead of the projection valued measures.

Acknowledgment

This paper is partly supported by the National Science Research Fund.

References

1. E. Nelson, Ann. Math. **70**(1959), 572.
2. T.S. Todorov, Math. Balkanica **10**(1966), 373.
3. R.T. Denchev, Coll. math. societ. Janos Bolyai **5**(1970), 113.
4. R. Denchev, Ch. Poushkarov and T. Todorov, Acta Ph. Austr. **45**(1976), 276.
5. G. Epifanio, T. Todorov and C. Trapani, Helvetica Ph. Acta, **65**(1992), 1.
6. ______, Jour. Math. Phys. **37**(1966), 1148.
7. ______, in Topics in Complex Analysis, Differential Geometry and Mathematical Physics, Proc. Third Intern. Workshop on Complex Structures and Vector Fields, eds. S. Demiev, K. Sekigawa, World Scientific, Singapore, 1997.

QUANTIZATION ON CLOSED MANIFOLDS

Y. OHNUKI

Nagoya Women's University
1302 Takamiya Tempaku Nagoya 468-8507, Japan
E-mail: ohnuki@nagoya-wu.ac.jp

Based the assumption that quantization for a system constrained to move on a manifold is provided with the algebra formed by position and displacement operators on it, we analyze the theory in the cases when the manifolds are S^D, the chiral manifold $SU_L(n) \times SU_R(n)/SU_V(n)$ and the Grassmann manifold $Gr(n+m, n; C)$, respectively. It is observed that the theory is automatically equipped with special types of gauge potential, whose properties are discussed in somewhat detail, especially, in the case for S^D. Some remarks are also made of related topics.

1 Preliminaries

To study quantization problems for a particle constrained to move on a closed manifold we have required in our previous paper[1] that the algebra characterizing the quantization be described with a set of commutation relations for self-adjoint position and displacement operators of the particle. As argued already this quantization scheme is a generalization of the canonical quantization which is known to be inapplicable to a system constrained on a close manifold. We call the algebra mentioned here the fundamental algebra for the system under consideration.

Thus in the case of manifold S^D we have assumed the following:

$$\begin{aligned} [\, x_\alpha,\, x_\beta\,] &= 0, \\ [\, x_\lambda,\, G_{\alpha\beta}\,] &= i\,(x_\alpha \delta_{\lambda\beta} - x_\beta \delta_{\lambda\alpha}), \\ [\, G_{\alpha\beta},\, G_{\lambda\mu}\,] &= i\,(\delta_{\alpha\lambda}\, G_{\beta\mu} - \delta_{\alpha\mu}\, G_{\beta\lambda} + \delta_{\beta\mu}\, G_{\alpha\lambda} - \delta_{\beta\lambda}\, G_{\alpha\mu}), \\ (\, \alpha,\, \beta,\, \lambda,\, \mu &= 1,\, 2, \cdots, D+1\,) \end{aligned} \tag{1.1}$$

with the constraint

$$\sum_{\alpha=1}^{D+1} x_\alpha^2 - r^2 = 0 \tag{1.2}$$

where the constant $r\,(>\,0)$ stands for the radius of S^D which has been embedded to R^{D+1} to make our argument simple. The operators x_α and $G_{\alpha\beta}(-G_{\beta\alpha})$, which obey the Lie algebra of $SO(D+1)$, are assumed to be self-adjoint. Since they play roles of the position and the displacement operators on S^D respectively, we may regard the set of relations given by (1) with constraint (2) as the fundamental algebra for a system on S^D. Thus we have to examine possible inequivalent irreducible representations of (1) to extract concrete properties of the system. It is noted that the algebra (1) generates a Lie group isomorphic to $(D+1)$-dimensional Euclidean group $E(D+1)$ that consists of translations and rotations in R^{D+1}. For simplicity in what follows the former group will also be denoted as $E(D+1)$. In the x-diagonal representation the inner product of two wave functions $\chi(x)$ and $\varphi(x)$ that belong to an irreducible representation of $E(D+1)$ is expressed as

$$< \chi \,|\, \varphi >= \int dx^{D+1}\, \delta(\textstyle\sum_{\alpha=1}^{D+1} x_\alpha^2 - r^2)\, \chi^*(x)\, \varphi(x), \tag{1.3}$$

which is seen to be invariant under $E(D+1)$. Main results obtained from the quantization on S^D will be summarized in Section 2.1.

Similarly we can introduce the fundamental algebra for a quantum mechanical system on the chiral manifold $SU_L(n)\times SU_R(n)/SU_V(n)$, coordinates on which are described, for example, in terms of $(n \times n)$-matrices $z = ||z^\alpha{}_\beta||$ and $z^\dagger = ||z^{\dagger\alpha}{}_\beta||$ (α, $\beta = 1, 2, \ldots, n$) that satisfy the unitary and unimodular conditions

$$\begin{aligned} z^\alpha{}_\gamma\, z^{\dagger\gamma}{}_\beta = \delta^\alpha_\beta, \text{ or equivalently} \quad zz^\dagger = \mathbf{1}_n, \\ \det z = 1, \end{aligned} \tag{1.4}$$

where $z^\alpha{}_\beta \in C$ and $z^{\dagger\alpha}{}_\beta = (z^\beta{}_\alpha)^*$, and $\mathbf{1}_n$ stands for the unit $(n\times n)$-matrix. Then we can postulate the following to be the fundamental algebra in this case:

$$\begin{aligned} [\, z^\alpha_{\ \beta},\, G^\mu_{L,\nu}\,] &= \delta^\alpha_\nu\, z^\mu_{\ \beta} - \frac{1}{n}\delta^\mu_\nu\, z^\alpha_{\ \beta}, \\ [\, z^\alpha_{\ \beta},\, G^\mu_{R,\nu}\,] &= -\delta^\mu_\beta\, z^\alpha_{\ \nu} + \frac{1}{n}\delta^\mu_\nu\, z^\alpha_{\ \beta} \end{aligned} \tag{1.5}$$

together with

$$\begin{aligned} [\, G^\mu_{L,\nu},\, G^\lambda_{L,\rho}\,] &= \delta^\mu_\rho G^\lambda_{L,\nu} - \delta^\lambda_\nu G^\mu_{L,\rho}, \\ [\, G^\mu_{R,\nu},\, G^\lambda_{R,\rho}\,] &= \delta^\mu_\rho G^\lambda_{R,\nu} - \delta^\lambda_\nu G^\mu_{R,\rho}, \\ [\, G^\mu_{L,\nu},\, G^\lambda_{R,\rho}\,] &= 0, \\ G^\mu_{L,\mu} = G^\mu_{R,\mu} &= 0, \end{aligned} \tag{1.6}$$

where $G^{\mu}_{L,\nu}(= G^{\nu\,\dagger}_{L,\mu})$ and $G^{\mu}_{R,\nu}(= G^{\nu\,\dagger}_{R,\mu})$ are generators of $SU_L(n)$ and $SU_R(n)$, respectively. Introducing infinitesimal hermitian $(n \times n)$-matrices $\epsilon_L = \|\epsilon^{\mu}_{L,\nu}\|$ and $\epsilon_R = \|\epsilon^{\mu}_{R,\nu}\|$ that satisfy $\mathrm{Tr}\,\epsilon_L = \mathrm{Tr}\,\epsilon_R = 0$, we define a unitary operator such that

$$U_{LR} = 1 + i\epsilon^{\mu}_{L,\nu} G^{\nu}_{L,\mu} + i\epsilon^{\nu}_{R,\mu} G^{\mu}_{R,\nu}. \tag{1.7}$$

One finds from (1.5) that it generates infinitesimal left and right chiral transformations for z, i.e.,

$$\begin{aligned} U^{\dagger}_{LR}\, z\, U_{LR} &= z + i(\epsilon_L z - z\epsilon_R) \\ &= e^{i\epsilon_L} z\, (e^{i\epsilon_R})^{\dagger}. \end{aligned} \tag{1.8}$$

Thus in this case $G^{\mu}_{L,\nu}$ and $G^{\mu}_{R,\nu}$ are also regarded as generators of displacements on the chiral space $SU_L(n) \times SU_R(n)/SU_V(n)$, while $z^{\alpha}{}_{\beta}$ are the position operators on it. Now let us write the wave function corresponding to the state vector $|\,\varphi>$ as $\varphi(z,\, z^{\dagger})$. Then the inner product of $|\,\chi>$ and $|\,\varphi>$ can be expressed as

$$<\chi\,|\,\varphi>= \int dz dz^{\dagger}\, \delta(zz^{\dagger} - \mathbf{1}_n)\, \delta(\arg(\det z))\, \chi^{*}(z,\, z^{\dagger})\, \varphi(z,\, z^{\dagger}), \tag{1.9}$$

where

$$dz\, dz^{\dagger} \equiv \prod_{\alpha,\,\beta=1}^{n} dx^{\alpha}_{\ \beta}\, dy^{\alpha}_{\ \beta} \tag{1.10}$$

with

$$z^{\alpha}_{\ \beta} = x^{\alpha}_{\ \beta} + i\, y^{\alpha}_{\ \beta} \qquad (x^{\alpha}_{\ \beta},\ y^{\alpha}_{\ \beta};\ \text{real}),$$

and

$$\delta(zz^{\dagger} - \mathbf{1}_n) \equiv \prod_{\alpha=\beta} \delta\left((zz^{\dagger})^{\alpha}_{\ \beta} - 1\right) \cdot \prod_{\alpha>\beta} \delta\left(\Re(zz^{\dagger})^{\alpha}_{\ \beta}\right) \delta\left(\Im(zz^{\dagger})^{\alpha}_{\ \beta}\right). \tag{1.11}$$

It can be shown by some calculations that the integral measure in (1.9) is invariant under the chiral transformation (1.8). With these preparations we will argue, in Section 2.2, the structure of the fundamental algebra given by (1.5) and (1.6) with the constraint (1.4).

Our quantization scheme stated in the beginning of this section can also be applied to a system on the Grassmann manifold $U(n+m)/U(n) \times U(m)$ $(n \leq m)$ which is defined[2] as

$$\begin{aligned} &Gr(\, n+m,\ n;\ \boldsymbol{C}\,) \\ &\qquad \equiv \{P \in M(n+m,\ n+m;\ \boldsymbol{C})\,|\, P^2 = P = P^{\dagger},\ \mathrm{Tr}\,P = n\,\}. \end{aligned} \tag{1.12}$$

We have denoted here a p-row and q-column matrix with complex components as $M(p,\, q;\, \boldsymbol{C})$. As seen from the above the matrix P can be represented with $2mn$ real parameters. Its explicit form is written, for instance, as

$$P = \begin{pmatrix} \dfrac{1}{\mathbf{1}_n + \xi^\dagger \xi} & \xi^\dagger \dfrac{1}{\mathbf{1}_m + \xi\, \xi^\dagger} \\ \xi \dfrac{1}{\mathbf{1}_n + \xi^\dagger \xi} & \xi \dfrac{1}{\mathbf{1}_n + \xi^\dagger \xi} \xi^\dagger \end{pmatrix}, \tag{1.13}$$

where

$$\xi = \|\xi^j{}_\beta\| \in M(m,\, n;\, \boldsymbol{C}), \qquad \xi^\dagger = \|\xi^{\dagger\beta}{}_j\| \in M(n,\, m;\, \boldsymbol{C}) \tag{1.14}$$
$$(\, \xi^{\dagger\beta}{}_j \equiv (\xi^j{}_\beta)^*; \quad j = 1',\, 2',\, \ldots,\, m'; \quad \beta = 1,\, 2,\, \ldots,\, n\,)^a.$$

In fact we can derive (1.13) by applying the boost transformation[2]

$$\alpha_\xi = \begin{pmatrix} \dfrac{1}{\sqrt{\mathbf{1}_n + \xi^\dagger\xi}} & -\xi^\dagger \dfrac{1}{\sqrt{\mathbf{1}_m + \xi\, \xi^\dagger}} \\ \xi \dfrac{1}{\sqrt{\mathbf{1}_n + \xi^\dagger\xi}} & \dfrac{1}{\sqrt{\mathbf{1}_m + \xi\, \xi^\dagger}} \end{pmatrix} \tag{1.15}$$

to a special point P_0 defined by

$$P_0 = \begin{pmatrix} \overbrace{\mathbf{1}_n}^{n} & \overbrace{\mathbf{0}}^{m} \\ \mathbf{0} & \mathbf{0} \end{pmatrix} \begin{matrix} \}\, n \\ \}\, m \end{matrix}\,, \tag{1.16}$$

because we can show that $P = \alpha_\xi P_0 \alpha_\xi^\dagger$ which satisfies $P^2 = P = P^\dagger$ and $\mathrm{Tr} P = n$. Now let us consider an infinitesimal transformation $g \in U(m+n)$. It can generally be expressed as

$$g = \mathbf{1}_{n+m} + i \begin{pmatrix} \nu & \lambda^\dagger \\ \lambda & \mu \end{pmatrix} \tag{1.17}$$

with infinitesimal matrices

$$\nu = \nu^\dagger \in M(n,\, n;\, \boldsymbol{C}), \qquad \mu = \mu^\dagger \in M(m,\, m;\, \boldsymbol{C}),$$
$$\lambda \in M(m,\, n;\, \boldsymbol{C}), \qquad \lambda^\dagger \in M(n,\, m;\, \boldsymbol{C}).$$

Then under this transformation the matrix P is transformed as

$$P \xrightarrow{\;g\;} P_g = g P g^\dagger. \tag{1.18}$$

---

[a] In the argument of the Grassmann manifold the Latin and Greek indices are assumed to run over $1',\, 2',\, \ldots,\, m'$ and $1,\, 2,\, \ldots,\, n$, respectively.

Thus we find, with the aid of (1.18), that the transformations for ξ and $\xi^\dagger$ are given by

$$\xi \xrightarrow{g} \xi_g = \xi + \Delta\xi, \qquad \xi^\dagger \xrightarrow{g} \xi_g^\dagger = \xi^\dagger + \Delta\xi^\dagger, \tag{1.19}$$

with

$$\Delta\xi = i(\lambda + \mu\xi - \xi\nu - \xi\lambda^\dagger\xi), \text{ and } \Delta\xi^\dagger = (\Delta\xi)^\dagger. \tag{1.20}$$

Corresponding to (1.17) we will introduce a unitary operator $U(g)$ that realizes the transformation (1.19) in the following manner:

$$U^\dagger(g)\,\xi\,U(g) = \xi_g, \qquad U^\dagger(g)\,\xi^\dagger\,U(g) = \xi_g^\dagger. \tag{1.21}$$

When expanded in terms of the infinitesimal matrices the operator $U(g)$ is written as

$$U(g) = 1 + i(\nu^\alpha_{\ \beta} G^\beta_\alpha + \mu^j_{\ k} G^k_j + \lambda^j_{\ \alpha} G^\alpha_j + \lambda^{\dagger\alpha}_{\ \ j} G^j_\alpha), \tag{1.22}$$

where

$$G^{\alpha\dagger}_\beta = G^\beta_\alpha, \qquad G^{j\dagger}_k = G^k_j, \qquad G^{j\dagger}_\alpha = G^\alpha_j, \tag{1.23}$$

and

$$\begin{gathered}
[G^\alpha_\beta, G^\lambda_\rho] = \delta^\alpha_\rho G^\lambda_\beta - \delta^\lambda_\beta G^\alpha_\rho, \qquad [G^j_k, G^p_q] = \delta^j_q G^p_k - \delta^p_k G^j_q, \\
[G^\alpha_k, G^j_\beta] = \delta^\alpha_\beta G^j_k - \delta^j_k G^\alpha_\beta, \\
[G^j_\lambda, G^\alpha_\beta] = -\delta^\alpha_\lambda G^j_\beta, \qquad [G^j_\lambda, G^p_q] = \delta^j_q G^p_\lambda, \\
[G^j_\lambda, G^k_\rho] = [G^j_k, G^\alpha_\beta] = 0.
\end{gathered} \tag{1.24}$$

Then with the aid of (1.19), (1.20) and (1.21) we obtain

$$\begin{gathered}
[G^\alpha_\beta, \xi^p_{\ \lambda}] = \delta^\alpha_\lambda \xi^p_{\ \beta}, \qquad [G^j_k, \xi^p_{\ \lambda}] = -\delta^p_k \xi^j_{\ \lambda}, \\
[G^j_\alpha, \xi^p_{\ \lambda}] = \xi^j_{\ \lambda} \xi^p_{\ \alpha}, \qquad [G^\alpha_j, \xi^p_{\ \lambda}] = -\delta^\alpha_\lambda \delta^p_j,
\end{gathered} \tag{1.25}$$

by means of which we can describe a displacement of the position ξ on the Grassmann manifold under consideration. Thus we see that the set of equations given by (1.23)∼(1.25) forms the fundamental algebra in this case. On the other hand, the integral measure invariant under the transformation (1.19) is known[2] to be of the form $d\xi d\xi^\dagger / \det(\mathbf{1}_n + \xi^\dagger\xi)^{n+m}$, where $d\xi d\xi^\dagger \equiv \prod_{p,\lambda}(\Re\xi^p_{\ \lambda} \cdot \Im\xi^p_{\ \lambda})$, and hence the inner product of two wave functions χ and φ is written as

$$<\chi\,|\,\varphi> = \int \frac{d\xi d\xi^\dagger}{\det(\mathbf{1}_n + \xi^\dagger\xi)^{n+m}} \chi^*(\xi, \xi^\dagger)\,\varphi(\xi, \xi^\dagger). \tag{1.26}$$

With these preparations we will examine, in Section 2.3, irreducible representations of the fundamental algebra for the Grassmann manifold. However the situation is somewhat complicated when compared the with the cases of S^D and the chiral manifold. In fact it will be shown that G^{α}_{β}, G^{j}_{k}, G^{j}_{β} and also the induced gauge potential A^{α}_{j} are generally represented in integral forms.

In the final section, some remarks will be made on related topics.

2 Irreducible Representations of the Fundamental Algebra

In quantization problem we have to determine, first of all, all possible inequivalent irreducible representations of the fundamental algebra for a system under consideration. Through this task we can construct irreducible Hilbert spaces, on each of which the position and displacement operators are explicitly formulated. To this end we will apply the induced representation technique[3] of group theory, which was originally introduced by Wigner in his classical paper[4] on unitary representations of the Poincaré group. As a result it will be shown, in this section, that there emerges special types of gauge potential[5] in quantizing the system.

In the following we will examine our quantization scheme in respective cases of S^D, $SU_L(n) \times SU_R(n)/SU_V(n)$ and $Gr(n+m,\, n;\, \boldsymbol{C})$.

2.1 Case of S^D

Under the rotation $R \in SO(D+1)$ the wave function $\varphi(x)$ belonging to an irreducible representation of the fundamental algebra is known to obey a transformation of the form

$$\varphi'(x) = Q(R,\, x)\, \varphi(R^{-1}x), \tag{2.1}$$

where $Q(R,\, x)$ is represented with a unitary matrix acting on the wave function. It plays a role of characterizing the irreducible representation and is sometimes called the Wigner rotation. Since a study of its determination has been made already[1], in what follows we merely summarize our main results obtained so far.

We consider an infinitesimal rotation

$$x'_\alpha = x_\alpha + \sum_{\beta=1}^{D+1} \omega_{\alpha\beta}\, x_\beta \tag{2.2}$$

with infinitesimal rotation angles $\omega_{\alpha\beta}\,(= -\omega_{\beta\alpha})$. Then the wave function

$\varphi(R^{-1}x)$ can be expanded as

$$\varphi(R^{-1}x) = \Big(1 + \frac{1}{2}\sum_{\alpha,\,\beta=1}^{D+1} \omega_{\alpha\beta}\,(x_\alpha\partial_\beta - x_\beta\partial_\alpha)\Big)\varphi(x) \tag{2.3}$$

with $\partial_\alpha \equiv \partial/\partial x_\alpha$. We denote the Wigner rotation corresponding to the infinitesimal rotation R as

$$Q(R,\,x) = 1 + \frac{i}{2}\sum_{\alpha,\,\beta=1}^{D+1} \omega_{\alpha\beta}\,f_{\alpha\beta}(x), \tag{2.4}$$

where $f_{\alpha\beta}(x)\,(= -f_{\beta\alpha}(x))$ are hermitian matrices that specify an irreducible representation of the fundamental algebra. Thus for an infinitesimal rotation R the transformation (2.1) is rewritten as

$$\varphi'(x) = \Big(1 + \frac{i}{2}\sum_{\alpha,\,\beta=1}^{D+1} \omega_{\alpha\beta}\,G_{\alpha\beta}\Big)\varphi(x) \tag{2.5}$$

with

$$G_{\alpha\beta} = -G_{\beta\alpha} = \frac{1}{i}(x_\alpha\partial_\beta - x_\beta\partial_\alpha) + f_{\alpha\beta}(x). \tag{2.6}$$

Now we write down the explicit form of $f_{\alpha\beta}(x)$ which uniquely corresponds to an irreducible representation of the fundamental algebra[1].

$D = 1$:

$$f_{12} = -f_{21} = \alpha, \qquad (\,0 \le \alpha < 1\,), \tag{2.7}$$

where α is a real constant that specifies an irreducible representation in a unique manner, that is, two different α's belonging to $[\,0,1\,)$ correspond to two inequivalent irreducible representations respectively, and vice versa.

$D = 2$:

$$f_{12} = -f_{21} = S, \quad f_{23} = -f_{32} = \frac{x_1}{r + x_3}S, \quad f_{31} = -f_{13} = \frac{x_2}{r + x_3}S, \tag{2.8}$$

$$(\,S = 0,\ \pm\frac{1}{2},\ \pm 1,\ \pm\frac{3}{2},\ \pm 2,\ \pm\frac{5}{2},\cdots).$$

Any unitary irreducible representation of $E(3)$ can uniquely be specified by S.

$\boldsymbol{D \geq 3}$:

$$\left\{\begin{array}{l} f_{jk} = -f_{kj} = S_{jk}, \\ f_{j\,D+1} = -f_{D+1\,j} = -\dfrac{1}{r + x_{D+1}} \displaystyle\sum_{k=1}^{D} S_{jk} x_k, \end{array}\right. \tag{2.9}$$

$$(j,\, k = 1,\, 2,\, \cdots,\, D),$$

where S_{jk} are irreducible hermitian generators of $SO(D)$;

$$[\, S_{jk},\, S_{lm}\,] = i\,(\,\delta_{jl} S_{km} - \delta_{jm} S_{kl} + \delta_{km} S_{jl} - \delta_{kl} S_{jm}\,). \tag{2.10}$$

Any unitary irreducible representation of $E(D+1)$ with $D \geq 3$ is uniquely specified by an irreducible representation of S_{jk}.

It is emphasized that all possible irreducible representations of the fundamental algebra are completely exhausted by the above, and for each value of D we have an infinite number of inequivalent descriptions of quantum mechanics unlike the case of canonical commutation relation, which is known to provide us with a unique irreducible representation. Furthermore, the quantization scheme mentioned above automatically induces a gauge potential in itself. It is expressed as

$$A_\alpha(x) = \frac{1}{r^2} \sum_{\beta=1}^{D+1} f_{\alpha\beta}\, x_\beta \qquad (\,\alpha = 1,\, 2,\cdots, D+1\,), \tag{2.11}$$

whose properties have been largely studied[1]. As a result the followings have been shown: 1) For $D = 1$ the induced gauge potential is of the Aharonov-Bohm type which is produced by a linear magnetic flux $\Phi = -2\pi\alpha e$ confined in a infinitely thin and infinitely long solenoid set on the z-axis. This gauge potential is known to be topologically non-trivial. 2) For $D = 2$ the induced gauge potential becomes identical with that produced by a magnetic monopole with magnetic charge $g = S/e$ which is located at the origin of the coordinate in the 3-dimensional space. Dirac's magnetic charge quantization just corresponds to the discrete values of S. The gauge potential in this case is topologically non-trivial. 3) For $D = 4$ the gauge potential takes the same form as the BPST instanton. This is topologically non-trivial. 4) For $D = 2n$ $(\,n = 3,\, 4,\, \ldots)$ the induced gauge potentials are shown to be gauge equivalent to the generalized BPST configurations which are also known to be topologically non-trivial. 5) For $D = 2n+1$ the induced gauge potentials have no special name. They are all topologically trivial.

It is to be noted that most of physically interesting gauge potentials with non-trivial topology are observed in this quantization.

Furthermore, as a byproduct[1] of our above study we can completely determine possible irreducible representations of the canonical variables in the Dirac quantization scheme[6] applied to a particle, which is constrained to move on S^D under the Hamiltonian $H = \sum_{\alpha=1}^{D+1} x_\alpha^2/2 + V(x)$. However, we will not enter this problem here.

2.2 Case of $SU_L(n) \times SU_R(n)/SU_V(n)$

Let us denote the left and right chiral transformations belonging to $SU_L(n) \times SU_R(n)$ as

$$u_L = L \times \mathbf{1}_n \quad \text{and} \quad u_R = \mathbf{1}_n \times R^\dagger \tag{2.12}$$

with unitary unimodular $(n \times n)$-matrices L and R. They are assumed to obey

$$\begin{cases} u_{L'}\, u_L \equiv L'L \times \mathbf{1}_n = u_{L'L}, \\ u_{R'}\, u_R \equiv \mathbf{1}_n \times R^\dagger R'^\dagger = \mathbf{1}_n \times (R'R)^\dagger = u_{R'R}, \\ u_L\, u_R \;= u_R\, u_L \equiv L \times R^\dagger = u_{LR} \end{cases} \tag{2.13}$$

which provide us with coordinate transformations such that

$$\begin{cases} z' = u_L \cdot z \equiv Lz & \text{for left transformation,} \\ z' = u_R \cdot z \equiv zR^\dagger & \text{for right transformation.} \end{cases} \tag{2.14}$$

We may summarize these transformations as

$$z' = u_{LR} \cdot z = LzR^\dagger. \tag{2.15}$$

The so called vector transformation that belongs to the subgroup $SU_V(n)$ is defined with

$$u_V = V \times V^\dagger, \tag{2.16}$$

where V is a unitary unimodular $(n \times n)$-matrix. It leads us to the coordinate transformation

$$z' = VzV^\dagger, \tag{2.17}$$

under which the point $z_0 \equiv \mathbf{1}_n$ is left unchanged. The subgroup $SU_V(n)$ is the little group or isotropy group in this case.

To apply the induced representation technique we introduce the boost transformation α_z which is an element of $SU_L(n) \times SU_R(n)$ and connects the point z_0 to an arbitrary point z on the manifold in such a way that

$$\alpha_z \cdot z_0 = z, \tag{2.18}$$

from which one finds

$$\alpha_z = \mathbf{1}_n \times (z^{-1})^\dagger \tag{2.19}$$

as a possible form[b] of α_z. It is known that the Wigner rotation $q(z,\, u_{LR})$ that corresponds to the chiral transformation u_{LR} is given by

$$q(z,\, u_{LR}) = \alpha_z^{-1}\, u_{LR}\, \alpha_{u_{LR}^{-1}\cdot z}. \tag{2.20}$$

Then using $\alpha_z^{-1} = \mathbf{1}_n \times z^\dagger$, $u_{LR} = L \times R^\dagger$ and $\alpha^{-1}_{u_{LR}\cdot z} = \mathbf{1}_n \times \{(L^{-1}zR)^{-1}\}^\dagger$ one easily obtains

$$\begin{aligned} q(z,\, u_{LR}) &= L \times \{zR(L^{-1}zR)^{-1}\}^\dagger \\ &= L \times L^\dagger, \end{aligned} \tag{2.21}$$

which is of course an element of the little group $SU_V(n)$. Thus corresponding to the chiral transformation u_{LR} we may write the representation of the Wigner rotation $q(z,\, u_{LR})$ as $Q(L)$ which denotes an irreducible representation of $L \in SU_L(n)$. In this connection it is to be noted that $Q(L)$ is z-independent and its form is asymmetric in left and right, though we are dealing with u_{LR}. Evidently the last property comes from an asymmetric form of the boost transformation (2.20). In fact, if we start with the boost transformation $\alpha_z = z \times \mathbf{1}_n$ in place of (2.19) we then obtain from (2.20)

$$q(z, u_{LR}) = R \times R^\dagger \in U_V(n), \tag{2.22}$$

in which the left chiral transformation L is missing. Accordingly the the representation of the Wigner rotation in this case is given by an irreducible representation $Q(R)$ of $R \in SU_R(n)$. However, it is not difficult to show that $Q(R)$ and $Q(L)$ lead us to those irreducible representations of the fundamental algebra which are unitary-equivalent to each other, and hence we can employ $Q(L)$ without loss of generality.

[b]More generally we may use for instance $\alpha_z = zw \times w^\dagger$, where $w \equiv w(z)$ is a unitary and unimodular $(n \times n)$-matrix that is a single-valued function of z. However the final result derived using this α_z is shown to be unitary-equivalent to that obtained from (2.19) with $w = z^{-1}$. Hence in the following we will use (2.19) for simplicity.

Thus under the chiral transformation u_{LR} the wave function $\varphi(z,\, z^\dagger)$ undergoes

$$\varphi'(z,\, z^\dagger) = Q(L)\,\varphi(u_{LR}^{-1}\cdot z,\, (u_{LR}^{-1}\cdot z)^\dagger), \tag{2.23}$$

where, by definition, $u_{LR}^{-1}\cdot z = L^{-1}zR$. For an infinitesimal transformation $u_{LR} = (\mathbf{1}_n + i\epsilon_L)\times(\mathbf{1}_n - i\epsilon_R)$ the Wigner rotation $Q(L)$ can be written as

$$Q(L) = 1 + i\epsilon^{\mu}_{L,\nu} S^{\nu}_{\mu} \tag{2.24}$$

where S^{ν}_{μ}'s are generators of the irreducible representation of $SU(n)$ that satisfy

$$\begin{cases} [\, S^{\mu}_{\nu},\, S^{\lambda}_{\rho}\,] = S^{\lambda}_{\nu}\,\delta^{\mu}_{\rho} - S^{\mu}_{\rho}\,\delta^{\lambda}_{\nu}, \\ (S^{\mu}_{\nu})^\dagger = S^{\nu}_{\mu}, \qquad S^{\mu}_{\mu} = 0. \end{cases} \tag{2.25}$$

Similarly, we can write $\varphi(u_{LR}^{-1}\cdot z,\, (u_{LR}^{-1}\cdot z)^\dagger)$ as

$$\begin{aligned} &\varphi(u_{LR}^{-1}\cdot z,\, (u_{LR}^{-1}\cdot z)^\dagger) \\ &= \left\{1 + i\epsilon^{\mu}_{L,\nu}(-z^{\nu}{}_{\lambda}\partial^{\lambda}{}_{\mu} + z^{\dagger\lambda}{}_{\mu}\partial^{\dagger\nu}{}_{\lambda}) + i\epsilon^{\mu}_{R,\nu}(z^{\lambda}{}_{\mu}\partial^{\nu}{}_{\lambda} - z^{\dagger\nu}{}_{\lambda}\partial^{\dagger\lambda}{}_{\mu})\right\}\varphi(z,\, z^\dagger), \end{aligned} \tag{2.26}$$

with $\partial^{\alpha}_{\beta} \equiv \partial/\partial z^{\beta}{}_{\alpha}$ and $\partial^{\dagger\alpha}{}_{\beta} \equiv \partial/\partial z^{\dagger\beta}{}_{\alpha}$. Then combining (2.25) and (2.27) we obtain

$$\varphi'(z,\, z^\dagger) = (1 + i\epsilon^{\mu}_{L,\nu}G^{\nu}_{L,\mu} + i\epsilon^{\mu}_{R,\nu}G^{\nu}_{R,\mu})\,\varphi(z,\, z^\dagger) \tag{2.27}$$

with

$$\begin{cases} G^{\nu}_{L,\mu} = -z^{\nu}{}_{\lambda}\partial^{\lambda}{}_{\mu} + z^{\dagger\lambda}{}_{\mu}\partial^{\dagger\nu}{}_{\lambda} - \dfrac{\delta^{\nu}_{\mu}}{n}(-z^{\rho}{}_{\lambda}\partial^{\lambda}{}_{\rho} + z^{\dagger\lambda}{}_{\rho}\partial^{\dagger\rho}{}_{\lambda}) + S^{\nu}_{\mu}, \\ G^{\nu}_{R,\mu} = z^{\lambda}{}_{\mu}\partial^{\nu}{}_{\lambda} - z^{\dagger\nu}{}_{\lambda}\partial^{\dagger\lambda}{}_{\mu} - \dfrac{\delta^{\nu}_{\mu}}{n}(z^{\lambda}{}_{\rho}\partial^{\rho}{}_{\lambda} - z^{\dagger\rho}{}_{\lambda}\partial^{\dagger\lambda}{}_{\rho}), \end{cases} \tag{2.28}$$

where the terms proportional to δ^{ν}_{μ}/n have been added in the right hand sides to guarantee the condition $G^{\mu}_{L,\mu} = G^{\mu}_{R,\mu} = 0$ though they make no contribution to Eq.(2.27) because of $\mathrm{Tr}\epsilon_L = \mathrm{Tr}\epsilon_R = 0$. Then one can easily check that $G^{\nu}_{L,\mu}$ and $G^{\nu}_{R,\mu}$ given here satisfy the fundamental algebra (1.5) and (1.6). Furthermore, as done usually it can be shown that $G^{\nu}_{L,\mu}$ and $G^{\nu}_{R,\mu}$ of (2.28) form an irreducible representation of the fundamental algebra together with z if and only if S^{ν}_{μ} are irreducible generators of the little group $SU_V(n)$, and any of irreducible representations of the fundamental algebra can uniquely be specified, through (2.28), by irreducible S^{ν}_{μ}.

We will now try to extract a gauge potential induced at an arbitrary point, say $\tilde{z}$, on the chiral manifold. In this respect it is expected that the gauge potential, if it exists, will appear in a covariant derivative produced by an infinitesimal transformation on the coset space $SU_L(n)\times SU_R(n)/SU_V(n)$. Thus to derive out an explicit form of the covariant derivative we will decompose the chiral transformation $u_{LR}=(\mathbf{1}_n+i\epsilon_L)\times(\mathbf{1}_n-i\epsilon_R)$ into two parts $u_{\tilde z}^T$ and $u_{\tilde z}^H$, where the former plays a role of giving rise to a pure displacement at the point $\tilde z$ on the coset space while the latter keeps $\tilde z$ invariant[c]. The covariant derivative is thought to be found in the transformation $u_{\tilde z}^T$ when it is applied to the wave function $\varphi(\tilde z,\,\tilde z^\dagger)$. Thus we have to find the explicit forms of $u_{\tilde z}^T$ and $u_{\tilde z}^H$, which, in what follows, will be denoted as

$$u_{\tilde z}^T=(\mathbf{1}_n+i\tilde\epsilon_L^T)\times(\mathbf{1}_n-i\tilde\epsilon_R^T)\quad\text{and}\quad u_{\tilde z}^H=(\mathbf{1}_n+i\tilde\epsilon_L^H)\times(\mathbf{1}_n-i\tilde\epsilon_R^H),\tag{2.29}$$

where $\tilde\epsilon_L^T$, $\tilde\epsilon_R^T$, $\tilde\epsilon_L^H$ and $\tilde\epsilon_R^H$ are infinitesimal hermitian matrices satisfying the traceless condition. They will be determined in the following manner. From $u_{LR}=u_{\tilde z}^T u_{\tilde z}^H$ we obtain

$$\epsilon_L=\tilde\epsilon_L^T+\tilde\epsilon_L^H\quad\text{and}\quad\tilde\epsilon_R=\tilde\epsilon_R^T+\tilde\epsilon_R^H.\tag{2.30}$$

In the case for $\tilde z=z_0(\equiv\mathbf{1}_n)$ the decomposition is given by

$$u_{z_0}^T=(\mathbf{1}_n+i\epsilon_T)\times(\mathbf{1}_n+i\epsilon_T)\quad\text{and}\quad u_{z_0}^H=(\mathbf{1}_n+i\epsilon_H)\times(\mathbf{1}_n-i\epsilon_H)\tag{2.31}$$

with

$$\epsilon_T\equiv\frac{1}{2}(\epsilon_L-\epsilon_R)\quad\text{and}\quad\epsilon_H\equiv\frac{1}{2}(\epsilon_L+\epsilon_R).\tag{2.32}$$

In fact we can see $u_{z_0}^T\cdot z_0=z_0+2i\epsilon_T$ and $u_{z_0}^H\cdot z_0=z_0$. Since $u_{z_0}^T$ and $u_{z_0}^H$ are to be obtained, respectively, from $u_{\tilde z}^T$ and $u_{\tilde z}^H$ by substituting z_0 for $\tilde z$, with application of the inverse boost transformation to $u_{\tilde z}^T$ and $u_{\tilde z}^H$ they must be reduced to those of the same type as (2.31). Then from

$$\alpha_{\tilde z}^{-1}\,u_{\tilde z}^T\,\alpha_{\tilde z}=(\mathbf{1}_n+i\epsilon_L^T)\times(\mathbf{1}_n-i\,\tilde z\,\epsilon_R^T\,\tilde z^{-1})$$

we obtain

$$\epsilon_L^T=-\tilde z\,\epsilon_R^T\,\tilde z^{-1}\tag{2.33}$$

and similarly from

$$\alpha_{\tilde z}^{-1}\,u_{\tilde z}^H\,\alpha_{\tilde z}=(\mathbf{1}_n+i\epsilon_L^H)\times(\mathbf{1}_n-i\,\tilde z\,\epsilon_R^H\,\tilde z^{-1})$$

we have

$$\epsilon_L^H=\tilde z\,\epsilon_R^H\,\tilde z^{-1}.\tag{2.34}$$

[c]The technique that determines the induced gauge potential from the expression of covariant derivative was also adopted[7] in examining the system constrained on S^D.

Hence with the aid of (2.30) we find

$$\begin{cases} \epsilon_L^T = \dfrac{1}{2}(\epsilon_L - \tilde{z}\,\epsilon_R\,\tilde{z}^{-1}) \quad \text{and} \quad \epsilon_R^T = \dfrac{1}{2}(\epsilon_R - \tilde{z}^{-1}\,\epsilon_L\,\tilde{z}), \\ \epsilon_L^H = \dfrac{1}{2}(\epsilon_L + \tilde{z}\,\epsilon_R\,\tilde{z}^{-1}) \quad \text{and} \quad \epsilon_R^H = \dfrac{1}{2}(\epsilon_R + \tilde{z}^{-1}\,\epsilon_L\,\tilde{z}). \end{cases} \tag{2.35}$$

Thus we have completed the decomposition of the infinitesimal chiral transformation u_{LR} into $u_{\tilde{z}}^T$ and $u_{\tilde{z}}^H$. Since under $u_{\tilde{z}}^T$ the wave function $\varphi(z,\, z^\dagger)$ is transformed into $\{1 + i(\epsilon_L^T)^\mu{}_\nu\, G^\nu_{L,\mu} + i(\epsilon_R^T)^\mu{}_\nu\, G^\nu_{R,\mu}\}\,\varphi(z,\, z^\dagger)$, we find in this case

$$\begin{aligned} \varphi'(\tilde{z},\, \tilde{z}^\dagger) &= \{1 + i(\epsilon_L^T)^\mu_\nu\, G^\nu_{L,\mu} + i(\epsilon_R^T)^\mu_\nu\, G^\nu_{R,\mu}\}\,\varphi(z,\, z^\dagger)|_{z=\tilde{z}} \\ &= \{1 + i(\epsilon_L - \tilde{z}\,\epsilon_R\,\tilde{z}^{-1})^\mu{}_\nu\}\,(X^\nu_\mu + \frac{1}{2}S^\nu_\mu)\,\varphi(\tilde{z},\, \tilde{z}^\dagger) \end{aligned} \tag{2.36}$$

with

$$X^\nu{}_\mu \equiv -\tilde{z}^\nu{}_\lambda \tilde{\partial}^\lambda{}_\mu + \tilde{z}^{\dagger\lambda}{}_\mu \tilde{\partial}^{\dagger\nu}{}_\lambda \tag{2.37}$$

and similarly for $u_{\tilde{z}}^V$

$$\begin{aligned} \varphi'(\tilde{z},\, \tilde{z}^\dagger) &= \{1 + i(\epsilon_L^V)^\mu{}_\nu\, G^\nu_{L,\mu} + i(\epsilon_R^V)^\mu{}_\nu\, G^\nu_{R,\mu}\}\,\varphi(z,\, z^\dagger)|_{z=\tilde{z}} \\ &= \frac{1}{2}\{1 + i(\epsilon_L + \tilde{z}\,\epsilon_R\,\tilde{z}^{-1})\}^\mu{}_\nu\, S^\nu_\mu\,\varphi(\tilde{z},\, \tilde{z}^\dagger). \end{aligned} \tag{2.38}$$

In deriving the above we have used (2.35) together with the identities

$$\begin{cases} (\tilde{z}\,\epsilon_R\,\tilde{z}^{-1})^\mu{}_\nu\, X^\nu_\mu = -\epsilon^\mu_{R,\nu} Y^\nu_\mu \\ (\tilde{z}^{-1}\,\epsilon_R\,\tilde{z})^\mu{}_\nu\, Y^\nu_\mu = -\epsilon^\mu_{L,\nu} X^\nu_\mu, \end{cases}$$

where

$$Y^\nu_\mu \equiv \tilde{z}^\lambda{}_\mu \tilde{\partial}^\nu{}_\lambda - \tilde{z}^{\dagger\nu}{}_\lambda \tilde{\partial}^{\dagger\lambda}{}_\nu.$$

It is noted that no derivative term emerges for the transformation $u_{\tilde{z}}^H$. The pure displacement at $\tilde{z}$ on the coset space is realized merely through the transformation $u_{\tilde{z}}^T$.

In order to rewrite X^ν_μ in a form of covariant derivative we introduce $n(n-1)/2$ real coordinates in the neighborhood of $\tilde{z}$, which will be denoted by $n \times n$-matrix t $((t^\alpha_\beta)^* = t^\beta_\alpha)$ with $\mathrm{Tr}\, t = 0$. In terms of this t we write the wave function in the neighborhood of $\tilde{z}$ as

$$\varphi_{\tilde{z},\, \tilde{z}^\dagger}(t) \equiv \varphi(e^{-2it}\tilde{z},\, \tilde{z}^\dagger e^{2it}). \tag{2.39}$$

Then we have

$$\frac{\partial}{\partial t^{\mu}_{\nu}}\varphi_{\tilde{z},\tilde{z}^{\dagger}}(t)|_{t=0} = \frac{\partial}{\partial t^{\mu}_{\nu}}\varphi((\mathbf{1}_n - 2it)\tilde{z},\, \tilde{z}^{\dagger}(\mathbf{1}_n + 2it))|_{t=0}$$

$$= -2i\frac{\partial}{\partial t^{\mu}_{\nu}}\left\{(t\tilde{z})^{\rho}{}_{\lambda}\tilde{\partial}^{\lambda}{}_{\rho} - (\tilde{z}^{\dagger}t)^{\lambda}{}_{\rho}\tilde{\partial}^{\dagger\rho}{}_{\lambda}\right\}\varphi(\tilde{z},\, \tilde{z}^{\dagger})$$

$$= 2iX^{\nu}_{\mu}\,\varphi(\tilde{z},\, \tilde{z}^{\dagger}), \tag{2.40}$$

where we have treated t^{μ}_{ν} as if they were independent variables in spite of the existence of the condition $\mathrm{Tr}\, t = 0$. However, this will cause no problem finally, because the constraints are commutable with generators of the chiral transformation. As a result we can write the right hand side of (2.36) as

$$rhs.\text{ of }(2.36) = \left\{1 + \frac{1}{2}(\epsilon_L - \tilde{z}\,\epsilon_R\,\tilde{z}^{-1})^{\mu}{}_{\nu}\,(\frac{\partial}{\partial t^{\mu}_{\nu}} + iS^{\nu}_{\mu})\right\}\varphi_{\tilde{z},\tilde{z}^{\dagger}}(t) \tag{2.41}$$

Thus we find that the gauge potential can be represented by a position independent matrix such that

$$A^{\mu}_{\nu} = -S^{\mu}_{\nu} \qquad ((A^{\nu}_{\mu})^{\dagger} = A^{\mu}_{\nu},\ A^{\mu}_{\mu} = 0). \tag{2.42}$$

In this connection we note the following: For $n = 2$ the chiral manifold $SU_L(2) \times SU_R(2)/SU_V(2) \sim SO(4)/SO(3)$ is equivalent to S^3. Hence the gauge potential $A_{\alpha}(x)$ ($\alpha = 1, 2, 3, 4$) obtained already in quantizing a system on S^3 must be gauge equivalent to the above A^{ν}_{μ} with $n = 2$. In fact we can show it explicitly with somewhat technical and lengthy calculations[8].

2.3 Case of Grassmann Manifold $Gr(n+m,\, n;\, \boldsymbol{C})$

Now let us start with determining the Wigner rotation in this case. For purpose we introduce the following quantities:

$$\rho_1 = \sqrt{\mathbf{1}_n + \xi^{\dagger}\xi + \Delta(\xi^{\dagger}\xi)} - \sqrt{\mathbf{1}_n + \xi^{\dagger}\xi},$$

$$\rho_2 = \sqrt{\mathbf{1}_m + \xi\xi^{\dagger} + \Delta(\xi\xi^{\dagger})} - \sqrt{\mathbf{1}_m + \xi\xi^{\dagger}} \tag{2.43}$$

with

$$\Delta(\xi^{\dagger}\xi) \equiv \Delta\xi^{\dagger}\cdot\xi + \xi^{\dagger}\cdot\Delta\xi$$
$$= i\{(\nu - \xi^{\dagger}\lambda)(\mathbf{1}_n + \xi^{\dagger}\xi) - (\mathbf{1}_n + \xi^{\dagger}\xi)(\nu - \lambda^{\dagger}\xi)\},$$

$$\Delta(\xi\xi^{\dagger}) \equiv \Delta\xi\cdot\xi^{\dagger} + \xi\cdot\Delta\xi^{\dagger}$$
$$= i\{(\mu - \xi\lambda^{\dagger})(\mathbf{1}_m + \xi\xi^{\dagger}) - (\mathbf{1}_m + \xi\xi^{\dagger})(\mu - \lambda\xi^{\dagger})\}, \tag{2.44}$$

where use has been made of (1.20). Then applying the boost transformation α_ξ given by (1.15) one finds after some calculation

$$\alpha_\xi^{-1} g \alpha_{\xi_{g^{-1}}} = \mathbf{1}_{n+m} + i \begin{pmatrix} \overbrace{\gamma_1}^{n} & \overbrace{\mathbf{0}}^{m} \\ \mathbf{0} & \gamma_2 \end{pmatrix} \begin{matrix} \}\, n \\ \}\, m \end{matrix} \in U(n) \times U(m) \tag{2.45}$$

where $\xi_{g^{-1}}$ is that obtained by substituting g^{-1} for g in (1.19), and γ_1 and γ_2 are given respectively as

$$\gamma_1 = \frac{1}{i}\rho_1 \frac{1}{\sqrt{\mathbf{1}_n + \xi^\dagger \xi}} + \sqrt{\mathbf{1}_n + \xi^\dagger \xi}\,(\nu + \lambda^\dagger \xi) \frac{1}{\sqrt{\mathbf{1}_n + \xi^\dagger \xi}},$$

$$\gamma_2 = \frac{1}{i}\rho_2 \frac{1}{\sqrt{\mathbf{1}_m + \xi \xi^\dagger}} + \sqrt{\mathbf{1}_m + \xi \xi^\dagger}\,(\mu - \lambda \xi^\dagger) \frac{1}{\sqrt{\mathbf{1}_m + \xi \xi^\dagger}}. \tag{2.46}$$

The hermiticity of γ_1 and γ_2 is recognized by showing that $\gamma_1^\dagger$ and $\gamma_2^\dagger$ are reduced, respectively, to γ_1 and γ_2 when use is made of (2.43) and (2.44). Thus to explicitly write down the Wigner rotation $\alpha_\xi^{-1} g \alpha_{\xi_{g^{-1}}}$ we have to express the matrices ρ_1 and ρ_2 in terms of the infinitesimal matrices μ, ν, λ and $\lambda^\dagger$ up to their first order. For this we will use a technical lemma which will be stated as

Lemma:
For a positive hermitian matrix Λ there hold the following identities:

$$\sqrt{\mathbf{1} + \Lambda + \delta} - \sqrt{\mathbf{1} + \Lambda} = \frac{1}{2}\int_0^1 dt \int_0^1 dx \frac{(\mathbf{1}+\Lambda)^{\frac{1-t}{2}}}{\mathbf{1} + x\Lambda}\, \delta\, \frac{(\mathbf{1}+\Lambda)^{-\frac{t}{2}}}{\mathbf{1} + x\Lambda}, \tag{I}$$

with infinitesimal matrix δ, and

$$\frac{\mathbf{1}}{\sqrt{\mathbf{1}+\Lambda}}\, B\, \frac{\mathbf{1}}{\sqrt{\mathbf{1}+\Lambda}}$$

$$= \frac{1}{2}\int_0^1 dt \int_0^1 dx \left[\frac{(\mathbf{1}+\Lambda)^{-\frac{t}{2}}}{\mathbf{1}+x\Lambda}\, B\, \frac{(\mathbf{1}+\Lambda)^{\frac{t}{2}}}{\mathbf{1}+x\Lambda} + \frac{(\mathbf{1}+\Lambda)^{\frac{t}{2}}}{\mathbf{1}+x\Lambda}\, B\, \frac{(\mathbf{1}+\Lambda)^{-\frac{t}{2}}}{\mathbf{1}+x\Lambda} \right], \tag{II}$$

where B is an arbitrary matrix.

A proof of the lemma is not a difficult task. It will be left to readers' exercise. Any way, with help of this lemma after some lengthy arguments we can finally arrive at the following expressions for γ_1 and γ_2:

$$\gamma_1 = \nu + \frac{1}{2}\int_0^1 dt \int_0^1 dx \frac{(\mathbf{1}_n + \xi^\dagger\xi)^{\frac{t}{2}}}{\mathbf{1}_n + x\,\xi^\dagger\xi}\, \xi^\dagger \lambda \frac{(\mathbf{1}_n + \xi^\dagger\xi)^{1-\frac{t}{2}}}{\mathbf{1}_n + x\,\xi^\dagger\xi}$$

$$+\frac{1}{2}\int_0^1 dt \int_0^1 dx\, \frac{(\mathbf{1}_n+\xi^\dagger\xi)^{1-\frac{t}{2}}}{\mathbf{1}_n+x\,\xi^\dagger\xi}\,\lambda^\dagger\xi\,\frac{(\mathbf{1}_n+\xi^\dagger\xi)^{\frac{t}{2}}}{\mathbf{1}_n+x\,\xi^\dagger\xi}, \tag{2.47}$$

$$\gamma_2=\mu-\frac{1}{2}\int_0^1 dt \int_0^1 dx\, \frac{(\mathbf{1}_m+\xi\xi^\dagger)^{\frac{t}{2}}}{\mathbf{1}_m+x\,\xi\xi^\dagger}\,\xi\lambda^\dagger\,\frac{(\mathbf{1}_m+\xi\xi^\dagger)^{1-\frac{t}{2}}}{\mathbf{1}_m+x\,\xi\xi^\dagger}$$

$$-\frac{1}{2}\int_0^1 dt \int_0^1 dx\, \frac{(\mathbf{1}_m+\xi\xi^\dagger)^{1-\frac{t}{2}}}{\mathbf{1}_m+x\,\xi\xi^\dagger}\,\lambda\xi^\dagger\,\frac{(\mathbf{1}_m+\xi\xi^\dagger)^{\frac{t}{2}}}{\mathbf{1}_m+x\,\xi\xi^\dagger}. \tag{2.48}$$

Now in order to connect our argument to the irreducible representation of the Wigner rotation we rewrite the matrices γ_1 and γ_2 with the aid of ortho-complete matrices $\hat{n}_\rho^\lambda$ and $\hat{m}_q^p$, whose $\alpha\beta$- and jk-components are given by

$$(\hat{n}_\rho^\lambda)_\beta^\alpha \equiv \delta_\beta^\lambda\,\delta_\rho^\alpha \quad \text{and} \quad (\hat{m}_q^p)_k^j \equiv \delta_k^p\,\delta_q^j \tag{2.49}$$

$$(\,\alpha,\,\beta,\lambda,\,\rho=1,\,2,\,\ldots,\,n \quad \text{and} \quad j,\,k,\,p,\,q=1',\,2',\,\ldots,\,m'\,).$$

Then we have

$$\gamma_1=\mathrm{Tr}(\,\gamma_1\hat{n}_\rho^\lambda\,)\,\hat{n}_\lambda^\rho \quad \text{and} \quad \gamma_2=\mathrm{Tr}(\,\gamma_2\hat{m}_q^p\,)\,\hat{m}_p^q$$

so that the Wigner rotation $\alpha_\xi^{-1}g\alpha_{\xi_{g^{-1}}}$ is rewritten as

$$\alpha_\xi^{-1}g\alpha_{\xi_{g^{-1}}}=\mathbf{1}_{n+m}+i\{\,\mathrm{Tr}(\,\gamma_1\hat{n}_\rho^\lambda\,)\,\hat{n}'^\rho_\lambda+\mathrm{Tr}(\,\gamma_2\hat{m}_q^p\,)\,\hat{m}'^q_p\,\} \tag{2.50}$$

with

$$\hat{n}'^\rho_\lambda=\begin{pmatrix}\hat{n}_\lambda^\rho & 0\\ 0 & 0\end{pmatrix} \quad \text{and} \quad \hat{m}'^q_p=\begin{pmatrix}0 & 0\\ 0 & \hat{m}_p^q\end{pmatrix}.$$

It is obvious that $\hat{n}'^\rho_\lambda$ and $\hat{m}'^q_p$ form a representation of the generators of $U(n)\times U(m)$ and obey the relations

$$(\hat{n}'^\lambda_\rho)^\dagger=\hat{n}'^\rho_\lambda, \qquad (\hat{m}'^p_q)^\dagger=\hat{m}'^q_p, \qquad [\,\hat{n}'^\lambda_\rho,\,\hat{m}'^p_q\,]=0,$$

$$[\,\hat{n}'^\lambda_\rho,\,\hat{n}'^\tau_\sigma\,]=\delta_\sigma^\lambda\,\hat{n}'^\tau_\rho-\delta_\rho^\tau\,\hat{n}'^\lambda_\sigma, \qquad [\,\hat{m}'^p_q,\,\hat{m}'^r_s\,]=\delta_s^p\,\hat{m}'^r_q-\delta_q^r\,\hat{m}'^p_s. \tag{2.51}$$

Corresponding to $\hat{n}'^\lambda_\rho$ and $\hat{m}'^p_q$ we introduce irreducible generators of $U(n)\times U(m)$, which will be denoted as N_ρ^λ and M_q^p, respectively. They satisfy the same relations as (2.51);

$$(N_\rho^\lambda)^\dagger=N_\lambda^\rho, \qquad (M_q^p)^\dagger=M_p^q, \qquad [\,N_\rho^\lambda,\,M_q^p\,]=0,$$

$$[\,N_\rho^\lambda,\,N_\sigma^\tau\,]=\delta_\sigma^\lambda\,N_\rho^\tau-\delta_\rho^\tau\,N_\sigma^\lambda, \qquad [\,M_q^p,\,M_s^r\,]=\delta_s^p\,M_q^r-\delta_q^r\,M_s^p. \tag{2.52}$$

Thus making the replacements $\hat{n}'^{\lambda}_{\rho} \to N^{\lambda}_{\rho}$ and $\hat{m}'^{p}_{q} \to M^{p}_{q}$ in (2.50) we obtain an irreducible representation of the Wigner rotation which is now denoted as

$$Q(g,\xi,\xi^{\dagger}) = 1 + i\{\,\mathrm{Tr}(\gamma_1 \hat{n}^{\lambda}_{\rho})\,N^{\rho}_{\lambda} + \mathrm{Tr}(\gamma_2 \hat{m}^{p}_{q})\,M^{q}_{p}\,\} \tag{2.53}$$

for the infinitesimal transformation g of (1.19). The wave function $\varphi(\xi,\,\xi^{\dagger})$ then undergoes the transformation

$$\begin{aligned}\varphi'(\xi,\,\xi^{\dagger}) &= Q(g,\,\xi,\xi^{\dagger})\,\varphi'(\xi_{g^{-1}},\,\xi^{\dagger}_{g^{-1}})\\ &= [\,1 + \frac{1}{i}\{(\,\lambda + \mu\xi - \xi\nu - \xi\lambda^{\dagger}\xi\,)^{j}_{\alpha}(\partial_{\xi^{\dagger}})^{\alpha}_{j}\\ &\qquad -(\,\lambda^{\dagger} + \xi^{\dagger}\mu - \nu\xi^{\dagger} - \xi^{\dagger}\lambda\xi)^{\alpha}_{j}(\partial_{\xi}\,)^{j}_{\alpha}\}\\ &\qquad +i\{\,\mathrm{Tr}(\,\gamma_1 \hat{n}^{\sigma}_{\rho}\,)\,N^{\rho}_{\sigma} + \mathrm{Tr}(\,\gamma_2\,\hat{m}^{p}_{q}\,)\,M^{q}_{p}\,\}]\varphi(\xi,\,\xi^{\dagger}),\end{aligned}$$

where $(\partial_{\xi})^{\alpha}_{j} \equiv \partial/\partial\xi^{j}_{\alpha}$ and $(\partial_{\xi^{\dagger}})^{j}_{\alpha} \equiv \partial/\partial\xi^{\dagger\alpha}_{j}$. In deriving the above we have used (1.19) and (1.20) to evaluate $\xi_{g^{-1}}$ and $\xi^{\dagger}_{g^{-1}}$. Thus applying (2.47) and (2.48) we finally arrive at

$$\begin{aligned}&\varphi'(\xi,\,\xi^{\dagger}) - \varphi(\xi,\,\xi^{\dagger})\\ &\quad = i\,\{\,\mathrm{Tr}[\,(\,\lambda^{\dagger} + \xi^{\dagger}\mu - \nu\xi^{\dagger} - \xi^{\dagger}\lambda\xi^{\dagger}\,)\partial_{\xi^{\dagger}}\,]\\ &\qquad -\mathrm{Tr}[\,(\,\lambda + \mu\xi - \xi\nu - \xi\lambda^{\dagger}\lambda\,)\partial_{\xi}\,]\\ &\qquad +\mathrm{Tr}(\,\nu\hat{n}^{\sigma}_{\rho}\,)\,N^{\rho}_{\sigma} + \mathrm{Tr}(\,\mu\hat{m}^{p}_{q}\,)\,M^{q}_{p}\\ &\qquad +[\,\mathrm{Tr}(\,\hat{f}^{\sigma\dagger}_{\rho}\lambda) \;-\; \mathrm{Tr}(\hat{g}^{p\dagger}_{q}\lambda)\,M^{q}_{p} + h.c.\,\}\,\varphi(\xi,\,\xi^{\dagger})\end{aligned} \tag{2.54}$$

with

$$\hat{f}^{\sigma}_{\rho} = \frac{1}{2}\xi\int_0^1 dt\int_0^1 dx\,\frac{(\mathbf{1}_n + \xi^{\dagger}\xi)^{\frac{t}{2}}}{\mathbf{1}_n + x\,\xi^{\dagger}\xi}\,\hat{n}^{\sigma}_{\rho}\,\frac{(\mathbf{1}_n + \xi^{\dagger}\xi)^{1-\frac{t}{2}}}{\mathbf{1}_n + x\,\xi^{\dagger}\xi} \tag{2.55}$$

and

$$\hat{g}^{p}_{q} = \frac{1}{2}\int_0^1 dt\int_0^1 dx\,\frac{(\mathbf{1}_m + \xi\xi^{\dagger})^{1-\frac{t}{2}}}{\mathbf{1}_m + x\,\xi\xi^{\dagger}}\,\hat{m}^{p}_{q}\,\frac{(\mathbf{1}_m + \xi\xi^{\dagger})^{\frac{t}{2}}}{\mathbf{1}_m + x\,\xi\xi^{\dagger}}\xi. \tag{2.56}$$

Since on the other hand we can write $\varphi'(\xi,\,\xi^{\dagger})$ as $U(g)\varphi(\xi,\,\xi^{\dagger})$ with the aid of (1.22), comparing it with (2.54) we find that the irreducible representation of the fundamental algebra (1.23)$\sim$(1.25) is given by

$$\begin{aligned}G^{\alpha}_{\beta} &= \xi^{j}_{\beta}(\partial_{\xi})^{\alpha}_{j} - \xi^{\alpha\dagger}_{j}(\partial_{\xi^{\dagger}})^{j}_{\beta} + N^{\alpha}_{\beta},\\ G^{j}_{k} &= -\xi^{j}_{\beta}(\partial_{\xi})^{\beta}_{k} + \xi^{\beta\dagger}_{k}(\partial_{\xi^{\dagger}})^{j}_{\beta} + M^{j}_{k},\end{aligned}$$

$$G_j^\alpha = -(\partial_\xi)_j^\alpha - \xi_j^{\beta\dagger}\xi_k^{\alpha\dagger}(\partial_{\xi^\dagger})_\beta^k + (\hat{f}_\rho^{\sigma\dagger})_j^\alpha N_\sigma^\rho - (\hat{g}_q^{p\dagger})_j^\alpha M_p^q,$$

$$G_\alpha^j = \xi_\alpha^k \xi_\beta^j (\partial_\xi)_k^\beta + (\partial_{\xi^\dagger})_\alpha^j (\hat{f}_\rho^\sigma)_\alpha^j N_\sigma^\rho - (\hat{g}_q^p)_\alpha^j M_p^q \tag{2.57}$$

with irreducible generators N_β^α and M_k^j of $U(n) \times U(m)$. According to the theory of induced representation all possible irreducible representations of the fundamental algebra are exhausted by the above.

We will then try to find out the gauge potential inherent in the representation (2.57). As done previously in the case of chiral manifold we write the infinitesimal transformation g to be a product of two transformations like $g = g_{\tilde{\xi}}^T g_{\tilde{\xi}}^H$, where $g_{\tilde{\xi}}^T$ is defined as to cause a pure displacement at $\tilde{\xi}$ on the coset space while $g_{\tilde{\xi}}^H$ makes $\tilde{\xi}$ invariant. For $\tilde{\xi} = 0$ the decomposition is obvious. It is written as $g = g_0^T g_0^H$ with

$$g_0^T = \mathbf{1}_{n+m} + i\begin{pmatrix} \mathbf{0} & \lambda^\dagger \\ \lambda & \mathbf{0} \end{pmatrix}, \qquad g_0^H = \mathbf{1}_{n+m} + i\begin{pmatrix} \nu & \mathbf{0} \\ \mathbf{0} & \mu \end{pmatrix}. \tag{2.58}$$

For a general case we write $g_{\tilde{\xi}}^T$ and $g_{\tilde{\xi}}^H$ as

$$g_{\tilde{\xi}}^T = \mathbf{1}_{n+m} + i\begin{pmatrix} \nu_T & \lambda_T^\dagger \\ \lambda_T & \mu_T \end{pmatrix} \quad \text{and} \quad g_{\tilde{\xi}}^H = \mathbf{1}_{n+m} + i\begin{pmatrix} \nu_H & \lambda_H^\dagger \\ \lambda_H & \mu_H \end{pmatrix}, \tag{2.59}$$

respectively. Then by definition there must hold

$$\nu = \nu_T + \nu_H, \quad \mu = \nu_T + \mu_H, \quad \lambda = \lambda_T + \lambda_H, \quad \lambda^\dagger = \lambda_T^\dagger + \lambda_H^\dagger. \tag{2.60}$$

As done previously we apply here the boost transformation $\alpha_{\tilde{\xi}}$ to obtain

$$g_{\tilde{\xi}}^T = \mathbf{1}_{n+m} + i\alpha_{\tilde{\xi}}\begin{pmatrix} \mathbf{0} & \tilde{\lambda}^\dagger \\ \tilde{\lambda} & \mathbf{0} \end{pmatrix}\alpha_{\tilde{\xi}}^{-1} \text{ and } g_{\tilde{\xi}}^H = \mathbf{1}_{n+m} + i\alpha_{\tilde{\xi}}\begin{pmatrix} \tilde{\nu} & \mathbf{0} \\ \mathbf{0} & \tilde{\mu} \end{pmatrix}\alpha_{\tilde{\xi}}^{-1}, \tag{2.61}$$

because $\alpha_{\tilde{\xi}}^{-1} g_{\tilde{\xi}}^T \alpha_{\tilde{\xi}}$ and $\alpha_{\tilde{\xi}}^{-1} g_{\tilde{\xi}}^H \alpha_{\tilde{\xi}}$ should be of the same form as g_0^T and g_0^H, respectively. Then Eqs. (2.59) and (2.61) lead us to

$$\nu_T = \frac{1}{\sqrt{\mathbf{1}_n + \tilde{\xi}^\dagger\tilde{\xi}}}(-\tilde{\xi}^\dagger\tilde{\lambda} - \tilde{\lambda}^\dagger\tilde{\xi})\frac{1}{\sqrt{\mathbf{1}_n + \tilde{\xi}^\dagger\tilde{\xi}}},$$

$$\mu_T = \frac{1}{\sqrt{\mathbf{1}_m + \tilde{\xi}\tilde{\xi}^\dagger}}(\tilde{\lambda}\tilde{\xi}^\dagger + \tilde{\xi}\tilde{\lambda}^\dagger)\frac{1}{\sqrt{\mathbf{1}_m + \tilde{\xi}\tilde{\xi}^\dagger}},$$

$$\lambda_T = \frac{1}{\sqrt{\mathbf{1}_m + \tilde{\xi}\tilde{\xi}^\dagger}}(\tilde{\lambda} - \tilde{\xi}\tilde{\lambda}^\dagger\tilde{\xi})\frac{1}{\sqrt{\mathbf{1}_n + \tilde{\xi}^\dagger\tilde{\xi}}},$$

$$\lambda_T^\dagger = \frac{1}{\sqrt{\mathbf{1}_n + \tilde{\xi}^\dagger\tilde{\xi}}}(\tilde{\lambda}^\dagger - \tilde{\xi}^\dagger\tilde{\lambda}\tilde{\xi}^\dagger)\frac{1}{\sqrt{\mathbf{1}_m + \tilde{\xi}\tilde{\xi}^\dagger}} \tag{2.62}$$

and

$$\begin{aligned}
\nu_H &= \frac{1}{\sqrt{\mathbf{1}_n + \tilde{\xi}^\dagger\tilde{\xi}}}(\tilde{\nu} + \tilde{\xi}^\dagger\tilde{\mu}\tilde{\xi})\frac{1}{\sqrt{\mathbf{1}_n + \tilde{\xi}^\dagger\tilde{\xi}}}, \\
\mu_H &= \frac{1}{\sqrt{\mathbf{1}_m + \tilde{\xi}\tilde{\xi}^\dagger}}(\tilde{\mu} + \tilde{\xi}\tilde{\nu}\tilde{\xi}^\dagger)\frac{1}{\sqrt{\mathbf{1}_m + \tilde{\xi}\tilde{\xi}^\dagger}}, \\
\lambda_H &= \frac{1}{\sqrt{\mathbf{1}_m + \tilde{\xi}\tilde{\xi}^\dagger}}(\tilde{\xi}\tilde{\nu} - \tilde{\mu}\tilde{\xi})\frac{1}{\sqrt{\mathbf{1}_n + \tilde{\xi}^\dagger\tilde{\xi}}}, \\
\lambda_H^\dagger &= \frac{1}{\sqrt{\mathbf{1}_n + \tilde{\xi}^\dagger\tilde{\xi}}}(\tilde{\nu}\tilde{\xi}^\dagger - \tilde{\xi}^\dagger\tilde{\mu})\frac{1}{\sqrt{\mathbf{1}_m + \tilde{\xi}\tilde{\xi}^\dagger}}.
\end{aligned} \tag{2.63}$$

Inserting (2.62) and (2.63) into the right hand sides of (2.60) and solving the resulting equations with respect to $\tilde{\nu}$, $\tilde{\mu}$, $\tilde{\lambda}$ and $\tilde{\lambda}^\dagger$ we obtain

$$\begin{aligned}
\tilde{\nu} &= \frac{1}{\sqrt{\mathbf{1}_n + \tilde{\xi}^\dagger\tilde{\xi}}}(\nu + \tilde{\xi}^\dagger\lambda + \lambda^\dagger\tilde{\xi} + \tilde{\xi}^\dagger\mu\tilde{\xi})\frac{1}{\sqrt{\mathbf{1}_n + \tilde{\xi}^\dagger\tilde{\xi}}}, \\
\tilde{\mu} &= \frac{1}{\sqrt{\mathbf{1}_m + \tilde{\xi}\tilde{\xi}^\dagger}}(\mu - \lambda\tilde{\xi}^\dagger - \tilde{\xi}\lambda^\dagger + \tilde{\xi}\nu\tilde{\xi}^\dagger)\frac{1}{\sqrt{\mathbf{1}_m + \tilde{\xi}\tilde{\xi}^\dagger}}, \\
\tilde{\lambda} &= \frac{1}{\sqrt{\mathbf{1}_m + \tilde{\xi}\tilde{\xi}^\dagger}}(\lambda - \tilde{\xi}\nu + \mu\tilde{\xi} - \tilde{\xi}\lambda^\dagger\tilde{\xi})\frac{1}{\sqrt{\mathbf{1}_n + \tilde{\xi}^\dagger\tilde{\xi}}}, \\
\tilde{\lambda}^\dagger &= \frac{1}{\sqrt{\mathbf{1}_n + \tilde{\xi}^\dagger\tilde{\xi}}}(\lambda^\dagger - \nu\tilde{\xi}^\dagger + \tilde{\xi}^\dagger\mu - \tilde{\xi}^\dagger\lambda\tilde{\xi}^\dagger)\frac{1}{\sqrt{\mathbf{1}_m + \tilde{\xi}\tilde{\xi}^\dagger}}.
\end{aligned} \tag{2.64}$$

Accordingly if use is made of these $\tilde{\nu}$, $\tilde{\mu}$, $\tilde{\lambda}$ and $\tilde{\lambda}^\dagger$ in the right hand sides of (2.62) and (2.63) we can describe $g^T_{\tilde{\xi}}$ and $g^H_{\tilde{\xi}}$ in terms of ν, μ, λ, $\lambda^\dagger$, $\tilde{\xi}$ and $\tilde{\xi}^\dagger$. Thus we have accomplished the decomposition $g = g^T_{\tilde{\xi}} g^H_{\tilde{\xi}}$.

Since under the transformation $g^T_{\tilde{\xi}}$ the wave function behaves as

$$\varphi'(\xi, \xi^\dagger) = \left[1 + i\{(\nu_T)^\alpha_\beta G^\beta_\alpha + (\mu_T)^j_k G^k_j + (\lambda_T)^j_\alpha G^\alpha_j + (\lambda^\dagger_T)^\alpha_j G^j_\alpha\}\right]\varphi(\xi, \xi^\dagger),$$

it becomes at the point $\tilde{\xi}$

$$
\begin{aligned}
&\varphi'(\tilde{\xi},\tilde{\xi}^\dagger)-\varphi(\tilde{\xi},\tilde{\xi}^\dagger)\\
&=i\{(\nu_T)^\alpha_\beta G^\beta_\alpha+(\mu_T)^j_k G^k_j+(\lambda_T)^j_\alpha G^\alpha_j+(\lambda^\dagger_T)^\alpha_j G^j_\alpha\}\varphi(\xi,\xi^\dagger)\Big|_{\substack{\xi=\tilde{\xi},\\ \xi^\dagger=\tilde{\xi}^\dagger}}\\
&=i\Big\{\mathrm{Tr}[\sqrt{\mathbf{1}_n+\tilde{\xi}^\dagger\tilde{\xi}}\;\tilde{\lambda}^\dagger\sqrt{\mathbf{1}_m+\tilde{\xi}\tilde{\xi}^\dagger}\;\partial_{\tilde{\xi}^\dagger}-\sqrt{\mathbf{1}_m+\tilde{\xi}\tilde{\xi}^\dagger}\;\tilde{\lambda}\sqrt{\mathbf{1}_n+\tilde{\xi}^\dagger\tilde{\xi}}\;\partial_{\tilde{\xi}}]\\
&\quad+\mathrm{Tr}[\tilde{\lambda}^\dagger\frac{1}{\sqrt{\mathbf{1}_m+\tilde{\xi}\tilde{\xi}^\dagger}}(-\tilde{\xi}\mathcal{N}+\mathcal{M}\tilde{\xi}+\tilde{\mathcal{K}}-\tilde{\xi}\tilde{\mathcal{K}}^\dagger\tilde{\xi})\frac{1}{\sqrt{\mathbf{1}_n+\tilde{\xi}^\dagger\tilde{\xi}}}]\\
&\quad+\mathrm{Tr}[\frac{1}{\sqrt{\mathbf{1}_n+\tilde{\xi}^\dagger\tilde{\xi}}}(-\mathcal{N}\tilde{\xi}^\dagger+\tilde{\xi}^\dagger\mathcal{M}+\tilde{\mathcal{K}}^\dagger-\tilde{\xi}^\dagger\tilde{\mathcal{K}}\tilde{\xi}^\dagger)\frac{1}{\sqrt{\mathbf{1}_m+\tilde{\xi}\tilde{\xi}^\dagger}}\tilde{\lambda}]\Big\}\\
&\qquad\qquad\times\varphi(\tilde{\xi},\tilde{\xi}^\dagger),
\end{aligned}
\tag{2.65}
$$

where

$$
\begin{cases}
\mathcal{N}\equiv\hat{n}^\alpha_\beta N^\beta_\alpha, & \mathcal{M}\equiv\hat{m}^j_k M^k_j,\\
\tilde{\mathcal{K}}\equiv\mathcal{K}|_{\xi=\tilde{\xi},\,\xi^\dagger=\tilde{\xi}^\dagger}, & \tilde{\mathcal{K}}^\dagger\equiv\mathcal{K}^\dagger|_{\xi=\tilde{\xi},\,\xi^\dagger=\tilde{\xi}^\dagger},\\
\mathcal{K}\equiv\hat{f}^\alpha_\beta N^\beta_\alpha-\hat{g}^j_k M^k_j, & \mathcal{K}^\dagger\equiv\hat{f}^{\dagger\alpha}_{\ \beta}N^\beta_\alpha-\hat{g}^{\dagger j}_{\ k}M^k_j
\end{cases}
\tag{2.66}
$$

and

$$
\begin{cases}
\mathrm{Tr}[A\mathcal{N}A']\equiv\mathrm{Tr}(A\hat{n}^\alpha_\beta A')\,N^\beta_\alpha,\\
\mathrm{Tr}[B\mathcal{M}B']\equiv\mathrm{Tr}(B\hat{n}^j_k B')\,M^k_j,\\
\mathrm{Tr}[A\tilde{\mathcal{K}}B]\equiv\{\mathrm{Tr}(B\hat{f}^{\dagger\alpha}_{\ \beta}A)\,N^\beta_\alpha-\mathrm{Tr}(A\hat{g}^j_k B)\,M^k_j\}|_{\substack{\xi=\tilde{\xi},\\ \xi^\dagger=\tilde{\xi}^\dagger}},\\
\mathrm{Tr}[B\tilde{\mathcal{K}}^\dagger A]\equiv\{\mathrm{Tr}(A\hat{f}^{\dagger\alpha}_{\ \beta}B)\,N^\beta_\alpha-\mathrm{Tr}(B\hat{g}^{\dagger j}_{\ k}A)\,M^k_j\}|_{\substack{\xi=\tilde{\xi},\\ \xi^\dagger=\tilde{\xi}^\dagger}}.
\end{cases}
\tag{2.67}
$$

$(A,\,A';\ (n\times n)\text{-matrices},\quad B,\,B';\ (m\times m)\text{-matrices})$

In deriving (2.65) we have used (2.62) together with(2.57).

On the other hand, for the transformation $g^H_{\tilde{\xi}}$ we have

$$
\begin{aligned}
&\varphi'(\tilde{\xi},\tilde{\xi}^\dagger)-\varphi(\tilde{\xi},\tilde{\xi}^\dagger)\\
&=i\{(\nu_H)^\alpha_\beta G^\beta_\alpha+(\mu_H)^j_k G^k_j+(\lambda_H)^j_\alpha G^\alpha_j+(\lambda^\dagger_H)^\alpha_j G^j_\alpha\}\varphi(\xi,\xi^\dagger)\Big|_{\substack{\xi=\tilde{\xi},\\ \xi^\dagger=\tilde{\xi}^\dagger}}
\end{aligned}
$$

$$= i\Big\{\mathrm{Tr}[\,\tilde{\nu}\frac{1}{\sqrt{\mathbf{1}_n+\tilde{\xi}^\dagger\tilde{\xi}}}(\mathcal{N}+\tilde{\xi}^\dagger\mathcal{M}\tilde{\xi}+\tilde{\xi}^\dagger\mathcal{K}+\mathcal{K}^\dagger\tilde{\xi})\frac{1}{\sqrt{\mathbf{1}_n+\tilde{\xi}^\dagger\tilde{\xi}}}\,]$$

$$+\mathrm{Tr}[\,\tilde{\mu}\frac{1}{\sqrt{\mathbf{1}_m+\tilde{\xi}\tilde{\xi}^\dagger}}(\mathcal{M}+\tilde{\xi}\mathcal{N}\tilde{\xi}^\dagger-\mathcal{K}\tilde{\xi}^\dagger-\tilde{\xi}\mathcal{K}^\dagger)\frac{1}{\sqrt{\mathbf{1}_m+\tilde{\xi}\tilde{\xi}^\dagger}}\,]\varphi(\tilde{\xi},\tilde{\xi}^\dagger). \tag{2.68}$$

As has been expected the derivatives $\partial_{\tilde{\xi}}$ and $\partial_{\tilde{\xi}^\dagger}$ appear only in the $g^T_{\tilde{\xi}}$ transformation (2.65).

Then by defining the gauge potentials as

$$\begin{cases} \mathcal{A}(\tilde{\xi},\tilde{\xi}^\dagger) = \dfrac{i}{\mathbf{1}_m+\tilde{\xi}\tilde{\xi}^\dagger}(-\tilde{\xi}\mathcal{N}+\mathcal{M}\tilde{\xi}+\tilde{\mathcal{K}}-\tilde{\xi}\tilde{\mathcal{K}}^\dagger\tilde{\xi})\dfrac{1}{\mathbf{1}_m+\tilde{\xi}^\dagger\tilde{\xi}}, \\ \mathcal{A}^\dagger(\tilde{\xi},\tilde{\xi}^\dagger) = \dfrac{i}{\mathbf{1}_n+\tilde{\xi}^\dagger\tilde{\xi}}(-\mathcal{N}\tilde{\xi}^\dagger+\tilde{\xi}^\dagger\mathcal{M}+\tilde{\mathcal{K}}^\dagger-\tilde{\xi}^\dagger\tilde{\mathcal{K}}\tilde{\xi}^\dagger)\dfrac{1}{\mathbf{1}_n+\tilde{\xi}\tilde{\xi}^\dagger} \end{cases} \tag{2.69}$$

we can rewrite (2.65) into

$$\varphi'(\tilde{\xi},\tilde{\xi}^\dagger)-\varphi(\tilde{\xi},\tilde{\xi}^\dagger)$$

$$= i\Big\{\mathrm{Tr}[\sqrt{\mathbf{1}_n+\tilde{\xi}^\dagger\tilde{\xi}}\;\tilde{\lambda}^\dagger\sqrt{\mathbf{1}_m+\tilde{\xi}\tilde{\xi}^\dagger}\;(\partial_{\tilde{\xi}^\dagger}-i\mathcal{A}(\tilde{\xi},\tilde{\xi}^\dagger))$$

$$-\sqrt{\mathbf{1}_m+\tilde{\xi}\tilde{\xi}^\dagger}\;\tilde{\lambda}\sqrt{\mathbf{1}_n+\tilde{\xi}^\dagger\tilde{\xi}}\;(\partial_{\tilde{\xi}}-i\mathcal{A}^\dagger(\tilde{\xi},\tilde{\xi}^\dagger))\,]\Big\}\varphi(\tilde{\xi},\tilde{\xi}^\dagger), \tag{2.70}$$

which clearly shows that the gauge potentials appear in forms of covariant derivative. Thus written in terms of components at the point $(\xi,\,\xi^\dagger)$ they are expressed as

$$\begin{cases} A^j_\alpha(\xi,\xi^\dagger) \\ \quad = (\dfrac{i}{\mathbf{1}_m+\xi\xi^\dagger})^j_k\,\{-\xi^k_\rho N^\rho_\beta+M^k_l\xi^l_\beta+K^k_\beta-(\xi K^\dagger\xi)^k_\beta\}(\dfrac{1}{\mathbf{1}_n+\xi^\dagger\xi})^\beta_\alpha, \\ A^{\dagger\alpha}_j(\xi,\xi^\dagger) \\ \quad = (\dfrac{-i}{\mathbf{1}_n+\xi^\dagger\xi})^\alpha_\rho\,\{-N^\rho_\beta\xi^{\dagger\beta}_k+\xi^{\dagger\rho}_l M^l_k+K^{\dagger\rho}_k-(\xi^\dagger K\xi^\dagger)^\rho_k\}(\dfrac{1}{\mathbf{1}_m+\xi\xi^\dagger})^k_j, \end{cases} \tag{2.71}$$

where $A^{\dagger\alpha}_j(\xi,\,\xi^\dagger) = (A^j_\alpha(\xi,\,\xi^\dagger))^\dagger$, and by virtue of (2.55),(2.56) and (2.66) $K^j_\alpha\,(=(K^{\dagger\alpha}_j)^\dagger)$ is represented in terms of irreducible generators of the little group $U(n) \times U(m)$ as

$$K_\alpha^j = \frac{1}{2}\int_0^1 dt \int_0^1 dx \Big[\big(\xi \frac{(\mathbf{1}_n + \xi^\dagger\xi)^{\frac{t}{2}}}{\mathbf{1}_n + x\xi^\dagger\xi}\big)^j_{\ \beta} N^\beta_{\ \rho} \big(\frac{(\mathbf{1}_n + \xi^\dagger\xi)^{1-\frac{t}{2}}}{\mathbf{1}_n + x\xi^\dagger\xi}\big)^\rho_{\ \alpha}$$

$$- \big(\frac{(\mathbf{1}_m + \xi\xi^\dagger)^{1-\frac{t}{2}}}{\mathbf{1}_m + x\xi\xi^\dagger}\big)^j_{\ k} M^k_{\ l} \big(\frac{(\mathbf{1}_m + \xi\xi^\dagger)^{\frac{t}{2}}}{\mathbf{1}_m + x\xi\xi^\dagger}\xi\big)^l_{\ \alpha}\Big]. \quad (2.72)$$

Hence we have determined the induced gauge potential in the case of Grassmann manifold.

The integration in (2.72) can easily be performed for $n = 1$ and $\xi \in \boldsymbol{C}_m$. Writing

$$\xi = \begin{pmatrix} \xi^1 \\ \xi^2 \\ \vdots \\ \xi^m \end{pmatrix} \qquad \text{and} \qquad |\xi|^2 = \sum_{j=1}^m \xi^{j*}\xi^j$$

we then obtain

$$A^j = \frac{1}{2i}\Big(\frac{1}{\mathbf{1}_m + \xi\xi^\dagger}\Big)^j_{\ k} (S\xi^k - M^k_{\ l}\xi^l)$$

$$+ \frac{1}{2i}\Big(1 - |\xi|^2 + \frac{2}{1 + \sqrt{1 + |\xi|^2}}\Big)\Big(M^j_{\ k}\xi^k - \frac{(\xi^\dagger M \xi)}{|\xi|^2}\xi^j\Big), \quad (2.73)$$

where $S\,(= N)$ is a real number.

Detailed properties of the gauge potential will be discussed elsewhere.

3 Remarks

We have examined, in the present note, fundamental algebras defined on the respective manifolds S^D, $SU_L(n) \times SU_R(n)/SU_V(n)$ and $Gr(n+m, n; \boldsymbol{C})$ and obtained all possible irreducible representations for them together with explicit form of the gauge potentials induced in quantizing the system. Unlike the case of canonical algebra we have observed that there exists an infinite number of inequivalent irreducible representations of the fundamental algebra for a given system. In this connection it is noted that the gauge potential obtained therein satisfies[9] the Yang-Mills equation defined on the manifold.

It is obvious that our quantization scheme may be extended to a system constrained on a manifold of the type G/H, in which the induced representation technique is explicitly applicable to determine irreducible representation of the fundamental algebra. However a generalization to a system on a manifold that has no geometric symmetry seems a difficult task. In fact, in this case the expression of the displacement between two points on the manifold will generally depend on the pass connecting them. Thus in order to set up a fundamental algebra with avoiding this difficulty it would be necessary for us to introduce further assumptions based on some physical requirements, though unfortunately at present time nothing is known about them.

Moreover, a generalization of our quantization scheme to field theory seems an interesting problem. However this has not yet been succeeded in because of infinite degrees of freedom of quantized field.

Any way a further detailed investigation of these problems would highly be desired.

References

1. Y. Ohnuki and S. Kitakado, J. Math. Phys. **A9**, 2827 (1993); Mod. Phys. Lett. **A7**, 2447 (1992); K. Fujii, S. Kitakado and Y. Ohnuki, Mod. Phys. Lett. **A10** 867 (1995); and references cited therein.
2. K. Fujii, T. Kashiwa and S. Sakoda, J. Math. Phys. **37**, 567 (1966).
3. G. W. Mackey, *Induced Representations of Groups and Quantum Mechanics* (Benjamin, N. Y. 1968).
4. E. P. Wigner, Ann. Math. **40**, 149 (1939); Y. Ohnuki, *Unitary Representations of the Poincaré Group* (World Scientific, 1998).
5. N. P. Landsman and N. Linden, Nucl. Phys. **B365**, 121 (1991).
6. P. A. M. Dirac, Can. J. Math. **2**, 129 (1950); *Lectures on Quantum Mechanics* (Belfer Graduate School of Science, Yeshiva University, New York, 1964).
7. Y. Ohnuki, *Proc. of Vth Int. School on Theoret. Phys. – Symmetry and Structural Properties of Condensed Matter –*, (ed. T. Lulek et al, World Scientific) in press.
8. Y. Ohnuki, to be published.
9. D. MacMullan and I. Tsutsui, Phys. Lett. **B320**, 287 (1994); Ann. Phys. **34**, 269 (1995).

INVARIANT PROJECTIONS AND COVARIANT DIVERGENCY OF THE ENERGY-MOMENTUM TENSORS

S. MANOFF

Bulgarian Academy of Sciences
Institute for Nuclear Research and Nuclear Energy
Department of Theoretical Physics,
Blvd. Tzarigradsko Chaussee 72
1784 Sofia, Bulgaria
E-mail: smanov@inrne.bas.bg

R. LAZOV

Bulgarian Academy of Sciences
Institute for Mathematics and Informatics
Acad. G. Bontchev Street, Bl. 8
1113 Sofia, Bulgaria
E-mail: lazov@math.bas.bg

The invariant projections of the energy-momentum tensors of Lagrangian densities for tensor fields over differentiable manifolds with contravariant and covariant affine connections and metrics [$(\overline{L}_n, g)$-*spaces*] are found by the use of an non-null (non-isotropic) contravariant vector field and its corresponding projective metrics. The notions of rest mass density, momentum density, energy current density and stress tensor are introduced as generalizations of these notions from the relativistic continuum media mechanics. The energy-momentum tensors are represented by means of the introduced notions and the corresponding identities are found. The notion of covariant differential operator along a contravariant tensor field is introduced. On its basis, as a special case, the notion of contravariant metric differential operator is proposed. The properties of the operators are considered. By the use of these operators the notion of covariant divergency of a mixed tensor field is determined. The covariant divergency of tensor fields of second rank of the types 1 and 2 is found. Invariant representations of the covariant divergency of the energy-momentum tensor are obtained by means of the projective metrics of a contravariant non-isotropic (non-null) vector field and the corresponding rest mass density, momentum density, and energy flux density.

1 Introduction

In the relativistic continuum media mechanics notions are introduced as generalizations of the same notions of the classical continuum media mechanics. This has been done by means of the projections of the (canonical, symmetric of Belinfante or symmetric of Hilbert) energy-momentum tensors along or orthogonal to a non-isotropic (non-null) contravariant vector field. In an analogous way as in (pseudo) Riemannian spaces without torsion (V_n-spaces), the different relations between the quantities with well known physical interpretations can be considered for physical systems described by means of mathematical models over differentiable manifold with affine connections and metrics [$(\overline{L}_n, g)$-spaces].

2 Invariant projections of a mixed tensor field of second rank

The energy-momentum tensors are obtained as mixed tensor fields of second rank of type 1 [by the use of the procedure on the grounds of the method of Lagrangians with covariant derivatives (MLCD) [1]]

$$G = G_\alpha{}^\beta \cdot e_\beta \otimes e^\alpha = G_i{}^j \cdot \partial_j \otimes dx^i \tag{1}$$

in contrast to the mixed tensor fields of second rank of type 2

$$\overline{G} = \overline{G}^\beta{}_\alpha \cdot e^\alpha \otimes e_\beta = \overline{G}^j{}_i \cdot dx^i \otimes \partial_j . \tag{2}$$

$\{e^\alpha\}$ and $\{e_\alpha\}$ are non-co-ordinate (non-holonomic) covariant and contravariant basic vector fields respectively, (α, $\beta = 1, \ldots, n$), $\{dx^i\}$ and $\{\partial_i\}$ are co-ordinate (holonomic) covariant and contravariant basic vector fields respectively (i, $j = 1, \ldots, n$), $\dim(\overline{L}_n, g) = n$.

The Kronecker tensor field Kr appears as a mixed tensor field of second rank of the type 1,

$$Kr = g^\alpha_\beta \cdot e_\alpha \otimes e^\beta = g^i_j \cdot \partial_i \otimes dx^j ,$$

and can be projected by means of the non-isotropic (non-null) contravariant vector field u and its projection metrics h_u and h^u [$h_u = g - \frac{1}{e} \cdot g(u) \otimes g(u)$, $h^u = \overline{g} - \frac{1}{e} \cdot u \otimes u$, $e = g(u, u) \neq 0$]

$$Kr = \varepsilon_{Kr} \cdot u \otimes g(u) + u \otimes g({}^{Kr}\pi) + {}^{Kr}s \otimes g(u) + ({}^{Kr}S)g,$$

where

$$\begin{aligned} \varepsilon_{Kr} &= \frac{1}{e^2} \cdot [g(u)](Kr)u = \frac{1}{e^2} . u_{\overline{\alpha}} \cdot u^{\overline{\alpha}} \\ &= \frac{1}{e^2} . g_{\overline{\alpha}\overline{\beta}} \cdot u^{\overline{\alpha}} \cdot u^\beta = \frac{1}{e^2} . f^\rho{}_\beta \cdot f^\sigma{}_\gamma \cdot f^\beta{}_\delta \cdot g_{\rho\sigma} \cdot u^\gamma \cdot u^\delta \\ &= \frac{1}{e^2} . u^{\overline{i}} \cdot u_{\overline{i}} = \frac{1}{e^2} \cdot g_{\overline{i}\overline{k}} \cdot u^{\overline{i}} \cdot u^k = \frac{1}{e} \cdot k, \end{aligned} \tag{3}$$

$$k = \frac{1}{e} \cdot [g(u)](Kr)u = \frac{1}{e} . u^{\overline{i}} \cdot u_{\overline{i}} \, , \quad u_{\overline{i}} = g_{\overline{i}\overline{j}} \cdot u^j , \tag{4}$$

$$
\begin{aligned}
{}^{Kr}\pi &= \frac{1}{e} \cdot [g(u)](Kr)h^u = \frac{1}{e} \cdot g_{\overline{\beta\gamma}} \cdot u^\gamma \cdot g^\beta_\delta \cdot h^{\overline{\delta}\alpha} \cdot e_\alpha \\
&= \frac{1}{e} \cdot g_{\overline{\beta\gamma}} \cdot h^{\overline{\beta}\alpha} \cdot u^\gamma \cdot e_\alpha = \frac{1}{e} \cdot u_{\overline{\beta}} \cdot h^{\overline{\beta}\alpha} \cdot e_\alpha = {}^{Kr}\pi^\alpha \cdot e_\alpha ,
\end{aligned} \tag{5}
$$

$$
\begin{aligned}
{}^{Kr}\pi^\alpha &= \frac{1}{e} \cdot g_{\overline{\beta\gamma}} \cdot u^\gamma \cdot h^{\overline{\beta}\alpha} = \frac{1}{e} \cdot u_{\overline{\beta}} \cdot h^{\overline{\beta}\alpha} , \\
{}^{Kr}\pi^i &= \frac{1}{e} \cdot g_{\overline{kl}} \cdot u^l \cdot h^{\overline{k}i} = \frac{1}{e} \cdot u_{\overline{k}} \cdot h^{\overline{k}i} ,
\end{aligned} \tag{6}
$$

$$
{}^{Kr}s = \frac{1}{e} \cdot h^u(g)(Kr)(u) = {}^{Kr}s^\alpha \cdot e_\alpha = {}^{Kr}s^i \cdot \partial_i , \tag{7}
$$

$$
{}^{Kr}s^\alpha = \frac{1}{e} \cdot h^{\alpha\beta} \cdot g_{\overline{\beta\delta}} \cdot u^{\overline{\delta}} , \quad {}^{Kr}s^i = \frac{1}{e} \cdot h^{ik} \cdot g_{\overline{kl}} \cdot u^{\overline{l}} , \tag{8}
$$

$$
{}^{Kr}S = h^u(g)(Kr)h^u = {}^{Kr}S^{\alpha\beta} \cdot e_\alpha \otimes e_\beta = {}^{Kr}S^{ij} \cdot \partial_i \otimes \partial_j , \tag{9}
$$

$$
{}^{Kr}S^{\alpha\beta} = h^{\alpha\gamma} \cdot g_{\overline{\gamma\delta}} \cdot h^{\overline{\delta}\beta} , {}^{Kr}S^{ij} = h^{ik} \cdot g_{\overline{kl}} \cdot h^{\overline{l}j} , \tag{10}
$$

$$
h^u[g(u)] = [g(u)](h^u) = 0, \quad h^u(g)h^u = h^u ,
$$

$$
\begin{aligned}
(Kr)\overline{g} &= \frac{k}{e} \cdot u \otimes u + u \otimes {}^{Kr}\pi + {}^{Kr}s \otimes u + {}^{Kr}S , \\
(Kr)\overline{g} &= g^{\overline{\alpha}\beta} \cdot e_\alpha \otimes e_\beta = g^{\overline{i}j} \cdot \partial_i \otimes \partial_j .
\end{aligned} \tag{11}
$$

Remark. The corresponding to the Kronecker tensor field mixed tensor field of the type 2, $\overline{Kr} = g(Kr)\overline{g} = g_{i\overline{k}} \cdot g^{\overline{k}j} \cdot dx^i \otimes \partial_j$, does not appear in general as a Kronecker tensor field.

Special case: $S = C : f^i{}_j = g^i_j$, $(f^\alpha{}_\beta = g^\alpha_\beta)$:

$$
\begin{aligned}
Kr &= \frac{1}{e} \cdot u \otimes g(u) + (h^u)g = \overline{g}(g), \quad (Kr)\overline{g} = \overline{g}, \\
k = 1, \quad \varepsilon_{Kr} &= \frac{1}{e}, \quad {}^{Kr}\pi = 0, \quad {}^{Kr}s = 0, \quad {}^{Kr}S = h^u .
\end{aligned} \tag{12}
$$

3 Energy-momentum tensors and the rest mass density

The covariant Noether identities (generalized covariant Bianchi identities) can be considered as identities for the components of mixed tensor fields of second rank of type 1. The second covariant Noether identity,

$$\overline{\theta}_\alpha{}^\beta - {}_sT_\alpha{}^\beta \equiv \overline{Q}_\alpha{}^\beta ,$$

can be written in the form $\theta - {}_sT \equiv Q$, where

$$\begin{aligned} \theta &= \overline{\theta}_\alpha{}^\beta \cdot e_\beta \otimes e^\alpha = \overline{\theta}_i{}^j \cdot \partial_j \otimes dx^i , \\ {}_sT &= {}_sT_\alpha{}^\beta \cdot e_\beta \otimes e^\alpha = {}_sT_i{}^j \cdot \partial_j \otimes dx^i , \\ Q &= \overline{Q}_\alpha{}^\beta \cdot e_\beta \otimes e^\alpha = \overline{Q}_i{}^j \cdot \partial_j \otimes dx^i . \end{aligned} \tag{13}$$

θ is the generalized canonical energy-momentum tensor (GC-EMT) of the type 1; ${}_sT$ is the symmetric energy-momentum tensor of Belinfante (S-EMT-B) of the type 1; Q is the variational energy-momentum tensor of Euler-Lagrange (V-EMT-EL) of the type 1.

The second covariant Noether identity for the energy-momentum tensors of the type 1 is called *second covariant Noether identity of type 1.*

By means of the non-isotropic contravariant vector field u and its corresponding projective metric the energy-momentum tensors can be represented in an analogous way as the mixed tensor fields of the type 1.

The structure of the generalized canonical energy-momentum tensor and the symmetric energy-momentum tensor of Belinfante for the metric and non-metric tensor fields has similar elements and they can be written in the form

$$G = {}_kG - L \cdot Kr : \quad \theta = {}_k\theta - L \cdot Kr, \quad {}_sT = T - L \cdot Kr, \tag{14}$$

where

$${}_k\theta = {}_k\overline{\theta}_\alpha{}^\beta \cdot e_\beta \otimes e^\alpha = {}_k\overline{\theta}_i{}^j \cdot \partial_j \otimes dx^i . \tag{15}$$

On the analogy of the notions of the continuum media mechanics ${}_kG$ is called *viscosity tensor field.*

G and ${}_kG$ can be written by means of u, h^u and h_u in the form

$$\begin{aligned} G &= \varepsilon_G \cdot u \otimes g(u) + u \otimes g({}^G\pi) + {}^Gs \otimes g(u) + ({}^GS)g, \\ {}_kG &= \varepsilon_k \cdot u \otimes g(u) + u \otimes g({}^k\pi) + {}^ks \otimes g(u) + ({}^kS)g. \end{aligned} \tag{16}$$

From the relation (14) the relations between the different projections of G and ${}_kG$ follow. If we introduce the abbreviation $\varepsilon_G = \rho_G$, then

$$\begin{aligned} \varepsilon_k = \rho_G + \frac{1}{e} \cdot L \cdot k, \quad {}^k\pi = {}^G\pi + L \cdot{}^{Kr} \pi, \quad {}^ks = {}^Gs + L \cdot{}^{Kr} s, \\ {}^kS = {}^GS + L \cdot{}^{Kr} S, \end{aligned} \tag{17}$$

and G can be written by means of (14) in the form

$$G = (\rho_G + \frac{1}{e} \cdot L \cdot k) \cdot u \otimes g(u) - L \cdot Kr + u \otimes g({}^k\pi) + {}^k s \otimes g(u) + ({}^k S)g, \tag{18}$$

where

$$\rho_G = \frac{1}{e^2} \cdot [g(u)](G)(u)$$

is the *rest mass density* of the energy-momentum tensor G of the type 1. This type of representation of a given energy-momentum tensor G by means of the projective metrics of u and ρ_G is called *representation of G by means of the projective metrics of the contravariant non-isotropic (non-null) vector field u and the rest mass density ρ_G.*

There are other possibilities for representation of G by means of u and its corresponding projective metrics.

If we introduce the abbreviation

$$p_G = \rho_G \cdot u + {}^G\pi, \tag{19}$$

where p_G is the *momentum density* of the energy-momentum tensor G of the type 1, then G can be written in the form

$$\begin{aligned} G &= u \otimes g(\rho_G \cdot u + {}^G\pi) + {}^G s \otimes g(u) + ({}^G S)g, \\ G &= u \otimes g(p_G) + {}^G s \otimes g(u) + ({}^G S)g. \end{aligned} \tag{20}$$

The representation of G by means of the last relation is called *representation of G by means of the projective metrics of the contravariant non-isotropic contravariant vector field u and the momentum density p_G.*

By the use of the relations

$$g({}^G\pi, u) = 0, \quad ({}^G S)[g(u)] = 0, \tag{21}$$

valid for every energy-momentum tensor $G \sim (\theta, {}_sT, Q)$, and the definitions

$$e_G = G(u) = (G)(u) = e \cdot (\rho_G \cdot u + {}^G s), \quad g(u,u) = e \neq 0, \tag{22}$$

where e_G is the *energy flux density* of the energy-momentum tensor G of the type 1, the tensor field G can be written in the form

$$\begin{aligned} G &= (\rho_G \cdot u + {}^G s) \otimes g(u) + u \otimes g({}^G\pi) + ({}^G S)g, \\ G &= \frac{1}{e} \cdot e_G \otimes g(u) + u \otimes g({}^G\pi) + ({}^G S)g. \end{aligned} \tag{23}$$

The representation of G by means of the last expression is called *representation of G by means of the projective metrics of the contravariant non-isotropic vector field u and the energy flux density e_G.*

Example. The generalized canonical energy-momentum tensor θ can be represented, in accordance with the above described procedure, by the use of the projective metrics of u and the rest mass density ρ_θ

$$\theta = {}_k\theta - L \cdot Kr, \quad {}_k\theta = \theta + L \cdot Kr, \tag{24}$$

$$\theta = (\rho_\theta + \frac{1}{e} \cdot L \cdot k) \cdot u \otimes g(u) - L \cdot Kr + u \otimes g({}^\theta\overline{\pi}) + {}^\theta\overline{s} \otimes g(u) + ({}^\theta\overline{S})g, \tag{25}$$

where

$$\rho_\theta = \frac{1}{e^2} \cdot [g(u)](\theta)(u), \quad k = \frac{1}{e} \cdot [g(u)](Kr)(u), \tag{26}$$

$$\rho_\theta = \frac{1}{e^2} \cdot g_{\overline{i}\overline{j}} \cdot u^j \cdot \overline{\theta}_k{}^i \cdot u^{\overline{k}} = \frac{1}{e^2} \cdot \overline{\theta}_k{}^i \cdot u_{\overline{i}} \cdot u^{\overline{k}}, \tag{27}$$

$$\varepsilon_{k\theta} = \rho_\theta + \frac{1}{e} \cdot L \cdot k, \tag{28}$$

$${}^\theta\overline{\pi} = \frac{1}{e} \cdot [g(u)]({}_k\theta)h^u = {}^\theta\overline{\pi}^\alpha \cdot e_\alpha = {}^\theta\overline{\pi}^i \cdot \partial_i, \tag{29}$$

$${}^\theta\overline{\pi}^\alpha = \frac{1}{e} \cdot {}_k\overline{\theta}_\gamma{}^\beta \cdot u_{\overline{\beta}} \cdot h^{\overline{\gamma}\alpha}, \quad {}^\theta\overline{\pi}^i = \frac{1}{e} \cdot {}_k\overline{\theta}_l{}^k \cdot u_{\overline{k}} \cdot h^{\overline{l}i}, \tag{30}$$

$${}^\theta\overline{s} = \frac{1}{e} \cdot h^u(g)({}_k\theta)(u) = {}^\theta\overline{s}^\alpha \cdot e_\alpha = {}^\theta\overline{s}^i \cdot \partial_i, \tag{31}$$

$${}^\theta\overline{s}^\alpha = \frac{1}{e} \cdot h^{\alpha\beta} \cdot g_{\overline{\beta}\overline{\gamma}} \cdot {}_k\overline{\theta}_\delta{}^\gamma \cdot u^{\overline{\delta}}, \quad {}^\theta\overline{s}^i = \frac{1}{e} \cdot h^{ij} \cdot g_{\overline{j}\overline{k}} \cdot {}_k\overline{\theta}_l{}^k \cdot u^{\overline{l}}, \tag{32}$$

$${}^\theta\overline{S} = h^u(g)({}_k\theta)h^u = {}^\theta\overline{S}^{\alpha\beta} \cdot e_\alpha \otimes e_\beta = {}^\theta\overline{S}^{ij} \cdot \partial_i \otimes \partial_j, \tag{33}$$

$${}^\theta\overline{S}^{\alpha\beta} = h^{\alpha\gamma} \cdot g_{\overline{\gamma}\overline{\delta}} \cdot {}_k\overline{\theta}_\kappa{}^\delta \cdot h^{\overline{\kappa}\beta}, \quad {}^\theta\overline{S}^{ij} = h^{ik} \cdot g_{\overline{k}\overline{l}} \cdot {}_k\overline{\theta}_m{}^l \cdot h^{\overline{m}j}. \tag{34}$$

In a co-ordinate basis, θ can be represented in the form

$$\overline{\theta}_i{}^j = (\rho_\theta + \frac{1}{e} \cdot L \cdot k) \cdot u_i \cdot u^j - L \cdot g_i^j + {}^\theta\overline{\pi}_i \cdot u^j + u_i \cdot {}^\theta\overline{s}^j + g_{i\overline{k}} \cdot {}^\theta\overline{S}^{jk}, \tag{35}$$

where

$$u_i = g_{i\overline{j}} \cdot u^j, \quad {}^\theta\overline{\pi}_i = g_{i\overline{k}} \cdot {}^\theta\overline{\pi}^k, \quad {}^\theta\overline{s}_i = g_{i\overline{l}} \cdot {}^\theta\overline{s}^l, \quad {}^\theta\overline{S}_{ij} = g_{i\overline{k}} \cdot {}^\theta\overline{S}^{kl} \cdot g_{\overline{l}j}. \tag{36}$$

The variational energy-momentum tensor of Euler-Lagrange Q can be represented in the standard manner by the use of the projective metrics h^u, h_u and the rest mass density ρ_Q in the form

$$Q = -\rho_Q \cdot u \otimes g(u) - u \otimes g({}^Q\pi) - {}^Q s \otimes g(u) - ({}^Q S)g, \tag{37}$$

where

$$\rho_Q = -\frac{1}{e^2} \cdot [g(u)](Q)(u), \tag{38}$$

$$\rho_Q = -\frac{1}{e^2} \cdot g_{\overline{i}\overline{j}} \cdot u^j \cdot \overline{Q}_k{}^i \cdot u^{\overline{k}} = -\frac{1}{e^2} \cdot \overline{Q}_k{}^i \cdot u_{\overline{i}} \cdot u^{\overline{k}}, \tag{39}$$

$${}^Q\pi = -\frac{1}{e} \cdot [g(u)](Q)h^u = {}^Q\pi^\alpha \cdot e_\alpha = {}^Q\pi^i \cdot \partial_i, \tag{40}$$

$${}^Q\pi^\alpha = -\frac{1}{e} \cdot \overline{Q}_\gamma{}^\beta \cdot u_{\overline{\beta}} \cdot h^{\overline{\gamma}\alpha}, \quad {}^Q\pi^i = -\frac{1}{e} \cdot \overline{Q}_l{}^k \cdot u_{\overline{k}} \cdot h^{\overline{l}i}, \tag{41}$$

$${}^Q s = -\frac{1}{e} \cdot h^u(g)(Q)(u) = {}^Q s^\alpha \cdot e_\alpha = {}^Q s^i \cdot \partial_i, \tag{42}$$

$${}^Q s^\alpha = -\frac{1}{e} \cdot h^{\alpha\beta} \cdot g_{\overline{\beta}\overline{\gamma}} \cdot \overline{Q}_\delta{}^\gamma \cdot u^{\overline{\delta}}, \quad {}^Q s^i = -\frac{1}{e} \cdot h^{ij} \cdot g_{\overline{j}\overline{k}} \cdot \overline{Q}_l{}^k \cdot u^{\overline{l}}, \tag{43}$$

$${}^Q S = -h^u(g)(Q)h^u = {}^Q S^{\alpha\beta} \cdot e_\alpha \otimes e_\beta = {}^Q S^{ij} \cdot \partial_i \otimes \partial_j, \tag{44}$$

$${}^Q S^{\alpha\beta} = -h^{\alpha\gamma} \cdot g_{\overline{\gamma}\overline{\delta}} \cdot \overline{Q}_\kappa{}^\delta \cdot h^{\overline{\kappa}\beta}, \quad {}^Q S^{ij} = -h^{ik} \cdot g_{\overline{k}\overline{l}} \cdot \overline{Q}_m{}^l \cdot h^{\overline{m}j}. \tag{45}$$

In a co-ordinate basis, Q can be represented in the form

$$\overline{Q}_i{}^j = -\rho_Q \cdot u_i \cdot u^j - {}^Q\pi_i \cdot u^j - u_i \cdot {}^Q s^j - g_{i\overline{k}} \cdot {}^Q S^{jk}. \tag{46}$$

The introduced abbreviations for the different projections of the energy-momentum tensors have their analogous forms in V_3- and V_4-spaces, where their physical interpretations have been proposed [3], [5] (S.383-385). The stress tensor in V_3-spaces has been generalized to the energy-momentum tensor ${}_sT$ in V_4-spaces. The viscosity stress tensor ${}_{sk}T$ appears as the tensor T in the structure of the symmetric energy-momentum tensor of Belinfante ${}_sT$.

On the analogy of the physical interpretation of the different projections, the following definitions can be proposed for the quantities in the representations of the different energy-momentum tensors $G \sim (\theta, {}_sT, Q)$:

Energy-momentum tensor of the type 1: $G \sim (\theta, {}_sT, Q)$

(a) Viscous energy-momentum tensor of the type 1. .... $_kG$
(b) Rest mass density of the energy-momentum tensor G. ...... ρ_G
(c) Conductive momentum density
of the energy-momentum tensor G. .... ${}^G\pi$
(d) Conductive energy flux density
of the energy-momentum tensor G. .... $e \cdot {}^Gs$
(e) Stress tensor of the energy-momentum tensor G. .... ${}^GS$

The projections of the energy-momentum tensors have properties which are due to their construction, the orthogonality of the projective metrics h_u and h^u correspondingly to the vector fields u and $g(u)$ $[h_u(u) = 0,\ h^u[g(u)] = 0]$

$$g(u, {}^\theta\overline{\pi}) = g({}^\theta\overline{\pi}, u) = 0, \quad g(u, {}^T\overline{\pi}) = 0, \quad g(u, {}^Q\pi) = 0, \tag{47}$$

$$g(u, {}^\theta\overline{s}) = g({}^\theta\overline{s}, u) = 0, \quad g(u, {}^T\overline{s}) = 0, \quad g(u, {}^Q s) = 0, \tag{48}$$

$$\begin{gathered} g(u)({}^\theta\overline{S}) = 0,\ ({}^\theta\overline{S})g(u) = 0,\ g(u)({}^T\overline{S}) = 0,\ ({}^T\overline{S})g(u) = 0, \\ g(u)({}^Q S) = 0, \quad ({}^Q S)g(u) = 0. \end{gathered} \tag{49}$$

From the properties of the different projections, it follows that the conductive momentum density π (or $\overline{\pi}$) is a contravariant vector field orthogonal to the vector field u. The conductive energy flux density $e \cdot s$ (or $e \cdot \overline{s}$) is also a contravariant vector field orthogonal to u. The stress tensor S (or $\overline{S}$) is orthogonal to u independently of the side of the projection by means of the vector field u.

The second covariant Noether identity $\theta - {}_sT \equiv Q$ can be written by the use of the projections of the energy-momentum tensors in the form

$$\begin{aligned} &(\rho_\theta - \rho_T + \rho_Q) \cdot u \otimes g(u) + u \otimes g({}^\theta\overline{\pi} - {}^T\overline{\pi} + {}^Q\pi) \\ &\qquad + ({}^\theta\overline{s} - {}^T\overline{s} + {}^Q s) \otimes g(u) + ({}^\theta\overline{S} - {}^T\overline{S} + {}^Q S)g \equiv 0. \end{aligned} \tag{50}$$

After contraction of the last expression consistently with u and $\overline{g}$ and taking into account the properties (47)$\div$(49) the second covariant Noether identity disintegrates in identities for the different projections of the energy-momentum tensors

$$\rho_\theta \equiv \rho_T - \rho_Q, \quad {}^\theta\overline{\pi} \equiv {}^T\overline{\pi} - {}^Q\pi, \quad {}^\theta\overline{s} \equiv {}^T\overline{s} - {}^Q s, \quad {}^\theta\overline{S} \equiv {}^T\overline{S} - {}^Q S\,. \tag{51}$$

If the covariant Euler-Lagrange equations of the type $\delta_v L/\delta V^A{}_B = 0$ are fulfilled for the non-metric tensor fields of a Lagrangian system and ${}_gQ = 0$, then the variational energy-momentum of Euler-Lagrange $Q = {}_vQ + {}_gQ$ is equal to zero. This fact leads to vanishing the invariant projections of Q

$(\rho_Q = 0,\, {}^Q\pi = 0,\, {}^Q s = 0,\, {}^Q S = 0)$. The equality which follows between θ and ${}_sT$ has as corollaries the identities

$$\rho_\theta \equiv \rho_T, \qquad {}^\theta\overline{\pi} \equiv {}^T\overline{\pi}, \qquad {}^\theta\overline{s} \equiv {}^T\overline{s}, \qquad {}^\theta\overline{S} \equiv {}^T\overline{S}. \tag{52}$$

From the first identity ($\rho_\theta \equiv \rho_T$) and the identity (51) for ρ, it follows that the covariant Euler-Lagrange equations of the type $\delta_v L/\delta V^A{}_B = 0$ for non-metric fields V and ${}_gQ = 0$ appear as sufficient conditions for the unique determination of the notion of rest mass density ρ for a given Lagrangian system.

Proposition. *The necessary and sufficient condition for the equality*

$$\rho_\theta = \rho_T \tag{53}$$

is the condition $\rho_Q = 0$.

Proof. It follows immediately from the first identity in (51). □

The condition $\rho_Q \neq 0$ leads to the violation of the unique determination of the notion of rest mass density and to the appearance of three different notions of rest mass density corresponding to the three different energy-momentum tensors for a Lagrangian system. Therefore, the violation of the covariant Euler-Lagrange equations $\delta_v L/\delta V^A{}_B = 0$ for the non-metric tensor fields or the existence of metric tensor fields in a Lagrangian density with ${}_gQ \neq 0$ induce a new rest mass density (a new rest mass respectively) for which the identity (51) is fulfilled.

The identity $\rho_\theta \equiv \rho_T - \rho_Q$ can be related to the physical hypotheses about the inertial, passive and active gravitational rest mass densities in models for describing the gravitational interaction. To every energy-momentum tensor a non-null rest mass density corresponds. The existence of the variational energy-momentum tensor of Euler-Lagrange is connected with the existence of the gravitational interaction in a Lagrangian system in Einstein's theory of gravitation [6] and therefore, with the existence of a non-null active gravitational rest mass density. When a Lagrangian system does not interact gravitationaly, the active gravitational rest mass density is equal to zero and the principle of equivalence between the inertial and the passive rest mass density is fulfilled [7]-[10].

The invariant representation of the energy-momentum tensors by means of the projective metrics h^u, h_u and the rest mass density allows a comparison of these tensors with the well known energy-momentum tensors from the continuum media mechanics (for instance, with the energy-momentum tensor of an ideal liquid in V_4-spaces:

$$ {}_sT_i^{\,j} = (\rho + \frac{1}{e} \cdot p) \cdot u_i \cdot u^j - p \cdot g_i^j, \qquad e = \text{const.} \neq 0, \qquad k = 1). \qquad (54) $$

It follows from the comparison that the Lagrangian invariant L can be interpreted as the pressure $p = L$ characterizing the Lagrangian system. This possibility for an other physical interpretation than the usual one (in the mechanics L is interpreted as the difference between the kinetic and the potential energy) allows a description of Lagrangian systems on the basis of phenomenological investigations determining the dependence of the pressure on other dynamical characteristics of the system. If these relations are given, then by the use of the method of Lagrangians with covariant derivatives (MLCD) the corresponding covariant Euler-Lagrange equations can be found as well as the energy-momentum tensors.

The use of a contravariant non-isotropic (non-null) vector field u and its corresponding projective metrics h^u and h_u is analogous to the application of a non-isotropic (time-like) vector field in the s. c. monad formalism [$(3+1)$-decomposition] in V_4-spaces for description of dynamical systems in Einstein's theory of gravitation (ETG) [11]-[21], [5], [22].

The $(n-1)+1$ decomposition (monad formalism) can be related to the method of Euler or to the method of Lagrange:

(a) *Method of Euler.* The vector field u is interpreted as the velocity vector field of an observer who describes a physical system with respect to his vector field (his velocity). This physical system is characterized by means of a Lagrangian system (Lagrangian density).

The motion of the observer (his velocity vector field) is given independently of the motion of the considered Lagrangian system.

(b) *Method of Lagrange.* The vector field u is interpreted as the velocity vector field of a continuum media with a co-moving with it observer. The last assumption means that the velocity vector field of the observer is identical with the velocity vector field of the media where he is situated.

The motion of the observer (his velocity vector field) is determined by the characteristics of the (Lagrangian) system. Its velocity vector field is, on the other side, determined by the dynamical characteristics of the system by means of equations of the type of the Euler-Lagrange equations.

4 Energy-momentum tensors and the momentum density

In the first section of this paper the notion of momentum density of an energy-momentum tensor G has been introduced as

$$ p_G = \rho_G \cdot u + {}^G\pi, $$

where $\rho_G.u$ is the convective momentum density of the energy-momentum tensor G; ${}^G\pi$ is the conductive momentum density of the energy-momentum tensor G. The tensor field G can be represented by means of the projective metrics h^u and h_u in the form

$$G = u \otimes g(p_G) + {}^G s \otimes g(u) + ({}^G S)g, \qquad p_G = \frac{1}{e} \cdot [g(u)](G)(\overline{g}).$$

The Kronecker tensor can be represented in the form

$$Kr = u \otimes g(p_{Kr}) + {}^{Kr}s \otimes g(u) + ({}^{Kr}S)g, \tag{55}$$

where

$$p_{Kr} = \frac{1}{e} \cdot k \cdot u + {}^{Kr}\pi, \quad L \cdot Kr = u \otimes g(L \cdot p_{Kr}) + L \cdot {}^{Kr}s \otimes g(u) + (L \cdot {}^{Kr}S)g. \tag{56}$$

The generalized canonical energy-momentum tensor θ will have then the form

$$\theta = u \otimes g(p_\theta) + {}^\theta s \otimes g(u) + ({}^\theta S)g, \tag{57}$$

where

$$p_\theta = \rho_\theta \cdot u + {}^\theta\pi = \rho_\theta \cdot u + {}^\theta\overline{\pi} - L \cdot {}^{Kr}\pi. \tag{58}$$

On the analogy of the notions in the continuum media mechanics one can introduce the following definitions connected with the notion momentum density of a given energy-momentum tensor $G \sim (\theta, {}_sT, Q)$:

(a) Momentum density of G

$$p_G = \frac{1}{e} \cdot [g(u)](G)(\overline{g}) = \rho_G \cdot u + {}^G\pi \,.$$

(b) Convective momentum density of G .... $\rho_G \cdot u$.

From the covariant Noether identities for the rest mass and the conductive momentum densities, it follows the identity for the momentum density of the different energy-momentum tensors

$$p_\theta \equiv p_T - p_Q. \tag{59}$$

If the Euler-Lagrangian equations $\delta_v L/\delta V^A{}_B = 0$ for the non-metric tensor fields are fulfilled and ${}_gQ = 0$, then

$$p_\theta = p_T. \tag{60}$$

The representations of the energy-momentum tensors of the type 2 is analogous to the representations of the energy-momentum tensors of the type 1 by the use of h^u, h_u and the momentum density.

5 Energy-momentum tensors and the energy flux density

In the first section, the notion of energy flux density has been introduced for a given energy-momentum tensor G as

$$e_G = G(u) = e \cdot (\rho_G \cdot u + {}^G s).$$

The energy-momentum tensors can be now represented by the use of the projective metrics h^u, h_u and the energy flux density in the forms

$$G = \frac{1}{e} \cdot e_G \otimes g(u) + u \otimes g({}^G\pi) + ({}^G S)g$$
$$= G_\alpha{}^\beta \cdot e_\beta \otimes e^\alpha = G_i{}^j \cdot \partial_j \otimes dx^i,$$

$$(G)\overline{g} = \frac{1}{e} \cdot e_G \otimes u + u \otimes {}^G\pi + {}^G S = G^{\alpha\beta} \cdot e_\alpha \otimes e_\beta = G^{ij} \cdot \partial_i \otimes \partial_j, \qquad (61)$$

$$g(G) = \frac{1}{e} \cdot g(e_G) \otimes g(u) + g(u) \otimes g({}^G\pi) + g({}^G S)g$$
$$= G_{\alpha\beta} \cdot e^\alpha \otimes e^\beta = G_{ij} \cdot dx^i \otimes dx^j. \qquad (62)$$

The Kronecker tensor Kr can also be represented by the use of e_{Kr}, where

$$e_{Kr} = Kr(u) = k \cdot u + e \cdot {}^{Kr}s = e \cdot (\frac{1}{e} \cdot k \cdot u + {}^{Kr}s). \qquad (63)$$

The structure of e_G allows the introduction of the abbreviations:
(a) Convective energy flux density of the energy-momentum tensor G .... $\rho_G \cdot e \cdot u$.
(b) Conductive energy flux density of the energy-momentum tensor G .... $e. \cdot {}^G s$.

On the analogy of the determined notions, the following abbreviations can be introduced for a given energy-momentum tensor $G \sim (\theta, {}_sT, Q)$:
(a) Energy flux density of G .... e_G.
(b) Convective energy flux density of G .... $\rho_G \cdot e \cdot u$.

By means of the covariant Noether identities for the rest mass density and for the conductive energy flux density the identity for the energy flux density follows in the form

$$e_\theta \equiv e_T - e_Q. \qquad (64)$$

The relation between the momentum density and the energy flux density follows from the structure of their definitions

$$p_G = \rho_G \cdot u + {}^G\pi, \qquad e_G = (\rho_G \cdot u + {}^G s) \cdot e, \qquad G \sim (\theta, {}_sT, Q).$$

From the last expressions the relations

$$p_G = \frac{1}{e} \cdot e_G + {}^G\pi - {}^Gs, \qquad e_G = (p_G + {}^Gs - {}^G\pi) \cdot e, \tag{65}$$

follow.

By the use of the different representations of the energy-momentum tensors the different physical processes can be investigated. The application of an representation will depend on the role of the considered quantity (rest mass density, momentum density or energy flux density) in the dynamical process.

The physical interpretation of the introduced notions has been used for describing Lagrangian systems in V_4-spaces, where the contravariant vector field u has been interpreted as a time-like vector field tangential to the trajectories of the moving in the space-time particles. A V_4-space is considered as a model of the space-time.

6 Covariant divergency of a mixed tensor field

The operation of the covariant differentiation along a contravariant vector field can be extended to covariant differentiation along a contravariant tensor field.

The Lie derivative $\pounds_\xi u$ of a contravariant vector field u along a contravariant vector field ξ can be expressed by the use of the covariant differential operators ∇_ξ and ∇_u in the form

$$\pounds_\xi u = \nabla_\xi u - \nabla_u \xi - T(\xi, u), \qquad u \in T(M),$$

where $T(\xi, u)$ is the contravariant torsion vector field

$$T(\xi, u) = T_{\alpha\beta}{}^{\gamma} \cdot \xi^\alpha \cdot u^\beta \cdot e_\gamma = T_{ij}{}^{k} \cdot \xi^i \cdot u^j \cdot \partial_k,$$

constructed by means of the components $T_{\alpha\beta}{}^{\gamma}$ (or $T_{ij}{}^{k}$) of the contravariant torsion tensor field T.

The Lie derivative $\pounds_\xi V$ of a contravariant tensor field $V = V^A \cdot e_A = V^A \cdot \partial_A \in \otimes^m(M)$ along a contravariant vector field ξ can be written on the analogy of the relation for $\pounds_\xi u$ and by the use of the covariant differential operator ∇_ξ and an operator ∇_V in the form

$$\pounds_\xi V = \nabla_\xi V - \nabla_V \xi - T(\xi, V), \quad \xi \in T(M), \quad V \in \otimes^m(M),$$

where

$$\nabla_V \xi = -\xi^\alpha{}_{/\beta} \cdot S_{B\alpha}{}^{A\beta} \cdot V^B \cdot e_A = -\xi^i{}_{;j} \cdot S_{Ci}{}^{Aj} \cdot V^C \cdot \partial_A,$$

$$T(\xi, V) = T_{B\gamma}{}^{A} \cdot \xi^\gamma \cdot V^B \cdot e_A = T_{Ck}{}^{A} \cdot \xi^k \cdot V^C \cdot \partial_A,$$

$$T_{B\gamma}{}^{A} = T_{\beta\gamma}{}^{\alpha} \cdot S_{B\alpha}{}^{A\beta}, \quad T_{Ck}{}^{A} = T_{jk}{}^{i} \cdot S_{Ci}{}^{Aj}.$$

(For the definition of $S_{Ci}{}^{Aj}$ see Appendix A.2 of [1])

$\nabla_V \xi$ appears as a definition of the action of the operator ∇_V on the vector field ξ. Let us now consider more closely this operator and its properties.

Let a mixed tensor field $K \in \otimes^k{}_l(M)$ be given in a non-co-ordinate (or co-ordinate) basis

$$K = K^C{}_D \cdot e_C \otimes e^D = K^{C_1\alpha}{}_D \cdot e_{C_1} \otimes e_\alpha \otimes e^D,$$
$$e_C = e_{C_1} \otimes e_\alpha, \qquad e^D = e^{\alpha_1} \otimes \cdots \otimes e^{\alpha_l}.$$

The action of the operator ∇_V on the mixed tensor field K can be defined on the analogy of the action of ∇_V on a contravariant vector field ξ

$$\nabla_V K = -K^{C_1\alpha}{}_{D/\beta} \cdot S_{B\alpha}{}^{A\beta} \cdot V^B \cdot e_{C_1} \otimes e^D \otimes e_A, \tag{66}$$

where ∇_V is the covariant differential operator along a contravariant tensor field V

$$\nabla_V : K \Rightarrow \nabla_V K, \quad K \in \otimes^k{}_l(M), \quad V \in \otimes^m(M), \quad \nabla_V K \in \otimes^{k-1+m}{}_l(M).$$

Remark. There is an other possibility for a generalization of the action of ∇_V on a mixed tensor field

$$\nabla_V K = -\sum_{m=1}^{k} K^{\alpha_1 \ldots \alpha_m \ldots \alpha_k}{}_{D/\beta} . S_{B\alpha_m}{}^{A\beta} . V^B . e_{\alpha_1} \otimes \cdots$$
$$\otimes e_{\alpha_{m-1}} \otimes e_{\alpha_{m+1}} \otimes \cdots \otimes e_{\alpha_k} \otimes e^D \otimes e_A.$$

The result of the action of this operator on a contravariant vector field ξ is identical with the action of the above defined operator ∇_V.

Remark. A covariant differential operator along a contravariant tensor field can also be defined through its action on mixed tensor fields in the form $\overline{\nabla}_V K = K^C{}_{D/\beta} . V^{A_1\beta} . e_C \otimes e^D \otimes e_{A_1}$.

The operator $\overline{\nabla}_V$ differs from ∇_V in its action on a contravariant vector field and does not appear as a generalization of the already defined operator by its action on a contravariant vector field. It appears as a new differential operator acting on mixed tensor fields.

The covariant differential operator ∇_V has the properties:

(a) Linear operator

$$\nabla_V(\alpha \cdot K_1 + \beta \cdot K_2) = \alpha \cdot \nabla_V K_1 + \beta \cdot \nabla_V K_2,$$
$$\alpha, \beta \in \mathbf{R} \text{ (or } \mathbf{C}), \qquad K_1, K_2 \in \otimes^k{}_l(M).$$

The proof of this property follows immediately from (66) and the linear property of the covariant differential operator along a basic contravariant vector field.

(b) Differential operator (not obeying the Leibniz rule)

$$\nabla_V(K \otimes S) = \nabla_{e_\beta} K \otimes \overline{S}^\beta + K \otimes \nabla_V S,$$
$$K = K^A{}_B \cdot e_A \otimes e^B, \qquad \nabla_{e_\beta} K = K^A{}_{B/\beta} \cdot e_A \otimes e^B,$$
$$S = \widetilde{S}^C{}_D \cdot e_C \otimes e^D = \widetilde{S}^{C_1\alpha}{}_D \cdot e_{C_1} \otimes e_\alpha \otimes e^D,$$
$$\overline{S}^\beta = -\widetilde{S}^{C_1\alpha}{}_D \cdot S_{E\alpha}{}^{F\beta} \cdot V^E \cdot e_{C_1} \otimes e^D \otimes e_F.$$

The proof of this property follows from the action of the defined in (66) operator ∇_V and the properties of the covariant derivative of the product of the components of the tensor fields K and S.

If the tensor field V is given as a contravariant metric tensor field $\overline{g}$, then the covariant differential operator ∇_V $(V = \overline{g})$ will have additional properties connected with the properties of the contravariant metric tensor field.

Definition. Contravariant metric differential operator $\nabla_{\overline{g}}$. Covariant differential operator ∇_V for $V = \overline{g}$.

By means of the relations

$$\begin{aligned} -S_{B\alpha}{}^{A\beta} \cdot g^B \cdot e_A &= (g^\sigma_\alpha \cdot g^{\beta\kappa} + g^\kappa_\alpha \cdot g^{\beta\sigma}).e_\sigma \otimes e_\kappa \\ &= (g^\sigma_\alpha \cdot g^{\beta\kappa} + g^\kappa_\alpha \cdot g^{\beta\sigma}).e_\sigma . e_\kappa, \end{aligned} \tag{67}$$

$$e_\sigma . e_\kappa = \frac{1}{2}(e_\sigma \otimes e_\kappa + e_\kappa \otimes e_\sigma), \qquad K^{C_1\alpha}{}_{D/\beta} \cdot g^\sigma_\alpha = K^{C_1\sigma}{}_{D/\beta}, \tag{68}$$

the action of the contravariant metric differential operator on a mixed tensor field K can be represented in the form

$$\begin{aligned} \nabla_{\overline{g}} K &= (K^{C_1\sigma}{}_{D/\beta} \cdot g^{\beta\kappa} + K^{C_1\kappa}{}_{D/\beta} \cdot g^{\beta\sigma}) \cdot e_{C_1} \otimes e^D \otimes e_\sigma \otimes e_\kappa \\ &= (K^{C_1\sigma}{}_{D/\beta} \cdot g^{\beta\kappa} + K^{C_1\kappa}{}_{D/\beta} \cdot g^{\beta\sigma}).e_{C_1} \otimes e^D \otimes e_\sigma . e_\kappa . \end{aligned} \tag{69}$$

The properties of the operator $\nabla_{\overline{g}}$ are determined additionally by the properties of the contravariant metric tensor field of second rank:

(a) $\nabla_{\overline{g}} : K \Rightarrow \nabla_{\overline{g}} K$, $K \in \otimes^k{}_l(M)$, $\nabla_{\overline{g}} K \in \otimes^{k+1}{}_l(M)$.

(b) Linear operator

$$\nabla_{\overline{g}}(\alpha \cdot K_1 + \beta \cdot K_2) = \alpha \cdot \nabla_{\overline{g}} K_1 + \beta \cdot \nabla_{\overline{g}} K_2.$$

(c) Differential operator (not obeying the Leibniz rule)

$$\nabla_{\overline{g}}(K \otimes S) = \nabla_{e_\beta} K \otimes \overline{S}^\beta + K \otimes \nabla_{\overline{g}} S,$$
$$K \in \otimes^k{}_l(M), \qquad S \in \otimes^m{}_r(M),$$
$$K = K^A{}_B \cdot e_A \otimes e^B, \qquad \nabla_{e_\beta} K = K^A{}_{B/\beta} \cdot e_A \otimes e^B, \tag{70}$$
$$S = \widetilde{S}^C{}_D \cdot e_C \otimes e^D = \widetilde{S}^{C_1\alpha}{}_D \cdot e_{C_1} \otimes e_\alpha \otimes e^D,$$
$$\overline{S}^\beta = (\widetilde{S}^{C_1\sigma}{}_D \cdot g^{\beta\kappa} + \widetilde{S}^{C_1\kappa}{}_D \cdot g^{\beta\sigma}) \cdot e_{C_1} \otimes e^D \otimes e_\sigma . e_\kappa .$$

Remark. The definition of $\nabla_{\overline{g}}$ in (69) differs from the definition in [8], where $\nabla_{\overline{g}} \equiv \overline{\nabla}_{\overline{g}}$, i.e. the contravariant metric differential operator is defined in the last case as a special case of the covariant differential operator $\overline{\nabla}_V$ for $V = \overline{g}$.

The notion of covariant divergency of a mixed tensor field has been used in V_4-spaces for the determination of conditions for the existence of local conserved quantities and in identities of the type of the first covariant Noether identity. Usually, the covariant divergency of a contravariant or mixed tensor field has been given in co-ordinate or non-co-ordinate basis in the form

$$\delta K = K^{A\beta}{}_{B/\beta} \cdot e_A \otimes e^B = K^{Ci}{}_{D;i} \cdot \partial_C \otimes dx^D, \tag{71}$$

where

$$K^{A\beta}{}_{B/\beta} = K^{A\beta}{}_{B/\gamma} \cdot g^\gamma_\beta, \qquad K^{Ci}{}_{D;i} = K^{Ci}{}_{D;k} \cdot g^k_i . \tag{72}$$

For full anti-symmetric covariant tensor fields (differential forms) the covariant divergency (called also codifferential) δ is defined by means of the Hodge operator $*$, its reverse operator $*^{-1}$ and the external differential operator ${}_a\overline{d}$ in the form [24], [25] (pp. 147–149)

$$\delta = *^{-1} \circ {}_a\overline{d} \circ * . \tag{73}$$

By the use of the contravariant metric differential operator $\nabla_{\overline{g}}$, the covariant metric tensor field g and the contraction operator one can introduce the notion of covariant divergency of a mixed tensor field K with finite rank.

Definition. Covariant divergency δK of a mixed tensor field K

$$\delta K = \frac{1}{2} . [\nabla_{\overline{g}} K] g = K^{A\beta}{}_{B/\beta} . e_A \otimes e^B = K^{Ci}{}_{D;i} . \partial_C \otimes dx^D,$$

where

$$K = K^{A\beta}{}_B \cdot e_A \otimes e_\beta \otimes e^B = K^{Ci}{}_D \cdot \partial_C \otimes \partial_i \otimes dx^D,$$
$$K \in \otimes^k{}_l(M), \qquad k \geq 1.$$

δ is called operator of the covariant divergency

$$\delta : K \Rightarrow \delta K, \qquad K \in \otimes^k{}_l(M), \qquad \delta K \in \otimes^{k-1}{}_l(M), \quad k \geq 1.$$

The properties of the operator of the covariant divergency δ are determined by the properties of the contravariant metric differential operator, the contraction operator and the metric tensor fields g and $\overline{g}$

(a) The operator of the covariant divergency δ is a linear operator

$$\begin{gathered}\delta(\alpha \cdot K_1 + \beta \cdot K_2) = \alpha \cdot \delta K_1 + \beta \cdot \delta K_2, \\ \alpha, \beta \in \mathrm{R} \text{ (or C)}, K_1, K_2 \in \otimes^k{}_l(M).\end{gathered} \tag{74}$$

The proof of this property follows immediately from the definition of the covariant divergency.

(b) Action on a tensor product of tensor fields

$$\delta(K \otimes S) = \overline{\nabla}_S K + K \otimes \delta S, \tag{75}$$

where

$$\begin{gathered}K = K^A{}_B \cdot e_A \otimes e^B, \quad S = S^{C\beta}{}_D \cdot e_C \otimes e_\beta \otimes e^D, \\ \overline{\nabla}_S K = K^A{}_{B/\beta} \cdot S^{C\beta}{}_D \cdot e_A \otimes e^B \otimes e_C \otimes e^D \quad \text{(see above } \overline{\nabla}_V\text{)}.\end{gathered} \tag{76}$$

The proof of this property follows from the properties of $\nabla_{\overline{g}}$ and from the relations

$$\frac{1}{2}[\nabla_{e_\beta} K \otimes \overline{S}^\beta]g = K^A{}_{B/\beta} \cdot S^{C\beta}{}_D \cdot e_A \otimes e^B \otimes e_C \otimes e^D, \tag{77}$$

$$\frac{1}{2}[K \otimes \nabla_{\overline{g}} S]g = K \otimes \frac{1}{2} \cdot [\nabla_{\overline{g}} S]g = K \otimes \delta S. \tag{78}$$

(c) Action on a contravariant vector field u

$$\delta u = \frac{1}{2} \cdot [\nabla_{\overline{g}} u]g = u^\beta{}_{/\beta} = u^i{}_{;i}. \tag{79}$$

(d) Action on the tensor product of two contravariant vector fields u and v

$$\delta(u \otimes v) = \nabla_v u + \delta u \cdot v, \qquad \overline{\nabla}_v u = \nabla_v u. \tag{80}$$

(e) Action on the product of an invariant function L and a mixed tensor field K

$$\delta(L.K) = \overline{\nabla}_K L + L \cdot \delta K, \tag{81}$$

$$\overline{\nabla}_K L = L_{/\beta} \cdot K^{A\beta}{}_B \cdot e_A \otimes e^B, \qquad \delta K = K^{A\beta}{}_{B/\beta} \cdot e_A \otimes e^B,$$
$$L_{/\beta} = e_\beta L, \qquad L_{;i} = L_{,i}, \tag{82}$$
$$K = K^{A\beta}{}_B \cdot e_A \otimes e_\beta \otimes e^B \in \otimes^k{}_l(M).$$

Special case: Action on the product of an invariant function L and the contravariant metric tensor $\overline{g}$:

$$\delta(L.\overline{g}) = (L_{/\beta} \cdot g^{\alpha\beta} + L \cdot g^{\alpha\beta}{}_{/\beta}) = (L_{,j} \cdot g^{ij} + L \cdot g^{ij}{}_{;j}). \tag{83}$$

(f) Action on an anti-symmetric tensor product of two contravariant vector fields u and v

$$\delta(u \wedge v) = \frac{1}{2} \cdot (\nabla_v u - \nabla_u v + \delta v \cdot u - \delta u \cdot v)$$
$$= -\frac{1}{2} \cdot [\pounds_u v + T(u,v) + \delta u \cdot v - \delta v \cdot u]. \tag{84}$$

(g) Action on a full anti-symmetric contravariant tensor field A of second rank

$$\delta A = \frac{1}{2} \cdot (A^{\alpha\beta} - A^{\beta\alpha})_{/\beta} \cdot e_\alpha = \frac{1}{2} \cdot (A^{ij} - A^{ji})_{;j} \cdot \partial_i,$$
$$A = A^{\alpha\beta} \cdot e_\alpha \wedge e_\beta = A^{ij} \cdot \partial_i \wedge \partial_j, \quad A^{\alpha\beta} = -A^{\beta\alpha}. \tag{85}$$

(h) Action on a tensor product of a contravariant vector field u, multiplied with an invariant function, and a covariant vector field $g(v)$ with the contravariant vector field v

$$\delta(\varepsilon \cdot u \otimes g(v)) = (u\varepsilon) \cdot g(v) + \varepsilon \cdot [\delta u \cdot g(v) + (\nabla_u g)(v) + g(\nabla_u v)]$$
$$= [u\varepsilon + \varepsilon \cdot \delta u] \cdot g(v) + \varepsilon \cdot [(\nabla_u g)(v) + g(\nabla_u v)], \tag{86}$$
$$\varepsilon \in C^r(M), \qquad \varepsilon'(x^{k'}) = \varepsilon(x^k), \qquad u, v \in T(M).$$

Special case: $v \equiv u$:

$$\delta(\varepsilon \cdot u \otimes g(u)) = [u\varepsilon + \varepsilon \cdot \delta u] \cdot g(u) + \varepsilon \cdot [(\nabla_u g)(u) + g(a)], \quad \nabla_u u = a. \tag{87}$$

Special case: $\varepsilon = 1$:

$$\delta(u \otimes g(v)) = \delta u \cdot g(v) + (\nabla_u g)(v) + g(\nabla_u v). \tag{88}$$

6.1 Covariant divergency of a mixed tensor field of second rank

From the definition of the covariant divergency δK of a mixed tensor field K, the explicit form of the covariant divergency of tensor fields of second rank of the type 1 or 2 follows as

$$\delta G = \frac{1}{2}[\nabla_{\overline{g}} G]g = G_\alpha{}^\beta{}_{/\beta} \cdot e^\alpha = G_i{}^j{}_{;j} \cdot dx^i, \tag{89}$$

$$\delta\overline{G} = \frac{1}{2}[\nabla_{\overline{g}}\overline{G}]g = \overline{G}^{\beta}{}_{\alpha/\beta} \cdot e^{\alpha} = \overline{G}^{j}{}_{i;j} \cdot dx^{i}. \tag{90}$$

By the use of the relations (79)÷(84), (86) ÷ (88), and the expression [see (75) ÷ (78)]

$$\overline{\nabla}_v(g(u)) = \nabla_v(g(u)) = (\nabla_v g)(u) + g(\nabla_v u), \tag{91}$$

$$\delta(g(u) \otimes v) = \delta v \cdot g(u) + (\nabla_v g)(u) + g(\nabla_v u), \tag{92}$$

$$\delta((^G S)g) = (g_{\alpha\overline{\gamma}} \cdot {}^G S^{\beta\gamma})_{/\beta} \cdot e^{\alpha}, \tag{93}$$

the covariant divergency of the representation of G by means of the rest mass density ρ_G $(\varepsilon_G = \rho_G)$

$$G = \rho_G \cdot u \otimes g(u) + u \otimes g(^G\pi) + {}^G s \otimes g(u) + (^G S)g,$$

can be found in the form $(\nabla_u u = a)$

$$\begin{aligned} \delta G =& \rho_G \cdot g(a) + (u\rho_G + \rho_G \cdot \delta u + \delta^G s) \cdot g(u) + \delta u \cdot g(^G\pi) \\ &+ g(\nabla_u {}^G\pi) + g(\nabla_{^G s} u) + \rho_G \cdot (\nabla_u g)(u) \\ &+ (\nabla_u g)(^G\pi) + (\nabla_{^G s} g)(u) + \delta((^G S)g). \end{aligned} \tag{94}$$

$\overline{g}(\delta G)$ will have the form

$$\begin{aligned} \overline{g}(\delta G) =& \rho_G \cdot a + (u\rho_G + \rho_G \cdot \delta u + \delta^G s) \cdot u + \delta u \cdot {}^G\pi \\ &+ \nabla_u {}^G\pi + \nabla_{^G s} u + \rho_G \cdot \overline{g}(\nabla_u g)(u) \\ &+ \overline{g}(\nabla_u g)(^G\pi) + \overline{g}(\nabla_{^G s} g)(u) + \overline{g}(\delta((^G S)g)). \end{aligned} \tag{95}$$

In a co-ordinate basis δG and $\overline{g}(\delta G)$ will have the forms

$$\begin{aligned} G_i{}^j{}_{;j} =& \rho_G \cdot a_i + (\rho_{G,j} \cdot u^j + \rho_G \cdot u^j{}_{;j} + {}^G s^j{}_{;j}) \cdot u_i + u^j{}_{;j} \cdot {}^G\pi_i \\ &+ g_{i\overline{j}} \cdot ({}^G\pi^j{}_{;k} \cdot u^k + u^j{}_{;k} \cdot {}^G s^k) \\ &+ g_{ij;k} \cdot (\rho_G \cdot u^k \cdot u^{\overline{j}} + u^k \cdot {}^G\pi^{\overline{j}} + {}^G s^k \cdot u^{\overline{j}}) + (g_{i\overline{k}} \cdot {}^G S^{jk})_{;j}, \end{aligned} \tag{96}$$

$$a_i = g_{i\overline{j}} \cdot a^j, \; a^i = u^i{}_{;j} \cdot u^j, \; \rho_{G;i} = \rho_{G,i}, \; u_i = g_{i\overline{k}} \cdot u^k, \; {}^G\pi_i = g_{i\overline{j}} \cdot {}^G\pi^j,$$

$$\begin{aligned} g^{i\overline{k}} \cdot G_k{}^j{}_{;j} =& \rho_G \cdot a^i + (\rho_{G,j} \cdot u^j + \rho_G \cdot u^j{}_{;j} + {}^G s^j{}_{;j}) \cdot u^i \\ &+ u^j{}_{;j} \cdot {}^G\pi^i + {}^G\pi^i{}_{;j} \cdot u^j + u^i{}_{;j} \cdot {}^G s^j \\ &+ g^{i\overline{l}} \cdot g_{lj;k} \cdot (\rho_G \cdot u^k \cdot u^{\overline{j}} + u^k \cdot {}^G\pi^{\overline{j}} + {}^G s^k \cdot u^{\overline{j}}) \\ &+ g^{i\overline{l}} \cdot (g_{lk} \cdot {}^G S^{jk})_{;j}. \end{aligned} \tag{97}$$

The relation between δG and $\delta\overline{G}$ follows from the relation between G and $\overline{G}$

$$\overline{G} = g(G)\overline{g} : \delta\overline{G} = \delta(g(G)\overline{g}) = \frac{1}{2}[\nabla_{\overline{g}}(g(G)\overline{g})]. \tag{98}$$

The covariant divergency of the Kronecker tensor field can be found in an analogous way as the covariant divergency of a tensor field of second rank of the type 1, since $Kr = g^{\alpha}_{\beta} \cdot e_{\alpha} \otimes e^{\beta} = g^{i}_{j} \cdot \partial_i \otimes dx^j$

$$\delta Kr = \frac{1}{2}[\nabla_{\overline{g}} Kr]g = g^{\beta}_{\alpha/\beta} \cdot e^{\alpha} = g^{j}_{i;j} \cdot dx^i. \tag{99}$$

If we use the representation of Kr

$$Kr = \frac{1}{e} \cdot k \cdot u \otimes g(u) + u \otimes g(^{Kr}\pi) + {}^{Kr}s \otimes g(u) + (^{Kr}S)g,$$

then the covariant divergency δKr can be written in the form

$$\begin{aligned} \delta Kr = & \frac{1}{e} \cdot k \cdot g(a) + [u(\frac{1}{e} \cdot k) + \frac{1}{e} \cdot k \cdot \delta u + \delta^{Kr} s] \cdot g(u) \\ & + \delta u \cdot g(^{Kr}\pi) + g(\nabla_u {}^{Kr}\pi) + g(\nabla_{^{Kr}s} u) \\ & + \frac{1}{e} \cdot k \cdot (\nabla_u g)(u) + (\nabla_u g)(^{Kr}\pi) + (\nabla_{^{Kr}s} g)(u) + \delta((^{Kr}S)g), \end{aligned} \tag{100}$$

7 Covariant divergency of the energy-momentum tensors and the rest mass density

The covariant divergency of the energy-momentum tensors can be represented by the use of the projective metrics h^u, h_u of the contravariant vector field u and the rest mass density for the corresponding energy-momentum tensor. In this case the representation of the energy-momentum tensor is in the form

$$G = (\rho_G + \frac{1}{e} \cdot L \cdot k) \cdot u \otimes g(u) - L \cdot Kr + u \otimes g(^k\pi) + {}^k s \otimes g(u) + (^k S)g,$$

where ${}^k\pi = {}^G\overline{\pi}$, ${}^k s = {}^G\overline{s}$, ${}^k S = {}^G\overline{S}$.

By the use of the relations

$$\begin{aligned} u(\rho_G + \frac{1}{e} \cdot L \cdot k) &= u\rho_G + L \cdot u(\frac{1}{e} \cdot k) + \frac{1}{e} \cdot k \cdot (uL) \\ &= [\rho_{G/\alpha} + L \cdot (\frac{1}{e} \cdot k)_{/\alpha} + \frac{1}{e} \cdot k \cdot L_{/\alpha}] \cdot u^{\alpha} \\ &= [\rho_{G,j} + L \cdot (\frac{1}{e} \cdot k)_{,j} + \frac{1}{e} \cdot k \cdot L_{,j}] \cdot u^j \end{aligned} \tag{101}$$

$$KrL = L_{/\alpha} \cdot e^{\alpha} = L_{,i} \cdot dx^i = \overline{\nabla}_{Kr} L, \quad \delta(L \cdot Kr) = KrL + L \cdot \delta Kr, \quad (102)$$

$$\delta(L \cdot Kr) = \frac{1}{2} \cdot [\nabla_{\overline{g}}(L.Kr)]g = (L \cdot g_i^j)_{;j} \cdot dx^i = (L_{,i} + L \cdot g_{i\ ;j}^{j}) \cdot dx^i, \quad (103)$$

$$\delta(u \otimes g({}^{G}\overline{\pi})) = \delta u \cdot g({}^{G}\overline{\pi}) + g(\nabla_u {}^{G}\overline{\pi}) + (\nabla_u g)({}^{G}\overline{\pi}), \quad (104)$$

$$\delta({}^{G}\overline{s} \otimes g(u)) = \delta^{G}\overline{s} \cdot g(u) + g(\nabla_{{}^{G}\overline{s}} u) + (\nabla_{{}^{G}\overline{s}} g)(u), \quad (105)$$

$$\delta(({}^{G}\overline{S})g) = (g_{\alpha\overline{\gamma}} \cdot {}^{G}\overline{S}^{\beta\gamma})_{/\beta} \cdot e^{\alpha} \text{ [see (93)]},$$

δG can be found in the form

$$\begin{aligned}
\delta G =& (\rho_G + \frac{1}{e} \cdot L \cdot k) \cdot g(a) \\
&+ [u(\rho_G + \frac{1}{e} \cdot L \cdot k) + \ (\rho_G + \frac{1}{e} \cdot L \cdot k) \cdot \delta u + \delta^{G}\overline{s}] \cdot g(u) \\
&- KrL - L \cdot \delta Kr + \delta u \cdot g({}^{G}\overline{\pi}) + g(\nabla_u {}^{G}\overline{\pi}) + g(\nabla_{{}^{G}\overline{s}} u) \\
&+ (\rho_G + \frac{1}{e} \cdot L \cdot k) \cdot (\nabla_u g)(u) + (\nabla_u g)({}^{G}\overline{\pi}) + (\nabla_{{}^{G}\overline{s}} g)(u) \\
&+ \delta(({}^{G}\overline{S})g).
\end{aligned} \quad (106)$$

In a co-ordinate basis δG will have the form

$$\begin{aligned}
G_i{}^j{}_{;j} =& (\rho_G + \frac{1}{e} \cdot L \cdot k) \cdot a_i \\
&+ [(\rho_G + \frac{1}{e} \cdot L \cdot k)_{,j} \cdot u^j + (\rho_G + \frac{1}{e} \cdot L \cdot k) \cdot u^j{}_{;j} + {}^{G}\overline{s}^j{}_{;j}] \cdot u_i \\
&- L_{,i} - L \cdot g_{i\ ;j}^{j} + u^j{}_{;j} \cdot {}^{G}\overline{\pi}_i + g_{i\overline{j}} \cdot ({}^{G}\overline{\pi}^j{}_{;k} \cdot u^k + u^j{}_{;k} \cdot {}^{G}\overline{s}^k) \\
&+ g_{ij;k} \cdot [(\rho_G + \frac{1}{e} \cdot L \cdot k) \cdot u^{\overline{j}} \cdot u^k + {}^{G}\overline{\pi}^{\overline{j}} \cdot u^k + u^{\overline{j}} \cdot {}^{G}\overline{s}^k] \\
&+ (g_{i\overline{k}} \cdot {}^{G}\overline{S}^{jk})_{;j}.
\end{aligned} \quad (107)$$

On the grounds of the representations of δG and $\overline{g}(\delta G)$ the representation of the different energy-momentum tensors θ, ${}_sT$ and Q can be found.

8 Covariant divergency of the energy-momentum tensors and the momentum density

In a previous section the energy-momentum tensors are represented by the use of the projective metrics h^u, h_u and the momentum density p. For this type of representation of a tensor field G of the type 1, which has the form

$$G = u \otimes g(p_G) + {}^{G}s \otimes g(u) + ({}^{G}S)g,$$

the covariant divergency δG can be calculated in the form

$$\begin{aligned}\delta G =& g(\nabla_u p_G) + \delta u \cdot g(p_G) + \delta^G s \cdot g(u) + g(\nabla_{{}^G s} u) \\ &+ (\nabla_u g)(p_G) + (\nabla_{{}^G s} g)(u) + \delta(({}^G S)g).\end{aligned} \tag{108}$$

In a co-ordinate basis δG will have the form

$$\begin{aligned}G_i{}^j{}_{;j} =& g_{i\overline{j}} \cdot (p^{\overline{j}}_{G;k} \cdot u^k + u^k{}_{;k} \cdot p^{\overline{j}}_G + {}^G s^k{}_{;k} \cdot u^{\overline{j}} + u^{\overline{j}}{}_{;k} \cdot {}^G s^k) \\ &+ g_{i\overline{j};k} \cdot (p^{\overline{j}}_G \cdot u^k + u^{\overline{j}} \cdot {}^G s^k) + (g_{i\overline{k}} \cdot {}^G S^{jk})_{;j}.\end{aligned} \tag{109}$$

On the other side, the covariant divergency of $(G)\overline{g}$ can be found. $(G)\overline{g}$ is represented in the form

$$(G)\overline{g} = u \otimes p_G + {}^G s \otimes u + {}^G S = G^{\alpha\beta} \cdot e_\alpha \otimes e_\beta = G^{ij} \cdot \partial_i \otimes \partial_j. \tag{110}$$

$\delta((G)\overline{g})$ will have the form

$$\delta((G)\overline{g}) = \nabla_u {}^G s + \nabla_{p_G} u + \delta p_G \cdot u + \delta u \cdot {}^G s + \delta^G S, \tag{111}$$

$$G^{\alpha\beta}{}_{/\beta} = {}^G s^\alpha{}_{/\beta} \cdot u^\beta + u^\alpha{}_{/\beta} \cdot p^\beta_G + p^\beta_{G/\beta} \cdot u^\alpha + u^\beta{}_{/\beta} \cdot {}^G s^\alpha + {}^G S^{\alpha\beta}{}_{/\beta}, \tag{112}$$

$$G^{ij}{}_{;j} = {}^G s^i{}_{;j} \cdot u^j + u^i{}_{;j} \cdot p^j_G + p^j_{G;j} \cdot u^i + u^j{}_{;j} \cdot {}^G s^i + {}^G S^{ij}{}_{;j}, \tag{113}$$

$$G^{\alpha\beta} = G_\gamma{}^\alpha \cdot g^{\overline{\gamma}\beta}, \qquad G^{ij} = G_k{}^i \cdot g^{\overline{k}j}.$$

9 Covariant divergency of the energy-momentum tensors and the energy flux density

In a previous section the notion of energy flux density e_G has been introduced. By means of it a mixed tensor field G of second rank and of type 1 can be represented in the form

$$G = \frac{1}{e} \cdot e_G \otimes g(u) + u \otimes g({}^G \pi) + ({}^G S)g.$$

The covariant divergency δG of a given through e_G tensor field G can be found by the use of the relation

$$\delta(\frac{1}{e} \cdot e_G \otimes g(u)) = [e_G(\frac{1}{e}) + \frac{1}{e} \cdot \delta e_G] \cdot g(u) + \frac{1}{e} \cdot [(\nabla_{e_G} g)(u) + g(\nabla_{e_G} u)] \tag{114}$$

and the property (75) of δ in the form

$$\begin{aligned}\delta G = {} & g(\nabla_u {}^G\pi) + \frac{1}{e}\cdot g(\nabla_{e_G} u) + [e_G(\frac{1}{e}) + \frac{1}{e}\cdot \delta e_G]\cdot g(u) + \delta u \cdot g({}^G\pi) \\ & + \frac{1}{e}\cdot (\nabla_{e_G} g)(u) + (\nabla_u g)({}^G\pi) + \delta(({}^G S)g),\end{aligned} \tag{115}$$

or in the form

$$\begin{aligned}\overline{g}(\delta G) = {} & \nabla_u {}^G\pi + \frac{1}{e}\cdot \nabla_{e_G} u + [e_G(\frac{1}{e}) + \frac{1}{e}\cdot \delta e_G]\cdot u + \delta u \cdot {}^G\pi \\ & + \frac{1}{e}\cdot \overline{g}(\nabla_{e_G} g)(u) + \overline{g}(\nabla_u g)({}^G\pi) + \overline{g}(\delta(({}^G S)g)).\end{aligned} \tag{116}$$

On the other side, the covariant divergency of $(G)\overline{g}$ can be found in the form

$$\delta((G)\overline{g}) = \frac{1}{e}\cdot \nabla_u e_G + \nabla_{{}^G\pi} u + [u(\frac{1}{e}) + \frac{1}{e}\cdot \delta u]\cdot e_G + \delta^G\pi \cdot u + \delta({}^G S). \tag{117}$$

In a co-ordinate basis δG and $\overline{g}(\delta G)$ will have the forms

$$\begin{aligned}G_i{}^j{}_{;j} = {} & g_{i\overline{j}}\cdot \{{}^G\pi^j{}_{;k}\cdot u^k + \frac{1}{e}\cdot u^j{}_{;k}\cdot e^k_G \\ & \qquad + [(\frac{1}{e})_{,k}\cdot e^k_G + \frac{1}{e}\cdot e^k_{G;k}]\cdot u^j + u^k{}_{;k}\cdot {}^G\pi^j\} \\ & + g_{ij;k}\cdot [\frac{1}{e}\cdot u^{\overline{j}}\cdot e^k_G + {}^G\pi^{\overline{j}}\cdot u^k] + (g_{i\overline{k}}\cdot {}^G S^{jk})_{;j},\end{aligned} \tag{118}$$

$$\begin{aligned}g^{i\overline{k}}\cdot G_k{}^j{}_{;j} = {} & {}^G\pi^i{}_{;j}\cdot u^j + \frac{1}{e}\cdot u^i{}_{;j}\cdot e^j_G + [(\frac{1}{e})_{,j}\cdot e^j_G + \frac{1}{e}\cdot e^j_{G;j}]\cdot u^i + u^j{}_{;j}\cdot {}^G\pi^i \\ & + g^{i\overline{l}}\cdot g_{lj;k}\cdot [\frac{1}{e}\cdot u^{\overline{j}}\cdot e^k_G + {}^G\pi^{\overline{j}}\cdot u^k] + g^{i\overline{l}}\cdot (g_{l\overline{k}}\cdot {}^G S^{jk})_{;j}, \quad (\frac{1}{e})_{;j} = (\frac{1}{e})_{,j}.\end{aligned} \tag{119}$$

The different possibilities for a representation of the covariant divergency of the energy-momentum tensors are related to the different possibilities for a representation of the first covariant Noether identity for a given Lagrangian system with its corresponding energy-momentum tensors.

The different representation of the energy-momentum tensors and their divergencies could be used for considering problems related to the dynamics of material systems over spaces with affine connections and metrics.

Acknowledgments

The authors are grateful to Prof. St. Dimiev from the Institute for Mathematics and Informatics of the Bulgarian Academy of Sciences and to Prof. K. Sekigawa from the Department of Mathematics of the Faculty of Science at the Niigata University (Japan) for their support of the presented topics. One of the authors (S.M.) is also grateful to Prof. A.N. Chernikov from the Bogoliubov Laboratory of Theoretical Physics at the Joint Institute for Nuclear Research (Russia) and to Prof. D.I. Kazakov for the kind hospitality at the Joint Institute for Nuclear Research. This work is supported in part by the National Science Foundation of Bulgaria under Grant No. F-642.

References

1. S. Manoff, *Lagrangian formalism for tensor fields*, In Topics in Complex Analysis, Differential Geometry and Mathematical Physics. eds. St. Dimiev and K. Sekigawa, World Sci. Publ., Singapore, 1997, pp. 177–218.
2. ———, *Mechanics of tensor fields. Kinematics of vector fields and Lagrangian theory of tensor fields over spaces with contravariant and covariant affine connections and metrics*, Preprint INRNE, 1997, pp. 1–650.
3. L.D. Landau and E.M. Lifshitz, *Gidrodinamika (Hydrodynamics)*, Nauka, Moscow, 1986 (in Russian).
4. ———, *Teorija uprugosti (Theory of elasticity)*, Nauka, Moscow, 1987 (in Russian), pp. 14–15.
5. E. Schmutzer, *Relativistische Physik (Klassische Theorie)*, B.G. Teubner Verlagsgesellschaft, Leipzig, 1968.
6. S. Manoff, *Einstein's theory of gravitation as a Lagrangian theory for tensor fields*, Intern. J. Mod. Phys. **13**(1998), 1941–1967.
7. ———, *Invariant projections of energy-momentum tensors for field theories in spaces with affine connection and metric*, J. Math. Phys. **32**(1991), 728–734.
8. ———, *On the energy-momentum tensors for field theories in spaces with affine connection and metric. Generalized Bianchi identities and different energy-momentum tensors*, Comm. JINR Dubna **E2-91-77**(1991), 1–16.
9. ———, *On the energy-momentum tensors for field theories in spaces with affine connection and metric. Conditions for existence of energy-momentum tensors*, Comm. JINR Dubna **E2-91-78**(1991), 1–15.
10. ———, *On the energy-momentum tensors for field theories in spaces with affine connection and metric*, In Proc. 9th Italian Conf. on Gen. Rel. and

Grav. Physics. Capri, Napoli, 25-28.09., Italy 1990. eds. Cianci R. et al., World Sci. Publ. Co., Singapore, 1991, pp. 476–480.

11. C. Eckart, Phys. Rev **58**(1940), 919.
12. A. Uhlmann, Zeitschrift der Friedrich-Schiller Universität Jena, Math.-Naturwiss. Reihe 1959/1960.
13. E.T. Newman and A.I. Janis, *Rigid Frames in Relativity*, Phys. Rev. **116**(1959), 1610–1614.
14. G. Dehnen, *Über allgemein-relativistischen Dynamik*, Wissensch. Zeitschrift der Friedrich-Schiller Universität Jena. Math. Naturw. Reihe. Jahrg. **15**(1966) 15–23, Translation in Einsteinovskii sbornik, Einstein's miscellany, 1969–1970, Energoatomizdat, Moscow, 1970 (in Russian), p. 140.
15. Ph.J. Greenberg, *Space-Like Congruences with an Application to Vorticity in Relativistic Hydrodynamics*, J. Math. Ann. Appl **30**(1970), 128–143.
16. E. Massa, *Space tensors in general relativity. I: Spatial tensor algebra and analysis*, Gen. Rel. Grav. **5**(1974), 555–572.
17. ______, *Space tensors in general relativity. II: Physical Applications*, Gen. Rel. Grav. **5**(1974), 573–591.
18. ______, *Space tensors in general relativity. III: The structural equations*, Gen. Rel. Grav. **5**(1974), 715–736.
19. V.I. Roditchev, *Theory of gravitation in an orthogonal reper*, Nauka, Moscow, 1974 (in Russian).
20. R.H. Gowdy, *Projection Tensor Hydrodynamics: Generalized Perfect Fluids*, Gen. Rel. and Grav. **10**(1979), 431–443.
21. J.S. Vladimirov, *Reference frames in the theory of gravitation*, Energoizdat, Moscow, 1982 (in Russian).
22. E. Schmutzer, *Symmetrien und Erhaltungssätze der Physik*, WTB Bd. 75. (Akademie-Verlag, Berlin, 1972); Grundlagen der Theoretischen Physik. Teil II., Wissenschaftsverlag Mannheim, Wien, Zürich, 1989, S. 1540–1541
23. V.G. Pavlenko, *Osnovii mechaniki jidkosti (Foundation of the fluid mechanics)*, Sudostroenie, Leningrad, 1988 (in Russian), pp. 18–20.
24. B.A. Dubrovin, S.P. Novikov and A.T. Fomenko, *Sovremennaja geometrija. Metodii i prilojenija (Modern geometry. Methods and applications)*, 2-nd ed. Nauka, Moscow, 1986 (in Russian).
25. C. von Westenholz, *Differential Forms in Mathematical Physics*, North-Holland Publ. Co., Amsterdam, 1978.
26. S. Manoff, *Local conserved quantities and equations for the gravitational field in the general theory of relativity*, In Proc. Second Marcel Grossman Meeting on (the Recent Developments of) General Relativity, North-Holland Publ. Co., 1982, pp. 641–646.

QUANTUM TELEPORTATION AND SPIN ECHO: A UNITARY SYMPLECTIC SPINOR APPROACH

ERNST BINZ

Lehrstuhl für Mathematik I, Universität Mannheim
D-68159 Mannheim - Deutschland
e-mail: binz@math.uni-mannheim.de

WALTER SCHEMPP

Lehrstuhl für Mathematik I, Universität Siegen
D-57068 Siegen - Deutschland
e-mail: schempp@mathematik.uni-siegen.de

It is well known that instantaneous information transfer is impossible. Due to the EPR paradox, the theory of quantum information is considered to be mysterious and counterintuitive for half a century. The concept of projective spinorial symmetry, however, definitely changed the perspective: Quantum holography is capable to unveil the mysteries of EPR singlets in terms of the theory of unitary symplectic spinors. For the quantum information transferred from the cryptographic sender to the decoding receiver, it is crucial to observe that the metaplectic group $\mathbf{Mp}(2,\mathbf{R})$ of simultaneous observables forms the uniquely determined connected two–fold covering of the symplectic group $\mathbf{Sp}(1,\mathbf{R})$. The elements of the metaplectic representation R of $\mathbf{Mp}(2,\mathbf{R})$ are viewed as intertwining operators of various incarnations of the linear Schrödinger representation of the Heisenberg nilpotent Lie group G. The mod 4 Maslov index allows an explicit construction of $\mathbf{Mp}(2,\mathbf{R})$. Its covering sheets, inherently unidentifiable by the cryptographic sender, actually prevent the phase information from being cloned by the decoding receiver. Due to the covariance of the automorphism group $\mathbf{Mp}(2,\mathbf{R})$ of the quantum logic which is associated to the unitary dual $\hat{G}$ of the Heisenberg group G, quantum holography provides an approach to the quantum teleportation phenomenon of photonics as well as the spin echo teleportation of nuclear magnetic resonance (NMR) spectroscopy, magnetic resonance imaging (MRI), and electron paramagnetic resonance spectroscopy (EPRS). Because it is based on the transition from ebits to the chirality exchange qubits via classical information, the spin echo teleportation forms the prototype of the quantum teleportation experiment which is targeted for the teleportation of the quantum state of a particle at a particular location, not the particle itself, to another destination. The classical message plays an important role for the perfect reconstruction problem of the quantum teleportation of photonics as well as NMR spectroscopy and MRI. It could not be otherwise, because any long range quantum cross–correlation between the input and the estimated output could be used to swap a superluminal phase coherent wavelet packet from the cryptographic sender to the decoding receiver. Due to the conjunction of qubits and classical bits, quantum information theory is not in conflict with special relativitytheory.

The spin echo is still striking to those who see it for the first time, as it probably was to Hahn. A nuclear magnetization, having disappear in a free induction decay, spontaneously reappears as a giant fluctuation. [J.S. Waugh (1992)]

Erwin Hahn*'s discovery of spin echoes surprised thescientific community because it seemed to be a violation of the second law of thermodynamics. The formation of the echo seemed to indicate a violation of time reversal. Violating causality is one of the strictest taboos of theoretical physics, and the formation of spin echoes seemed to break this taboo.* [M. Weger (1992)]

There was NMR *before* Erwin Hahn*'s 1950 original spin echo paper, there is* NMR *after. It is not the same* NMR. [Anatole Abragam (1992)]

We could argue that Bloch *and* Hahn *set the stage for all of modern spin choreography.* [Ray Freeman (1997)]

MRI *will become the most important clinical imaging modality by the year* 2010. *It will be the major morphological staging method for the abdomen and pelvis.* MR*-guided surgery will be commonplace, and thermal ablation of disc and soft-tissue tumours will be carried out under* MR *control. Morphology, early stroke detection, and even emergency trauma cases will done by* MR. *Brain imaging will no longer make use of* CT *by* 2010. [Gustav K. von Schulthess (1998)]

The irreversibility of time is a topic generally associated with fundamental physics. Imaging techniques, on the other hand, are usually related to engineering sciences. Remarkably, however, these two domains can be fruitfully connected to yield many applications in medicine, undersea communications, hydrodynamics, and material analysis, as well as elegant experiments in pure physics. [Mathias Fink (1997)]

An unknown quantum state can be disassembled into, then later reconstructed from, purely classical information and purely nonclassical Einstein–Podolsky–Rosen (EPR) *correlations. To do so the sender and the receiver must prearrange the sharing of an* EPR*-correlated pair of particles. Teleportation defies no physical laws. It requires, among other things, sending a classical message. The net result of teleportation is completely prosaic: The only remarkable feature is that, in the interim, the information has been cleanly separated into classical and nonclassical parts.* [Charles H. Bennett (1993)]

Quantum teleportation is a striking application of the holistic nature of the physical world revealed by quantum mechanics. The entanglement is like an invisible cat's cradle connecting the 'handset' particles, and it is delicate – the particles must well isolated from their environments to remain in a coherent entangled state. [Tony Sudbery (1997)]

A look at the physics underpinning quantum teleportation and quantum information theory in general takes us into the realm of some of the most astonishing scientific discoveries that have ever been made. It shows that the kind of quantum manipulations needed to perform quantum computations are done on a daily basis in hospitals and laboratories around the world. Although modern computers already exploit some quantum phenomena they do not make use of the full repertoire of quantum phenomena that nature provides. The devices that will be built in future will not only exceed the capabilities of their classical counterparts but they will serve as a test-bed to illuminate much of the basic physics underlying quantum phenomena. [Colin P. Williams (1998)]

The area lies at the crossroads of mathematics and physics. This adds greatly to its interest but increases the difficulty of presentation. [Michael Atiyah (1989)]

Formal perfection, in mathematics, is never negligible; it is even a necessary ingredient for progress. [Jean–Marie Souriau (1997)]

Auch für die Physik gilt der Satz, daß man nicht selig wird ohne Glauben, zum mindesten den Glauben an eine gewisse Realität außer uns. [Max Planck (Berliner Rektoratsrede 1913)]

Over the past 30 years, there has been a dramatic miniaturization in the electronic circuitry of computer technology. The chips of a computer memory still behave in accordance with the laws of classical physics which describe *local* interactions of forces. In particular, electromagnetic interactions tend to weaken the farther they are interacting from the source.

The current trend in computer technology, however, is that the chips will be the size of individual atoms. At the atomic scale, the new field of quantum computing emerges which is based on *non-local* quantum phenomena such as quantum teleportation. Quantum computing which is consistent with quantum physics has arisen in anticipation of the step to replace classical circuitry by quantum circuitry.

The young field of quantum computing presupposes the preparation of the states of a *quantum logic*. The quantum logic associated to coherent entangled singlets is determined by the ebits implemented by the unitary dual $\hat{G}$ of the Heisenberg nilpotent Lie group G. Indeed, $\hat{G}$ consists of *pairs* of equivalence classes of irreducible unitary linear representations of G. These equivalence classes admit a natural contragredient pairing. The group of automorphisms of the quantum logic determine the *projective spinorial symmetries* of the two-state quantum system ([51]), and the geometric quantization procedure involving the coadjoint orbit visualization of $\hat{G}$ reveals the fundamental insight that classical information is sandwiched by quantum information.

The sandwich configuration is formed by a pair of *entangled* planar coadjoint orbits of G which encode the ebits of entangled quantum states ([55]). It can be glued together to form the principal *sandwich bundle* over the compact 2-sphere $\mathbf{S}_2 \hookrightarrow \mathbf{R}^3$ and allows to model the long range cross-correlation between the coadjoint orbits $\mathcal{O}_U \hookrightarrow \mathrm{Lie}(G)^\star$ which encode the qubits of *chirality* exchange. The non-local correlation is based on the projective spinorial symmetries of the principal sandwich bundle over the compact 2-sphere $\mathbf{S}_2 \hookrightarrow \mathbf{R}^3$ which operate unmediated, instantaneous, and unaffected by the nature of the intervening medium on quantum states. The projective spinorial symmetries of quantum states are the hallmark of a *non-local* influence. The sandwich bundle geometrically implements the *interface* of quantum and classical information which allows the faithful transition from ebits to the chirality exchange qubits via classical information. In particular, *none* of the four known forces in the universe are actually interacting with quantum information which *solely* is governed by the projective spinorial symmetries of quantum states. Indeed, one of the fundamental achievements of quantum physics is that its predictions are independent of the electromagnetic, the gravitational, the strong nuclear, and the weak nuclear forces which are all local. The non–local interactions are not mediated, not limited to acting at the speed of light, and do not drop off in strength with distance from the source. Even the *collapse* of the quantum state vector involves *no* forces of any kind. The collapse is governed by the projective spinorial symmetries of the principal sandwich bundle, too. Also in the realm of classical information channels, temporal symmetry principles are now well recognized as an important consideration in the design of NMR pulse sequences.

The young field of quantum information exploits the entangled quantum states for new types of communication and information processing ([55]). In photonics or laser optics, there exist only a few experimental methods to generate pairs of coherent entangled photons. The most reliable method to create an Einstein-Podolsky-Rosen (EPR) singlet is the type II degenerate parametric down-conversion via a non-linear crystal. In the preparation part of the quantum teleportation experiment ([35]), an ultraviolet (UV) laser pulse with a duration of 200 fs at a wavelength of $\lambda = 490$nm is shining onto a crystal of β-barium borate (BBO) so that an incoming pump photon can decay spontaneously into a pair of coherent entangled photons at equal wavelength and orthogonal polarization. The geometric quantization of the extraordinary and ordinary Heisenberg helices provides photons of horizontal and vertical polarizations, respectively. The phase shift arises from the crystal birefringence. The quantization allows to superimpose a photon in its initial quantum state with one of the coherent entangled photons at a polarizing beam splitter (PBS) to classically detect the triple counts of photon state vector collapses by means of a coincidence logic ([25,30,6,54]).

In nuclear magnetic resonance (NMR) spectroscopy and electron paramagnetic resonance spectroscopy (EPRS), the acquired signal is very rich in measurable characteristics that reflect the nature of a population of atoms and electrons, the structure of their environment, and the way in which they interact with this environment. NMR spectroscopists have demonstrated the unitary symplectic spinor character of spin-$\frac{1}{2}$ nuclei ([48,49]). In NMR spectroscopy and EPRS, pairs of coherent entangled spins are created as ebits, and chirality qubits are exchanged via classical information in the *spin echo* teleportation experiment. The spin echo teleportation experiment transcended in its conceptual originality and elegance all that had preceded it. Today it forms one of the most versatile tools available to the NMR spectroscopist to overcome the phase information *loss* caused by decoherence. Actually the spin echo teleportation experiment solves the longitudinal decoherence problem of NMR spectroscopy and EPRS consisting in the signal degradation of magnetic spin isochromats of *no* set phase difference by clever nuclear spin manipulations at the interface of classical and quantum physics.

The manipulation of the external magnetic environment in space and time allows to modify the NMR signal without significantly affecting material structure. Because spin systems take a longer time period to lose transversal phase coherence by the irreversible spin state coupling to other spins of the environment, there is time enough to perform longitudinal coherence improving spin manipulations. For this reason, the spin echo teleportation concept pervades the whole of the NMR spectroscopists' methodology.

As described in the spin echo classic [20], the original formation of the Hahn spin echo teleportation produces the primary and stimulated spin echo teleportation by a $(\frac{\pi}{2}, \frac{\pi}{2}, \frac{\pi}{2})$-pulse sequence, composite of *two* successive radiofrequency pulse excitations after the initial $\frac{\pi}{2}$-flip. When the radiofrequency pulse sequence is extinguished, spin echoes persist for a time period of the order of the spin-spin relaxation time. Besides the ordinary primary spin echoes, an analysis of the stimulated secondary spin echoes reveals that a wide variety of spin echo responses are available for experiment, all of them directly related to the time intervals between the successive perturbations and therefore accessible to the observer's control ([17]). In the quantum teleportation experiments, which are targeted for the teleportation of an *unknown* pure quantum state via (*not* through) EPR communication channels and the complete reconstruction by means of two bits of classical information through a dual classical information channel, it has not been possible to produce coherent entangled states of more than two photons. Thus the manipulation of coherent entangled photon states is limited.

Rumour has it that Erwin Louis Hahn discovered the spin echo teleportation experiment by accident when his radiofrequency excitation pulse unit malfunctioned ([31]). However, it was his genius to appreciate that he should *not* force the delicate formation of the spin echo teleportation to the model of interference beat phenomena derived from a linear superposition of the individual responses for an ensemble of oscillators having either slightly different frequencies, or the same frequency but loosely coupled together. One day, in the late autumn of 1949, Hahn realized that he had discovered an excitation pulse sequence which provided a response at time intervals controlled by the investigator, in contrast to the usual beat interferences which are controlled by frequency differences specific to the sample and so do not give rise to a variable parameter. The brilliant conceptual step in allowing a second radiofrequency pulse excitation after the initial $\frac{\pi}{2}$-flip to initiate a re-phasing of the magnetic spin system in order to compensate for the phase information loss of the maximally mixed state could account for the spin echo teleportation phenomenon ([1]). The phase information loss which is produced by the fanning out of dispersed magnetic spin isochromats is implemented by the $\mathbf{O}(2, \mathbf{R})$-invariance of the tracial frame functions on the compact unit sphere $\mathbf{S}_2 \hookrightarrow \mathbf{R}^3$. Because they produce the Kepplerian *integrality condition*, they are a crucial tool in the Gleason geometric quantization procedure ([18]). The spin echo teleportation technique of NMR spectroscopy became a reality.

The conventional spin echo process is conceptually easier to follow. It generates the spin echo teleportation phenomenon by the Carr-Purcell technique of a $(\frac{\pi}{2}, \pi)$-sequence ([10]). In a letter from Hahn to Carr dated November 14, 1950, Hahn noted that the $(\frac{\pi}{2}, \pi)$-sequence was the one he always used in explaining the spin echo phenomenon ([31]) by an application of the "pancake flip".

The Carr-Purcell-Meiboom-Gill (CPMG) spin echo technique is based on a coherent $(\frac{\pi}{2}, \pi)$-pulse train with the modification of introducing an additional $\frac{\pi}{2}$-phase shift ([32]). All of these various radiofrequency excitation pulse trains stress the significance of the second perturbation at a specified *time* in contrast to a specified distance. The spin echo teleportation re-establishes a coherent state of the magnetic spin system, therefore, at a time controlled by the observer. Due to the time control over the two-fold covered tori, timing is everything in NMR spectroscopy, magnetic resonance imaging (MRI), and EPRS ([13,27,42,43]).

The magnetic moments of nuclei are typically 1000 times smaller and more difficult to detect than those of electron spins ([13]). The nuclear spin echoes are weak signals and are liable to be swamped by the lingering effects of the radiofrequency excitation pulse trains. There are several ways in which these lingering effects can manifest themselves. For instance, the amplifier following the detector and the power supplies feeding the amplifier may not have recovered. These problems are the major source of difficulties that limit the effectiveness with which magnetic spin echoes can be detected after short $\frac{\pi}{2}$- to π-pulse delay times. Overall, it is very helpful to employ a linear gate, implemented by a matched filter, to help suppress the remaining effects of the radiofrequency excitation pulse train.

It is worth emphasizing that the spin echo teleportation is especially valuable in that it engineers a response from the magnetic spin system when there are no sources of interference from the oscillating magnetic fields that interact with the magnetic spin isochromats. Hence the signal-to-noise ratio can be close to its ideal value ([33]).

A whole new science has grown up, based on the *manipulation* of magnetic spin isochromats, but the inspiration clearly comes from the early work of Hahn ([20]), Carr and Purcell ([10]), and Meiboom and Gill ([32]). What has followed is essentially a set of clever extensions of their designs which makes nuclear spin choreography a relatively straightforward reprogramming operation.

The non-local emitter-absorber transaction of quantum physics ([12]) admits a mathematical formulation in terms of Fourier analysis on the Heisenberg nilpotent Lie group G ([40]). The Heisenberg group G implements the Heisenberg uncertainty principle at the Lie algebra level Lie(G). The point is to include a reconstructing identification process into the emergence of the interference pattern and its read-out procedure. In quantum holography, the interference phenomena are described by the non-local quantum cross-correlations which include the phase information as an essential ingredient ([26]). Applications *geometrically* implemented by the principal sandwich bundle over the compact 2-sphere $\mathbf{S}_2 \hookrightarrow \mathbf{R}^3$ are the Aharonov–Bohm effect of electron holography, and the fractional quantum Hall effect of condensed matter physics ([4]).

Notice that the full phase information needed to completely reconstruct the state of an object can be divided into two parts, quantum *and* classical information. The former can be transferred *instantaneously*, but it cannot be used without the latter, which can only be transmitted at the speed of light or even slower ([50]). As a consequence, instantaneous information transfer is impossible. Taking advantage of the holistic aspect of quantum physics, both parts, quantum communication channels and classical information channels, have to be modelled group representationally by the semantic filter bank concept which is basic for the subband coding of quantum holography as well as the theory of quantum information transfer ([15]).

The concept of virtual image in optical holography suggests to introduce the important concept of *contragredient* group representation into quantum holography. The contragredient representation $\check{U}$ is deduced from the conjugate representation $\bar{U}$ by the canonical anti-automorphism of the representation Hilbert space associated with the unitary group representation U of central differential i onto its vector space dual ([46,47]). Whereas $\bar{U}$ acts on the anti-dual and has the complex conjugate central differential $-i$, $\check{U}$ acts on the dual of the representation Hilbert space associated with the unitary group representation U and has the same central differential $-i$. The linkage of the mirrored pair

$$(U, \check{U})$$

has a cohomology theoretic background. Indeed, the irreducible unitary linear representation U of the nilpotent Lie group G appears in the L^2-cohomology of the homogeneous line bundle over the associated coadjoint orbit $\mathcal{O}_U \hookrightarrow \mathrm{Lie}(G)^\star$ exactly when $\check{U}$ admits non–trivial Lie algebra cohomology ([39]). The quantum teleportation phenomenon suggests a geometrical realization of the discrete series of G on the L^2-cohomology by the compact 2–sphere $\mathbf{S}_2 \hookrightarrow \mathbf{R}^3$ with antipodal points identified.

The concept of "entanglement" is in correspondence to Erwin Schrödinger's concept of "Verschränktheit". Both are related to the Greek "συμπλέκειν" which is at the origin of Hermann Weyl's concept of "symplectic" space. In terms of the theory of symplectic spinors ([24,19,4]), the coherent path entanglement is described by the irreducible tensor product representation ([42])

$$U \hat{\otimes} \check{U}.$$

The irreducibility allows to consider this representation as a *singlet* state of a particle instead of two separate particles carrying the *twin aspects* of the quantum teleportation phenomenon ([2,3,50,55]). Under coherent entanglement, the original particle is not available in its original state. The irreducibility, however, allows to mathematically realize the coherent superposition principle which is basic for quantum physics, and to cohomologically classify the *automorphisms* of the quantum logic associated to the two–state quantum system and the EPR communication channels. In this way, the metaplectic representation R adopts a quantum informational meaning. Note however, that the metaplectic representation is a singular unitary representation and that its existence has *not* been recaptured by any systematic procedure.

The existence of the metaplectic representation R is predicted by geometric quantization ([24]), and has its roots in the quantization rules of the early quantum mechanics. It is based on two observations. One, the Stone-von Neumann theorem which asserts that for the Heisenberg nilpotent Lie group G, given a central character χ, there exists up to unitary equivalence a unique irreducible unitary representation U with χ as its central character. It is reasonable to regard square integrability as an essential part of the Stone-von Neumann theorem, because a unitary representation of a nilpotent Lie group is determined by its central character χ if and only if it is square integrable modulo center. The *strong* Stone-von Neumann theorem implies unicity plus square integrability. Second, the symplectic group $\mathbf{Sp}(1,\mathbf{R})$ is the group of continuous automorphisms $\eta(s)$ of G which fix the center of G pointwise.

The metaplectic representation R has a strong impact on the quantum teleportation phenomenon because the covariance identity

$$R(s)^{-1} \circ U \circ R(s) = U \circ \eta(s)$$

of the intertwiner R ([52]) and the covering projection η performs the transition to the long range cross-correlation between EPR communication channels by *transferring* the action of the metaplectic group

$$\mathbf{Mp}(2,\mathbf{R})$$

from the irreducible unitary representation U onto the contragredient representation $\check{U}$. Thus, the Stone-von Neumann theorem, or the uniqueness of the representation of the canonical Heisenberg commutation relations, in its *strong* group representational formulation, reappears in the quantum holographic regime. In this regime, the preparation of a replica of the original quantum state can be contragrediently *reversed*, but it cannot be duplicated. There is *no* collision of quantum teleportation with relativity theory.

The covariance projection η is at the fundament of the non-cloning principle ([56]) and therefore illustrates an *essential* feature of quantum information theory. The quantum information can be swapped from one system to another, but it *cannot* be cloned. In this regard, quantum information is unlike classical information which actually can be duplicated. The non–locality inherent to quantum information theory therefore is not in a conflict to special relativity theory.

Conversely, the covariance identity characterizes, up to unitary equivalence, the metaplectic representation R and the contragredient representation $\check{R}$ of $\mathbf{Mp}(2,\mathbf{R})$. Notice the analogy to Wigner's theorem concerning the group

$\mathrm{Gal}_{\mathbf{R}}\mathbf{C}$ of order $[\mathbf{C}:\mathbf{R}] = 2$ and its action on transition probabilities ([35,51]). The quantum-mechanical formalism makes extensive use of the operation of complex conjugation and square integrability because the probability is equal to the absolute square of state vectors rather than to the absolute value, or to the real part as Max Born first suggested.

In the context of quantum teleportation, the EPR entangled pair of particles serve as two ends of a quantum communication channel, one particle being retained by the person wishing to teleport the quantum state and the other by the person wishing to receive it. Thus, in order to teleport a quantum state, the sender and receiver must each already possess one member of a pair of EPR entangled particles.

The quantum teleportation experiment can be looked at as a quantum informational transfer realized by the tensor product representation $R \otimes \check{R}$ of the metaplectic representation R and its contragredient $\check{R}$. The experimental verification of this realization via the *simultaneous* coincidence counts of photon state vector collapses ([12]) is included in quantum holography.

It is a remarkable fact that, prior to the experiments establishing the quantum teleportation phenomenon, the non–isomorphic unitary linear representations U and $\check{U}$ have been called "physically equivalent" ([51]). In order to *represent* the non-local phenomena as demonstrated by the quantum teleportation experiments ([5,6]), however, it is absolutely essential to make a chirality *distinction* between the unitary linear representation U and the *unitarily inequivalent* contragredient representation $\check{U}$.

Theorem (Deterministic Entanglement Theorem). *Let G denote a connected, simply connected nilpotent Lie group with real Lie algebra* Lie(G), *dual* Lie$(G)^\star$ *of the vector space* Lie(G), *and unitary dual $\hat{G}$. The following conditions are pairwise equivalent*

(i) The irreducible unitary representation $U \in \hat{G}$ is square integrable modulo its kernel.

(ii) The coadjoint orbit $\mathcal{O}_U \hookrightarrow$ Lie$(G)^\star$ *associated with the equivalence class of U is an affine–linear symplectic manifold.*

(iii) The one-dimensional identity representation 1 of G associated with the collapsed coadjoint orbit $\{0\}$ is weakly contained as a completely entangled state in the mixing tensor product of the reconstructing read-out process

$$U \hat{\otimes} V,$$

where $V \in \hat{G}$, if and only if

$$V = \check{U}$$

is the contragredient representation of U *in the set* $\hat{G}$ *of equivalence classes of irreducible unitary linear representations of* G. *The coherent entangled representation*

$$U\hat{\otimes}\check{U}$$

is of Hilbert–Schmidt class.

The proof depends upon the smooth direct integral decomposition of representations of nilpotent Lie groups commuting with Lie algebra cohomology ([39]).

For the Heisenberg nilpotent Lie group G, the deterministic entanglement theorem is a consequence of the strong Stone-von Neumann theorem and therefore at the heart of quantum physics. Its group representational aspects are geometrically implemented by the projective spinorial symmetries of the principal sandwich bundle over the compact 2-sphere $\mathbf{S}_2 \hookrightarrow \mathbf{R}^3$ which encodes the ebits of entangled quantum states. As a special case, the deterministic entanglement theorem implies Gleason's geometric quantization procedure ([18]) which asserts that every state of a standard quantum logic can be described in the canonical way by the *trace class* operators of a complex Hilbert space.

Theorem (Quantization Theorem). *Let* $\mathcal{W}$ *denote the convex set of all bounded, self-adjoined, non-negative, trace-class operators of unit trace acting on the standard complex Hilbert space* $L^2(\mathbf{R})$. *Then the mapping*

$$U \rightsquigarrow p_U$$

defines a convex isomorphism of $\mathcal{W}$ *onto the convex set of all states of the standard logic of* $L^2(\mathbf{R})$. *For each closed subspace* $M \hookrightarrow L^2(\mathbf{R})$ *with orthogonal projection*

$$P_M : L^2(\mathbf{R}) \longrightarrow M,$$

the tracial identity

$$p_U(M) = \mathrm{tr}\,(P_M \circ U) \qquad (U \in \mathcal{W})$$

holds. A state p *in the logic of* $L^2(\mathbf{R})$ *is pure if and only if*

$$p = p_u$$

holds for a unit vector $u \in L^2(\mathbf{R})$, *where*

$$p_u(M) = < P_M u | u > = \|P_M u\|^2,$$

and

$$p_u = p_v$$

if and only if

$$v = c.u \qquad (c \in \mathbf{T})$$

holds. Therefore the pure states p in the logic of the standard complex Hilbert space $L^2(\mathbf{R})$ *are in a natural one-to-one correspondence with the elements of the complex projective space* $\mathbf{P}L^2(\mathbf{R})$.

Although Gleason's geometric quantization procedure in its *projective* formulation ([51]) is a cornerstone of the mathematical foundations of quantum physics outlined in the spirit of the Mackey theory, and one of the most profound theorems in this subject, the majority of physicists is not familiar with this result. Therefore, some comments are in order.

The three-dimensional Heisenberg nilpotent Lie group G is neither commutative nor compact, as can be seen from the global parametrization

$$\begin{pmatrix} 1 & x & z \\ 0 & 1 & y \\ 0 & 0 & 1 \end{pmatrix} \qquad (((x,y),z) \in (\mathbf{R} \oplus \mathbf{R}) \times \mathbf{R}),$$

but its non-trivial irreducible unitary linear representations $U \in \hat{G}$ are square integrable modulo center. Equivalently, the coadjoint orbits of dimension > 1 are *flat* manifolds ([40]). Notice that if a connected, simply connected nilpotent Lie group has one such square integrable representation, then almost all, with respect to Plancherel measure, irreducible unitary representations are square integrable and belong to the *reduced* dual. Thus the dual and the reduced dual are equal.

The coadjoint orbit method encompasses many of the deepest results of the representation theory of Lie groups. For the coadjoint orbit visualization of $\hat{G}$, the *weak containment* is the key concept already present in the subband decomposition performed by Keppler's conchoid construction ([41]). It leads to the hull-kernel topology of group representation theory ([16]) and ensures the approximation property of *positive* Schwartz kernels in $\mathcal{S}'(\mathbf{R}\oplus\mathbf{R})$ ([46,47]). Due to the coadjoint orbit visualization of the representation $1 \in \hat{G}$ and Gleason's geometric quantization theorem, the *prearranged* weak containment of the positive Schwartz kernel forming a completely entangled state permits the differentiation between the two-states of quantum holography and classical optical holography.

In the emitter-absorber transaction the projective spinorial symmetries of quantum states mean time-reversal invariance *combined* with spatial reciprocity. In mathematical terms, it means invariance under the orientation reversing conjugation transfer

$$U \rightsquigarrow \check{U}$$

which cohomologically models the basic projective spinorial symmetry of the non-local cross-correlation associated with the coherent entanglement experiment. This long range cross-correlation is in terms of phase and spatial frequency, and is actually *independent* of the spatial distances involved in the coherent entanglement experiment for ebit encoding. Indeed, the cryptographic sender knows neither the quantum state to be transferred to the destination nor the location of the intended receiver.

In the preceding theorem, note that the tensor product representation

$$\sigma_U = U \hat{\otimes} \check{U}$$

of $G \times G$ is *irreducible*. Due to Schur's lemma, σ_U defines a *pure* quantum state. Moreover, σ_U is in a natural way unitarily isomorphic to the coherent entangled representation of $G \times G$ given by the assignment

$$\sigma_U : T \rightsquigarrow U \circ T \circ \check{U}$$

with Hilbert-Schmidt operators T and the associated kernel functions ([52]). The isomorphism arises via the standard procedure of linear extension of dyads by continuity from the class of operators of finite rank to the class of Hilbert-Schmidt operators (von Neumann-Schatten 2-class).

For the realization of states in a standard quantum logic it should be noticed that the product of two Hilbert-Schmidt operators is trace class (von Neumann-Schatten 1-class), and conversely, every trace class operator factorizes into the product of two Hilbert-Schmidt operators. The Hilbert-Schmidt operators form a complex Hilbert space under the standard trace scalar product. They admit extensions to Schwartz kernels, and the associated kernel functions extend with respect to the symplectic convolution product

$$(\varphi, \psi) \rightsquigarrow \varphi \natural \psi$$

to factorized kernel distributions.

Consider the symplectically factorized distribution of positive type

$$D = \varphi \natural \check{\bar{\varphi}},$$

where φ is square integrable modulo center. Then D is a tempered distribution modulo center and defines a *positive* Schwartz kernel. Note that a trace class operator $U(D)$ only implies a factorization of the type $D = \psi_1 \natural \psi_2$ with factors that are square integrable modulo center.

By realizing σ_U in terms of the Dirac measure $\varepsilon_{(.,.)}$, the irreducible representation σ_U yields the left/right symplectic shift identity

$$U((\varepsilon_{(-x,-y)} \natural \varphi) \natural (\check{\varphi} \natural \varepsilon_{(x,y)})) = U(\exp_G(x,y))^{-1} \circ U(D) \circ U(\exp_G(x,y))$$

for $(x, y) \in \mathbf{R} \oplus \mathbf{R}$. The operator $U(\exp_G(x, y))$ is of course *unitary*, and conjugating by a unitary operator preserves *simultaneous* observability, *positivity*, and is *isometric*. The coordinates (x, y) depend on the digital computer controlled gradient switching protocol. For the reconstruction process, they have to be transferred through classical information channels. Quantum communication channels afforded by ebits are used in conjunction with classical information channels. Hence the classical message plays an important role for the perfect reconstruction problem of the quantum teleportation performed by MRI.

It follows for the trace

$$\operatorname{tr} U(D) = D(0).$$

In terms of symbols, the symplectic Fourier transform

$$\hat{D} = \frac{1}{2} \cdot D \natural 1$$

gives rise to the trace class operator $U(\hat{D}) = \frac{1}{2} \cdot U(D \natural 1)$. The trace is given by the orbital integral

$$\operatorname{tr} U(\hat{D}) = 2 \int_{\mathbf{R} \oplus \mathbf{R}} D \, \mathrm{d}x \wedge \mathrm{d}y.$$

The distributional extensions give rise to a phase history summation by a Feynman path integral ([38]).

By restriction to the diagonal of $G \times G$, the internal tensor product representation $U \hat{\otimes} \check{U}$ of G contains U as well as $\check{U}$. In the context of quantum information theory, the prearranged *weak* containment of the one-dimensional identity representation 1 as a completely entangled state implies the non-local, distance independent resonance conditions for the coherent entangled states. After the transmission of quantum states via a EPR communication channel, the weak containment of the one-dimensional identity representation 1 allows a *recombination* at the intended destination in order to *reconstruct* the original quantum state.

In the case of the three-dimensional Heisenberg nilpotent Lie group G the Hilbert-Schmidt kernels associated with T define a *semantic filter bank*.

Note also that the one-dimensional identity representation 1 of the Heisenberg group G induces as a completely entangled state the distributional *reproducing* kernel ([42,46,47])

$$1\hat{\otimes}1$$

with respect to a rotating frame of symplectic coordinates attached to the planar coadjoint orbit $\mathcal{O}_U \hookrightarrow \mathrm{Lie}(G)^\star$. Thus in quantum holography, the two-fold covered coadjoint orbit $\mathcal{O}_U$ provides the affine symplectic *state projection* plane $\mathbf{R} \oplus \mathbf{R}$ for the four EPR-Bell states

$$\{J^0, J^1, J^2, J^3\}$$

of quantum teleportation ([2]), and the polyphase holograms generated by phase coherent wavelet interference as the circular grating arrays of magnetic resonance imaging (MRI). The EPR-Bell states of polarized photons

$$\{\text{identity}, \text{phase shift}, \text{bit flip}, \text{combined phase shift/bit flip}\}$$

are in correspondence to the 4–periodicity of the ordinary one–dimensional Fourier cotransform

$$\bar{\mathcal{F}}_{\mathbf{R}} = R(J)$$

acting on the standard complex Hilbert space $L^2(\mathbf{R})$ ([52]) and are given by the dual string of k-vectors

$$\{\bar{\mathcal{F}}^0_{\mathbf{R}}, \bar{\mathcal{F}}^1_{\mathbf{R}}, \bar{\mathcal{F}}^2_{\mathbf{R}}, \bar{\mathcal{F}}^3_{\mathbf{R}}\}.$$

The affine symplectic state projection plane carries $\mathbf{R} \oplus \mathbf{R}$ the irreducible non–local information content of quantum physics ([7]) and allows constructive as well as destructive interferences of the dual string.

The planar coadjoint orbits $\mathcal{O}_U \hookrightarrow \mathrm{Lie}(G)^\star$ are transversal to the propagation direction C of quantum teleportation, and the central direction C of the driving oscillator of MRI, respectively. Their symplectic structure is given by the skew-symmetric wedge product of linear differential forms

$$(\alpha, \beta) \rightsquigarrow \alpha \wedge \beta$$

with respect to an orthonormal basis $\{P, Q\}$ of the symplectic plane $\mathbf{R} \oplus \mathbf{R}$. The Weyl symplectic matrix associated to the rotational curvature form is given by ([52])

$$J = \begin{pmatrix} 0 & -1 \\ 1 & 0 \end{pmatrix} \in \mathbf{SO}(2, \mathbf{Z}).$$

Notice that the matrix J implements the phase sensitive quadrature detection used in Keppler's strategy of physical astronomy ([45]). The *central* reflection matrix

$$J^2 = \begin{pmatrix} -1 & 0 \\ 0 & -1 \end{pmatrix} \in \mathbf{SO}(2, \mathbf{Z})$$

of bit flipping corresponds to the squared Fourier cotransform

$$\bar{\mathcal{F}}_{\mathbf{R}}^2 = R(J)^2$$

embodied by the polyphase hologram. The identity

$$J^2 = -J^0$$

shows that J defines a complex structure of the plane $\mathbf{R} \oplus \mathbf{R}$. Finally, the chirality exchange qubit matrix

$$J^3 = \begin{pmatrix} 0 & 1 \\ -1 & 0 \end{pmatrix} \in \mathbf{SO}(2, \mathbf{Z})$$

of phase shift/bit flip ebits corresponds to the direct Fourier transform

$$\mathcal{F}_{\mathbf{R}} = R(J)^{-1}$$

of the real line $\mathbf{R}$.

In contrast to the special orthogonal group $\mathbf{SO}(2, \mathbf{R})$ which is isomorphic to the one-dimensional compact torus group

$$\mathbf{T} = \mathbf{R}/\mathbf{Z},$$

the orthogonal group $\mathbf{O}(2, \mathbf{R})$ acting on the symplectic plane $\mathbf{R} \oplus \mathbf{R}$ is non-commutative. By interchanging the rows of J, the orientation reversing *line* reflection matrix

$$\begin{pmatrix} 1 & 0 \\ 0 & -1 \end{pmatrix} \in \mathbf{O}(2, \mathbf{Z}) - \mathbf{SO}(2, \mathbf{Z})$$

with respect to the orthonormal basis $\{P, Q\}$ of the symplectic plane $\mathbf{R} \oplus \mathbf{R}$ arises. An interchange of the columns of J provides the orientation reversing *conjugate* or complementary line reflection matrix

$$\begin{pmatrix} -1 & 0 \\ 0 & 1 \end{pmatrix} \in \mathbf{O}(2, \mathbf{Z}) - \mathbf{SO}(2, \mathbf{Z}).$$

The line reflection identities

$$\mathbf{O}(2, \mathbf{R}) - \mathbf{SO}(2, \mathbf{R}) = \mathbf{SO}(2, \mathbf{R}) \cdot \begin{pmatrix} 1 & 0 \\ 0 & -1 \end{pmatrix},$$

and

$$\mathbf{O}(2,\mathbf{R}) - \mathbf{SO}(2,\mathbf{R}) = \mathbf{SO}(2,\mathbf{R}) \,.\, \begin{pmatrix} -1 & 0 \\ 0 & 1 \end{pmatrix}$$

follow. The orientation reversing orthogonal matrices perform the transition of chirality exchange qubits via classical information to the ebit encoding planar coadjoint orbits. Because the exchanged coadjoint orbit $\mathcal{O}_{\check{U}}$ of the principal sandwich bundle is associated to the contragredient representation $\check{U}$ of G, and the line reflections should not be confused with the powers of J displayed above. A careful chirality distinction between the orientation preserving rotations of the EPR-Bell states string, and the orientation reversing line reflections induced by the contragredient cover transition actually is the key to their *simultaneous* coincidence detection performed in quantum teleportation experiments aiming to experimentally test the chirality by photons of a double refracting polarizer ([43,44]). In the actual experimental setup, the reflections are performed by the backprojection of calcite crystals which are transmitted through the BBO crystal ([5]). To align the system, the beam splitter and the back reflecting mirror acting on the path b_2 are mounted on a micrometrical stage which allows an incrementation of step size 0.1μm by means of a digital computer controlled stepping procedure.

The symplectic structure J of the planar coadjoint orbit n $\mathcal{O}_U \hookrightarrow \mathrm{Lie}(G)^\star$ is also given by the two transverse Lagrangian lines

$$\{\mathbf{R}, i.\mathbf{R}\}$$

of linear polarization which subtend the planar $\frac{\pi}{2}$-rotation of the orthonormal basis $\{P, Q\}$ of the symplectic plane $\mathbf{R} \oplus \mathbf{R}$, or equivalently, the symplectic basis $\{1, i\}$ of the complex plane $\mathbf{C}$. Indeed, the field $\mathbf{C}$ forms the Clifford algebra of the one-dimensional real vector space $\mathbf{R}$ under the quadratic form

$$x \rightsquigarrow -x^2,$$

and the action of $i \in \mathbf{C}$ is given by the real automorphism J of $\mathbf{R} \oplus \mathbf{R}$.

Notice that the Lagrangian lines $L \hookrightarrow \mathbf{C}$ are in bijection to the abelian polarizing subalgebras $\mathcal{P}$ of the complexified Heisenberg Lie algebra $\mathrm{Lie}(G) \otimes_{\mathbf{R}} \mathbf{C}$ ([24]). Then $\dim_{\mathbf{C}} \mathcal{P} = 2$ and $\mathcal{P}$ acts by multiples of a homogeneous first–order polynomial differential operator on the complex vector space of $\mathcal{C}^\infty$-vectors for U and $\check{U}$ which actually coincides with the complex Schwartz space $\mathcal{S}(\mathbf{R})$ of rapidly decaying functions on the real line $\mathbf{R}$. If $\mathcal{S}'(\mathbf{R})$ denotes the complex vector space of tempered distributions on the real line $\mathbf{R}$, the inclusions

$$\mathcal{S}(\mathbf{R}) \hookrightarrow L^2(\mathbf{R}) \hookrightarrow \mathcal{S}'(\mathbf{R})$$

hold.

A computation of the associated Lie algebra cohomology groups allows an identification of the two non–vanishing cohomology groups $H^q(\mathcal{P} \cap \mathbf{C}, \mathcal{S}(\mathbf{R}))$ of degree *or* negativity index $q = 0$ and $q = 1$ with the symplectic plane $\mathbf{R} \oplus \mathbf{R}$ endowed with its complex structure J ([39]).

Associated with any three Lagrangian lines L_1, L_2, L_3 in $\mathbf{C}$ is the quadratic form

$$L_1 \oplus L_2 \oplus L_3 \ni (w_1, w_2, w_3) \rightsquigarrow \det(w_1, w_2) + \det(w_2, w_3) + \det(w_3, w_1).$$

Its signature gives rise to the mod 4 *Maslov index* which is defined according to the equation ([36])

$$\tau_0(L_1, L_2, L_3) = p - q \in \mathbf{Z}.$$

The transverse Lagrangian lines L consist of four *oriented* Lagrangian lines isomorphic to $\mathbf{R}$ inside the complex plane $\mathbf{C}$ of the skew-symmetric bilinear form

$$(u, v) \rightsquigarrow -\Im u\bar{v}.$$

The four EPR-Bell states are in correspondence to oriented Lagrangian lines and form a realization of the $\frac{\pi}{2}$-phase shift characteristic for the CPMG spin echo technique of NMR spectroscopy ([23]).

The symplectic structure gives rise to the symplectic Fourier transform by the left convolution structure, transferred from G via a cross-section to the one-dimensional center $C \hookrightarrow G$. Notice that the symplectic Fourier transform has periodicity 2 instead of the ordinary two-dimensional Fourier transform. The Schwartz kernel ([46,47]) of the symplectic Fourier transform is the one-dimensional identity representation 1. The convolution structure includes the tempered distributional sweep $1 \hat{\otimes} 1$ of the symplectic plane G/C ([42]).

Examples of the projective spinorial symmetries of entangled quantum states ([12]) include:

(i) The quantum teleportation experiments using coherent entanglement of separate horizontally and vertically photon states. The experimental setup consists not only of the arms of a balanced Mach-Zehnder interferometer but of two closed entangled photonic loops which are equipped with symmetrically adjusted taps for the 'handset' photons. The spatial cross-correlation between the linear polarizations of the coherent entangled pair of photons is converted to a long range cross-correlation between their paths ([50]). The outcome of the quantum teleportation experiment is the *non-locality* of the *simultaneously* detected handset photons in their horizontal and vertical polarizations, or the coincidence of the line reflections of the two-state quantum system.

Thus quantum teleportation permits the transfer and reconstruction over arbitrary distances of the quantum states. Experimental quantum teleportation establishes that the instantaneous influence of an initial quantum state at a distant location which Einstein once called "spooky action-at-a-distance" effect actually is a groupn representationally predicted "telepathic" reality the *non–locality* of which is *not* in a conflict to special relativity theory. Due to the projective spinorial symmetries of quantum states which are implemented by the Heisenberg nilpotent Lie group G, there is nothing speculative about the physical basis for quantum teleportation. The "action-at-a-distance" effect is real and reality really is, due to the projective spinorial symmetries, non–local, but *not* spooky.

> *"Es erscheint hart, dem Herrgott in die Karten zu gucken. Aber daß er würfelt und sich sogar telepathischer Mittel bedient, wie es ihm von der gegenwärtigen Quantentheorie zugemutet wird, kann ich keinen Augenblick glauben."*

Einstein's statement is in agreement with Newton's quote. The point, however, is that in quantum teleportation not the particle itself but its quantum state is transferred and that, due to the non-cloning principle of quantum information theory, the transfer of quantum information has to be separated into a EPR communication channel and a classical information channel. Thus, the swapping of quantum information is unlike classical information which actually can be duplicated. The non-locality of quantum information theory therefore is not in a conflict to special relativity theory.

"That one body may act upon another at a distance through a vacuum without the mediation of anything else... is to me so great an absurdity, that I believe no man, who has in philosophical matters a competent faculty for thinking, can ever fall into."

(ii) Magnetic resonance imaging (MRI) of medical diagnostics for cross-sectional imaging the human body in-viva. The emitter corresponds to nuclear magnetic induction (Bloch's method), the absorber corresponds to nuclear magnetic absorption (Purcell's method): East meets West. The conjugation corresponds to the spin echo concept ([41,42,44]). It follows that the spin echo phenomenon is a quantum teleportation phenomenon. Notice, however, that even a *soft* X-ray photon contains over 10^{12} times the energy of one photon in clinical MRI.

(iii) Nuclear magnetic resonance microscopy involves the acquisition of phase coherent wavelets in the presence of linear magnetic field gradients ([9,28]).

(iv) Confocal scanning laser microscopy (CSLM) for cross-sectional imaging optical transparent specimen without physically slicing them into thin planar sections ([11,42]).

(v) Time reversed acoustics. In the experiments performed by Mathias Fink, the time-reversal operation is performed on a limited angular area, called a time-reversal mirror. Applications include compensation for multipath propagation in waveguide channels or cavities. In the medical area, one of the first clinical applications of the pulse echo time-reversal mirror is the destruction of stones in the kidneys and gallbladders. To locate a reflecting target such as a kidney stone in its environment of other stones and organ walls, one illuminates the region of interest (ROI) with a few elements of the transducer matrix. The reflected signals are then recorded by the whole matrix and time-reversed. When the process is iterated several times, the ultrasonic beam converges towards the most reflective area of a stone. The time-reversal iteration selects one of those spots. Once the spot has been reliably located, intermittent amplified pulses can be applied to shatter the stone. As breathing moves the stone, the process is repeated to track it in real time.

It sometimes transpires that mathematics and physics, pursuing quite different agendas, find that their intellectual wanderings have converged upon the same fundamental ideas and that, once it is recognized that this mysterious convergence has occurred, each field breathes new life into the other. The classic example is the symbiosis between quantum physics and symplectic geometry. As clinical MRI amply attests, the results of such an interaction can be not only spectacular but even extremely applicable and useful for the cogniscenti in radiology ([21]).

The symmetry group of the planar coadjoint orbits $\mathcal{O}_U \hookrightarrow \mathrm{Lie}(G)^\star$ of G under their natural symplectic structure $\wedge$ of matrix representation J is given by the metaplectic group $\mathbf{Mp}(2,\mathbf{R})$, a two-fold cover η of the unimodular group $\mathbf{SL}(2,\mathbf{R})$. Indeed, $\mathbf{SL}(2,\mathbf{R})$ coincides with the symplectic group $\mathbf{Sp}(1,\mathbf{R})$ which acts as a group of automorphisms of $\mathrm{Lie}(G)$ and hence on G. Each fiber of the canonical surjection

$$\eta : \mathbf{Mp}(2,\mathbf{R}) \longrightarrow \mathbf{Sp}(1,\mathbf{R})$$

consists of two points ([52]). Thus the fibres of the covering projection η correspond to the element

$$(\mathrm{id}, -\mathrm{id}) \in \mathbf{Z}/2\mathbf{Z}$$

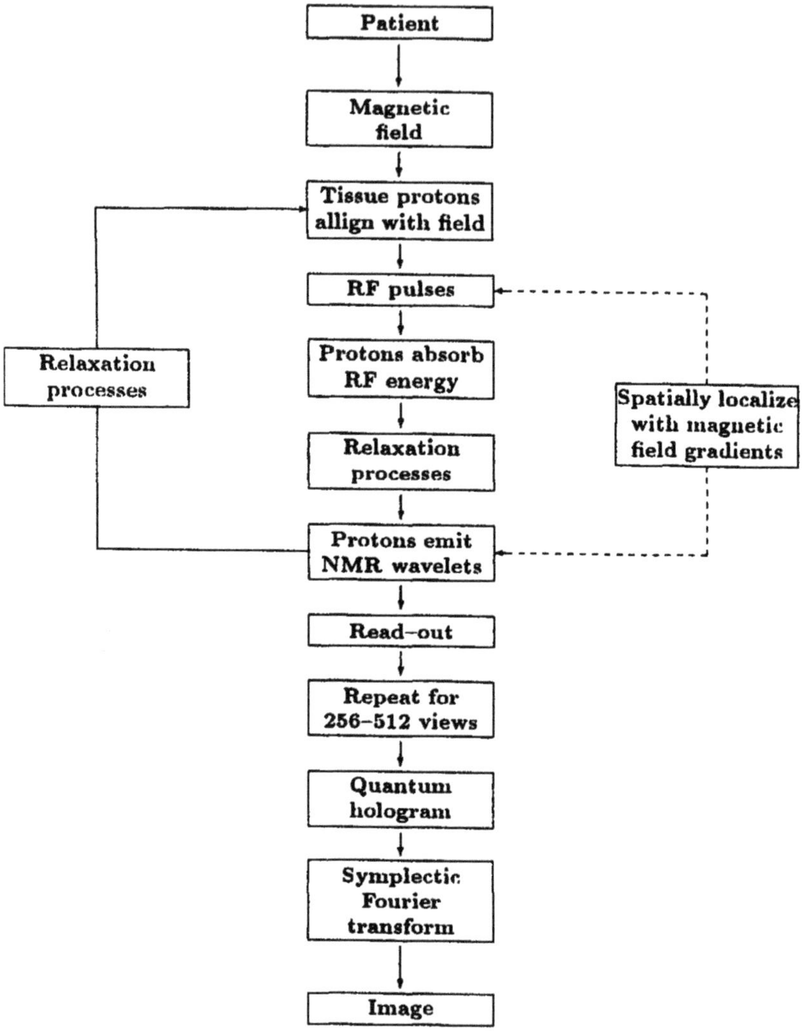

Figure 1. Diagram displaying the main principles on which a clinical MRI scanner is based. RF means radiofrequency. RF excitation pulse trains are used to induce the spin echo teleportation phenomenon. In the read-out procedure, the symplectic Fourier transform algorithm reflects the basic structural feature of the MRI modality. The classical message plays an important role for the perfect reconstruction problem of the quantum teleportation of MRI.

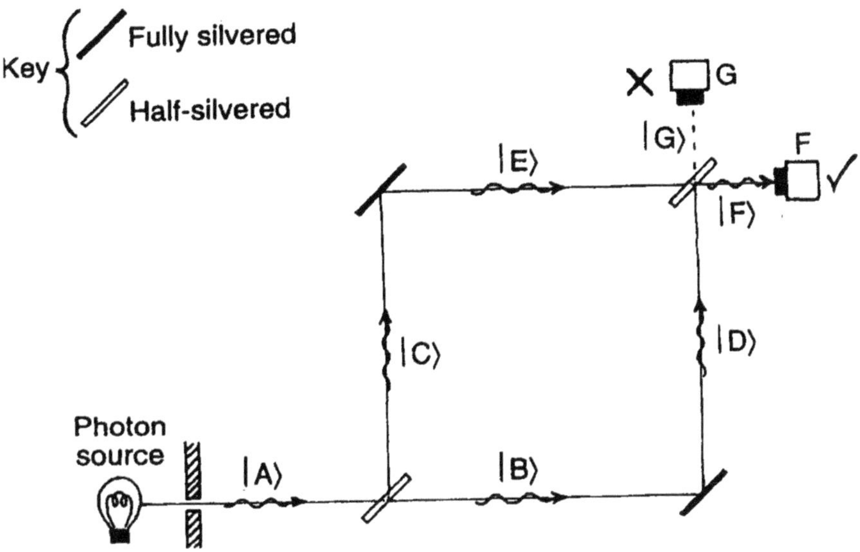

Figure 2. The beam splitter experiment: The two parts of the photon state are mixed together by two fully silvered mirrors, so as to encounter each other at a final half-silvered mirror. The quantum interference is performed in such a way that the entire state emerges in state **F** which is detected in the photon's initial motion , and the detector at **G** cannot receive the photon. This quantum magic outcome arises due to the prearranged weak containment of the one-dimensional identity representation 1 of collapsed coadjoint orbit $\{0\}$ as a completely entangled state in the mixing of irreducible unitary linear representation of G. Thus quantum holography allows to describe the concept of coherent entanglement by phase conjugation and its experimental verification by the simultaneous coincidence counts of photon state vector collapses. The phase conjugation is performed with respect to the symplectic affine structure attached to the coadjoint orbit leaf of the equivalence class of the irreducible unitary representation of G. Actually, the invariance of the coherent entanglement symmetry group $\mathbf{Z}/4\mathbf{Z}$ of the symplectic affine coadjoint orbit is broken by the symmetry group $\mathrm{Gal}_{\mathbf{R}}\mathbf{C}$ of order $[\mathbf{C}:\mathbf{R}]=2$ of the balanced Mach-Zehnder interferometer. Provided an effective phase shifter is inserted into the arm of the interferometer which is transverse to the photon's initial direction pointing to the detector at **F**, the detector at **C** can actually receive the photon. In this way, the projectively symmetrized setup of a delayed-choice experiment arises. Unfortunately, most treatments of the beam splitter experiment are oversimplified in explaining wrongly the beam splitter effect by Niels Bohr's principle of wave-particle complementarity and ignoring the cohomological double covering phenomenon geometrically displayed by the projective spinorial symmetries of the principal sandwich bundle over the compact 2−sphere $\mathbf{S}_2 \hookrightarrow \mathbf{R}^3$.

consisting of two bits. It follows that the sequence

$$\{1\} \longrightarrow \mathbf{Z}/2\mathbf{Z} \longrightarrow \mathbf{Mp}(2,\mathbf{R}) \stackrel{\eta}{\longrightarrow} \mathbf{Sp}(1,\mathbf{R}) \longrightarrow \{1\}$$

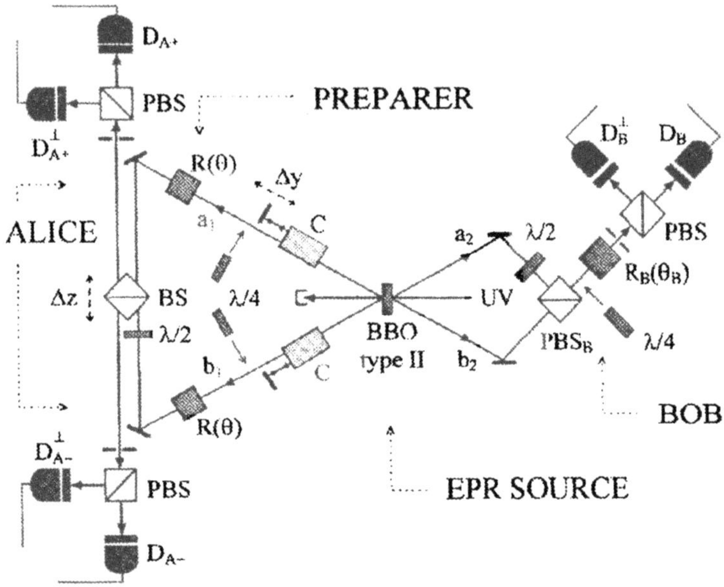

Figure 3. The quantum teleportation experiment of Francesco de Martini and associates to simultaneously detect in phase sensitive quadrature the four EPR-Bell states within two symmetric closed beam splitter photonic loops. The EPR-Bell state ordering is performed by unitary operators. Calcite crystals (**C**) acting as double refracting polarizers separate horizontally and vertically polarized photons. The spatial cross-correlation between the two opposite polarizations of the coherent entangled pair of photons is converted to a long range cross-correlation between their EPR communication channels. The interference pattern of quantum holography, independently generated by the two polarizations, is emergent in the ordinary 50:50 beam splitter (BS) of two input ports. On the receiver side, the phase sensitive quadrature detector located in the direction orthogonal to the emitter-absorber transaction is used for alignment purposes. Because the coadjoint orbit visualization of the unitary dual $\widehat{G}$ of three-dimensional Heisenberg nilpotent Lie group G provides the transition to the spatial cross-correlation and the affine symplectic state projection plane $\mathcal{O}_U$, this experimental setup has a much higher proportion of successes than the unitarily equivalent triple coincidence detection of photon state vector collapses. The standard Dirac formalism of quantum physics does not provide the geometric transition to the two-fold covering phenomenon which is at the basis of the complete quantum teleportation experiment. In covering phenomenon which is at the basis of the complete quantum teleportation experiment. In particular, the Dirac formalism does not predict the interference of single photons with themselves.

forms an exact sequence of Lie groups and $\mathbf{Mp}(2,\mathbf{R})$ as the same Lie algebra $\mathrm{sl}(2,\mathbf{R})$ as $\mathbf{SL}(2,\mathbf{R})$ itself. The covariance identity of the metaplectic

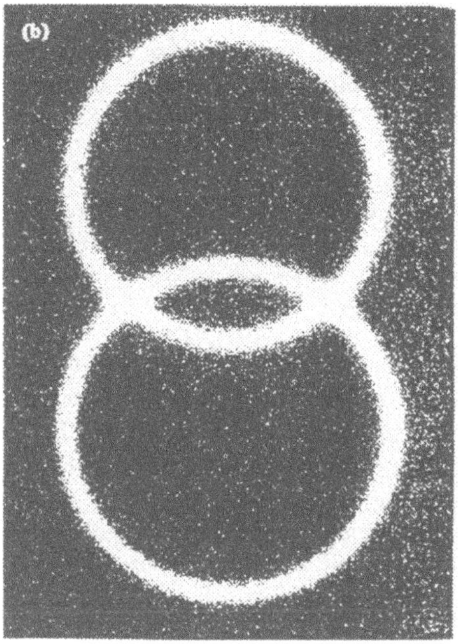

Figure 4. In the quantum teleportation experiments of photonics, photons are produced in coherent entangled singlets. The figure displays the inherent projective spinorial symmetry. A photon on the top circle traced in the coadjoint orbit $\mathcal{O}_U \hookrightarrow \text{Lie}(G)^*$ is horizontally polarized while its contragredient partner traced in the bottom circle is vertically polarized. However, only the photons corresponding to the intersections of the two circles in completely symmetric projective positions of $\mathcal{O}_U \hookrightarrow \text{Lie}(G)^*$ are coherent entangled photons. the offset of the concentric circles is due to the birefringence or double refracting polarization property of the BBO crystal which permits type II degenerate parametric down-conversion. The concentric circles are the planar traces of the extraordinary and ordinary Heisenberg helices and indicate the rotational curvature form of the affine symplectic state projection plane $\mathcal{O}_U \hookrightarrow \text{Lie}(G)^*$. Due to the two-fold covering of the toral traces, it is in principle impossible to verify the chirality determined by the orientations of the extraordinary and ordinary Heisenberg helices and the associated Coriolis phases. It must be so according to the non-cloning principle of quantum information theory. However, there is no interference trace of photons projecting on a coherent state symmetrically positioned within the coadjoint orbit $\mathcal{O}_U \hookrightarrow \text{Lie}(G)^*$.

representation R implies that only the *liftings* of the elements

$$g = \begin{pmatrix} a & b \\ c & d \end{pmatrix} \qquad \det g = ad - bc = 1$$

of the underlying Lie group $\mathbf{SL}(2,\mathbf{R})$ are *simultaneous observables*. The transition from $\mathbf{SL}(2,\mathbf{R})$ to the two-fold cover $\mathbf{Mp}(2,\mathbf{R})$ needs to transmit the additional information of two bits through a classical information channel in order to reconstruct the quantum pixels (qupixels) encrypted by a quantum hologram.

Theorem (Observability Theorem). *Provided two bits of classical external information are available, each single coadjoint orbit of the coherent entangled pair of planar coadjoint orbits $(\mathcal{O}_U, \mathcal{O}_{\check{U}})$ of G consists of an ensemble of simultaneously detectable qupixels.*

Therefore the *planar* coadjoint orbits of G are suitable substrata for the encoding of quantum information. Due to the Stone-von Neumann theorem, however, the qupixels of *both* planar coadjoint orbits $(\mathcal{O}_U, \mathcal{O}_{\check{U}})$ are *not* simultaneously detectable because the irreducible unitary linear representations U and $\check{U}$ of G are *not* isomorphic. Quantum teleportation is therefore distinct from classical information processing, which leaves the original intact and only creates an approximate replica and hypothetically perfect cloning which would result in two identical versions of the original.

Theorem (Exclusion Theorem). *The simultaneous detection of the ensemble of qupixels encrypting one of the planar coadjoint orbits $(\mathcal{O}_U, \mathcal{O}_{\check{U}})$ excludes the simultaneous detectability of the qupixels encrypting the other one.*

The exclusion theorem actually has *no* counterpart in classical information processing. It points out the importance of the principal sandwich bundle over the compact 2-sphere $\mathbf{S}_2 \hookrightarrow \mathbf{R}^3$ for quantum information processing. Due to the action of the metaplectic group $\mathbf{Mp}(2,\mathbf{R})$, its importance for the theory of quantum information is superior over the importance of the Hopf bundle $\mathbf{S}_1 \longrightarrow \mathbf{S}_3 \longrightarrow \mathbf{S}_2$.

The importance of the two-fold cover $\mathbf{Mp}(2,\mathbf{R})$ comes from the fact that the action of the unitary projective cocycle representation R of $\mathbf{SL}(2,\mathbf{R})$ on the *rays* of the infinite dimensional projective geometry $\mathbf{P}L^2(\mathbf{R})$ *lifts* to the action of an "ordinary" unitary linear representation of $\mathbf{Mp}(2,\mathbf{R})$ on the standard complex Hilbert space $L^2(\mathbf{R})$ with the multiplier removed. Notice that the linear Schrödinger representation U of central character $\chi = 1$ can be regarded as a representation of the semi-direct product $\mathbf{Mp}(2,\mathbf{R}) \times G$ on $L^2(\mathbf{R})$. The vector space of C^∞-vectors is the same whether U and $\check{U}$ are regarded as representations of G or $\mathbf{Mp}(2,\mathbf{R}) \times G$.

Notice that $\mathbf{T}$ forms a maximal compact subgroup $H \hookrightarrow \mathbf{Sp}(1, \mathbf{R})$. It is worth emphasizing that the 2-cocycle c of G with values in $\mathbf{T}$ in the cohomology group $\mathrm{H}^2(G, \mathbf{T})$ associated to R provides phase information. Notice that this multiplier of modulus one or phase factor includes the 2-cocycle

$$\tau : \mathbf{Sp}(1, \mathbf{R}) \times \mathbf{Sp}(1, \mathbf{R}) \ni (g, g') \rightsquigarrow \tau_0(L, g.L, g.g'.L)$$

with respect to a preassigned abelian polarizing subalgebra of the complexified Heisenberg Lie algebra $\mathrm{Lie}(G) \otimes_{\mathbf{R}} \mathbf{C}$ ([42]). With $\tau \in \mathrm{H}^2(\mathbf{Sp}(1, \mathbf{R}), \mathbf{Z})$ define the 2-cocycle

$$c = e^{\frac{\pi i \tau}{4}}.$$

Then $c \in \mathrm{H}^2(\mathbf{Sp}(1, \mathbf{R}), \mathbf{T})$ and $\tau \in \mathrm{H}^2(\mathbf{Sp}(1, \mathbf{R}), \mathbf{Z})$ are independent on the choice of the prefixed Lagrangian line $L \hookrightarrow \mathbf{C}$.

The mod 4 Maslov index of closed loops plays a decisive role in the Gleason geometric quantization procedure which nicely illustrates the eight-ball echo trajectory of the spin echo teleportation. The non-local character of the entangled loop configuration permits intananeous enforcement of long range quantum correlations across spatial distances but does not allow signalling. Trying to draw Feynman diagrams with tree structure rather than closed loops does not lead to a experimental setup performing accurate long range quantum teleportation.

Theorem (Non-Cloning Theorem). *The complete reconstruction of quantum information needs to transmit at least two bits of external classical information through a classical information channel. Thus quantum information and classical information cannot be separated. By cohomological obstruction reasons, quantum information theory is* not *in conflict to special relativity theory.*

The abuse of notation and not distinguish between R as a unitary projective cocycle representation of $\mathbf{Sp}(1, \mathbf{R})$ satisfying

$$R(g.g') = c(g, g').R(g) \circ R(g') \qquad g, g' \in \mathbf{SL}(2, \mathbf{R})$$

and as a unitary representation of $\mathbf{Mp}(2, \mathbf{R})$ is disastrous in the theory of quantum information transfer. The rays of the infinite dimensional projective geometry $\mathbf{P}L^2(\mathbf{R})$ provides the directions of the beam splitter experiment and its late-choice version.

The Lie group $\mathbf{Mp}(2, \mathbf{R})$ acts via R by automorphisms on the three-dimensional Heisenberg nilpotent Lie group G and leaves the one-dimensional center $C \hookrightarrow G$ pointwise fixed [52,40]. Notice that within quantum physics it is impossible to decide on which of the two covers the automorphism group $\mathbf{Mp}(2, \mathbf{R})$ is actually acting because the pure quantum state to be teleported is *unknown* within the quantum logic associated to the two-state quantum system. Observing the non-cloning principle of quantum information theory, two additional bits of classical information actually allow to communicate through a dual classical information channel the decision inherent to the two-fold covered logic in order to overcome the cohomological obstruction [39] associated to the central group extension

$$C \lhd G \longrightarrow \mathbf{R} \oplus \mathbf{R}$$

and to complete the perfect reconstruction process of the quantum teleportation experiment. In this way, a dual classical information channel compensates the phase information loss in the EPR communication channels which is caused by the experimental transition to the dual of the teleportation phenomenon. Reliable teleportation of a two-state quantum system cannot be achieved with a classical information channel of less than two bits of capacity. The EPR communication channel and the classical channel have to be used *in tandem* for accurate long range teleportation of a two-state quantum state. There is no collision of quantum teleportation with relativity theory.

Note that the mathematical foundations of quantum physics in the spirit of the Mackey theory is in terms of the cohomology theory of group extensions. The oriented Lagrangian lines in $\mathbf{C}$ give rise to a function

$$s : \mathbf{Sp}(1, \mathbf{R}) \longrightarrow \mathbf{T}$$

which satisfies the coboundary condition

$$c^2 = \mathrm{d}s.$$

The exact sequence of abelian groups

$$\{1\} \longrightarrow \mathbf{Z}/2\mathbf{Z} \longrightarrow \mathbf{T} \xrightarrow{(.)^2} \mathbf{T} \longrightarrow \{1\}$$

admits the exact sequence of dual groups

$$\{0\} \longrightarrow \mathbf{Z} \xrightarrow{2.} \mathbf{Z} \longrightarrow \mathbf{Z}/2\mathbf{Z} \longrightarrow \{0\},$$

where $(.)^2$ is just the square and 2. denotes multiplication by 2. It induces a long exact sequence of cohomology groups

$$\begin{aligned} \longrightarrow \mathrm{H}^1(\mathbf{Sp}(1,\mathbf{R}), \mathbf{T}) \longrightarrow \mathrm{H}^2(\mathbf{Sp}(1,\mathbf{R}), \mathbf{Z}/2\mathbf{Z}) \longrightarrow \\ \longrightarrow \mathrm{H}^2(\mathbf{Sp}(1,\mathbf{R}), \mathbf{T}) \longrightarrow \mathrm{H}^2(\mathbf{Sp}(1,\mathbf{R}), \mathbf{T}) \longrightarrow \end{aligned}$$

which makes clear that the class of c is the image of an element $\gamma \in \mathrm{H}^1(\mathbf{Sp}(1,\mathbf{R}),\ \mathbf{Z}/2\mathbf{Z})$. Due to the identity $\mathrm{H}^1(\mathbf{Sp}(1,\mathbf{R}), \mathbf{T}) =$

$\mathrm{Hom}(\mathbf{Sp}(1,\mathbf{R}),\mathbf{T}) = \{0\}$, the element γ is well defined. It follows that the metaplectic group $\mathbf{Mp}(2,\mathbf{R})$ is the extension η of $\mathbf{Sp}(1,\mathbf{R})$ by $\mathbf{Z}/2\mathbf{Z}$ associated with γ.

Theorem (Chirality Theorem). *There exist exactly two diffeomorphisms*

$$h_\pm : \mathbf{SU}(2,\mathbf{C}) - \{\mathrm{id}, -\mathrm{id}\} \longrightarrow \mathbf{S}_2 \times \mathbf{S}_1$$

which are compatible with the fiber projections

$$\pi_1 : \mathbf{SU}(2,\mathbf{C}) \longrightarrow \mathbf{SO}(3,\mathbf{R}), \qquad \pi_2 : \mathbf{S}_2 \times \mathbf{S}_1 \longrightarrow \mathbf{SO}(3,\mathbf{R}),$$

to the basis group $\mathbf{SO}(3,\mathbf{R}) = \mathbf{P}_3(\mathbf{R})$ *in the sense that the identities*

$$\pi_2 \circ h_\pm = \pi_1$$

hold on $\mathbf{SU}(2,\mathbf{C}) - \{\mathrm{id}, -\mathrm{id}\}$. *Thus the two-fold coverings* $\mathbf{S}_2 \times \mathbf{S}_1$ *and* $\mathbf{SU}(2,\mathbf{C}) = \mathbf{Spin}(3,\mathbf{R})$ *of the special orthogonal group* $\mathbf{SO}(3,\mathbf{R})$ *define equivalent chirality of unitary symplectic spinors.*

Let $\mathbf{Sp}(1,\mathbf{R})\times_c\mathbf{T}$ denote the extension of $\mathbf{Sp}(1,\mathbf{R})$ by $\mathbf{T}$ corresponding to the 2-cocycle c. This means that the Cartesian product $\mathbf{Sp}(1,\mathbf{R}) \times \mathbf{T}$ is equipped with the law

$$(g,t).(g',t') = (g.g', t.t'.c(g,g')).$$

The mapping

$$(g,t) \rightsquigarrow s(g).t^2$$

is a homomorphism of $\mathbf{Sp}(1,\mathbf{R})\times_c\mathbf{T}$ into $\mathbf{T}$ with kernel $\mathbf{Mp}(2,\mathbf{R})$. Therefore $\mathbf{Mp}(2,\mathbf{R})$ admits the explicit form

$$\mathbf{Mp}(2,\mathbf{R}) = \{(g,t) \in \mathbf{Sp}(1,\mathbf{R}) \times \mathbf{T} \mid s(g).t^2 = 1\}.$$

Let H denote a Lie group acting on $\mathbf{R} \oplus \mathbf{R}$. If the action respects the non-degenerate alternating bilinear form det, then there exists a morphism $\kappa : H \longrightarrow \mathbf{Sp}(1,\mathbf{R})$. Let $\tilde{H}$ denote the inverse image of η under κ. Then

$$\tilde{H} = \{(g,t) \in H \times \mathbf{T} \mid s(\kappa(g)).t^2 = 1\}$$

In the following diagram, the projection $\tilde{H} \longrightarrow H$ is onto the first coordinate. The non-trivial element of its kernel is the pair $\{\mathrm{id}, -\mathrm{id}\}$.

$$\begin{array}{ccc} \tilde{H} & \longrightarrow & \mathbf{Mp}(2,\mathbf{R}) \\ \downarrow & & \downarrow \\ H & \xrightarrow{\kappa} & \mathbf{Sp}(1,\mathbf{R}) \end{array}$$

The uniqueness of the metaplectic group $\mathbf{Mp}(2,\mathbf{R})$ as an extension of $\mathbf{Sp}(1,\mathbf{R})$ by $\mathbf{Z}/2\mathbf{Z}$ is important for the application to the theory of quantum information.

Let exp denote the exponential map $sl(2,\mathbf{R}) \to \mathbf{SL}(2,\mathbf{R})$. Similarly let Exp be the exponential map $sl(2,\mathbf{R}) \to \mathbf{Mp}(2,\mathbf{R})$. Let

$$\widetilde{\mathbf{T}} = \frac{\mathbf{R}}{\{\mathrm{id}, -\mathrm{id}\} \times \mathbf{Z}}$$

denote the two-fold covered torus group. The circular polarization of chirality is the given by the tori

$$\mathbf{T} = \exp(\mathbf{R}.J)$$

and

$$\widetilde{\mathbf{T}} = \mathrm{Exp}(\mathbf{R}.J)$$

where

$$\ker(\exp : \mathbf{R}J \longrightarrow \mathbf{T}) = 2\pi\mathbf{Z}.J$$

and

$$\ker(\mathrm{Exp} : \mathbf{R}J \longrightarrow \widetilde{\mathbf{T}}) = 4\pi\mathbf{Z}.J\,.$$

Note the 4-periodicity factor $4\pi\mathbf{Z}$ in front of the Weyl symplectic matrix J which arises from the cycles of the mod 4 Maslov index τ_0. The metaplectic representation R of $\mathbf{Mp}(2,\mathbf{R})$ acts on U via the intertwining operator

$$R\,(\mathrm{Exp}(\theta J))\,.$$

It generates independently of the distance the circular polarization rotation

$$\begin{pmatrix} \cos\theta & -\sin\theta \\ \sin\theta & \cos\theta \end{pmatrix} \in \mathbf{SO}(2,\mathbf{R})$$

through the simultaneously observable phase angle $\theta \in \mathbf{R}$. The associated line reflection with respect to the axis admitting the angle $\frac{\theta}{2}$ to the line $\mathbf{R} \times \{0\}$ is given by the matrix

$$\begin{pmatrix} \cos\theta & \sin\theta \\ \sin\theta & -\cos\theta \end{pmatrix} \in \mathbf{O}(2,\mathbf{R}) \; - \; \mathbf{SO}(2,\mathbf{R}).$$

For the phase angles $\theta = 0$ and $\theta = \pi$, the line reflection matrices

$$\begin{pmatrix} 1 & 0 \\ 0 & -1 \end{pmatrix} \in \mathbf{O}(2,\mathbf{Z}) \; - \; \mathbf{SO}(2,\mathbf{Z})$$

and

$$\begin{pmatrix} -1 & 0 \\ 0 & 1 \end{pmatrix} \in \mathbf{O}(2,\mathbf{Z}) - \mathbf{SO}(2,\mathbf{Z})$$

arise. In the special case of the Weyl symplectic matrix J, the covariance identity for the intertwiner

$$R(\mathrm{Exp}(\theta J)) \otimes R(\mathrm{Exp}(-\theta J))$$

which simultaneously acts on the coherent entangled representation

$$U \hat{\otimes} \check{U}$$

shows that the phase changes θ of the oriented Lagrangian lines, swapped to and detected in phase sensitive quadrature by the decoding receiver, represent simultaneous observables for the cryptographic sender, and his phase sensitive quadrature detectors. In the actual experimental setup, the phase change on the receiver side is realized by a Fresnel rhomb photon polarization rotator ([5]).

The one-dimensional torus $\mathbf{T}$ of circular polarization realizes under its natural quotient topology the principal $\mathbf{SL}(2,\mathbf{R})$-bundle Λ of all Lagrangian lines L of the planar coadjoint orbit $\mathcal{O}_U \hookrightarrow \mathrm{Lie}(G)^\star$. Its two-fold covering group $\tilde{\mathbf{T}}$ represents the principal $\mathbf{Mp}(2,\mathbf{R})$-bundle $\tilde{\Lambda}$ of all oriented Lagrangian lines in the planar coadjoint orbit $\mathcal{O}_U$. The number of EPR-Bell states

$$\sharp(\mathbf{O}(2,\mathbf{Z})/(\mathbf{O}(1,\mathbf{Z}) \times \mathbf{O}(1,\mathbf{Z}))) = 4$$

is the minimal number of generating quantum states which is necessary to verify accurate quantum teleportation by oriented Lagrangian line ordering and phase sensitive quadrature detection. The $\mathbf{O}(2,\mathbf{R})$-manifold $\mathbf{O}(2,\mathbf{R})/(\mathbf{O}(1,\mathbf{Z}) \times \mathbf{O}(1,\mathbf{Z}))$ can be represented by the matrices of the outcome phase measurements

$$\left\{ \begin{pmatrix} -1 & 0 \\ 0 & -1 \end{pmatrix}, \begin{pmatrix} 0 & 1 \\ 1 & 0 \end{pmatrix}, \begin{pmatrix} 0 & -1 \\ 1 & 0 \end{pmatrix}, \begin{pmatrix} -1 & 0 \\ 0 & 1 \end{pmatrix} \right\}.$$

Because the associated phase angles are given by $\theta = \pi$ and $\theta = \frac{\pi}{2}$, a transition from the orthonormal basis $\{P, Q\}$ to the diagonal orthogonal basis

$$\left\{ \frac{1}{2}\sqrt{2}(P+Q), \frac{1}{2}\sqrt{2}(P-Q) \right\}$$

of the two-dimensional commutative subalgebra $\mathcal{C}\ell_2^+(\mathbf{R}) = \mathbf{R} \oplus \wedge^2 \mathbf{R}^2$ of the real Clifford algebra $\mathcal{C}\ell_2(\mathbf{R})$ graded over $\mathbf{Z}/2\mathbf{Z}$, or to the orthonormal basis

$$\left\{ \frac{1}{2}\sqrt{2}(1+i), \frac{1}{2}\sqrt{2}(1-i) \right\}$$

of the complex plane $\mathbf{C}$, shows that the photon polarization states can be experimentally performed by a suitable combination of quarter-wave plates ([2]). The phase sensitive quadrature detectors filter out the circular polarization states. Considered as a quantum computer, the experimental setup allows to filter out the polarization state $\begin{pmatrix} -1 & 0 \\ 0 & -1 \end{pmatrix}$ which is unitarily equivalent to the quantum state

$$J_4 = -1 \begin{pmatrix} -1 & 0 \\ 0 & -1 \end{pmatrix}$$

associated to the identity representation.

The special unitary group

$$\mathbf{SU}(2, \mathbf{C}) = \left\{ -i \begin{pmatrix} -x_3 + ix_4 & x_1 + ix_2 \\ x_1 - ix_2 & x_3 + ix_4 \end{pmatrix} \middle| \, (x_1, x_2, x_3, x_4) \in \mathbf{S}_3 \right\}$$

consisting of unitary matrices of order 2 and determinant 1 is diffeomorphic to the three-dimensional compact manifold formed by the unit sphere $\mathbf{S}_3 \hookrightarrow \mathbf{R}^4$. Therefore the semi-simple Lie group $\mathbf{SU}(2, \mathbf{C})$ can be diffeomorphically embedded into the compact Lie group of rotations $\mathbf{SO}(4, \mathbf{R})$ of the four-dimensional real vector space $\mathbf{R}^4 = \mathbf{H}$ of Hamiltonian quaternions. Each rotation in $\mathbf{SO}(4, \mathbf{R})$ can be represented by two antipodal elements in $\mathbf{S}_3 \times \mathbf{S}_3$ so that the compact group $\mathbf{S}_3 \times \mathbf{S}_3$ forms a two-fold covering group of

$$\mathbf{SO}(4, \mathbf{R}) = \frac{\mathbf{S}_3 \times \mathbf{S}_3}{\{(\mathrm{id}, \mathrm{id}), (-\mathrm{id}, -\mathrm{id})\}},$$

and the mapping

$$\mathbf{SU}(2, \mathbf{C}) \longrightarrow \mathbf{SO}(4, \mathbf{R})$$

is a faithful representation. The tangent space

$$\mathrm{T}_{J_4}(\mathbf{SU}(2, \mathbf{C}))$$

at the neutral element $J_4 \in \mathbf{SU}(2, \mathbf{C})$ identifies with the semi-simple Lie algebra $\mathrm{Lie}(\mathbf{SU}(2, \mathbf{C}))$ of complex matrices X which are traceless and skew-hermitian. Hence

$$\mathrm{Lie}(\mathbf{SU}(2, \mathbf{C})) = \left\{ -i \begin{pmatrix} -x_3 & x_1 + ix_2 \\ x_1 - ix_2 & x_3 \end{pmatrix} \middle| \, (x_1, x_2, x_3) \in \mathbf{R}^3 \right\}$$

is the image of the traceless Clifford mapping

$$\mathbf{R}^3 \hookrightarrow \mathrm{End}_{\mathbf{R}}(\mathbf{C} \oplus \mathbf{C}).$$

By setting $x_1 = 1$, $x_2 = x_3 = 0$, and $x_1 = 0$, $x_1 = 1$, $x_3 = 0$, and $x_1 = x_2 = 0$, $x_3 = 1$, the other quantum states X can be identified with the Hamiltonian quaternions associated with the Pauli *spin matrices*

$$J_1 = -i\begin{pmatrix} 0 & 1 \\ 1 & 0 \end{pmatrix}, \quad J_2 = -1\begin{pmatrix} 0 & -1 \\ 1 & 0 \end{pmatrix}, \quad J_3 = -i\begin{pmatrix} -1 & 0 \\ 0 & 1 \end{pmatrix}.$$

It follows by an easy computation

$$J_j{}^2 = -J_4, \qquad J_j.J_k = -J_k.J_j \qquad (1 \leq j, k \leq 3),$$

and

$$J_1.J_2 = J_3, \qquad J_2.J_3 = J_1, \qquad J_3.J_1 = J_2,$$

so that the commutators in Lie($\mathbf{SU}(2,\mathbf{C})$) admit the form

$$[J_1, J_2] = 2J_3, \qquad [J_2, J_3] = 2J_1, \qquad [J_3, J_1] = 2J_1.$$

These relations characterize the canonical basis

$$\{J_1, J_2, J_3, J_4\}$$

of the skew-field **H** of Hamiltonian quaternions, considered as a real division algebra of dimension 4 over the field **R** of real numbers [the Acts 12,4: "he put him in prison, and delivered him to four quaternions of soldiers to keep him."]. Actually **H** forms the Clifford algebra of any two-dimensional vector subspace of the three-dimensional real vector space of *pure* quaternions. Consequently, the Hamiltonian quaternions $q \in \mathbf{H}$ can be considered as unitary symplectic *spinors* ([37]).

The Clifford group

$$\{\pm J_1, \pm J_2, \pm J_3, \pm J_4\}$$

forms the *finite* group of Hamiltonian quaternions [29]. Obviously it is a non-commutative subgroup of the Clifford group $\mathbf{H} - \{0\}$. The order of the finite group of quaternions is 8 and, with the exception of $\pm J_4$, every element of the group has order 4.

The commutators in Lie($\mathbf{SU}(2,\mathbf{C})$) are cyclic with respect to the basis

$$\left\{\frac{1}{2}J_1, \frac{1}{2}J_2, \frac{1}{2}J_3, \frac{1}{2}J_4\right\}$$

of **H**. Omitting the single element $\frac{1}{2}J_4$, the Clifford isomorphy between $\mathbf{R}^3$ under the standard vector product and Lie($\mathbf{SU}(2,\mathbf{C})$) is now obvious. As a consequence, it follows from the spectral theory of normal endomorphisms

([51]) that every automorphism of the skew-field $\mathbf{H}$ is an *inner* automorphism and that the Hamiltonian quaternions

$$\exp_{\mathbf{SU}(2,\mathbf{C})}\begin{pmatrix} 2\pi i\theta & 0 \\ 0 & -2\pi i\theta \end{pmatrix} = \begin{pmatrix} e^{2\pi i\theta} & 0 \\ 0 & e^{-2\pi i\theta} \end{pmatrix} \in T \qquad (\theta \in [0,1])$$

consisting of diagonal matrices of contragredient tori are in one-to-one correspondence with the conjugacy classes in $\mathbf{H}$ ([51]). Clearly the phase $\theta \in [0,1]$ is an Eulerian angle of the compact unit sphere

$$\mathbf{S}_3 = \{q \in \mathbf{H} \mid |q| = 1\} \hookrightarrow \mathbf{R}^4$$

which forms a two-fold covering group of the special orthogonal group

$$\mathbf{SO}(3,\mathbf{R}) = \frac{\mathbf{S}_3}{\{\mathrm{id}, -\mathrm{id}\}}.$$

The group of rotations $\mathbf{SO}(3,\mathbf{R}) = \mathbf{P}_3(\mathbf{R})$ which is given by the factor group of the spin group $\mathbf{Spin}(3,\mathbf{R}) = \mathbf{SU}(2,\mathbf{C})$ modulo its center $\{\mathrm{id}, -\mathrm{id}\}$ is not simply connected, its fundamental group is isomorphic to $\mathbf{Z}/2\mathbf{Z}$. By considering the complex quaternions as unitary symplectic spinors, this implies the chirality theorem. It emphasizes the unitary symplectic spinorial aspect of the Hopf fibration $\mathbf{S}_1 \longrightarrow \mathbf{S}_3 \longrightarrow \mathbf{S}_2$.

Theorem (Central Circle Resonance Theorem). *The central circle $\mathbf{S}_1$ of the Hopf bundle $\mathbf{S}_1 \longrightarrow \mathbf{S}_3 \longrightarrow \mathbf{S}_2$ is embedded by the sandwich configuration into the singular plane of the unitary dual $\hat{G}$ of the Heisenberg group G. In the laboratory frame of coordinates it is in transversal position to the Dirac string of the Hopf fibration $\mathbf{S}_1 \longrightarrow \mathbf{S}_3 \longrightarrow \mathbf{S}_2$. Under resonance with the stabilized maximal torus $T \hookrightarrow \mathbf{SU}(2,\mathbf{C})$, the central circle $\mathbf{S}_1$ of collapsed quantum state vectors projects stereographically onto contragrediently rotating frames of symplectic coordinates attached to the entangled coadjoint orbit inside* $\mathrm{Lie}(G)^\star$ *of the representation $U \hat{\otimes} \check{U}$ of G.*

The subgroup $T \hookrightarrow \mathbf{SU}(2,\mathbf{C})$ is itself conjugated by the Weyl symplectic matrix J which twists the contragredient tori inside the diagonal matrices:

$$T \ni \begin{pmatrix} e^{2\pi i\theta} & 0 \\ 0 & e^{-2\pi i\theta} \end{pmatrix} \rightsquigarrow \begin{pmatrix} e^{-2\pi i\theta} & 0 \\ 0 & e^{2\pi i\theta} \end{pmatrix} \in T$$

Thus T forms its own centralizer in $\mathbf{SU}(2,\mathbf{C})$ and therefore represents a maximal abelian subgroup of $\mathbf{SU}(2,\mathbf{C})$. Obviously, T is the stabilizer of

$$\begin{pmatrix} 2\pi i n & 0 \\ 0 & -2\pi i n \end{pmatrix} \in \mathrm{Lie}(\mathbf{SU}(2,\mathbf{C}))^\star \qquad (n \neq 0)$$

for $n \in \mathbf{Z} - \{0\}$.

Theorem (Stroboscopic Synchronization Theorem). *The stroboscopic qubit measurement of synchronization by the stabilized maximal torus* $T \hookrightarrow \mathbf{SU}(2,\mathbf{C})$ *of collapsed quantum states is uniquely determined if and only if the resonance frequencies* $\neq 0$ *of the rotating frames of symplectic coordinates attached to the projectively completed planar coadjoint orbits* $\mathbf{P}(\mathbf{R}\times\mathcal{O}_U)$ *inside* $\mathbf{P}(\mathbf{R} \times \mathrm{Lie}(G)^\star)$ *are integral. The integrality forms a manifestation of the Dirac* quantization *condition of magnetic monopoles.*

Because the function

$$X \rightsquigarrow \det X = \sum_{1\leq j\leq 3} x_j^2$$

is invariant under the coadjoint action of $\mathbf{SU}(2,\mathbf{C})$ on the vector space dual $\mathrm{Lie}(\mathbf{SU}(2,\mathbf{C}))^\star \cong \mathrm{Lie}(\mathbf{SU}(2,\mathbf{C}))$, the coadjoint orbits of $\mathbf{SU}(2,\mathbf{C})$ are the compact spheres $\mathcal{O}_r \hookrightarrow \mathbf{R}^3$ defined by the equation

$$\sum_{1\leq j\leq 3} x_j^2 = r^2 \qquad (r \geq 0).$$

Synchronization by *resonance* of the characters of $T \hookrightarrow \mathbf{SU}(2,\mathbf{C})$ provides the Kepplerian integrality condition

$$r = 2\pi n \qquad (n \in \mathbf{N} - \{0\})$$

for the radii $r > 0$ of the compact spheres $\mathcal{O}_r \hookrightarrow \mathbf{R}^3$. This *filtering* procedure represents the coadjoint orbit *reduction* process as a manifestation of the Dirac quantization condition of magnetic monopoles. The $\mathbf{SU}(2,\mathbf{C})$-homogeneous manifold $\mathbf{S}_2$ obtained by the quotient $\mathbf{SU}(2,\mathbf{C})/T$ to the stabilized maximal torus $T \hookrightarrow \mathbf{SU}(2,\mathbf{C})$ keeping the north-pole $J_3 \in \mathbf{S}_3$ fixed, provides the Hopf bundle projection

$$\mathbf{S}_3 \longrightarrow \mathbf{S}_2.$$

In the principal sandwich bundle, the base 2-sphere $\mathbf{S}_2 \hookrightarrow \mathbf{R}^3$ is regarded as the union of *two* copies of the complex plane $\mathbf{C}$ coherently arranged so that their positive real axes are parallel, with identifications arising from two *entangled* stereographic projections of $\mathbf{S}_2$ from the poles $\pm J_3$. The two ebit encoding copies of the complex plane $\mathbf{C}$ are in correspondence to the planar coadjoint orbits $\mathcal{O}_U$ and $\mathcal{O}_{\bar{U}}$ of G.

In the Hopf fibration $\mathbf{S}_1 \longrightarrow \mathbf{S}_3 \longrightarrow \mathbf{S}_2$, the inverse image of any point of $\mathbf{S}_2 \hookrightarrow \mathbf{R}^3$ forms a great circle in the compact 3-sphere $\mathbf{S}_3 \hookrightarrow \mathbf{R}^4$, and the inverse image of any circle on $\mathbf{S}_2$ in $\mathbf{S}_3$ forms a flat torus in the real vector space $\mathbf{R}^3$. All but two of the fibres of the Hopf bundle $\mathbf{S}_1 \longrightarrow \mathbf{S}_3 \longrightarrow \mathbf{S}_2$ lie

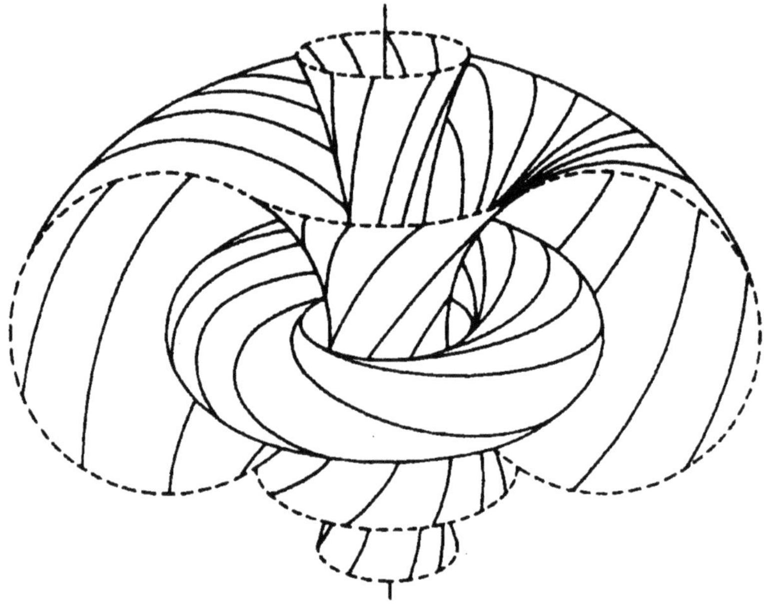

Figure 5. The Hopf fibration $\mathbf{S}_1 \longrightarrow \mathbf{S}_3 \longrightarrow \mathbf{S}_2$. The compact unit sphere $\mathbf{S}_3 \hookrightarrow \mathbf{R}^4$ is conformally represented according to the stereographic projection of $\mathbf{S}_2$. The Clifford parallels filling the whole of the compact 3-sphere $\mathbf{S}_3$ appear as a bundle of mutually linked circles $\mathbf{S}_1$ and one straight line which is the Dirac string in transversal position to the central circle $\mathbf{S}_1$. The circles $\mathbf{S}_1$ lie in a twisted fashion on a bundle of nested tori. In the context of computational geometry, nested means that these flat tori in the real vector space $\mathbf{R}^3$ lie threaded within one another in continuous succession.

on these flat tori in $\mathbf{R}^3$. The exceptional fibres are the Dirac string and, in transversal position to the Dirac string, the central circle $\mathbf{S}_1$.

An adaption of Keppler's conchoid construction to the sandwich configuration gives rise to a matched filter bank

$$v_{\mathbf{SU}(2,\mathbf{C})} : \begin{pmatrix} 2\pi i\theta & 0 \\ 0 & -2\pi i\theta \end{pmatrix} \rightsquigarrow \frac{1}{2}\sum_{n\in\mathbf{Z}}(e^{2\pi in\theta} + e^{-2\pi in\theta}) = \sum_{n\in\mathbf{Z}} \cos 2\pi n\theta$$

on the reduced coadjoint orbits of the compact Lie group $\mathbf{SU}(2,\mathbf{C})$ ([41]). Then the distributional Poisson-Plancherel formula for $\mathbf{SU}(2,\mathbf{C})$ can be derived from the rotational curvature form of the subband encoded filter bank

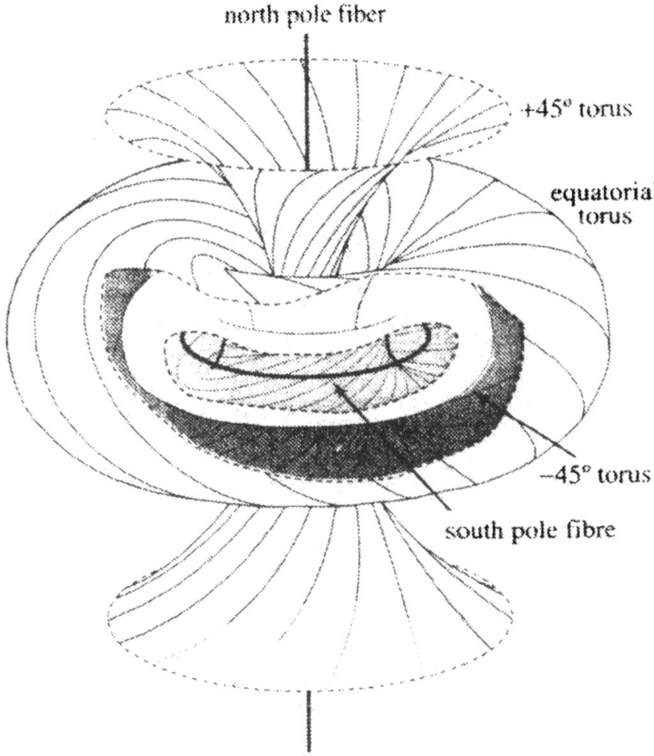

Figure 6. The Hopf bundle $\mathbf{S}_1 \longrightarrow \mathbf{S}_3 \longrightarrow \mathbf{S}_2$. Picturing the compact unit sphere $\mathbf{S}_3 \hookrightarrow \mathbf{R}^4$ minus the north pole as $\mathbf{R}^3$ via stereographic projection , the perspective view displays an accurate cutaway sight in $\mathbf{R}^3$ of the fibres on the three flat tori lying over the equator, and the $\pm 45^0$ parallels of $\mathbf{S}_2$. The fibres of the north and south poles, namely, the Dirac string and, in transversal position to the Dirac string, the central circle, are also shown. The complementary subset of $\mathbf{R}^3 - \{$Dirac string, south pole fibre$\}$ is foliated by the flat tori lying over the circles of latitude on the compact 2-sphere $\mathbf{S}_2 \hookrightarrow \mathbf{R}^3$. All but the two fibres lie on the flat tori in $\mathbf{R}^3$. On each torus the fibres run around once in each direction. The base 2-sphere $\mathbf{S}_2$ may be regarded as the space of fibres of the Hopf fibration $\mathbf{S}_1 \longrightarrow \mathbf{S}_3 \longrightarrow \mathbf{S}_2$ in the sense that the bundle projection establishes a bijection between the set of fibres and the base manifold. A cross-sectional disc for the solid torus meets every fibre inside the flat torus just once. The fibres on the flat torus meet each disc once in its boundary and provides a diffeomorphism between theses two circles. Thus the space of fibres is made up of two 2-discs glued together by a diffeomorphism of their boundaries, which is just the 2-sphere $\mathbf{S}_2$ topologically fit together.

channels. Indeed, denoting by

$$\frac{1}{2\pi}\sigma_n$$

the canonical Liouville measure on the reduced coadjoint orbit $\mathcal{O}_n \hookrightarrow \mathbf{R}^3$ of radius $r = 2\pi n$ ($n \in \mathbf{N}$) of $\mathbf{SU}(2,\mathbf{C})$, the identity

$$\int_{\mathrm{Lie}(\mathbf{SU}(2,\mathbf{C}))^*} v_{\mathbf{SU}(2,\mathbf{C})} = \sum_{n\in\mathbf{N}} \frac{n}{2\pi}\int_{\mathcal{O}_n} \sigma_n$$

holds in the sense of complex tempered distributions. For the proof of the distributional Poisson-Plancherel formula for $\mathbf{SU}(2,\mathbf{C})$, it should be observed that the action of the stabilized maximal torus $T \hookrightarrow \mathbf{SU}(2,\mathbf{C})$ on the reduced coadjoint orbit $\mathcal{O}_n \hookrightarrow \mathbf{R}^3$ has the *integral* fixed point subset of the Dirac string

$$\{(0,0,n),(0,0,-n)\}.$$

An application of the calculus of residues to the differential form

$$e^{2\pi i x_3 \theta}\,\frac{\sigma_n}{2\pi}$$

which is *exact* on the fixed point free subset of $\mathcal{O}_n \hookrightarrow \mathbf{R}^3$ provides the character formula

$$\frac{e^{2\pi i n\theta}}{2\pi i n\theta} - \frac{e^{-2\pi i n\theta}}{2\pi i n\theta} = \frac{1}{2\pi}\int_{\mathcal{O}_n} e^{2\pi i x_3\theta}\,\sigma_n \qquad (n \in \mathbf{N}-\{0\})$$

as a manifestation of the Dirac quantization condition of magnetic monopoles. Evaluation at the gauge function

$$\varphi : x_3 \rightsquigarrow e^{2\pi i x_3\theta} \qquad (\theta \in \mathbf{R})$$

of the derivative of the compactly supported localization distribution

$$K : \varphi \rightsquigarrow \begin{cases} \frac{1}{2\pi}\int_{\mathcal{O}_r} \varphi\,\sigma_r & \text{for} \quad r > 0, \\ 1 & \text{for} \quad r = 0, \\ -\frac{1}{2\pi}\int_{\mathcal{O}_{|r|}} \varphi\,\sigma_{|r|} & \text{for} \quad r < 0 \end{cases}$$

on $\mathrm{Lie}(G)$ along the radius $r \in \mathbf{R}$ of the reference circle provides the conchoid function

$$(\lambda, r) \rightsquigarrow \frac{1}{2}(e^{2\pi i\lambda r} - e^{-2\pi i\lambda r}) = \cos 2\pi\lambda r$$

associated to the conchoid construction ([41]). For $r \in \mathbf{R}$ deduce from the conchoid function

$$\Theta_r : \mathrm{Lie}(T)^\star \ni \begin{pmatrix} 2\pi i\lambda & 0 \\ 0 & -2\pi i\lambda \end{pmatrix} \rightsquigarrow \cos 2\pi\lambda r.$$

Then an application of the one-dimensional inverse Fourier transform provides the matched filter bank identity

$$\bar{\mathcal{F}}_{\mathbf{R}}\Theta_r = (\frac{\partial}{\partial r}K)(e^{2\pi x_3 \theta}) \qquad (r \in \mathbf{R}).$$

Specifically,

$$\bar{\mathcal{F}}_{\mathbf{R}}\Theta_0 = 1.$$

Finally, the Kepplerian stroboscopic synchronization strategy for the radii

$$r = 2\pi n \qquad (n \in \mathbf{N} - \{0\})$$

of the compact spheres $\mathcal{O}_r \hookrightarrow \mathbf{R}^3$ establishes as a manifestation of the Dirac quantization condition of magnetic monopoles the aforementioned distributional Poisson-Plancherel formula for $\mathbf{SU}(2, \mathbf{C})$ ([45]).

Theorem (Magnetic Monopole Theorem). *The tempered distribution*

$$v_G = \sum_{n \in 2\pi\mathbf{Z}} \Theta_n$$

obtained from the conchoid function by stroboscopic synchronization forms a matched filter bank. It admits a distributional expansion into Dirac magnetic monopoles by the distributional Poisson-Plancherel formula for $\mathbf{SU}(2, \mathbf{C})$.

An application of Keppler's stroboscopic synchronization strategy to establish the distributional Poisson-Plancherel formula for the category of semisimple Lie groups is also possible.

Returning from the compact case to the Heisenberg nilpotent Lie group G, which does not need a filtering procedure for coadjoint orbit reduction, the constructive and destructive interferences of the Lagrangian lines belonging to the real Grassmann manifolds Λ and $\widetilde{\Lambda}$ allow a Fourier analysis interpretation of the interferometric measurements via the orthogonality relations of the characters associated to the tori $\mathbf{T}$ and $\widetilde{\mathbf{T}}$, respectively. Regardless of the unknown pure quantum state to be teleported, the four phase measurement outcomes are equally likely, each occurring with probability $\frac{1}{4}$.

It should be noted that $\mathbf{SL}(2, \mathbf{R})$ contains the rotation group $\mathbf{SO}(2, \mathbf{R})$ of the affine symplectic state projection plane $\mathcal{O}_U$ as the maximal compact torus subgroup $\mathbf{T}$ of circular polarization generated by the Weyl symplectic matrix J. The symplectic form $dx \wedge dy$ of matrix J defines the rotational curvature form of the affine state projection plane $\mathcal{O}_U \hookrightarrow \mathrm{Lie}(G)^\star$.

Free precession allows each magnetic spin isochromat to go its own way at its own pace, so there is a fanning out of the magnetic isochromatic vectors in the symplectic plane $\mathbf{R} \oplus \mathbf{R}$ due to the spatial inhomogeneity of the applied magnetic flux density or as a result of any magnetic field gradient applied synchronously. The coadjoint orbit visualization of the Hahn spin echo requires some effort because the echo trajectories are the result of the fanning out of the magnetic isochromatic vectors which precesses into a "figure of eight" pattern on the compact unit sphere $\mathbf{S}_2 \hookrightarrow \mathbf{R}^3$ ([20]). The fanning out of magnetic spin isochromats forms a maximally mixed state which gives *no* phase information about the input state. In the teleportation of an eight-ball echo trajectory the points precess at different rates but with fixed points at the north and south poles. Points of the trajectory near the north pole are fast-moving, whereas those near the south pole move slowly. The Gleason geometric quantization procedure visualizes via the mod 4 Maslov index τ_0 of the metaplectic representation R of $\mathbf{Mp}(2,\mathbf{R})$ the dynamics of the principal $\mathbf{Mp}(2,\mathbf{R})$-bundle $\widetilde{\Lambda}$ of oriented Lagrangian lines. Indeed, taking account of the stationarity of the two poles, the planar tracial frame functions of Gleason's geometric quantization procedure admit the same periodicity $4\pi\mathbf{Z}$ as the mod 4 Maslov index τ_0 indicated above.

The motion of oriented Lagrangian lines of the planar *traces* of the extraordinary and ordinary Heisenberg helices generate two-fold covered tori $\widetilde{\mathbf{T}}$. The sequence

$$\{1\} \longrightarrow \mathbf{Z}/2\mathbf{Z} \longrightarrow \widetilde{\mathbf{T}} \longrightarrow \mathbf{T} \longrightarrow \{1\}$$

is an exact sequence of tori. In view of the fact that the extraordinary and ordinary Heisenberg helices and the associated Coriolis phases are wrapped along the energy cylinders, the motion of oriented Lagrangian lines of the planar traces provides a spin entanglement purification. Note that Gleason's geometric quantization procedure relies on a central projection of the energy cylinders onto the compact 2-sphere $\mathbf{S}_2 \hookrightarrow \mathbf{R}^3$ with antipodal points identified, and is based on an integrality condition manifesting the Dirac quantization of magnetic monopoles. Note also that the longitudinal Coriolis component allows to include the flip angle for contrast control and faster data acquisition.

It follows that the "halved" symplectic Fourier transform of period 2 is sufficient to read out the polyphase holograms of clinical MRI from the two-fold covered affine symplectic state projection plane $\mathcal{O}_U$ ([42]). Because only half of the data along the phase history lines in the polyphase holograms is acquired, the half Fourier reconstruction technique allows a reduction of the data acquisition times. The HASTE (half Fourier acquisition single-shot FSE) technique exploits this symmetry ([14]).

Once Hahn had shown the way, the idea that magnetic spin isochromats could be manipulated in quite sophisticated ways gradually gained acceptance in NMR spectroscopy and created the new field of nuclear spin choreography. Clinical MRI scanners use quantum methods such as the spin echo teleportation, and classical methods such as nuclear spin manipulations by protocols of digital computer controlled radiofrequency excitation pulse trains and *simultaneous* linear gradient switching *in tandem* ([43]).

Clinical MRI scanners have three separate physical magnetic field gradient coils, each capable of creating a linear gradient along one of three orthogonal directions. A given linear gradient becomes the slice select gradient by virtue of the fact that it is activated during the $\frac{\pi}{2}$ slice select radiofrequency pulse. Another linear gradient becomes the read-out gradient because it is turned on during the read-out process.

Whereas in experimental quantum teleportation a narrow-bandwidth filter ensures a coherence time much longer than the pump pulse length and in this way makes it sure that the coherent entangled photons cannot be distinguished by their arrival times ([6]), timing is everything in clinical MRI, which is why the activation of these options is called a pulse sequence timing diagram ([27]). It is this time dependence which forces to projectively complete the coadjoint orbit

$$\mathcal{O}_U \hookrightarrow \mathrm{Lie}(G)^\star$$

by the stratigraphic time line $\mathbf{R}$. The projective completion

$$\mathbf{P}(\mathbf{R} \times \mathcal{O}_U) \hookrightarrow \mathbf{P}(\mathbf{R} \times \mathrm{Lie}(G)^\star)$$

of the affine symplectic state projection plane $\mathcal{O}_U$ allows the direction of the magnetic moment to escape to infinity. It can be homeomorphically realized by the compact unit sphere

$$\mathbf{S}_2 = \mathbf{SO}(3, \mathbf{R})/\mathbf{SO}(2, \mathbf{R}) = \mathbf{O}(3, \mathbf{R})/\mathbf{O}(2, \mathbf{R})$$

with antipodal points identified ([42]), as in the "projective" proof of Gleason's geometric quantization theorem. Because it needs two bits of information through a classical information channel to distinguish the centres $\{\mathrm{id}\}$ and $\{\mathrm{id}, -\mathrm{id}\}$ of the compact Lie group $\mathbf{SO}(3, \mathbf{R}) = \mathbf{P}_3(\mathbf{R})$ and its covering group $\mathbf{O}(3, \mathbf{R})$, respectively, transferred via EPR communication channels, the antipodal spherical visualization of the two-fold covered symplectic state

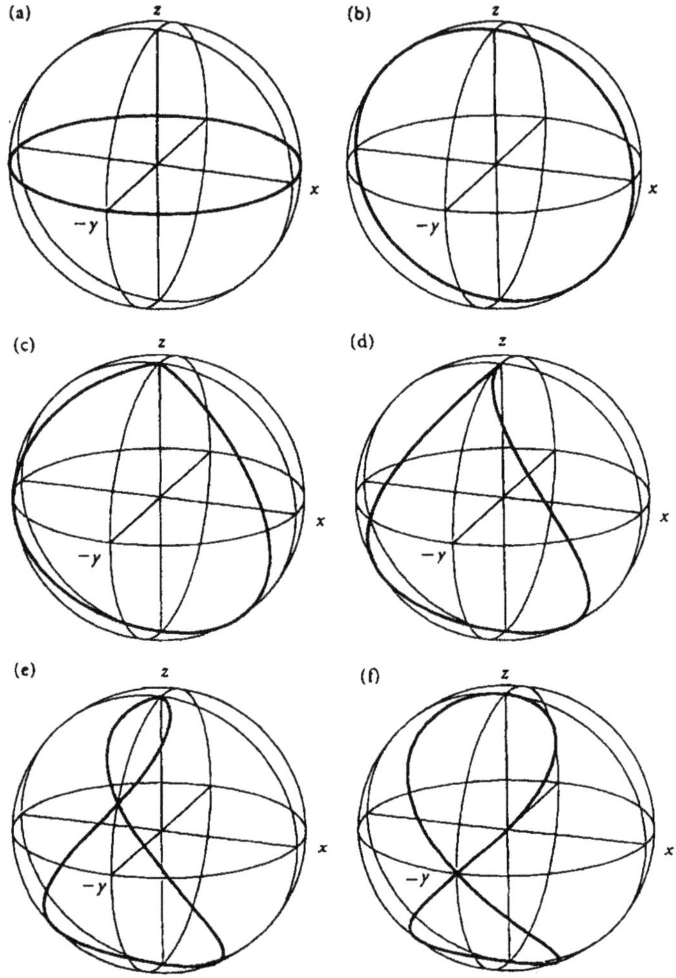

Figure 7. The formation of the Hahn spin echo teleportation. Visualization of an eight-ball echo trajectory in a rotating frame of symplectic co-ordinates attached to the projectively completed coadjoint orbit $\mathbf{P}(\mathbf{R} \times \mathcal{O}_U)$. The coadjoint orbit visualization displays the oriented Lagrangian lines of the planar *traces* of the extraordinary and ordinary Heisenberg helices. The magnetic isochromatic vectors are allowed to untwist before the second radiofrequency excitation pulse is applied. The planar tracial frame functions implementing the integrability conditions of Gleason's geometric quantization procedure admit the same periodicity $4\pi\mathbf{Z}$ as the mod 4 Maslov index τ_0.

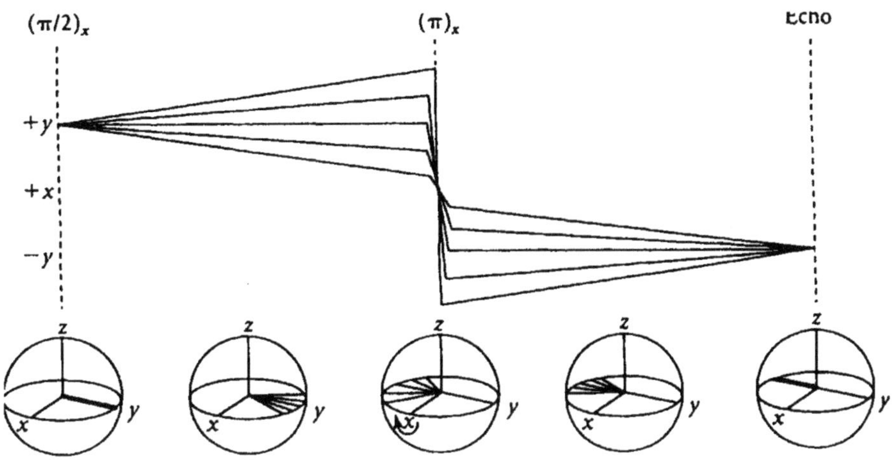

Figure 8. The formation of a spin echo teleportation by the Carr-Purcell technique. Visualization of the trajectories in a rotating frame of symplectic co-ordinates attached to the projectively completed coadjoint orbit $\mathbf{P}(\mathbf{R} \times \mathcal{O}_U)$. Initially, the magnetic isochromatic vectors are in their equilibrium position parallel to the central direction $C \hookrightarrow G$ of the strong external magnetic flux density.

projection plane $\mathcal{O}_U$ by the real Grassmann manifold

$$\mathbf{SO}(3,\mathbf{R})/\left(\{\mathrm{id},-\mathrm{id}\} \times \mathbf{SO}(2,\mathbf{R})\right) = \mathbf{O}(3,\mathbf{R})/\left(\mathbf{O}(1,\mathbf{R}) \times \mathbf{O}(2,\mathbf{R})\right)$$

provides Purcell's spin echo model of the teleportation of an eight-ball echo trajectory and illustrates Gleason's geometric quantization procedure. It provides via the mod 4 Maslov index τ_0 of the metaplectic representation R of $\mathbf{Mp}(2,\mathbf{R})$ a dimple on the antipodal unit sphere by destructive interference in the immersed Grassmann manifold $\widetilde{\Lambda}$ ([20,10]). The erasing of oriented Lagrangian lines which is visualized by the dimple is a characteristic of the quantum information transfer.

Taking into account of the longitudinal relaxation in the central direction C, the homeomorphic realization of $\mathbf{P}(\mathbf{R} \times \mathcal{O}_U)$ obtained from the ball $\mathbf{B}_2$ by identifying each point of the circle $\mathbf{S}_1$ with its antipodal point, should replace the realization of $\mathbf{P}(\mathbf{R} \times \mathcal{O}_U)$ by the compact unit sphere $\mathbf{S}_2 \hookrightarrow \mathbf{R}^3$ with antipodal points identified.

In the conventional spin echo teleportation sequence, slice selective linear gradients are applied simultaneously with frequency selective pulses. The negative lobe refocuses magnetic spins that dephased during the $\frac{\pi}{2}$-flip. A fre-

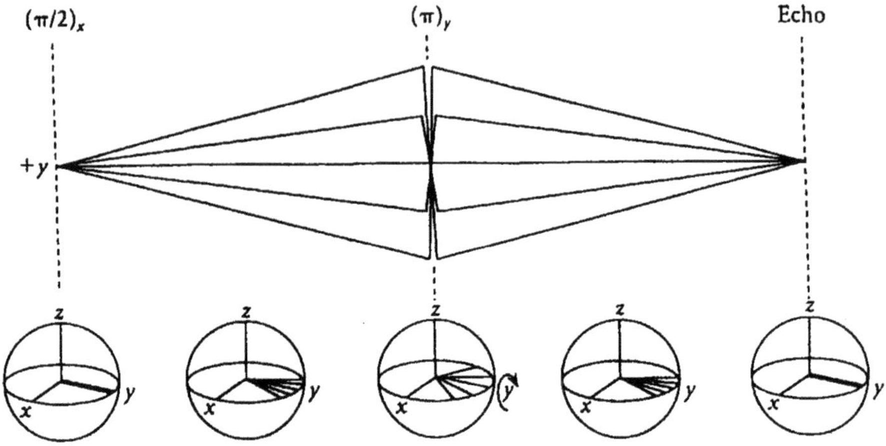

Figure 9. The formation of a spin echo teleportation by the CPMG technique. The two spin echo traces in protons H_2O were produced under identical conditions except for the introduction of an additional $\frac{\pi}{2}$-phase shift

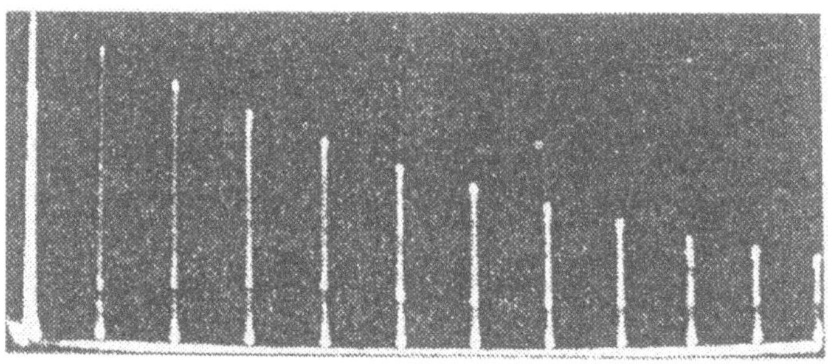

Figure 10. Stimulated spin echo amplitudes for protons in H_2O. Notice that the nuclear magnetization, having disappear in a free induction decay, spontaneously reappears in the quantum holographic regime under pulse repetitions from maximally mixed states. Hahn appreciated that he should not force the formation of the spin echo teleportation to the model of interference beat phenomena derived from a linear superposition of the individual response for an ensemble of oscillators having either slightly different frequencies, or the same frequency but loosely coupled together.

quency encoding linear gradient is applied during sampling of the spin echo. The additional linear gradient preceding data sampling refocuses the center of the echo at the center of the ADC window. The gradient following the data sampling period spoils any remaining transverse magnetization, preventing the formation of unwanted spin echo teleportations in later sampling periods. It is the gradient switching protocol which has to be transferred to the decoding receiver through classical information channels in order to decode the polyphase hologram generated by the clinical MRI scanner and to reconstruct the image from the sampled data set.

For early excitation pulse trains, data acquisition for a single scan required minutes or even tens of minutes. A great deal of research has targeted more rapid clinical MRI methods.

Due to the fact that the conventional spin echo teleportation technique is time consuming, its data acquisition has been made faster by acquiring multiple lines of the polyphase hologram after a single excitation. Fast spin echo (FSE) techniques, also known as turbo spin echo (TSE) imaging methods, do not fall under the category of ultrafast imaging in the sub-second range. However, the excellent image quality and ease of contrast manipulation possible with this class of radiofrequency excitation pulse trains combined with a significant time savings by a factor of 8 to 16 compared with conventional spin echo teleportation has led to the widespread clinical use of the FSE sequence ([14,8]). Recently, the FSE modality has been used in image-guided neurosurgery ([22]).

The FSE procedure does not collect all the spin echoes teletransported for image formation in a single shot. Instead, 8 to 16 spin echoes, each echo encoded at a different phase encoding level, are collected in every repetition time interval. Unlike in conventional spin echo imaging, the amplitude of echoes encoded with different *phase* gradient varies. Spin echoes teletransported early in the radiofrequency excitation pulse train have higher amplitude than the later echoes due to T_2 relaxation decay. The gradual falling off of amplitudes represents a low pass filter which ultimately results in a blurred image. An inversion recovery pulse can be added to an FSE pulse train.

In an effort to further reduce the data acquisition times, the HASTE (half Fourier acquisition single-shot FSE) sequence has been popularized. The HASTE technique combines the TSE procedure with the half Fourier reconstruction technique ([14]).

An ortant factor to note is that the *phase* encoding gradient is rewound after each echo so that no residual phase is added onto the phase encoding gradient applied to the later echoes of the same excitation pulse train. Relaxation weighting is readily obtained by a phase reordering scheme ([27,42]).

In Keppler's physical astronomy, a linear gradient leads from the phase encoding reference circle to the elliptic planetary orbit. Conversely, in clinical MRI the switching of linear gradients allows for a localization. The gradient switching must be *simultaneously* performed with the radiofrequency excitation pulse train ([41,45]). Therefore, timing is everything in clinical MRI protocols.

An analysis of the conchoid construction contained in the *Astronomia Nova* of 1609 reveals that the Kepplerian strategy is closer to Fourier and Feynman than to Newton ([41,42,45]). Indeed, the extraordinary and ordinary Heisenberg helices wrapped along the energy cylinders form the geodesics with respect to the Carnot-Carathéodory metric associated with the principal symbol of the left-invariant sub-Laplacian differential operator

$$\mathcal{L}_G = -\frac{1}{2}\left(\left(\frac{\partial}{\partial x} - y\frac{\partial}{\partial z}\right)^2 + \left(\frac{\partial}{\partial y} + x\frac{\partial}{\partial z}\right)^2\right).$$

Then the Feynman path integral on phase plane can be expressed in terms of $\mathcal{L}_G$ and the mod 4 Maslov index τ_0 of the metaplectic representation R of $\mathbf{Mp}(2, \mathbf{R})$ ([38]).

The left-invariant sub-Laplacian differential operator $\mathcal{L}_G$ provides a differential geometric approach to the spin echo teleportation concept and the non-cloning principle of quantum information theory. Indeed, due to the symplectic structure of the planar coadjoint orbits $\mathcal{O}_U \hookrightarrow \mathrm{Lie}(G)^\star$ of G, the sub-Laplacian differential operator $\mathcal{L}_G$ is sensitive to the chirality determined by the orientation of the extraordinary and ordinary Heisenberg helices and the circular polarization orientations.

If, however, $\mathcal{L}_G$ is assumed to be invariant under the orientation reversing line reflection

$$\begin{pmatrix} 1 & 0 \\ 0 & -1 \end{pmatrix} \in \mathbf{O}(2, \mathbf{Z}) - \mathbf{SO}(2, \mathbf{Z}),$$

which induces the isometry under the Carnot-Carathéodory metric of G given by the mapping

$$\begin{pmatrix} 1 & x & z \\ 0 & 1 & y \\ 0 & 0 & 1 \end{pmatrix} \rightsquigarrow \begin{pmatrix} 1 & x & -z \\ 0 & 1 & -y \\ 0 & 0 & 1 \end{pmatrix},$$

it reduces to the Laplacian differential operator given by

$$\Delta_{\mathbf{R}^3} = -\frac{1}{2}\left(\frac{\partial^2}{\partial x^2} + \frac{\partial^2}{\partial y^2} + \frac{\partial^2}{\partial z^2}\right).$$

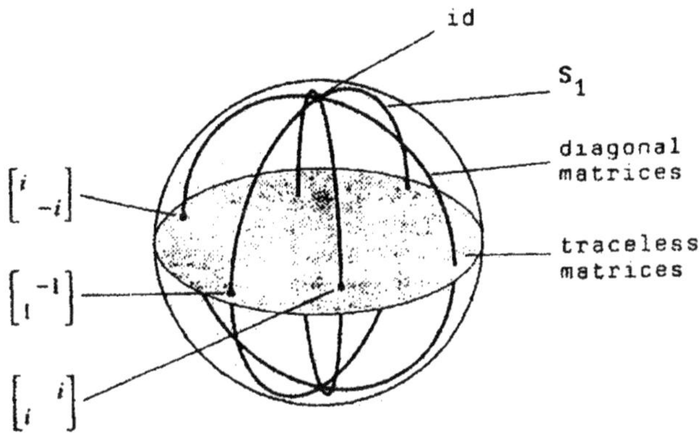

Figure 11. Projection of the 3-sphere $\mathbf{S}_3 \hookrightarrow \mathbf{R}^4$ onto the unit 2-disc in $\mathbf{R}^2$ displaying a degree of latitude and several meridians of $\mathbf{S}_3$. The degrees of latitude of $\mathbf{S}_3$ are 2-spheres embedded into $\mathbf{R}^4$ representing conjugacy classes of $\mathbf{SU}(2, \mathbf{C})$, whereas the meridians are great circles of the 3-sphere $\mathbf{S}_3$. The Dirac string and the central circle $\mathbf{S}_1$ in transversal position are exceptional fibres.

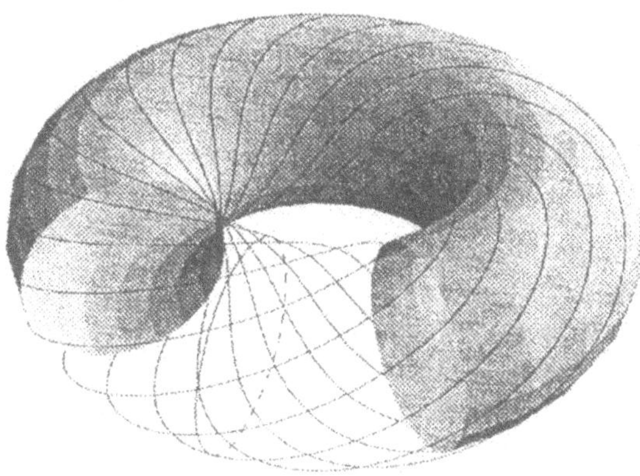

Figure 12. A flat torus singled out from the Hopf bundle $\mathbf{S}_1 \to \mathbf{S}_3 \to \mathbf{S}_2$. On the flat torus the fibres run around once in each direction. The transversal displays the chirality of unitary symplectic spinors which is not completely testable by quantum experiments. Due to the non-cloning principle of quantum information theory, a complete test of the swapped information needs at least two bits of classical information. The non-locality of quantum information theory therefore is not in a conflict to special relativity theory.

The Carnot-Carathéodory metric associated with the principal symbol of $\mathcal{L}_G$ reduces to the Euclidean metric of the vector space $\mathbf{R}^3$. Due to the Clifford isomorphism

$$\mathbf{R} \times \mathbf{R}^3 \longrightarrow \mathbf{H}$$

and the fact that the *center* of the skew-field $\mathbf{H}$ is isomorphic to the field $\mathbf{R}$, the negative definite Laplacian differential operator $\Delta_{\mathbf{R}^3}$ allows for a central *extension* to the bi-invariant Laplacian differential operator $\Delta_{\mathbf{R}^4}$ of the four-dimensional real vector space

$$\mathbf{R}^4 = \mathbf{C} \oplus \mathbf{C} = \mathbf{C} \otimes_{\mathbf{R}} \mathbf{C} = \mathbf{H}$$

of unitary symplectic spinors ([4]) which embeds the compact 3-sphere

$$\mathbf{S}_3 = \mathbf{SO}(4, \mathbf{R})/\mathbf{SO}(3, \mathbf{R}).$$

It is the Laplacian differential operator $\Delta_{\mathbf{R}^4}$ which actually governs the *harmonic* analysis of the compact Gelfand pair

$$(\mathbf{SO}(4, \mathbf{R}), \mathbf{SO}(3, \mathbf{R})),$$

hence of the compact spin group $\mathbf{SU}(2, \mathbf{C}) = \mathbf{Spin}(3, \mathbf{R})$ which is diffeomorphic to $\mathbf{S}_3 \hookrightarrow \mathbf{R}^4$. The traceless Clifford mapping transfers the second-rank antisymmetric magnetic tensor of the Maxwell equations

$$\begin{pmatrix} 0 & B_3 & B_2 \\ B_3 & 0 & -B_1 \\ -B_2 & B_1 & 0 \end{pmatrix}$$

into the unitary symplectic spinor

$$-i \begin{pmatrix} -B_3 & B_2 - iB_1 \\ B_2 + iB_1 & B_3 \end{pmatrix}$$

of $\mathbf{SU}(2, \mathbf{C})$ for the programming of temporally symmetric NMR pulse sequences of nuclear spin choreography. The isomorphy of the Clifford algebra of the negative definite quadratic space $\mathbf{R}^{0,3}$ to $\mathbf{H} \oplus \mathbf{H}$ makes sure that the *non-locality* caused by the projective spinorial symmetries of the principal sandwich bundle over the compact 2-sphere $\mathbf{S}_2 \hookrightarrow \mathbf{R}^3$ is *not* in a conflict to special relativity.

Notice that the metaplectic representation R is a privileged example of the deep relation between group representations and symplectic geometry. Quantum holography provides the affine symplectic state projection planes $\mathcal{O}_U \hookrightarrow \mathrm{Lie}(G)^\star$ to provide a realization of various features of the affine symplectic geometry of quantum physics by photonics, clinical MRI, and EPRS.

Although neither the type II degenerate parametric down-conversion nor the magnetic spin systems are linear, the affine symplectic state projection planes $\mathcal{O}_U$ provide *linear* encoding and decoding procedures. Despite the strong non-linearities of the magnetic spin systems, the concepts of *linear* response theory for filter banks hold exactly and are extremely useful for the application to clinical MRI.

The Fourier encoding of the coadjoint orbits $\mathcal{O}_U \hookrightarrow \mathrm{Lie}(G)^\star$ is the most natural encoding procedure of the MRI modality. If, however, the sinc wavelet basis is replaced by more general wavelet bases, the excitation pulse shape has to be adapted to the wavelet packet associated with the mother wavelet ([34]).

Acknowledgments

The first author owes much to valuable discussions with the late Professor Werner H. Greub. Moreover, he would like to thank Professor Shlomo Sternberg (Harvard) for his generous advice. The second author is grateful to Professors George L. Farre (Georgetown), Maurice de Gosson (Yale), and Dr. Peter J. Marcer (Bristol) for encouraging discussions on quantum holography and its historical roots. An observation made by Professor Cees R. Traas (Enschede) during the lectures at the Morningside Center of Mathematics (Beijing) has led to a differential geometric approach to the spin echo teleportation concept and the non-cloning principle of quantum information theory. The second author's thanks also go to Professor James M.S. Hutchinson (Aberdeen) for his introduction into the practice of 15 MHz spectroscopy, as well as to Dr. Francis W. Smith (Woodend Hospital) for his advice in the clinical practice.

References

1. D.M.S. Bagguley, *The echo phenomenon*, in Pulsed Magnetic Resonance-NMR, ESR, and Optics: A Recognition of E.L. Hahn, ed. D.M.S. Bagguley, Oxford University Press, Oxford, New York, Toronto, 1992, pp. 5–53.
2. C.H. Bennett, G. Brassard, C. Crépeau, R. Jozsa, A. Peres, and W.K. Wootters, *Teleporting an unknown quantum state via dual classical and Einstein-Podolsky-Rosen channels*, Phys. Rev. Lett. **70**(1993), 1895–1899.
3. C.H. Bennett, G. Brassard, S. Popescu, B. Schumacher, J.A. Smolin and W.K. Wootters, *Purification of noisy entanglement and faithful teleportation via noisy channels*, Phys. Rev. Lett. **76**(1996), 722–725.

4. E. Binz, W. Schempp, *The projective quantization of spinors*, Manuscript 1998 (to appear).
5. D. Boschi, S. Branca, F. de Martini, L. Hardy and S. Popescu, *Experimental realization of teleporting an unknown pure quantum state via dual classical and Einstein-Podolsky-Rosen channels*, Phys. Rev. Lett. **80**(1998), 1121–1125.
6. D. Bouwmeester, J.-W. Pan, K. Mattle, M. Eibl, H. Weinfurter and A. Zeilinger, *Experimental quantum teleportation*, Nature **390**(1997), 575–579.
7. S.I. Braunstein, H.J. Kimble, *Teleportation of continuous quantum variables*, Phys. Rev. Lett. **80**(1998), 869–873.
8. A. Bücker, *Fast imaging techniques for* MR-*guided biopsies*, in Interventional Magnetic Resonance Imaging, eds. J.F. Debatin, G. Adam, Springer-Verlag, Berlin, Heidelberg, New York, 1998, pp. 113–120.
9. P.T. Callaghan, *Principles of nuclear magnetic resonance microscopy*, Oxford University Press, Oxford, New York, Athens, 1991.
10. H.Y. Carr, E.M. Purcell, *Effects of diffusion on free precession in nuclear magnetic resonance experiments*, Phys. Rev. **94**(1954), 630–638.
11. T.R. Corle, G.S. Kino, *Confocal scanning optical microscopy and related imaging systems*, Academic Press, San Diego, London, Boston, 1996.
12. J.G. Cramer, *The transactional interpretation of quantum mechanics*, Rev. Mod. Phys. **58**(1986), 647–687.
13. E.R. Davies, *Electronics, noise, and signal recovery*, Academic Press, London, San Diego, New York, 1993.
14. J.F. Debatin, G.C. McKinnon, *Ultrafast* MRI: *Techniques and applications*, Springer-Verlag, Berlin, Heidelberg, New York, 1998.
15. G.L. Farre, *The semantic filter and the structure of observation*, in Cybernetics and Systems '98, ed. R. Trappl, Vol. I, Austrian Society for Cybernetic Studies, Vienna, 1998, pp. 146–151.
16. J.M.G. Fell, *Weak containment and Kronecker products of group representations*, Pacific J. Math. **13**(1963), 503–510.
17. R. Freeman, *Spin Choreography: Basic Steps in High Resolution* NMR, Spektrum Academic Publishers, Oxford, 1997.
18. A.M. Gleason, *Measures on the closed subspaces of a Hilbert space*, J. Math. Mech. **6**(1957), 885–893.
19. V. Guillemin, *Symplectic spinors and partial differential equations*, in Géométrie Symplectique et Physique Mathématique, Colloques Internationaux du Centre National de la Recherche Scientifique, No. 237, Editions du Centre National de la Recherche Scientifique, Paris 1975, pp. 217–252.

20. E.L. Hahn, *Spin echoes*, Phys. Rev. **80**(1950), 580–594.
21. R. Jäger, I. Moseley, *An introduction to magnetic resonance imaging of the brain*, Chapman & Hall Medical, London, Weinheim, New York, 1998.
22. F.A. Jolesz, J. Kettenbach, and R. Kikinis, *Image-guided neurosurgery with intraoperative* MRI, in Interventional Magnetic Resonance Imaging, eds. J.F. Debatin, G. Adam, Springer-Verlag, Berlin, Heidelberg, New York, 1998, pp. 253–260.
23. J.A. Jones, M. Mosca, and R.H. Hansen, *Implementation of a quantum research algorithm on a quantum computer*, Nature **393**(1998), 344–346.
24. B. Kostant, *Symplectic spinors*, in Geometria Simplettica e Fisica Matematica. Symposia Mathematica, Vol. XIV, Academic Press, London, New York, 1974, pp. 139–152.
25. P.G. Kwiat, K. Mattle, H. Weinfurter, A. Zeilinger, A.V. Sergienko and Y. Shih, *New high-intensity source of polarization-entangled photon pairs*, Phys. Rev. Lett. **75**(1995), 4337–4341.
26. C. Leichtle, W.P. Schleich, I.Sh. Averbukh and M. Shapiro, *Quantum state holography*, Phys. Rev. Lett. **80**(1998), 1418–1421.
27. *The* MRI *manual*, Second edition, ed. R.B. Lufkin, Mosby-Year Book, St. Louis, Baltimore, Boston, 1998.
28. NMR *imaging*, Proceedings of a Royal Society Discussion Meeting held on 5 and 6 June 1990, eds. P. Mansfield, E.L. Hahn, The Royal Society, London, 1990. Also in: Philosophical Transactions of the Royal Society of London, Series A, Volume 333(1990), pp. 401–572.
29. M. Manthey, *A combinatorial bit bang leading to quaternions*, Acta Polytechnica Scandinavica **91**(1998), 122–144.
30. K. Mattle, H. Weinfurter, P.G. Kwiat and A. Zeilinger, *Dense coding in experimental quantum communication*, Phys. Rev. Lett. **76**(1996), 4656–4659.
31. J. Mattson, M. Simon, *The pioneers of* NMR *and magnetic resonance in medicine: The story of* MRI, Bar-Ilan University Press, Ramat Gan, 1996.
32. S. Meiboom, D. Gill, *Modified spin-echo method for measuring nuclear relaxation times*, Rev. Sci. Instrum. **29**(1958), 688–691.
33. J.P. Mugler III, J.R. Brookeman, *The optimum data sampling period for maximun signal-to-noise ratio in* MR *imaging*, Rev. Magn. Reson. Med. **3**(1988), 3–51.
34. L.P. Panych, *Theoretical comparison of Fourier and wavelet encoding in magnetic resonance imaging*, IEEE Trans. Med. Imag. **15**(1996), 141–153.
35. A. Peres, *Quantum theory: Concepts and methods*, Kluwer Academic

Publishers, Dordrecht, Boston, London, 1993.
36. R. Ranga Rao, *The Maslov index on the simply connected covering group and the metaplectic represenation*, J. Funct. Anal. **107**(1992), 211–233.
37. M. Riesz, *Clifford numbers and spinors*, Kluwer Academic Publishers, Dordrecht, Boston, London, 1993.
38. J. Robbin, D. Salamon, *Phase functions and path integrals*, in Symplectic Geometry, ed. D. Salamon, Cambridge University Press, Cambridge, 1993, pp. 203–226.
39. J. Rosenberg, *Realization of square-integrable representations of unimodular Lie groups on L^2-cohomology spaces*, Trans. Amer. Math. Soc. **261**(1980), 1–32.
40. W. Schempp, *Harmonic analysis on the Heisenberg nilpotent Lie group, with applications to signal theory*, Pitman Research Notes in Mathematics Series, Vol. 147, Longman Scientific and Technical, London, 1986.
41. ———, *Zu Kepplers Conchoid-Konstruktion*, Result Math. **32**(1997), 352–390.
42. ———, *Magnetic resonance imaging: Mathematical foundations and applications*, Wiley-Liss, New York, Chichester, Weinheim, 1998.
43. ———, *Quantum computing by* NMR *spectroscopy*, in Cybernetics and Systems '98, ed. R. Trappl, Vol. I, Austrian Society for Cybernetic Studies, Vienna, 1998, pp. 107–112.
44. ———, *Quantum teleportation and spin echo*, Acta Polytechnica Scandinavica **91**(1998), 99–121.
45. ———, *Die Kepplerschen Strategien und ihre Anwendungen auf geometrische Spinorquantisierung*, Manuscript (to appear).
46. L. Schwartz, *Sous-espaces hilbertiens d'espaces vectoriels topologiques et noyaux associés (noyaux reproduisants)*, J. Analyse Math. **13**(1964), 115–256.
47. ———, *Sous-espaces hilbertiens et noyaux associés; applications aux représentations des groupes de Lie*, in Deuxième Colloq. L'Anal. Fonct., Centre Belge Recherches Mathématiques, Librairie Universitaire, Louvain, 1964, pp. 153–163.
48. M.E. Stoll, A.J. Vega and R.W. Vaughan, *Explicit demonstration of spinor character for a spin-$\frac{1}{2}$ nucleus via* NMR *interferometry*, Phys. Rev. **A 16**(1977), 1521–1524.
49. D. Suter, A. Pines and M. Mehring, *Indirect phase detection of* NMR *spinor-transitions*, Phys. Rev. Lett. **57**(1986), 242–244.
50. T. Sudbery, *The fastest way from* A *to* B, Nature **390**(1997), 551–552.
51. V.S. Varadarajan, *Geometry of quantum theory. Vol. I and II*, Van Nostrand Reinhold Company, New York, Cincinnati, Toronto, 1970.

52. A. Weil, *Sur certains groupes d'opérateurs unitaires*, Acta Math. **111**(1964), 143–211.
53. ———, *Œuvres Scientifiques, Collected Papers, Volume III (1964-1978)*, Corrected Sec. Printing, Springer-Verlag, New York, Heidelberg, Berlin, 1979, pp. 1–69.
54. H. Weinfurter, D. Bouwmeester, K. Mattle, J.-W. Pan, M. Eibl, A. Zeilinger, J. Brendel, W. Tittel, H. Zbinden, N. Gisin, J.G. Rarity, and P.R. Tapster, *Quantum communication and entanglement*, in Cybernetics and Systems '98, ed. R. Trappl, Vol. I, Austrian Society for Cybernetic Studies, Vienna, 1998.
55. C.P. Williams, S.H. Clearwater, *Explorations in Quantum Computing*, Springer-Verlag, New York, Berlin, Heidelberg; TELOS, The Electronic Library of Science, Santa Clara, California, 1998.
56. W.K. Wootters, W.H. Zurek, *A single quantum cannot be cloned*, Nature **299**(1982), 802–803.

www.ingramcontent.com/pod-product-compliance
Ingram Content Group UK Ltd.
Pitfield, Milton Keynes, MK11 3LW, UK
UKHW020400150926
8624IPUK00004B/268

9 789810 238681